W9-AVZ-365

DESIGN OF REINFORCED CONCRETE

Central Library, University of California, San Diego
(Courtesy of Glasheen Graphics of La Jolla)

DESIGN OF REINFORCED CONCRETE

Jack C. McCormac
Clemson University

Thomas Y. Crowell
Harper & Row Publishers
New York Hagerstown San Francisco London

Sponsoring Editor: Charlie Dresser
Project Editor: Penelope Schmukler
Designer: T. R. Funderburk
Production: Marion Palen
Compositor: Syntax International Pte., Ltd.
Printer and Binder: The Maple Press Company
Art Studio: Vantage Art, Inc.

DESIGN OF REINFORCED CONCRETE
Copyright © 1978 by Jack C. McCormac

All rights reserved. Printed in the United States of America. No part of this book may be used or reproduced in any manner whatsoever without written permission except in the case of brief quotations embodied in critical articles and reviews. For information address Harper & Row, Publishers, Inc., 10 East 53rd Street, New York, N.Y. 10022.

Library of Congress Cataloging in Publication Data

McCormac, Jack C.
Design of reinforced concrete.
Series in civil engineering
1. Reinforced concrete construction. I. Title.
TA683.2.M3 624'.1834 78-15882
ISBN 0-7002-2523-4

Preface

The purpose of this book is to introduce the fundamentals of reinforced concrete design in such a manner as to stimulate interest in the subject. Although the material was prepared primarily for the student just beginning the study of reinforced concrete, it is hoped that it will be of some use to practicing engineers as well.

After a discussion of several general topics is presented in Chapter 1, a brief introduction to *working stress design* (alternate design method) is given in Chapters 2 and 3. The main body of the book (Chapters 4–14) is devoted to the *strength design method* applied to beams, columns, footings, retaining walls, and floor slabs. Prestressed concrete is introduced in Chapter 15, while formwork is the topic of the final chapter (16).

Most students seem to like reinforced concrete design very much. Perhaps its because they get the feeling that they are actually determining the sizes of beams, columns, and slabs and not just picking sizes from a restricted list in a manufacturer's catalogue. The author hopes that through this textbook he will be able to promote this fondness for reinforced concrete.

Jack C. McCormac

To Mary

Contents

Chapter 9 DESIGN OF MEMBERS SUBJECT TO AXIAL LOAD AND BENDING 203

Chapter 10 FOOTINGS 236

Chapter 11 RETAINING WALLS 273

Chapter 12 CONTINUOUS REINFORCED CONCRETE STRUCTURES 309

Chapter 13 TORSION 346

Chapter 14 DESIGN OF TWO-WAY SLABS 365

Chapter 15 PRESTRESSED CONCRETE 404

Chapter 16 FORMWORK 442

Acknowledgments

The author gratefully acknowledges the aid he has received from several sources. He is indebted to G. Batson, J. E. Clark, R. E. Elling, D. P. Gustafson, C. R. Mitchell, L. Spiegel, L. A. Traina, and O. Ural, who have directly contributed to the preparation of the manuscript by their suggestions and criticisms, and to his own professors M.J.Holley, Jr., and B. B. Williams, who patiently instructed him in their classes. The books used in their classes and in the classes taught by the author have naturally influenced the text material to some extent. These books include *Design of Concrete Structures* by G. Winter and A. H. Nilson and *Reinforced Concrete Design* by C. K. Wang and C. G. Salmon. Finally, thanks are due to Mrs. Janice Ruelle who typed the manuscript.

SERIES IN CIVIL ENGINEERING

Series Editor
Russell C. Brinker
New Mexico State University

Brinker and Wolf
ELEMENTARY SURVEYING
6th edition

Clark, Viessman, and Hammer
WATER SUPPLY AND POLLUTION CONTROL
3rd edition

McCormac
STRUCTURAL ANALYSIS
3rd edition

McCormac
STRUCTURAL STEEL DESIGN
2nd edition

Meyer
ROUTE SURVEYING AND DESIGN
4th edition

Mikhail
OBSERVATIONS AND LEAST SQUARES

Moffitt
PHOTOGRAMMETRY
2nd edition

Moffitt and Bouchard
SURVEYING
6th edition

Salmon and Johnson
STEEL STRUCTURES: DESIGN AND BEHAVIOR

Spangler and Handy
SOIL ENGINEERING
3rd edition

Viessman, Harbaugh, Knapp, and Lewis
INTRODUCTION TO HYDROLOGY
2nd edition

Wang and Salmon
REINFORCED CONCRETE DESIGN
3rd edition

DESIGN OF REINFORCED CONCRETE

CHAPTER 1

Introduction

1.1 CONCRETE AND REINFORCED CONCRETE

Concrete is a mixture of sand, gravel, crushed rock, or other aggregates held together in a rocklike mass with a paste of cement and water. Sometimes one or more admixtures are added to change certain characteristics of the concrete such as its workability, durability, and time of hardening.

As with most rocklike substances, concrete has a high compressive strength and a very low tensile strength. *Reinforced concrete* is a combination of concrete and steel wherein the steel reinforcement provides the tensile strength lacking in the concrete. Steel reinforcing is also capable of resisting compression forces and is used in columns as well as in other situations to be described later.

1.2 ADVANTAGES OF REINFORCED CONCRETE AS A STRUCTURAL MATERIAL

Reinforced concrete may be the most important material available for construction. It is used in one form or another for almost all structures, great or small–buildings, bridges, pavements, dams, retaining walls, tunnels, viaducts, drainage and irrigation facilities, tanks, and so on.

The tremendous success of this universal construction material can be understood quite easily if its numerous advantages are considered. These include the following:

1. It has considerable compressive strength as compared to most other materials.
2. Reinforced concrete has great resistance to the actions of fire and water and, in fact, is the best structural material available for situations where water is present. During fires of average intensity, members with a satisfactory cover of concrete over the reinforcing bars suffer only surface damage without failure.
3. Reinforced concrete structures are very rigid.
4. It is a low-maintenance material.

Peachtree Center Hotel, The Tallest Hotel in the World,
Atlanta, Georgia (Courtesy Symons Corporation).

5. As compared with other materials, it has a very long service life. Under proper conditions reinforced concrete structures can be used indefinitely without reduction of their load-carrying abilities. This can be explained by the fact that the strength of concrete does not decrease with time but actually increases over a very long period, measured in years, due to the lengthy process of the solidification of the cement paste.
6. It is usually the only economical material available for footings, basement walls, piers, and similar applications.
7. A special feature of concrete is its ability to be cast into an extraordinary variety of shapes from simple slabs, beams, and columns to great arches and shells.

8. In most areas concrete takes advantage of inexpensive local materials (sand, gravel, and water) and requires relatively small amounts of cement.
9. A lower grade of skilled labor is required for erection as compared to other materials such as structural steel.

1.3 DISADVANTAGES OF REINFORCED CONCRETE AS A STRUCTURAL MATERIAL

To successfully use concrete the designer must be completely familiar with its weak points as well as with its strong ones. Among its disadvantages are the following:

1. Concrete has a very low tensile strength, requiring the use of tensile reinforcing.
2. Forms are required to hold the concrete in place until it hardens sufficiently. In addition, falsework or shoring may be necessary to keep the forms in place for roofs, walls, and similar structures until the concrete gains the necessary strength.
3. The low strength per unit of weight of concrete leads to heavy members. This becomes an increasingly important matter for long-span structures where concrete's large dead weight has a great effect on bending moments.
4. Similarly, the low strength per unit of volume of concrete means members will be relatively large, an important consideration for tall buildings and long-span structures.
5. The properties of concrete vary widely due to variations in its proportioning and mixing. Furthermore, the placing and curing of concrete is not as carefully controlled as is the production of other materials such as structural steel and laminated wood.

Two other characteristics that can cause problems are concrete's shrinkage and creep. The *shrinkage* of concrete is the lessening of volume of the concrete as time goes by which is not due to the application of load (thus due to insufficient moisture). The amount of shrinkage is heavily dependent on the type of exposure. For instance, if concrete is subjected to a considerable amount of wind during curing, its shrinkage will be large. In a related fashion, a humid atmosphere means smaller shrinkages while a dry one means larger ones.

Under sustained loads concrete will continue to deform for long periods of time. This additional deformation is called *creep* or *plastic flow*. If a compressive load is applied to a concrete member, an immediate or elastic shortening occurs. If the load is left in place for a long time, the member will continue to shorten over a period of several years and the final deformation will usually be two to three times the initial deformation. It is obviously quite difficult to distinguish between shrinkage and creep.

1.4 HISTORICAL BACKGROUND

Though the average person thinks that concrete has been in common use for many centuries, such is not the case. Although the Romans made cement—called pozzolana—before the birth of Christ by mixing slaked lime with a volcanic ash from Mount Vesuvius and used it to make concrete for buildings, the art was lost during the Dark Ages and was not revived until the eighteenth and nineteenth centuries. A deposit of natural cement rock was discovered in England in 1796 and was sold as "Roman cement." Various other deposits of natural cement were discovered in both Europe and America and were used for several decades.

Twisted Bar (Courtesy Clemson University Communications Center).

The real breakthrough for concrete occurred in 1824 when an English bricklayer named Joseph Aspdin, after long and laborious experiments, obtained a patent for a cement which he called "portland cement" because its color was quite similar to that of the stone quarried on the Isle of Portland off the English coast. He made his cement by taking certain quantities of clay and limestone, pulverizing them, burning them in his kitchen stove, and grinding the resulting clinker into a fine powder. During the early years after its development, his cement was primarily used in stuccos.[1] This wonderful product was very slowly adopted by the building industry and was not even introduced into the United States until 1868; the first portland cement was not manufactured in the United States until the 1870s.

The first uses of reinforced concrete are not very well known. Much of the early work was done by two Frenchmen, Joseph Lambot and Joseph Monier. In about 1850 Lambot built a concrete boat reinforced

[1] Kirby, R. S., and Laurson, P. G., 1932, *The Early Years of Modern Civil Engineering* (New Haven: Yale University Press), p. 266.

with a network of parallel wires or bars. Credit is usually given to Monier, however, for the invention of reinforced concrete. In 1867 he received a patent for the construction of concrete basins or tubs and reservoirs reinforced with a mesh of iron wire. His stated goal in working with this material was to obtain lightness without sacrificing strength.[2]

From 1867 to 1881 Monier received patents for reinforced concrete railroad ties, floor slabs, arches, footbridges, buildings, and other items in both France and Germany. Another Frenchman, Francois Coignet, built simple reinforced concrete structures and developed basic methods of design. In 1861 he published a book in which he presented quite a few applications. He was the first person to realize that the addition of too much water in the mix greatly reduced concrete strength. Other Europeans who were early experimenters with reinforced concrete included the Englishmen William Fairbairn and William B. Wilkinson, the German G. A. Wayss, and another Frenchman, Francois Hennebique.[3,4]

William E. Ward, built the first reinforced concrete building in the United States in Port Chester, N.Y., in 1875. In 1883 he presented a paper before the American Society of Mechanical Engineers in which he claimed that he got the idea of reinforced concrete by watching English laborers in 1867 trying to remove hardened cement from their iron tools.[5]

Thaddeus Hyatt, an American, was probably the first person to correctly analyze the stresses in a reinforced concrete beam, and in 1877 he published a 28-page book on the subject, entitled *An Account of Some Experiments with Portland Cement Concrete, Combined with Iron as a Building Material.* In this book he praised the use of reinforced concrete and said that "rolled beams have to be taken largely on faith." Hyatt put a great deal of emphasis on the high fire resistance of concrete.[6]

E. L. Ransome of San Francisco is supposed to have used reinforced concrete in the early 1870s and was the originator of deformed (or twisted) bars, for which he received a patent in 1884. These bars which were square were cold-twisted with one complete turn in a length of not more than 12 times the bar diameter.[7] In 1890 in San Francisco, Ransome built the Leland Stanford Jr. Museum. It is a reinforced concrete building 312 feet

[2] Kirby, R. S., and Laurson, P. G., 1932, *The Early Years of Modern Civil Engineering* (New Haven: Yale University Press), pp. 273–275.

[3] Straub, H., 1964, *A History of Civil Engineering* (Cambridge: The M.I.T. Press), pp. 205–215. Translated from the German *Die Geschickte der Bauingenieurkunst*, Verlog Birkhauser, Basle, 1949.

[4] Kirby, R. S., and Laurson, P. G., 1932, *The Early Years of Modern Civil Engineering* (New Haven: Yale University Press), pp. 273–275.

[5] Ward, W. E., 1883, "Béton in Combination with Iron as a Building Material," American Society of Mechanical Engineers (*Transactions*, vol. IV), pp. 388–403.

[6] Kirby, R. S., and Laurson, P. G., 1932, *The Early Years of Modern Civil Engineering* (New Haven: Yale University Press), p. 275.

[7] American Society for Testing Materials, 1911, *Proceedings* (vol. XI), pp. 66–68.

long and two stories high in which discarded wire rope from a cable-car system was used as tensile reinforcing. This building experienced little damage in the 1906 earthquake. Since 1890 the development and use of reinforced concrete in the United States has been very rapid.[8,9]

1.5 COMPARISON OF REINFORCED CONCRETE AND STRUCTURAL STEEL FOR BUILDINGS AND BRIDGES

When a particular type of structure is being considered, the student may be perplexed by the question, "Should reinforced concrete or structural steel be used?" There is much joking on this point, with the proponents of reinforced concrete referring to steel as that material which rusts, while those favoring structural steel refer to concrete as that material which when overstressed tends to return to its natural state—that is, sand and gravel.

There is no simple answer to this question as both of these materials have many excellent characteristics that can be utilized successfully for so many types of structures. In fact, they are often used together in the same structures with wonderful results.

The selection of the structural material to be used for a particular building depends on the height and span of the structure, the material market, foundation conditions, local building codes, and architectural considerations. For buildings of less than 4 stories, reinforced concrete, structural steel, and wall-bearing construction are competitive. From 4 to about 20 stories, reinforced concrete and structural steel are economically competitive, with steel having taken most of the jobs above 20 stories in the past. Today, however, reinforced concrete is becoming increasingly competitive above 20 stories, and there are a number of reinforced concrete buildings of greater height around the world. Water Tower Place in Chicago is the tallest reinforced concrete building in the world. It has 74 stories with a total height of 859 ft and 2 ins.

Foundation conditions can very often affect the selection of the material to be used for the structural frame. If foundation conditions are poor, a lighter structural steel frame may be desirable. The building code in a particular city may be favorable to one material over the other. For instance, many cities have fire zones in which only fireproof structures can be erected—a very favorable situation for reinforced concrete. Finally, the time element favors structural steel frames, as they can be erected

[8] Wang, C. K., and Salmon, C. G., 2nd ed., 1973, *Reinforced Concrete Design* (New York: Intext), p. 4.

[9] "The Story of Cement, Concrete and Reinforced Concrete," *Civil Engineering*. November 1977. ASCE: New York. (pp. 63–65).

much more quickly than reinforced concrete ones. The time advantage, however, is not as great as it might seem at first glance because if the structure is to have any type of fire rating, the builder will have to cover the steel with some kind of fireproofing material after it is erected.

To make a decision about using concrete or steel for a bridge will involve several factors, such as span, foundation conditions, loads, architectural considerations, and others. In general, concrete is an excellent compression material and normally will be favored for short-span bridges and for cases where rigidity is required (as, perhaps, for railway bridges).

1.6 MECHANICAL PROPERTIES OF REINFORCED CONCRETE

A thorough knowledge of the properties of concrete is necessary for the student before he or she begins to design reinforced concrete structures. An introduction to several of these properties is presented in this section.

Compressive Strength

The compressive strength of concrete is determined by testing 28-day-old 6-in. by 12-in. concrete cylinders at a specified rate of loading. Although concretes are available with 28-day ultimate strengths from 2,500 psi up to as high as 10,000 or 11,000 psi, most of the concretes used fall into the 3000- to 7000-psi range. For ordinary applications, 3000- and 4000-psi concretes are used, while for prestressed construction, 5000- and 6000-psi strengths are common.

It is quite feasible to move from 3000-psi concrete to 5000-psi concrete without requiring excessive amounts of labor or cement. The approximate increase in cost for such a strength increase is 15% to 20%. To move above 5000- or 6000-psi concrete, however, requires very careful mix designs and considerable attention to such details as mixing, placing, and curing. These requirements cause relatively larger increases in cost.

Several comments are made throughout the text regarding the relative economy of using different strength concretes for different applications, such as for beams, columns, or prestressed members.

The stress-strain curves of Figure 1.1 represent the results obtained from compression tests of sets of 28-day-old standard cylinders of varying strengths. You should carefully study these curves as they bring out several significant points:

(a) The curves are roughly straight while the load is increased from zero to about one-third to one-half the concrete's ultimate strength.
(b) Beyond this range the behavior of concrete is nonlinear. This lack of linearity of concrete stress-strain curves at higher stresses causes some

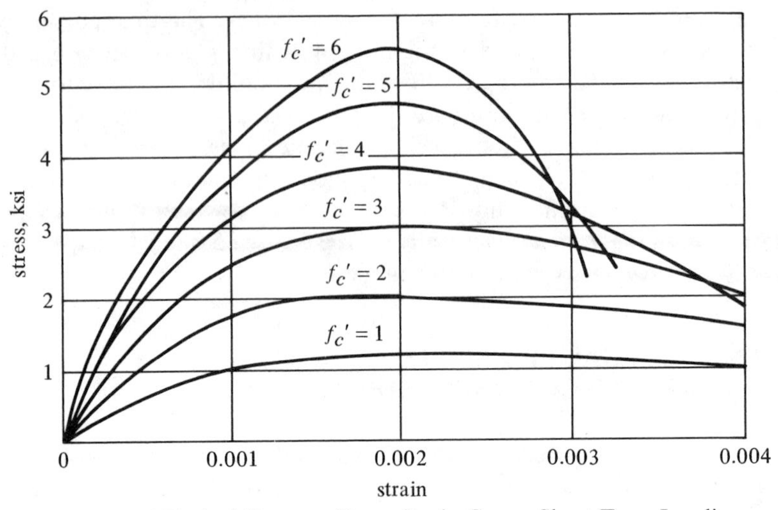

Figure 1.1 Typical Concrete Stress-Strain Curve, Short-Term Loading

problems in the structural analysis of concrete structures because their behavior is also nonlinear at higher stresses.

(c) Of particular importance is the fact that regardless of strengths all the concretes reach their ultimate strengths at strains of about 0.002.

(d) Concrete does not have a definite yield strength; rather, the curves run smoothly on to the point of rupture at strains of from 0.003 to 0.004. It will be assumed for the purpose of future calculations in this text that concrete fails at 0.003.

(e) Many tests have clearly shown that stress-strain curves of concrete cylinders are almost identical with those for the compression sides of beams.

(f) It should be further noticed that the weaker grades of concrete are less brittle than the stronger ones—that is, they will take larger strains before breaking.

Modulus of Elasticity

Truthfully speaking, concrete has no clearcut modulus of elasticity. Its value varies with different concrete strengths, concrete age, type of loading, and with the characteristics of the cement and aggregates. Furthermore, there are several different definitions of the modulus:

(a) The *initial modulus* is the slope of the stress-strain diagram at the origin of the curve.

(b) The *tangent modulus* is the slope of a tangent to the curve at some point along the curve, for instance, at 50% of the ultimate strength of the concrete.

(c) The slope of a line drawn from the origin to a point on the curve from 25% to 50% of the compressive strength is referred to as the *secant modulus*.
(d) In addition, there is another modulus which is called the *apparent modulus* or the *long-term modulus*. It is determined by using the stresses and strains obtained after the load has been applied for a certain length of time.

The American Concrete Institute (ACI) in Section 8.5 of their building code (to be discussed in Section 1.10 of this chapter) suggests that the following expression can be used for calculating the modulus of elasticity of concretes weighing from 90 to 155 #/ft³:

$$E_c = w_c^{1.5} 33\sqrt{f_c'}$$

In this expression w_c is the weight of the concrete in pounds per cubic foot and f_c' is its 28-day compressive strength in pounds per square inch. This is actually a secant modulus with the line (whose slope equals the modulus) drawn from the origin to a point on the stress-strain curve corresponding approximately to the stress that would occur under the estimated dead and live loads the structure has to support.

For normal-weight concrete weighing approximately 145 #/ft³, the ACI Code states that the following simplified version of the previous expression may be used to determine the modulus:

$$E_c = 57{,}000\sqrt{f_c'}$$

Table A.1 (see the Appendix of this text) shows values of E_c for different strength concretes.

Poisson's Ratio

As a concrete cylinder is subjected to compressive loads, it not only shortens in length but also expands laterally. The ratio of this lateral expansion to the longitudinal shortening is referred to as *Poisson's ratio*. Its value varies from about 0.11 for the higher-strength concretes to as high as 0.21 for the weaker-grade concretes, with average values at about 0.16.

Tensile Strength

The tensile strength of concrete varies from about 10% to 15% of its compressive strength. This strength, which is rather difficult to measure accurately, is determined by the so-called *split-cylinder test*. In this test a standard 6-in. × 12-in. cylinder is placed on its side in the testing machine and a compressive load is applied uniformly along the length of the cylinder with support supplied along the bottom for the cylinder's

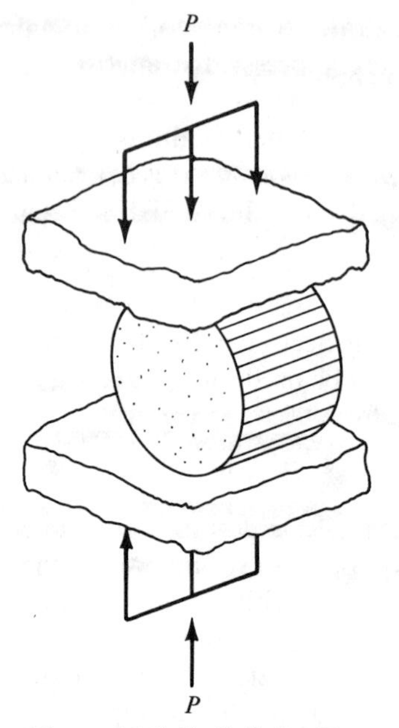

Figure 1.2 Split-Cylinder Test

full length. (See Figure 1.2.) The cylinder will split in half from end to end when its tensile strength is reached. The tensile stress at which splitting occurs is referred to as the split-cylinder strength and can be calculated by the expression to follow, in which P is the maximum compressive force, L is the length, and D is the diameter of the cylinder:

$$f_t = \frac{2P}{\pi LD}$$

Though the tensile strength of concrete is normally neglected in design calculations, nevertheless, it is an important property that affects the sizes and extent of the cracks that occur. Furthermore, the tensile strength of concrete members has a definite reduction effect on their deflections.

You might wonder why concrete is not assumed to resist a portion of the tension in a flexural member and the steel the remainder. The reason is that concrete cracks at such small tensile strains that the low stresses in the steel up to that time would make its use uneconomical.

Shear Strength

It is extremely difficult in testing to obtain pure shear failures unaffected by other stresses. As a result, the tests of concrete shearing strengths through the years have yielded values all the way from one-third to four-fifths of the ultimate compressive strengths. You will learn in Chapter 7 that you do not have to worry about these inconsistent shear strength tests because the pure shear strength of concrete is normally not of importance.

1.7 REINFORCING STEEL

The reinforcing used for concrete structures may be in the form of bars or welded wire fabric. Reinforcing bars are referred to as being *plain* or *deformed*. The deformed bars, which have ribbed projections rolled onto their surfaces (patterns differing with different manufacturers) to provide better bonding between the concrete and the steel, are used for almost all applications. Instead of rolled-on deformations, the deformed wire has indentations pressed into it. Plain bars are not used very often except for wrapping around longitudinal bars, primarily in columns. Such bars will be referred to later in this text as ties and spirals.

Plain round bars are indicated by their diameters in fractions of an inch as $\frac{3}{8}''\phi$, $\frac{1}{2}''\phi$ and $\frac{5}{8}''\phi$. Deformed bars are round and vary in sizes from #3 to #11 with two very large sizes, #14 and #18, also available. For bars up to and including #8, the number of the bar coincides with the bar diameter in eighths of an inch. Bars were formerly manufactured in both round and square cross sections, but today the larger bars are round and are manufactured with areas equivalent to certain square sizes. The #9, #10, and #11 bars have diameters that provide areas equal to the areas of 1 in. × 1 in. square bars, $1\frac{1}{8}$ in. × $1\frac{1}{8}$ in. square bars and $1\frac{1}{4}$ in. × $1\frac{1}{4}$ in. square bars, respectively. Similarly the #14 and #18 bars correspond to $1\frac{1}{2}$ in. × $1\frac{1}{2}$ in. square bars and 2 in. × 2 in. square bars, respectively. Table A-2 (see Appendix) provides details as to areas, diameters, and weights of reinforcing bars. Although #14 and #18 bars are shown in this table, the designer should check the suppliers to see if they have these very large sizes in stock.

Welded wire fabric is also frequently used for reinforcing slabs, pavements and shells, and places where there is normally not sufficient room for providing the necessary concrete cover required for regular reinforcing bars. The mesh is made of cold-drawn wires running in both directions and welded together at the points of intersection. The sizes and spacings of the wire may be the same in both directions or may be different, depending on design requirements.

Table A.3(a) in the Appendix provides information concerning certain styles of welded wire fabric which have been recommended by the Wire Reinforcement Institute as common stock styles (thus normally carried in stock at the mills or at warehousing points and thus usually immediately available). Table A.3(b) in the Appendix provides detailed information as to diameters, areas, weights and spacings of quite a few wire sizes which are normally used to manufacture welded wire fabric.

Smooth wire is denoted by the letter "W" followed by a number which equals the cross-sectional area of the wire in hundredths of a square inch. Deformed wire is denoted by the letter "D" followed by a number giving the area.

The fabric is usually indicated on drawings by the letters WWF followed by the spacings of the longitudinal wires and the transverse wires and then the wire areas in hundredths of a sq in. For instance WWF6 × 12—W16 × 8 represents smooth welded wire fabric with a 6-in. longitudinal and a 12-in. transverse spacing with cross sectional areas of 0.16 in.^{2}/ft and 0.08 in.^{2}/ft, respectively.

1.8 GRADES OF REINFORCING STEEL

Reinforcing bars may consist of billet steel, axle steel, or rail steel. Billet steel is newly manufactured steel, while axle and rail steels are made from the steel in old axles and rails. These latter steels generally have less ductility than do the new billet steels.

There are several types of reinforcing bars designated by the American Society for Testing and Materials (ASTM), which are listed at the end of this paragraph. In this listing grade 40 means the steel has a specified yield point of 40,000 psi; grade 50 means 50,000 psi; and so on.

1. ASTM A615, billet steel, grades 40 and 60.
2. ASTM A616, rail steel, grades 50 and 60.
3. ASTM A617, axle steel, grades 40 and 60.
4. ASTM A706, low-alloy steel, grade 60.

Grade 60 steel is the commonly used steel in reinforced concrete practice. Grade 75 steel is an appreciably higher cost steel and is usually not available from stock and probably has to be specially ordered from the steel mills. This means that they may have to have a special rolling to supply the steel, and thus its use could not be economically justified unless at least 50 to 100 tons were ordered.

The wire used for welded wire fabric may be smooth or deformed. Smooth welded wire fabric must conform to ASTM A185 and have a minimum yield point of 65 000 psi. Deformed welded wire fabric must conform to ASTM A497 and have a minimum yield point of 70 000 psi. Unless otherwise specified smooth welded wire fabric conforming to ASTM 185 will be furnished.

The modulus of elasticity for nonprestressed steels is considered to be equal to 29×10^6 psi. For prestressed steels it varies somewhat from manufacturer to manufacturer, with a value of 27×10^6 psi being fairly common.

1.9 COMPATABILITY OF CONCRETE AND STEEL

Concrete and steel work together beautifully in reinforced concrete structures. The advantages of each material seem to compensate for the disadvantages of the other. For instance, the great shortcoming of concrete is its lack of tensile strength; but tensile strength is one of the great advantages of steel. Reinforcing bars have tensile strengths equal to approximately 100 times that of the usual concretes used.

The two materials bond together very well so there is no slippage between the two, and thus they will act together as a unit in resisting forces. The excellent bond obtained is due to the chemical adhesion between the two materials, the natural roughness of the bars, and the closely spaced rib-shaped deformations rolled on the bar surfaces.

Reinforcing bars are subject to corrosion, but the concrete surrounding them provides them with excellent protection. The strength of exposed steel subject to the temperatures reached in fires of ordinary intensity is nil, but the enclosure of the reinforcement in concrete produces very satisfactory fire ratings. Finally, concrete and steel work very well together in relation to temperature changes because their coefficients of thermal expansion are quite close to each other (0.0000065 for steel and about 0.0000055 for concrete).

1.10 DESIGN CODES

The most important code in the United States for reinforced concrete design is the *Building Code Requirements for Reinforced Concrete* (*ACI* 318–77). This code, which is primarily used for the design of buildings, is followed for the majority of the examples given in this text. Frequent references are made to this document and section numbers are provided in the majority of cases. There is another ACI publication, entitled *Commentary on Building Code Requirements for Reinforced Concrete* (*ACI* 318–77), which is useful in providing a better background and understanding of the code.

The ACI Code is not in itself a legally enforceable document. It is merely a statement of current good practice in reinforced concrete design. It is, however, written in the form of a code or law so that various public bodies such as city councils can easily vote it into their local building codes, and as such it becomes legally enforceable in that area. In this manner the ACI Code has been incorporated into law by countless government organizations throughout the United States. It is also widely accepted in Canada and Mexico and has tremendous influence on concrete codes of all countries throughout the world.

Other well-known concrete specifications are those of the American Association of State Highway and Transportation Officials (AASHTO) and the American Railway Engineering Association (AREA).

1.11 CALCULATION ACCURACY

A most important point, which many students with their pocket calculators have difficulty in understanding, is that reinforced concrete design is not an exact science for which answers can be confidently calculated to six or eight places. The reasons for this statement should be quite obvious: The analysis of structures is based on partly true assumptions; the strengths of materials used vary widely; and maximum loadings can only be approximated. With respect to this last sentence, how many users of this book could estimate to within 10% the maximum load in pounds per square foot that will ever occur on the building floor they are now occupying? Calculations to more than two significant figures are obviously of little value and may actually be dangerous because they may mislead the student by giving him a fictitious sense of accuracy.

CHAPTER 2

Analysis of Beams by the Working-Stress Method

2.1 WORKING STRESS AND STRENGTH DESIGN METHODS

From the early 1900s until the early 1960s, nearly all reinforced concrete design in the United States was performed by the working stress design method (also called allowable stress design or straight-line design). In this method, frequently referred to as WSD, the dead and live loads to be supported, called working loads or service loads, were first estimated. Then the members of the structure were proportioned so that stresses calculated by an elastic analysis did not exceed certain permissible or allowable values.

Placing Concrete Slab (Courtesy Bethlehem Steel Corporation).

Since 1963 the ultimate strength design method has rapidly gained popularity. With this method (now called strength design) the working dead and live loads are multiplied by certain load factors (equivalent to safety factors). The members are then selected so that they theoretically will fail under the factored loads. Chapters 2 and 3 are devoted to WSD (now called the Alternate Design Method by the ACI Code as described in Chapter 3), while Chapters 4 through 14 are devoted to the strength design method.

2.2 INTRODUCTION TO WSD

Although the large percentage of existing reinforced concrete structures in the United States were designed by WSD, the situation is rapidly changing for several reasons (which are discussed in Chapter 4). Today the large proportion of designs are made by the strength design method.

Even though most of the concrete structures you will face after graduation will be designed by the strength design method, you should nevertheless be familiar with WSD for several reasons. First, the design of many highway structures is handled by WSD, although the 1973 AASHTO specifications permit the use of strength design[1] in a very similar manner to that of the ACI Code. The current AREA specifications are based on WSD only but are in the process of being revised to permit strength design. The next reason for being familiar with WSD is that many designers still use the method and will continue to do so in the foreseeable future despite the several advantages of strength design. Their attitude is expressed by the statement: "I've got a method that works and I like it." For this reason the change of practicing designers from WSD to strength design has been rather slow.

It is recognized today that the design of reinforced concrete structures should take advantage of the best features of both the strength and working stress methods. For instance, if the strength method alone is used to design a member, it may have excessive deflections and cracks under service loads even though its factor of safety against collapse is more than adequate. Thus to produce satisfactory structures, the members may often be initially proportioned by strength design, while their deflections and crack widths at service loads are checked with WSD.

2.3 ASSUMPTIONS MADE FOR WSD

The accurate estimation of the stresses in reinforced concrete members under working- or service-load conditions is very difficult because of the

[1] *Standard Specifications for Highway Bridges*, 11th ed., 1973 (Washington, D.C.: Association of State Highway Officials), pp. 75–95.

effects of shrinkage, tensile cracking, creep, and so on. As a result, it is concluded that conditions at failure provide a better measure of performance than does WSD. Nevertheless, this chapter and the next are used to present the basic theory used in WSD.

The following assumptions are made for this discussion:

1. A plane section before bending remains a plane section after bending.
2. Stress is proportional to strain; that is, Hooke's law applies to this nonhomogeneous material of concrete and steel.
3. The tensile strength of concrete is negligible and tensile forces are carried completely by steel reinforcing.
4. The concrete and steel bond together perfectly so that no slip occurs.

With regard to the third assumption, concrete does have a little tensile strength in bending but it is a very small percentage of its compression strength. Thus a plain concrete flexural member would fail in tension well before the strength of the concrete on the compression side of the beam was utilized. It is assumed, therefore, under service loads that the concrete has cracked on the tensile side. A brief discussion of concrete's tensile strength as it affects the behavior of beams is presented in the next section of this chapter.

The assumptions listed are fairly good with a notable exception for the second one. Stress is proportional to strain as long as the concrete compression stress is less than about one-half of its 28-day compressive strength. In this regard you should again examine Figure 1.1.

2.4 INTRODUCTION TO BENDING OF CONCRETE BEAMS

If a relatively long reinforced concrete beam has a load applied to it that is gradually increased, the beam will go through three distinct stages before collapse occurs. These are (1) the uncracked concrete stage, (2) the concrete cracked-elastic stresses stage, and (3) the ultimate strength stage. Each of these stages is briefly introduced in this section; the first two are considered in detail in this chapter while the third stage is presented in Chapter 4. A relatively long beam is used so that shear will not be a factor in this initial discussion.

1.) Uncracked Concrete Stage

At small loads when the tensile stresses are less than the *modulus of rupture* (the bending tensile stress at which the concrete begins to crack), the entire cross section of the beam resists bending, with compression on one side and tension on the other. Figure 2.1 shows the variation of stresses

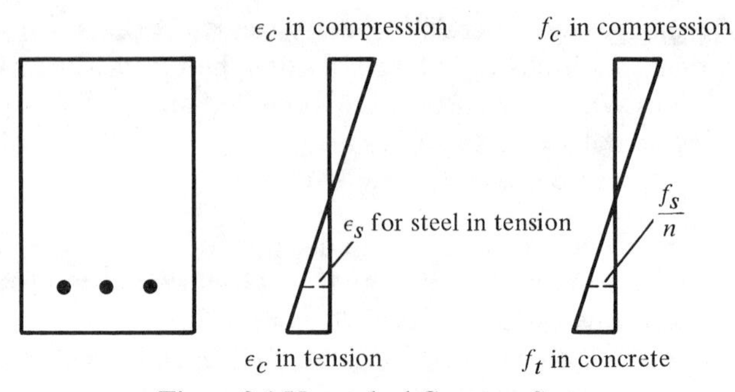

Figure 2.1 Uncracked Concrete Stage

and strains for these small loads; a numerical example of this type is presented in Section 2.5.

Concrete Cracked-Elastic Stresses Stage

As the load is increased after the modulus of rupture of the beam is exceeded, cracks begin to develop in the bottom of the beam. The moment at which these cracks begin to form—that is, when the tensile stress in the bottom of the beam equals the modulus of rupture—is referred to as the *cracking moment*. As the load is further increased, these cracks quickly spread up to the vicinity of the neutral axis, and then the neutral axis begins to move upward. Such a cracked section is shown in Figure 2.2(a).

Now that the bottom has cracked, another stage is present because the concrete in the cracked zone obviously cannot resist tensile stresses—the steel must do it. This stage will continue as long as the compression in the top fibers is less than about one-half of the concrete's 28-day strength

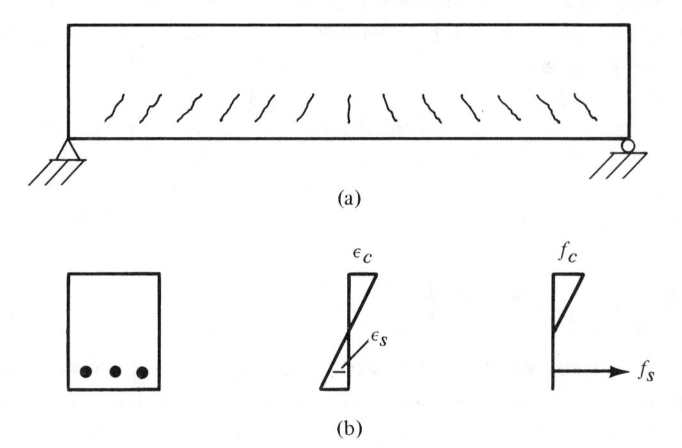

Figure 2.2 Concrete Cracked-Elastic Stresses Range

and as long as the steel stress is less than its yield point. The stresses and strains for this range are shown in Figure 2.2(b). In this stage the compressive stresses vary linearly with the distance from the neutral axis or as a straight line (thus the method is sometimes called "straight-line design" instead of WSD).

The straight-line stress-strain variation normally occurs in reinforced concrete beams under normal service-load conditions because at those loads the stresses are generally less than $0.50f_c'$. To compute the concrete and steel stresses in this range, the transformed area method (to be presented in Sections 2.5 and 2.6) is used.

Ultimate Strength Stage

As the load is increased further so that the compressive stresses are greater than one-half of the concrete's 28-day strength, the tensile cracks move upward, as does the neutral axis, and the concrete stresses begin to change appreciably from a straight line. The stress variation is much like that shown in Figure 2.3. You should again relate this figure to Figure 1.1. (This discussion is continued in more detail in Chapter 4.)

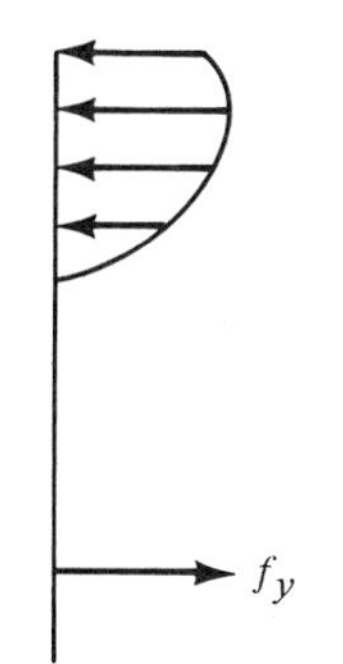

Figure 2.3 Ultimate Strength Stage

2.5 MODULAR RATIO

Example 2.1 presents sample stress calculations for a concrete beam whose tensile stresses are less than its modulus of rupture. As a result, no tensile cracks are present and the stresses are similar to those in a beam constructed with a homogeneous material.

There is, however, another material present in reinforced concrete beams: the reinforcing. An assumption of perfect bond between the two materials is made, and thus the strain in the concrete and the steel will be the same at equal distances from the neutral axis. But if the strains in the two materials at a particular point are the same, their stresses cannot be the same as they have different moduli of elasticity. Thus their stresses are in proportion to the ratio of their moduli of elasticity. The ratio of the steel modulus to the concrete modulus is called the *modular ratio n*:

$$n = \frac{E_s}{E_c}$$

If the modular ratio for a particular beam is 10, the stress in the steel will be 10 times the stress in the concrete at the same distance from the neutral axis. Another way of saying this is that 1 in.^{2} of steel will carry the same total force as 10 in.^{2} of concrete.

In Example 2.1 each square inch of steel is replaced with an equivalent area (nA_s) of fictitious concrete; this area is referred to as the *transformed area.* The resulting revised cross section or transformed section is handled by the usual methods for elastic homogeneous beams.

EXAMPLE 2.1

(a) For the beam of Figure 2.4, compute the bending stresses in the concrete top and bottom and in the reinforcing bars for a moment of 25 ft-k. The modulus of rupture of the concrete is 450 psi and n is 9.
(b) Compute the cracking moment of the section.

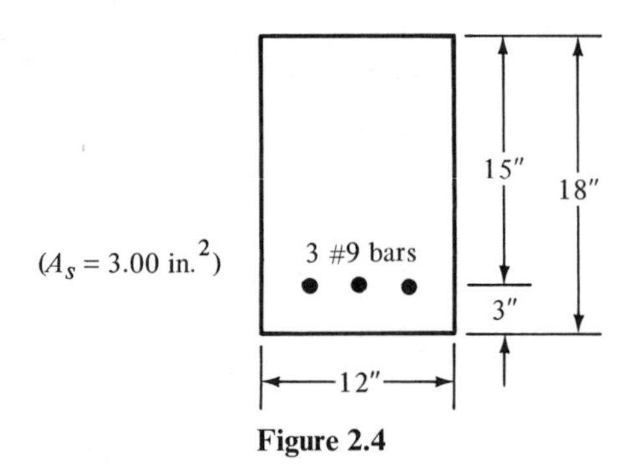

Figure 2.4

SOLUTION

Assuming the tensile bending stress is less than the modulus of rupture, the concrete and steel are replaced with a transformed section consisting only of concrete, as shown in Figure 2.5. The transformed area of the steel bars is nA_s, but as the holes in the concrete where the bars are located have an area equal to A_s, the area added to the basic concrete rectangle is $nA_s - A_s = (n - 1)A_s$. The moment of inertia of the bars about their own centroidal axis is assumed to be negligible.

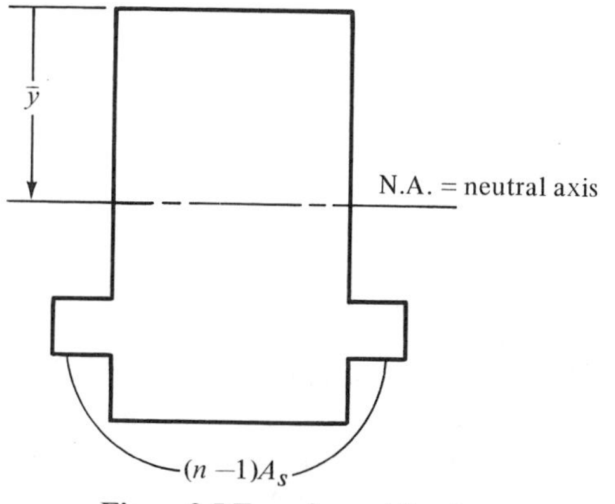

Figure 2.5 Transformed Section

Properties of section

$$A = (12)(18) + (9 - 1)(3.00) = 240 \text{ in.}^2$$

$$\bar{y} = \frac{(18)(12)(9) + (24)(15)}{240} = 9.60''$$

$$I_x = (\tfrac{1}{3})(12)(9.60^3 + 8.40^3) + (24)(5.40)^2 = 6610 \text{ in.}^4$$

(a) Bending stresses.

$$f_{ct} = \text{concrete stress in tension}$$

$$= \frac{Mc}{I} = \frac{(12)(25{,}000)(8.40)}{6{,}610} = 381 \text{ psi} < 450 \text{ psi}$$

Since this stress is less than the tensile strength or modulus of rupture of the concrete, no tensile cracks have developed, and the calculation of stresses by the uncracked transformed section is satisfactory.

$$f_{cc} = \text{concrete stress in compression}$$

$$= \frac{Mc}{I} = \frac{(12)(25{,}000)(9.60)}{6{,}610} = 436 \text{ psi}$$

f_s = steel stress in tension

= n times the stress in the concrete at the same distance from the neutral axis

$$= n\frac{Mc}{I} = (9)\frac{(12)(25{,}000)(5.40)}{6{,}610} = 2{,}206 \text{ psi}$$

(b) Cracking moment. The ACI Code (9.5.2.3) suggests that I_{gross} of the section, not counting the reinforcing, be used for calculating the cracking moment, but the author, to be consistent with the calculation of part (a) of this example, uses the transformed I for this one example.

$$M_{cr} = \frac{f_r I}{c} = \frac{(450)(6{,}610)}{8.40} = 354{,}100''\# = 29.5 \text{ ft-k}$$

2.6 ELASTIC STRESSES—CONCRETE CRACKED

When the bending moment is sufficiently large to cause the tensile stress in the extreme fiber of the concrete to be greater than the modulus of rupture, it is assumed that all the concrete on the tensile side of the beam is cracked and must be neglected in flexure calculations. As illustrated in Figure 2.6, the steel bars are replaced with an equivalent area of fictitious concrete (nA_s), which supposedly can resist tension. Also in the figure is a diagram showing the stress variation in the beam. On the tensile side a dotted line is shown because the diagram is discontinuous. There the concrete is assumed to be cracked and unable to resist tension. The value shown opposite the steel is the fictitious stress in the concrete if it could carry tension. This value is shown as f_s/n because it must be multiplied by n to give the steel stress f_s.

Examples 2.2, 2.3, and 2.4 are transformed area problems that illustrate the calculations necessary for determining the stresses and resisting moments for reinforced concrete beams. The first step to be taken in each of these problems is to locate the neutral axis, which is assumed to be located a distance x from the compression surface of the beam. The moment of the compression area of the beam cross section about the neutral axis must equal the moment of the tensile area about the neutral axis. The resulting quadratic equation can be solved by completing the squares.

After the neutral axis is located, the moment of inertia of the transformed section is calculated, and the stresses in the concrete and the steel are computed with the flexure formula.

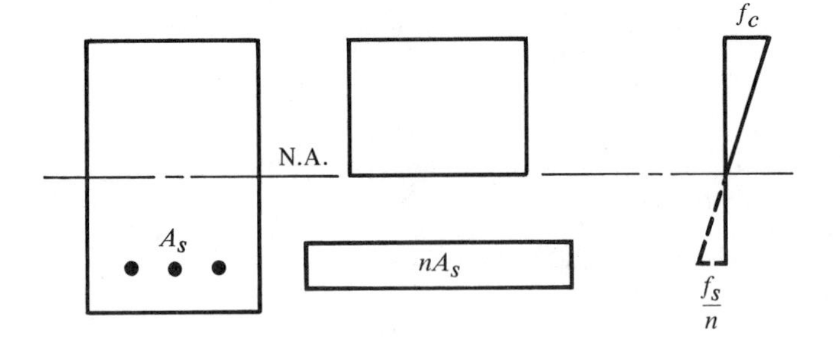

Figure 2.6

Arch Bridge, Pittsburgh, Pennsylvania (Courtesy Portland Cement Association).

EXAMPLE 2.2

Calculate the bending stresses in the beam shown in Figure 2.7 by using the transformed area method; $n = 9$ and $M = 70$ ft-k.

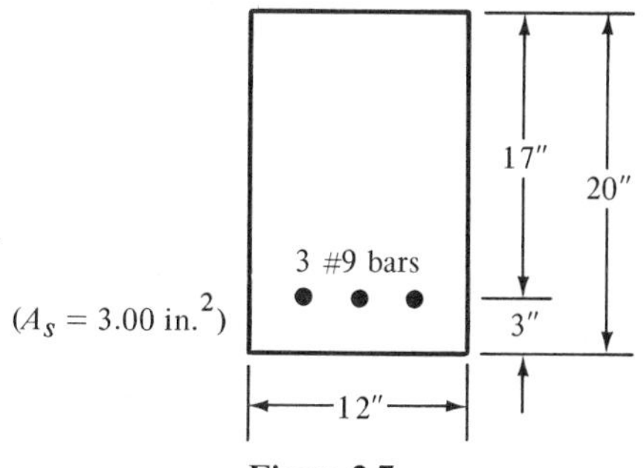

Figure 2.7

SOLUTION

Taking moments about neutral axis (referring to Figure 2.8)

$$(12x)\left(\frac{x}{2}\right) = (9)(3.00)(17 - x)$$

$$6x^2 = 459 - 27.00x$$

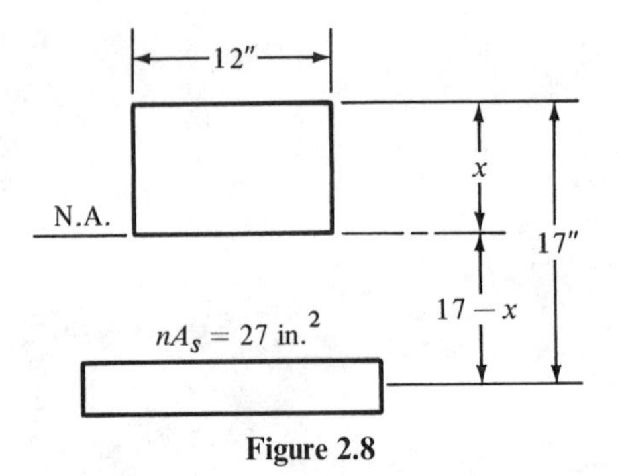

Figure 2.8

Solving by completing the square

$$6x^2 + 27.00x = 459$$
$$x^2 + 4.50x = 76.5$$
$$(x + 2.25)(x + 2.25) = 76.5 + (2.25)^2$$
$$x + 2.25 = \sqrt{76.5 + (2.25)^2} = 9.03''$$
$$x = 6.78''$$

Moment of inertia

$$I = (\tfrac{1}{3})(12)(6.78)^3 + (27.00)(10.22)^2 = 4{,}067 \text{ in.}^4$$

Bending stresses

$$f_c = \frac{Mc}{I} = \frac{(12)(70{,}000)(6.78)}{4{,}067} = 1{,}400 \text{ psi}$$

$$f_s = n\frac{Mc}{I} = (9)\frac{(12)(70{,}000)(10.22)}{4{,}067} = 18{,}998 \text{ psi}$$

EXAMPLE 2.3

Determine the resisting moment of the beam of Example 2.2 if the allowable stresses are $f_c = 1{,}350$ psi and $f_s = 20{,}000$ psi.

SOLUTION

$$M_c = \frac{f_c I}{c} = \frac{(1{,}350)(4{,}067)}{6.78} = 809{,}800''\# = \underline{67.5 \text{ ft-k}}$$

$$M_s = \frac{f_s I}{nc} = \frac{(20{,}000)(4{,}067)}{(9)(10.22)} = 884{,}323''\# = 73.7 \text{ ft-k}$$

Discussion

For a given beam the concrete and steel will not usually reach their maximum allowable stresses at exactly the same bending moments. Such is the case for this example beam, where the concrete reaches its maximum permissible stress at 67.5 ft-k while the steel does not reach its maximum value until 73.7 ft-k is applied. The resisting moment of the section is 67.5 ft-k because if that value is exceeded, the concrete becomes overstressed even though the steel stress is less than its allowable.

EXAMPLE 2.4

Compute the bending stresses in the beam shown in Figure 2.9 by using the transformed area method; $n = 8$ and $M = 110$ ft-k.

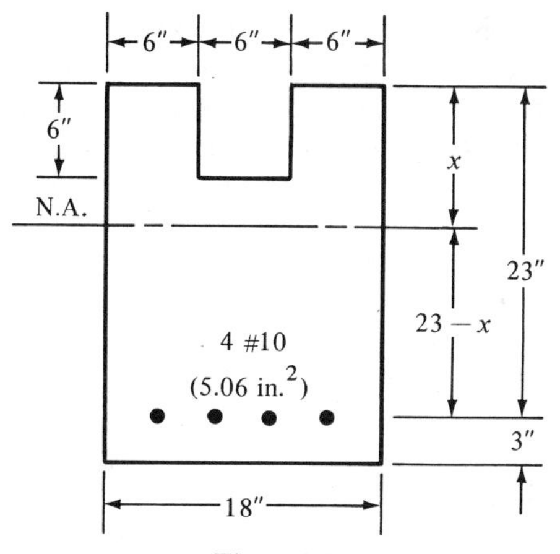

Figure 2.9

SOLUTION

Locating neutral axis

$$(18x)\left(\frac{x}{2}\right) - (6)(6)(x - 3) = (8)(5.06)(23 - x)$$

$$9x^2 - 36x + 108 = 931 - 40.48x$$

$$9x^2 + 4.48x = 823$$

$$x^2 + 0.50x = 91.44$$

$$(x + 0.25)(x + 0.25) = 91.44 + (0.25)^2 = 91.50$$

$$x + 0.25 = \sqrt{91.50} = 9.57$$

$$x = 9.32''$$

Moment of inertia

$$I = (\tfrac{1}{3})(6)(9.32)^3(2) + (\tfrac{1}{3})(6)(3.32)^3 + (8)(5.06)(13.68)^2 = 10{,}887 \text{ in.}^4$$

Computing stresses

$$f_c = \frac{(12)(110{,}000)(9.32)}{10{,}887} = 1{,}130 \text{ psi}$$

$$f_s = (8)\frac{(12)(110{,}000)(13.68)}{10{,}887} = 13{,}269 \text{ psi}$$

Example 2.5 illustrates the determination of the stresses in a T beam by the transformed area method. A detailed discussion of the uses, flange widths, and other information concerning such beams is presented in Chapter 5.

In Example 2.5 it is not known initially whether the neutral axis falls in the flange or down in the stem or web of the beam. An initial assumption is made that it is in the flange, and the usual quadratic equation is written. The computed value of x is greater in this case than the flange thickness, so the neutral axis falls in the web. The equation is rewritten to properly show the shape of the compression side of the beam and another value of x is determined. From a practical standpoint there is usually not a great deal of difference between the two values.

You should notice that the stresses in the compression side of T beams are usually quite small. That side of the beam is in effect much stronger than the other side. It is rather similar to a steel beam with a tremendous cover plate on top. The stresses in the top will be rather small compared to those on the tensile side of the beam.

EXAMPLE 2.5

Compute the bending stresses in the T beam shown in Figure 2.10; $n = 9$ and $M = 110$ ft-k.

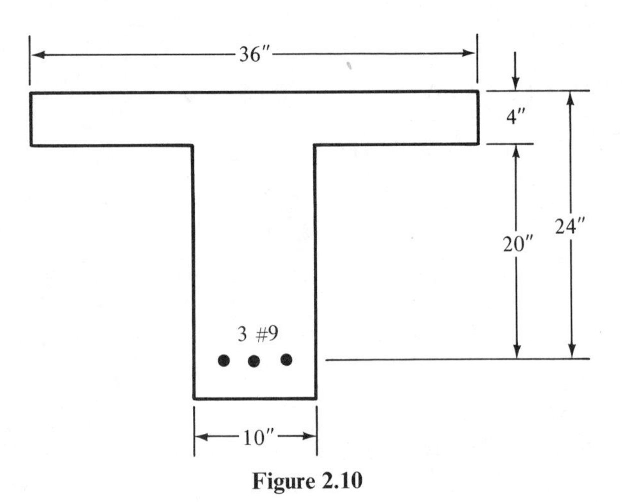

Figure 2.10

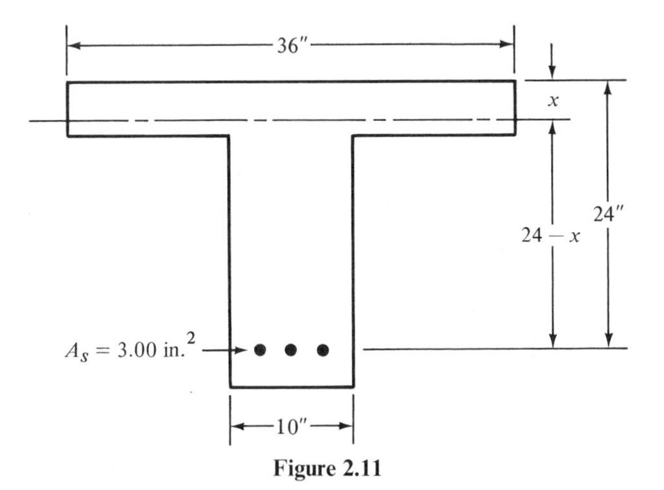

Figure 2.11

SOLUTION

Assuming neutral axis in flange (Figure 2.11)

$$(36x)\left(\frac{x}{2}\right) = (9)(3.00)(24 - x)$$

$$18x^2 = 648 - 27x$$

$$18x^2 + 27x = 648$$

$$x^2 + 1.50x = 36$$

$$(x + 0.75)(x + 0.75) = 36 + (0.75)^2 = 36.56$$

$$x + 0.75 = \sqrt{36.56} = 6.05''$$

$$x = 5.30'' > 4.00''$$

Assuming neutral axis in web (Figure 2.12)

$$(36)(4)(x - 2) + (10)(x - 4)\left(\frac{x - 4}{2}\right) = (9)(3.00)(24 - x)$$

$$144x - 288 + 5x^2 - 40x + 80 = 648 - 27x$$

$$5x^2 + 131x = 856$$

$$x^2 + 26.2x = 171.2$$

$$(x + 13.1)(x + 13.1) = 171.2 + (13.1)^2 = 342.8$$

$$x + 13.1 = \sqrt{342.8} = 18.52$$

$$x = 5.42'' > 4''$$

$\therefore$ N.A. is in web as assumed

Moment of Inertia

$$I = (\tfrac{1}{3})(36)(5.42)^3 - (\tfrac{1}{3})(26)(1.42)^3 + (9)(3.00)(18.58)^2 = 11{,}207 \text{ in.}^4$$

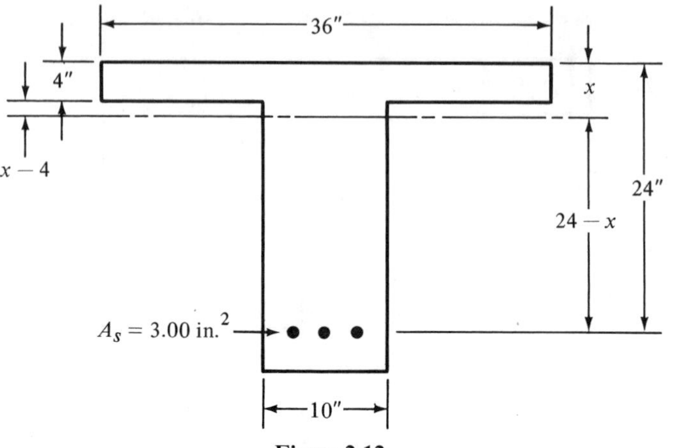

Figure 2.12

Computing stresses

$$f_c = \frac{(12)(110{,}000)(5.42)}{11{,}207} = 638 \text{ psi}$$

$$f_s = (9)\frac{(12)(110{,}000)(18.58)}{11{,}207} = 19{,}696 \text{ psi}$$

2.7 DOUBLY REINFORCED BEAMS

Example 2.6 illustrates the analysis of a doubly reinforced concrete beam, that is one that has compression steel as well as tensile steel. Compression steel is generally thought to be uneconomical, but there are occasional situations where its use is quite advantageous.

Compression steel will permit the use of appreciably smaller beams than those which make use of tensile steel only. Reduced sizes can be very important where space or architectural requirements limit the sizes of beams. Compression steel is quite helpful in reducing long-term deflections, and such steel is useful for positioning stirrups or shear reinforcing, a subject to be discussed in Chapter 7. A detailed discussion of doubly reinforced beams is presented in Chapter 5.

The creep or plastic flow of concrete was briefly mentioned in Section 1-3. Should the compression side of a beam be reinforced, the long-term stresses in that reinforcing will be greatly affected by the creep in the concrete. As time goes by, the compression concrete will compact more tightly, leaving the reinforcing bars (which themselves have negligible creep) to carry more and more of the load.

As a consequence of this creep in the concrete, the stresses in the compression bars are assumed to double as time goes by. In Example 2.6

the transformed area of the compression bars is assumed to equal $2n$ times their area A_s'.

On the subject of "hair splitting," it will be noted in the example that the compression steel area is really multiplied by $2n - 1$. The transformed area of the compression side equals the compression area of the concrete plus $2nA_s'$ minus the area of the holes in the concrete ($1A_s'$), which theoretically should not have been included in the concrete part. This equals the compression concrete area plus $(2n - 1)A_s'$. Similarly, $2n - 1$ is used in the moment of inertia calculations. The stresses in the compression bars are determined by multiplying $2n$ times the stresses in the concrete located at the same distance from the neutral axis.

EXAMPLE 2.6

Compute the bending stresses in the beam shown in Figure 2.13; $n = 10$ and $M = 118$ ft-k.

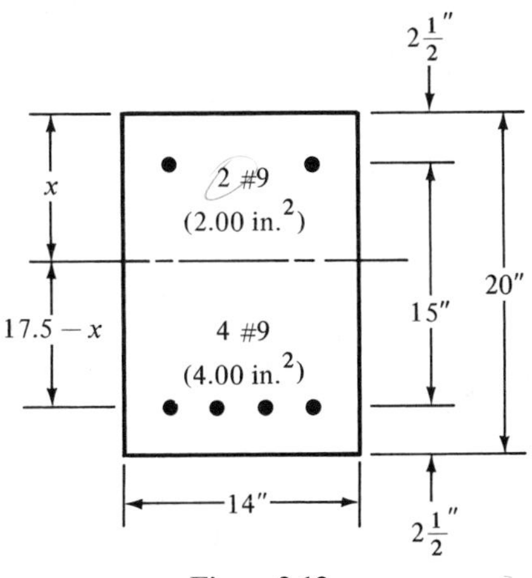

Figure 2.13

SOLUTION

Locating neutral axis

$$(14x)\left(\frac{x}{2}\right) + (20 - 1)(2.00)(x - 2.5) = (10)(4.00)(17.5 - x)$$

$$7x^2 + 38x - 95 = 700 - 40x$$

$$7x^2 + 78x = 795$$

$$x^2 + 11.14x = 113.57$$

$$x + 5.57 = \sqrt{113.57 + (5.57)^2} = 12.02$$

$$x = 6.45 \text{ in.}$$

Moment of inertia

$$I = (\tfrac{1}{3})(14)(6.45)^3 + (20 - 1)(2.00)(3.95)^2 + (10)(4.00)(11.05)^2 = 6{,}729 \text{ in.}^4$$

Bending stresses

$$f_c = \frac{(12)(118{,}000)(6.45)}{6{,}729} = 1{,}357 \text{ psi}$$

$$f_s' = 2n\frac{Mc}{I} = (2)(10)\frac{(12)(118{,}000)(3.95)}{6{,}729} = 16{,}624 \text{ psi}$$

$$f_s = (10)\frac{(12)(118{,}000)(11.05)}{6{,}729} = 23{,}253 \text{ psi}$$

2.8 FORMULAS FOR REVIEW OF RECTANGULAR BEAMS

The transformed area method presented in the last few pages is for the analysis of reinforced concrete beams of different shapes. It is possible to develop a set of formulas for a particular shaped beam with which the analysis can be quickly carried out. These formulas, which are derived by transformed area, yield exactly the same stresses. The general transformed area method is applicable to any shape of concrete section in flexure, whereas a special set of equations would have to be developed for each different shape of beam cross section. In this section a set of such equations is developed for rectangular beams with tensile steel only. Similar expressions can be derived for T beams, doubly reinforced beams, I beams, and others by the same process. For this discussion reference is made to Figure 2.14, and the analysis is made by transformed area purely in formula fashion. In the figure the letter d is used to represent the distance from the compression face of the beam to the center of gravity of the tensile reinforcing. This distance is referred to as the effective depth of the beam. It will also be noted that x is replaced with kd in the figure.

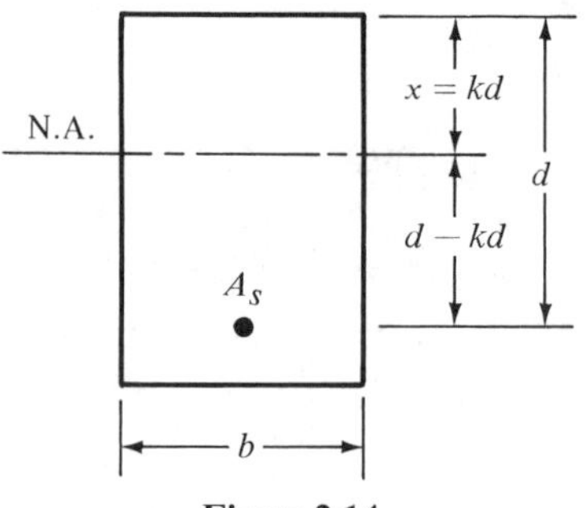

Figure 2.14

Placing Welded Wire Fabric, Cramlington, Northumberland, England (Courtesy Cement and Concrete Association).

Taking moments about the neutral axis

$$bkd\left(\frac{kd}{2}\right) = nA_s(d - kd)$$

Letting p = percentage of steel = A_s/bd; thus $A_s = pbd$:

$$\frac{bk^2d^2}{2} = n\rho bd^2 - n\rho bd^2k$$

$$k^2 = 2\rho n - 2\rho nk$$

$$k^2 + 2\rho nk = 2\rho n$$

$$(k + \rho n)(k + \rho n) = 2\rho n + (\rho n)^2$$

$$k + \rho n = \sqrt{2\rho n + (\rho n)^2}$$

$$k = \sqrt{2\rho n + (\rho n)^2} - \rho n \tag{2.1}$$

The internal forces (C = total compression and T = total tension) shown in Figure 2.15 are now considered. C is located at the center of gravity (c.g.) of the compression stress triangle that is a distance $kd/3$ from the top of the beam, and T is located at the center of gravity of the steel bars. The distance between C and T is shown as jd.

Solving for the value of *j*

$$jd = d - \frac{kd}{3}$$

$$j = 1 - \frac{k}{3}$$

The moment of the couple Cjd or Tjd must equal the external moment M, and from these expressions values for f_s and f_c can be obtained.

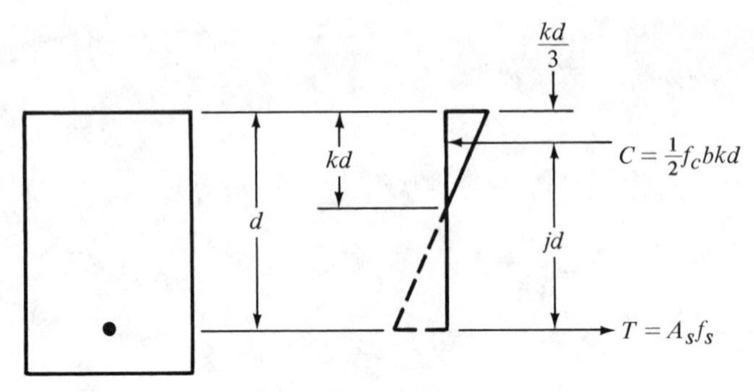

Figure 2.15

For the steel

$$Tjd = M$$
$$A_s f_s jd = M$$
$$f_s = \frac{M}{A_s jd} \tag{2.3}$$

For the concrete

$$Cjd = M$$
$$\frac{f_c}{2} bkdjd = M$$
$$f_c = \frac{2M}{bd^2kj} \tag{2.4}$$

EXAMPLE 2.7

Repeat Example 2.2 but use the review formulas developed in this section.

SOLUTION

$$\rho = \frac{A_s}{bd} = \frac{3.00}{(12)(17)} = 0.0147$$

$$\rho n = (0.0147)(9) = 0.1324$$
$$(\rho n)^2 = (0.1323)^2 = 0.01750$$
$$k = \sqrt{2\rho n + (\rho n)^2} - \rho n = \sqrt{(2)(0.1323) + 0.01750} - 0.1323 = 0.399$$

$$j = 1 - \frac{k}{3} = 1 - \frac{0.399}{3} = 0.867$$

$$f_c = \frac{2M}{bd^2kj} = \frac{(2)(12)(70{,}000)}{(12)(17)^2(0.399)(0.867)} = 1{,}400 \text{ psi}$$

$$f_s = \frac{M}{A_s jd} = \frac{(12)(70{,}000)}{(3.00)(0.867)(17)} = 18{,}997 \text{ psi}$$

2.9 SI UNITS

The International Bureau of Weights and Measures has as its goal the establishment of a rational and coherent worldwide system of units. In 1960 they named this system the "International System of Units," with the abbreviation SI in all languages. SI units, which are currently being adopted by quite a few countries including most English-speaking nations (Britain, Australia, Canada, South Africa, and New Zealand), differ from the metric system now being used in most European countries and in many other parts of the world.

The SI system has the very important advantage that only one unit is given for each physical quantity, such as the meter (m) for length, the kilogram (kg) for mass, the second (s) for time, the newton (N) for force, and so on. From these basic units other units are derived as follows:

area	square meter (m^2)
acceleration	meter per second squared (m/s^2)
force	kilogram meter per second squared ($kg \cdot m/s^2$) = newton (N)
stress	newton per square meter (N/m^2) = pascal (Pa)

The multiples of these units are in the decimal system and are expressed in powers of ten that are multiples of three, as shown in Table 2.1, where length units are illustrated. In other words, multiple and submultiple steps of one thousand are recommended. It will be noted that the centimeter is not shown because its value (which would be 10^{-2} in the table) is inconsistent with the theory of having prefixes that are ternary powers of 10.

Table 2.1 SI Decimal Multiples

SI Symbol	Name	Multiplier	Example
G	giga	1 000 000 000	Gm = 1 000 000 000 meters
M	mega	1 000 000	Mm = 1 000 000 meters
k	kilo	1 000	km = 1 000 meters
m	milli	0.001	mm = 0.001 meters
μ	micro	0.000 001	μm = 0.000 001 meters
n	nano	0.000 000 001	nm = 0.000 000 001 meters

In many countries of the world the comma is used to indicate a decimal; thus to avoid confusion in the SI system, spaces rather than commas are used. For a number having four or more digits, the digits are separated into groups of threes, counting both right and left from the decimal. For example, 3,245,621 is written as 3 245 627 and 2,015.3216 is written as 2 015.321 6.

When units are to be multiplied together, a dot is used to separate them. In the SI examples of this text, a common unit is the newton meter. It is to be written as N·m (not mN, which represents millinewton). There is a great premium in the SI system on symbology. As a result, the designer must be exceptionally careful to use the correct symbols. For example, it is correct to write 100 meters, but in abbreviation 100 m must be used and not 100 ms, which would be 100 milliseconds.

Table 2.2 gives conversion values for some customary measurements to the SI system. Perhaps if you were to memorize a few approximate values from this table, you would begin to get a feel for the relative values between the two systems. For instance, 25 mm is approximately equal to 1 in., and 300 mm is approximately equal to 1 ft.

Table 2.2 Common Values for Some Common Units

Customary Units	SI Units
1 in.	25.400 mm = 0.025 400 m
1 in.^{2}	645.16 mm^2 = 6.451 600 m^2 × 10^{-4}
1 ft	304.800 mm = 0.304 800 m
1 lb	4.448 222 N
1 kip	4 448.222 N = 4.448 222 kN
1 psi	6.894 757 kN/m^2 = 0.006 895 MN/m^2 = 0.006 895 N/mm^2
1 psf	47.880 N/m^2 = 0.047 880 kN/m^2
1 ksi	6.894 757 MN/m^2 = 6.894 757 MPa
1 in.-lb	0.112 985 N·m
1 ft-lb	1.355 818 N·m
1 in.-k	112.985 N·m
1 ft-k	1 355.82 N·m = 1.355 82 kN·m

The unit of stress in the SI system is the newton per square meter (N/m^2), which is also called a pascal (Pa). To have a feel for the magnitude of this number, a megapascal (MPa) is equal to 0.145 kip per square inch. Thus, 20 MPa is equal to 2.9 ksi or 2900 psi, and 30 MPa is equal to 4350 psi. You might think of 20- and 30-MPa concretes as being approximately equivalent to 3000- and 4000-psi concretes (actual values are 20.7 and 27.6 MPa).

In the same fashion, the terms in.-k and ft-k have been constantly used. In the SI system they are usually expressed in kN·m. One kN·m is equal to 0.738 ft-k. From Table 2.2 it can be seen that 1 pound is equal to 4.448 222 newtons. Thus, 1 newton is equal to about one-fourth of a pound, which is roughly equal to the weight of one apple.

Example 2.8 illustrates the calculation of stresses in a reinforced concrete beam using SI units. For convenience, millimeters are used for locating the neutral axis and calculating the moment of inertia. The stresses are obtained in newtons per square millimeter or megapascals and are then converted to pounds per square inch by dividing them by 0.006 895 from the conversion table. The last several problems at the end of this and most of the remaining chapters are to be solved by using SI units. It will be noted that Table A.2 (see Appendix) gives bar diameters, areas, and weights in both customary and SI units.

EXAMPLE 2.8

Compute the bending stresses in the beam shown in Figure 2.16 by using the transformed area method; $n = 10$ and $M = 100$ kN·m.

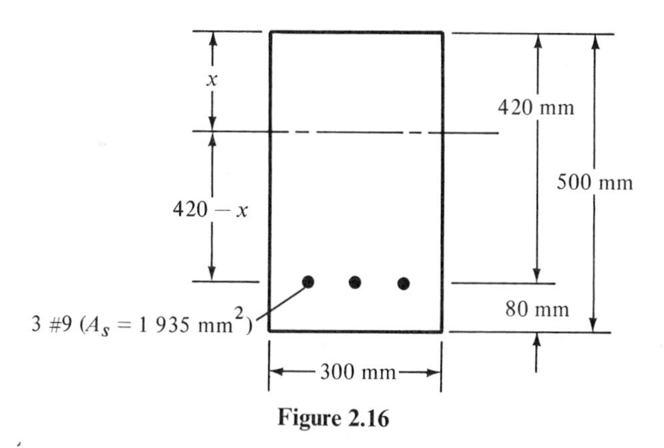

Figure 2.16

SOLUTION

Locating neutral axis

$$(300x)\left(\frac{x}{2}\right) = (10)(1\,935)(420 - x)$$

$$150x^2 = 8\,127\,000 - 19\,350x$$

$$x^2 + 129x = 54\,180$$

$$(x + 64.5)(x + 64.5) = 54\,180 + (64.5)^2 = 58\,340$$

$$x + 64.5 = \sqrt{58\,340} = 241.5$$

$$x = 177 \text{ mm}$$

Moment of inertia

$$I = (\tfrac{1}{3})(300)(177)^3 + (10)(1\,935)(243)^2 = 1.697 \times 10^9 \text{ mm}^4$$

Computing stresses (note: 10^6 is for units)

$$f_c = \frac{(100)(10^6)(177)}{1.697 \times 10^9} = 10.43 \text{ N/mm}^2 = 10.43 \text{ MPa} = 1\,513 \text{ psi}$$

$$f_s = (10)\frac{(100)(10^6)(243)}{1.697 \times 10^9} = 143.19 \text{ N/mm}^2 = 143.19 \text{ MPa} = 20\,767 \text{ psi}$$

PROBLEMS

2.1 to **2.24** Using the transformed area method, compute the bending stresses in the concrete and steel for the beams shown except for Problems 2.4, 2.5, 2.10, 2.12, 2.18, and 2.22, where other information is asked.

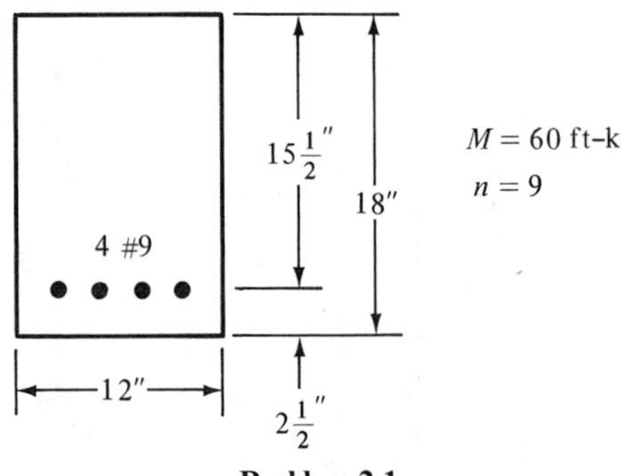

Problem 2.1

2.2 Repeat Problem 2.1 if 4 #7 bars are used.

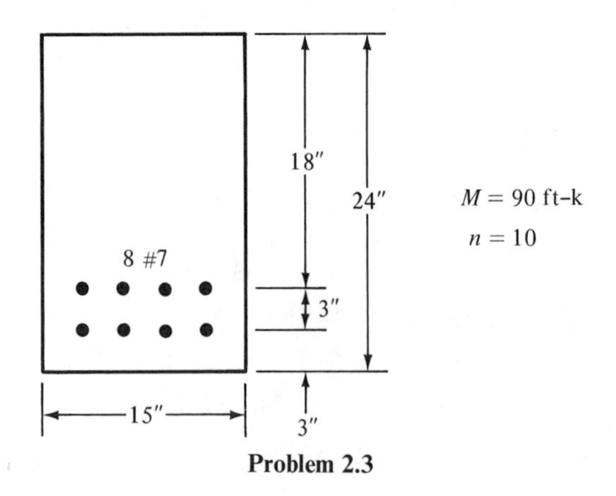

Problem 2.3

2.4 Compute the resisting moment of the beam of Problem 2.3 if $f_s = 20{,}000$ psi and $f_c = 1{,}125$ psi.

2.5 Compute the resisting moment of the beam of Problem 2.3 if 8 #5 bars are used and if $n = 10$, $f_s = 20{,}000$ psi and $f_c = 1{,}125$ psi.

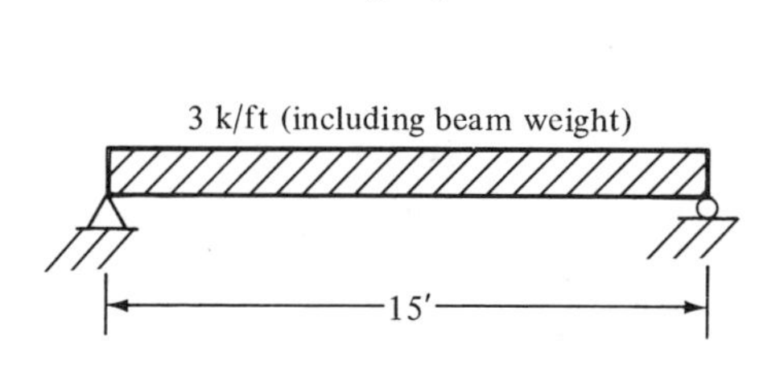

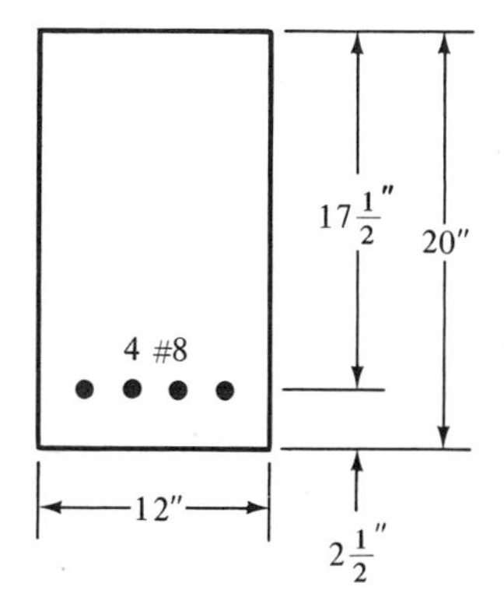

Problem 2.6

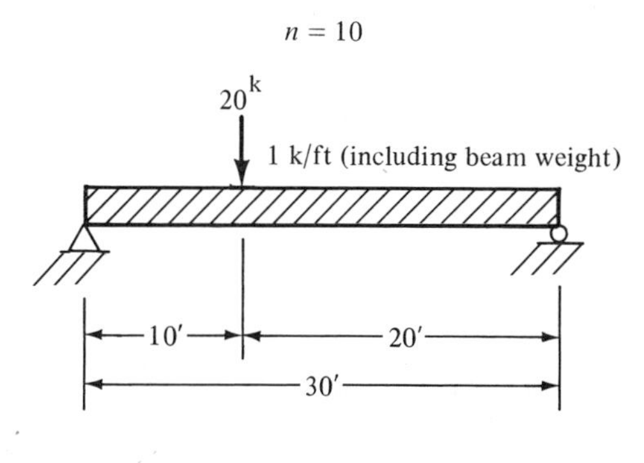

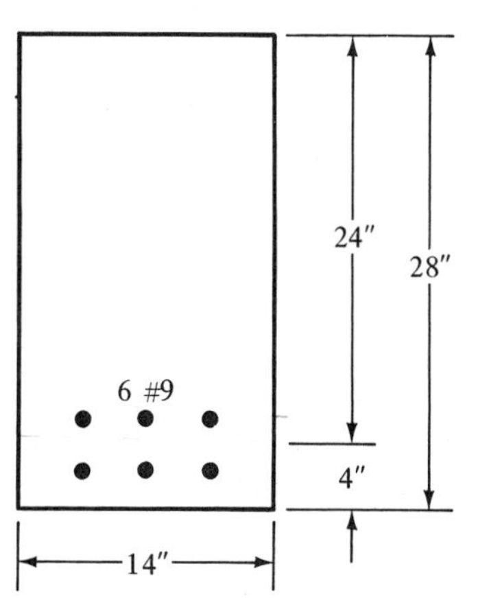

Problem 2.7

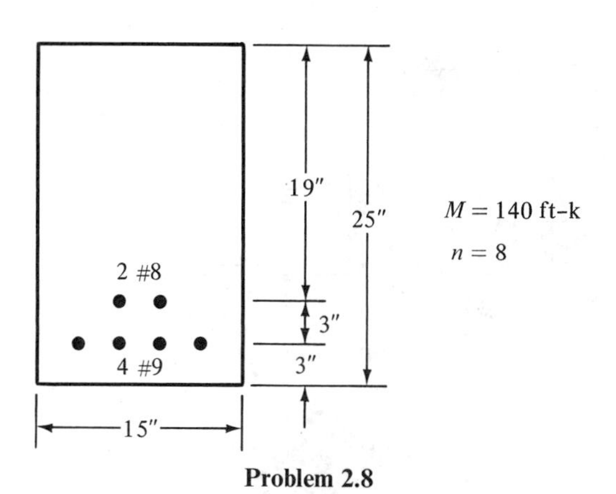

$M = 140$ ft-k

$n = 8$

Problem 2.8

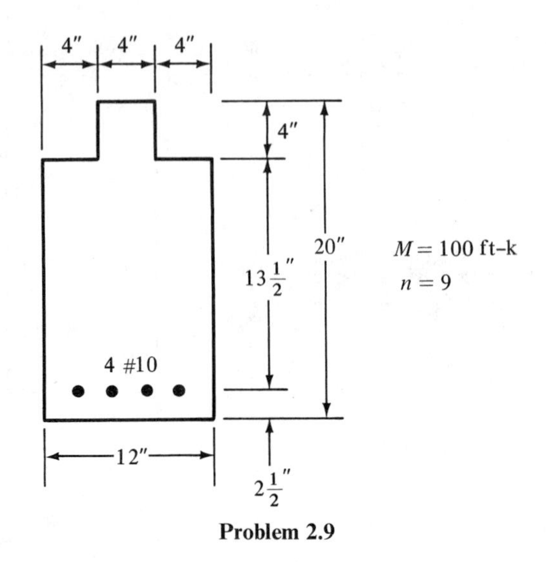

Problem 2.9

2.10 What allowable uniform load can the beam of Problem 2.9 support in addition to its own weight for a 24-ft simple span? Concrete weight = 150 #/ft^3, $f_s = 24{,}000$ psi, and $f_c = 1{,}350$ psi.

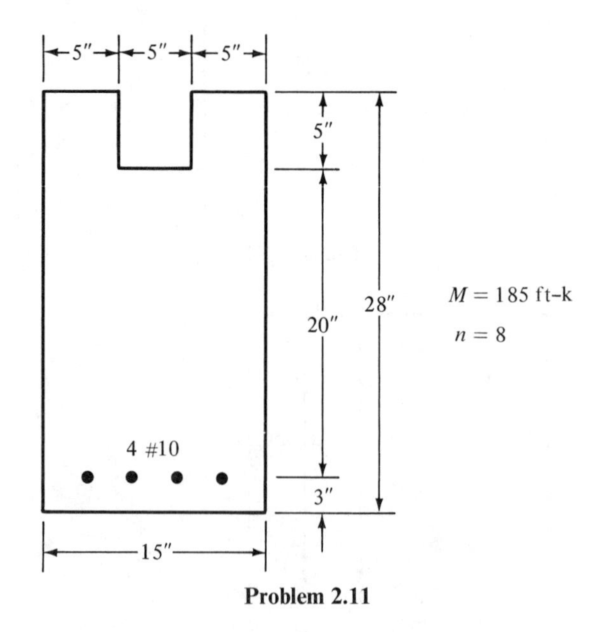

Problem 2.11

2.12 Compute the resisting moment of this section if $f_c = 1{,}350$ psi, $f_s = 24{,}000$ psi, and $n = 9$.

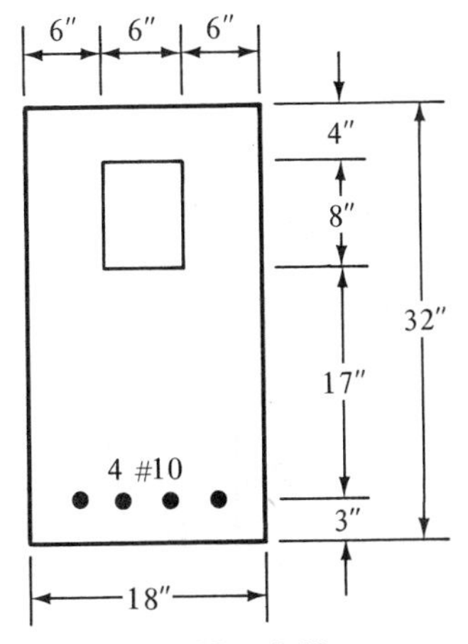

Problem 2.12

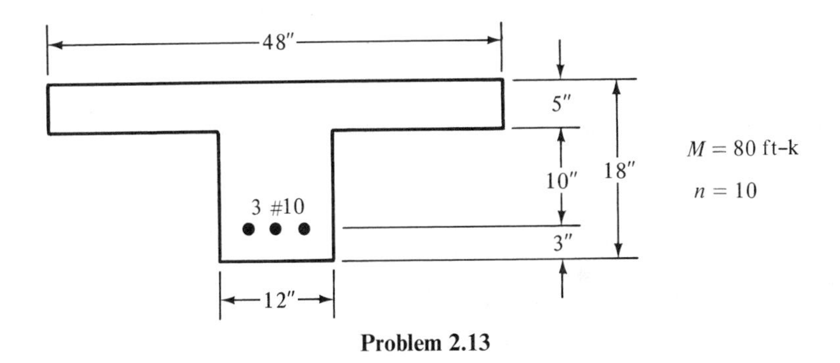

Problem 2.13

2.14 Repeat Problem 2.13 if the flange width is reduced to 24 in. and the moment to 70 ft k.

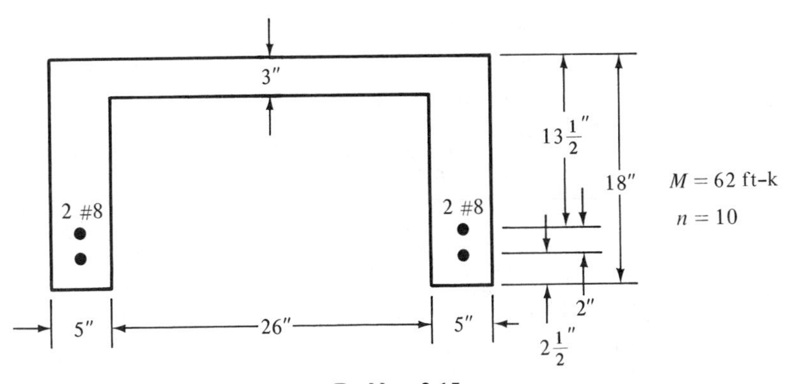

Problem 2.15

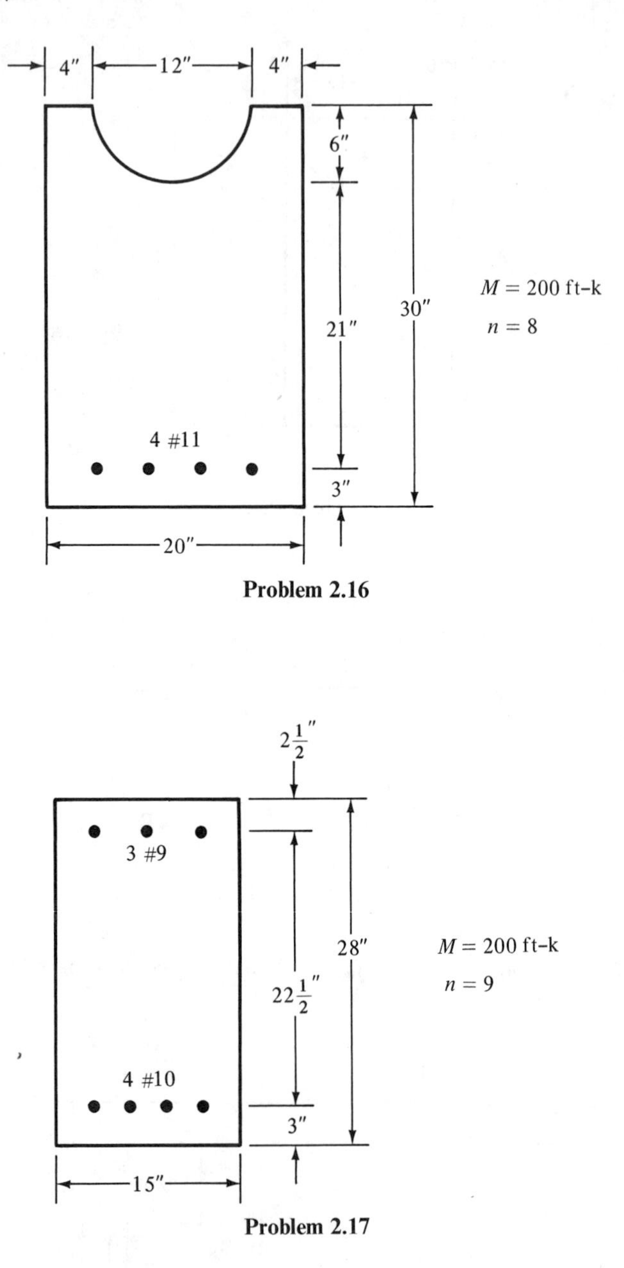

Problem 2.16

Problem 2.17

2.18 Compute the resisting moment of the beam of Problem 2.17 if $f_c = 1{,}350$ psi and $f_s = f_s' = 20{,}000$ psi.

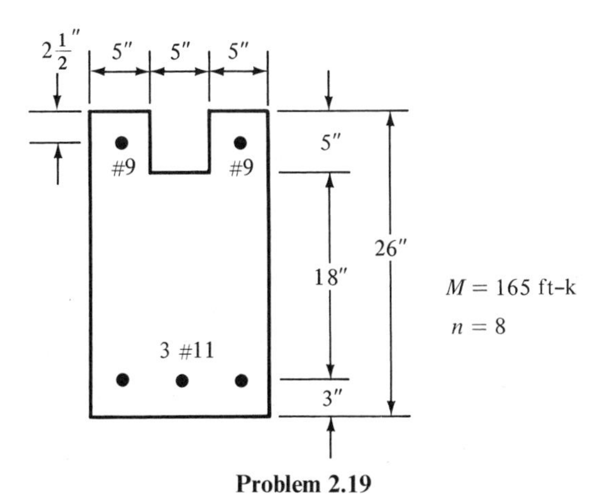

Problem 2.19

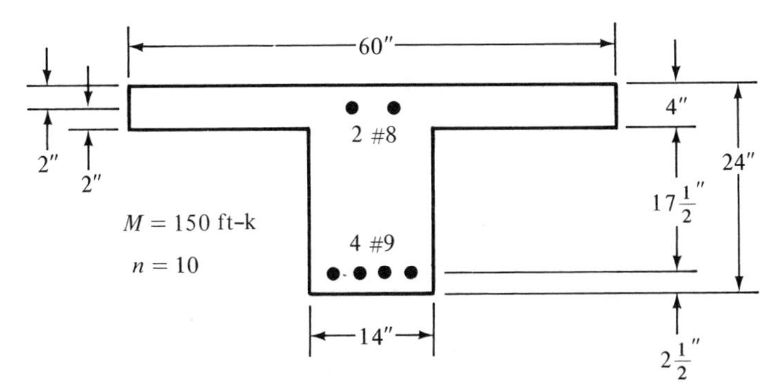

Problem 2.20

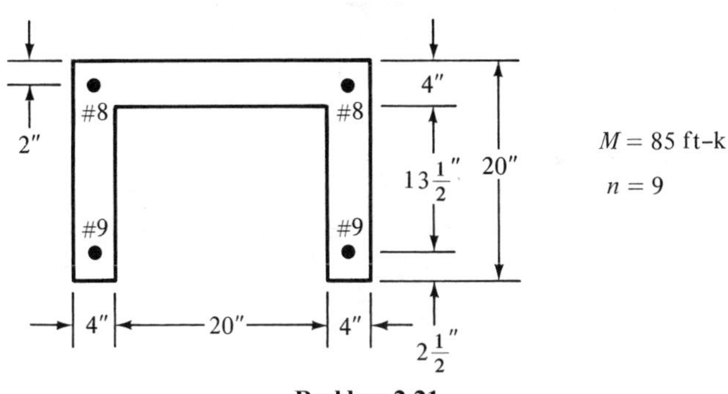

Problem 2.21

2.22 Compute the resisting moment of this section if $f_c = 1{,}800$ psi, $f_s = f_s' = 24{,}000$ psi, and $n = 8$.

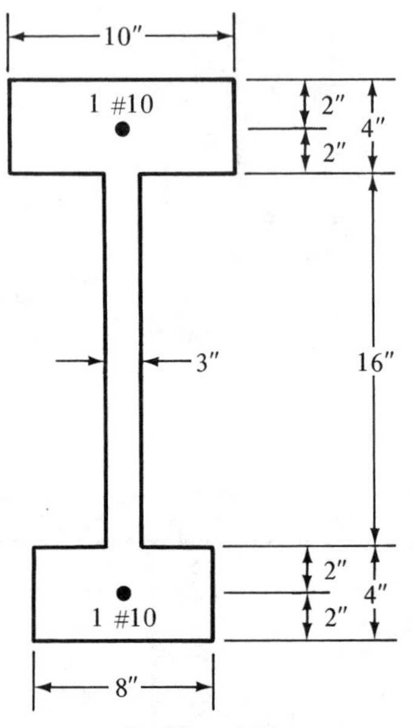

Problem 2.22

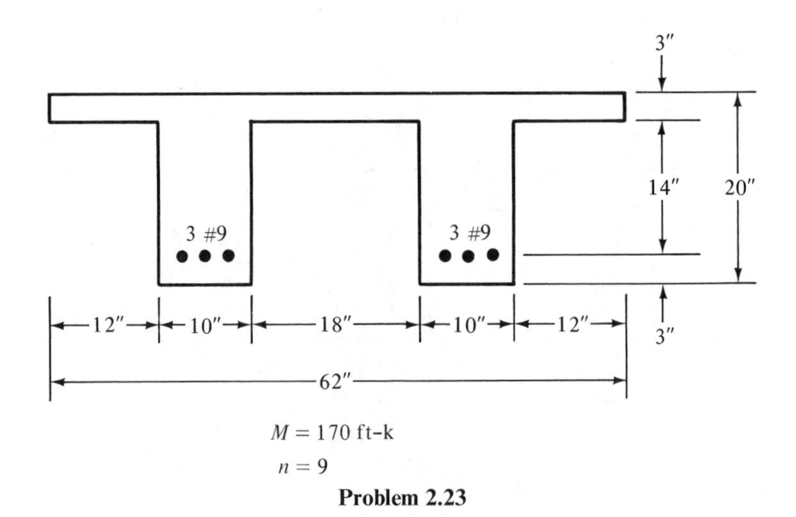

$M = 170$ ft-k

$n = 9$

Problem 2.23

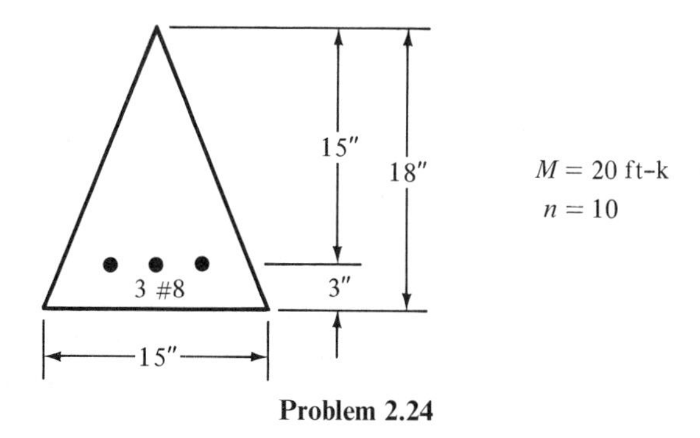

Problem 2.24

2.25 to **2.29** Compute the flexural stresses in the concrete and steel for the beams shown.

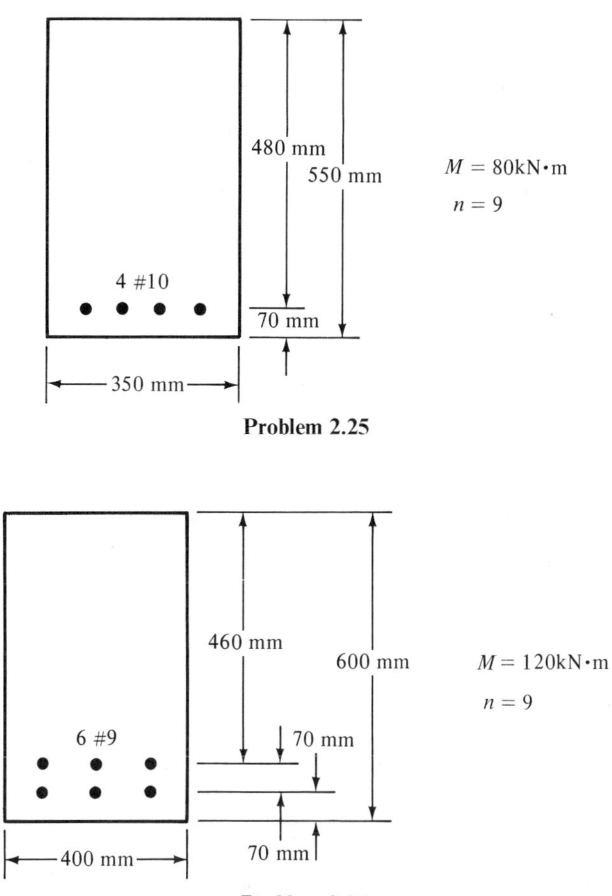

Problem 2.25

Problem 2.26

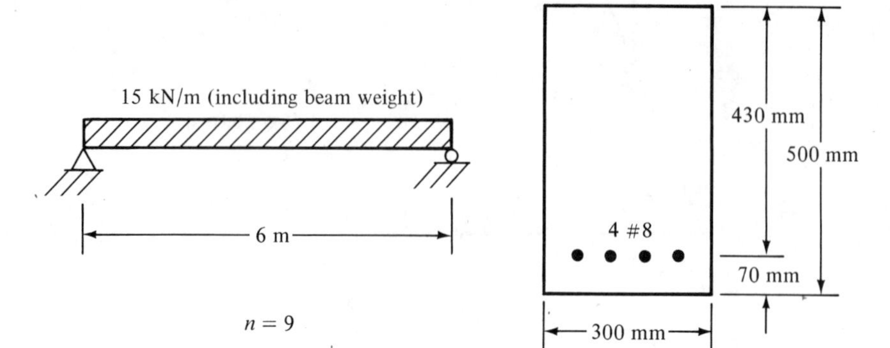

Problem 2.27

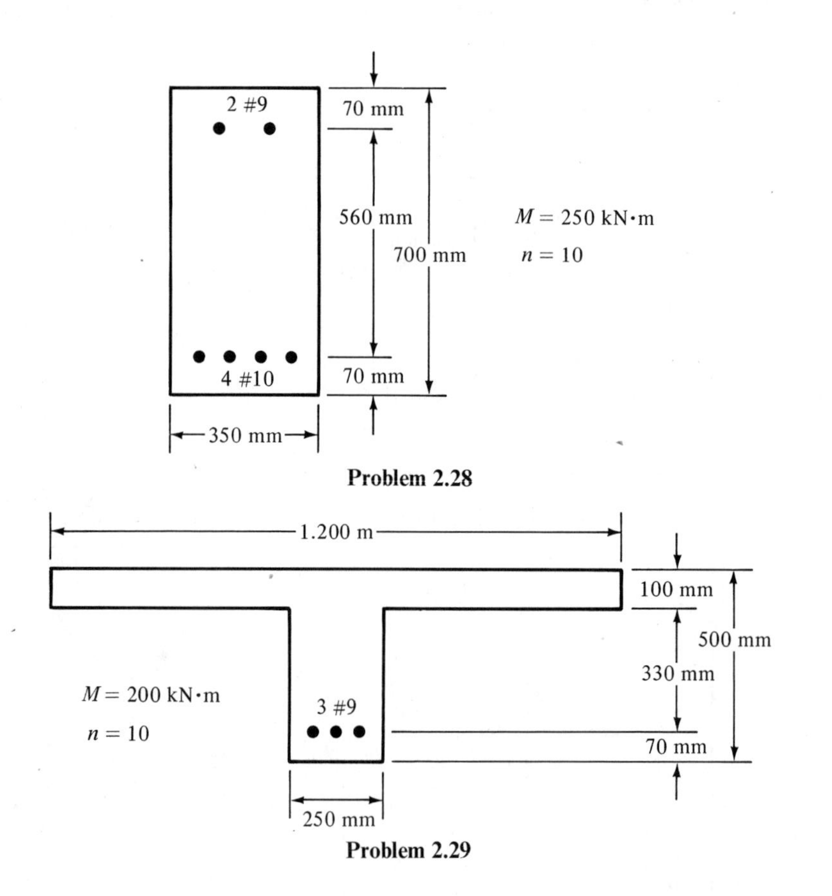

Problem 2.28

Problem 2.29

2.30 Compute the resisting moment of the beam of Problem 2.26 in kilonewton meters if $f_c = 9.3$ MPa and $f_s = 137.9$ MPa.

2.31 Compute the resisting moment of the beam of Problem 2.28 in kilonewton meters if $f_c = 7.8$ MPa and $f_s = f_s' = 137.9$ MPa.

CHAPTER 3

Working Stress Design (The Alternate Design Method)

3.1 INTRODUCTION

In working stress design a margin of safety is provided by permitting calculated flexure stresses to reach only a certain percentage of the ultimate strength of the concrete or of the yield strength of the reinforcing. These percentages are sufficiently small so that an approximately linear relation exists between the stress and strain in the concrete as well as in the reinforcing.

The Code (8.1.2) permits the design of reinforced concrete members using service loads and the working stress design method, except the method is now referred to as the *Alternate Design Method*. This method is the same as WSD for flexural members but some changes have been made for other designs. For instance, the permissible capacity of members subject to axial load and bending equals 40 percent of the value obtained by the strength design method.

Appendix B of the Code presents in some detail the allowable stresses and other conditions to be used for the alternate design method. The

Continuous Girder Bridge Over Chattahoochie River, Atlanta, Georgia (Courtesy Portland Cement Association).

placing of this information in an appendix may very well be a prelude to its complete deletion from the next edition of the Code in the 1980s.

From Appendix B of the Code, the maximum permissible concrete compressive stress in the extreme fiber of a member is $0.45f_c'$. The maximum allowable tensile stress in the reinforcing is 20,000 psi for grades 40 and 50 steels and 24,000 psi for grade 60 steels and steels with higher yield strengths. These values are summarized in Table A.7 (see Appendix of this text).

Should reference be made to the AASHTO or AREA bridge specifications for reinforced concrete, it will be found that those groups are more conservative than the ACI for obvious reasons. Bridge structures are, in general, subjected to more severe weather conditions than are buildings because of their more exposed locations; in addition, they must support more violent loadings from trucks or trains. The allowable stresses given in the 1973 AASHTO specification are $0.40f_c'$ for concrete and 20,000 psi for grade 40 steel and 24,000 psi for grade 60 steel.

3.2 DERIVATION OF WSD DESIGN FORMULAS

This section presents the derivation of the formulas needed for the design of rectangular beams with tensile steel only. Reference is made to Figure 3.1 for the beam cross section and the terms that will be used in the derivation. The steel is once again transformed into an equivalent area of fictitious concrete that can resist tension. This area is nA_s.

In WSD the most economical design possible is referred to as *balanced design.* A beam designed by this method will, under full service load, have its extreme fibers in compression stressed to their maximum permissible value f_c and its reinforcing bars stressed to their maximum

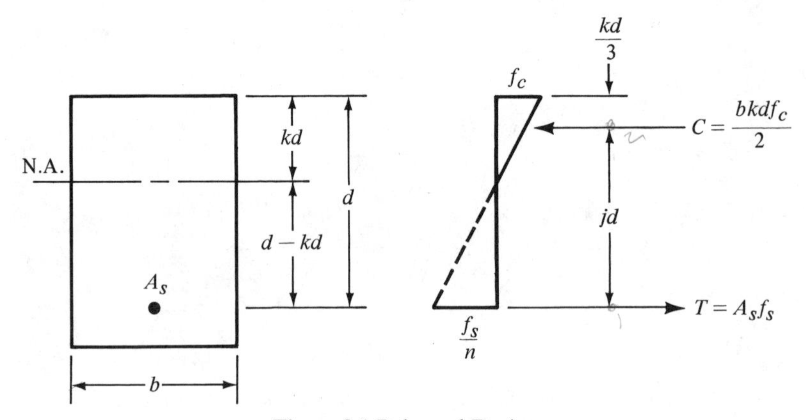

Figure 3.1 Balanced Design

permissible value f_s. Balanced design is the situation assumed for the beam and stress diagram shown in Figure 3.1.

In WSD the usual objective is to have a balanced design because it will often represent the most economical proportions. (Should a beam be designed in which the tensile steel reaches its maximum allowable tensile stress before the concrete fibers reach their maximum allowable compression stress, the beam is said to be *underreinforced*. Subsequent chapters concerning the strength design method will elaborate on this definition.)

The design formulas are derived on the basis of a consideration of the internal couple consisting of the two forces C and T. Once again, the total compression C equals the compression area bkd times the average compression stress $f_c/2$ and T equals $A_s f_s$. The sum of the horizontal forces in a beam in equilibrium is obviously zero and thus $C = T$. The internal moment can be written as Cjd or Tjd, and these are equated to the external moment M and the resulting expressions are solved for the beam dimensions and the steel area required.

As a straight-line variation of stress is assumed from f_c to f_s/n, the following ratio can be written and from it the design value of k obtained:

$$\frac{kd}{d} = \frac{f_c}{f_c + (f_s/n)}$$

$$k = \frac{f_c}{f_c + (f_s/n)} \tag{3.1}$$

In a similar manner j is determined.

$$jd = d - \frac{kd}{3}$$

$$j = 1 - \frac{k}{3} \tag{3.2}$$

Now using the internal couples:

$$M = Cjd$$

$$M = \frac{bkdf_c}{2} jd$$

$$bd^2 = \frac{2M}{f_c kj} \tag{3.3}$$

$$M = Tjd$$

$$M = A_s f_s jd$$

$$A_s = \frac{M}{f_s jd} \tag{3.4}$$

These equations (3.1 through 3.4) were derived for rectangular sections and they do not apply to sections where the compression area is not rectangular or to sections with compression reinforcing.

3.3 EXAMPLE DESIGN PROBLEMS

There are two general types of formulas that are presented for beams: those used for review and those used for design. In *review problems* the physical dimensions of the beam and the number, sizes, and placement of bars are given. It is desired to determine the flexural stresses for certain moments or the permissible moment a section can carry for certain allowable stresses. *Design problems* are those in which the member sizes are to be determined based on certain external moments and certain allowable steel and concrete fiber stresses. It is very important to realize that some of the formulas do not apply to both review and design. For this reason the formulas for the review and design of rectangular beams are repeated here.

Zimmer Nuclear Power Station Near Moscow, Ohio (Courtesy Symons Corporation).

For review of rectangular beams:

$$k = \sqrt{2pn + (pn)^2} - pn$$

$$j = 1 - \frac{k}{3}$$

$$f_c = \frac{2M}{bd^2kj}$$

$$f_s = \frac{M}{A_s jd}$$

For design of rectangular beams:

$$k = \frac{f_c}{f_c + (f_s/n)}$$

$$j = 1 - \frac{k}{3}$$

$$bd^2 = \frac{2M}{f_c kj}$$

$$A_s = \frac{M}{f_s jd}$$

Before the design of an actual beam is attempted, several miscellaneous topics need to be discussed. These include the following:

1. Estimated beam weight. The weight of the beam to be selected must be included in the calculation of the bending moment to be resisted, as the beam must support itself as well as the external loads. The weight estimates for the beams selected in this chapter are very close because the author was able to perform a little preliminary paperwork before making his estimates. You are not expected to be able to glance at a problem and estimate exactly the weight of the beam required. Following the same procedure as did the author, however, you can do a little figuring on the side and make a very reasonable estimate. For instance, you could calculate the moment due to the external loads only, select a beam size, and calculate its weight. From this beam size, you should be able to make a very good estimate of the weight of the final beam section.

2. Beam proportions. Unless architectural or other requirements dictate the proportions of reinforced concrete beams, the most economical beam sections are usually obtained when the ratio of b to d is in the range of $\frac{1}{2}$ to $\frac{2}{3}$. In Example 3.1 (to follow) the required bd^2 is 6323. Different values of b are assumed and the corresponding values of d computed. A pair of values of b and d are selected that are in the economical range given.

It will be noticed that the overall beam dimensions are selected to whole inches. This is done for simplicity in constructing forms or for the rental of forms, which are usually available in 1- or 2-in. increments.

3. Selection of bars. After the required reinforcing area is calculated, Table A.4 (see Appendix) can be used to select bars that provide the necessary area. For the usual situations bars of sizes #11 and smaller are practical. It is usually convenient to use bars of one size only in a beam, although occasionally two sizes will be used.

4. Cover. The reinforcing for concrete members must be protected from the surrounding environment; that is, fire and corrosion protection needs to be provided. To do this the reinforcing is located at certain minimum distances from the surface of the concrete so that a protective layer of concrete, called *cover*, is provided. In addition, the cover improves the bond between the concrete and the steel. In Section 7.7 of the ACI Code, minimum permissible cover is given for reinforcing bars under different conditions. Values are given for concrete beams, columns, and slabs, for cast-in-place members, for precast members, for prestressed members, for members exposed to earth and weather, for members not so exposed, and so on. The beam designed in Example 3.1 is assumed to be located inside a building and thus protected from the weather. For this case the Code requires a minimum cover of $1\frac{1}{2}$ in. of concrete outside of any reinforcement.

In Chapter 7 you will learn that vertical stirrups are used in most beams for shear reinforcing. Such a stirrup is shown for a beam in Figure 3.2 and a $1\frac{1}{2}$-in. cover is assumed to be required outside of the stirrup. Stirrups are commonly #3 bars and thus a $\frac{3}{8}$-in. diameter is assumed in this figure to determine the minimum edge distance of the bars.

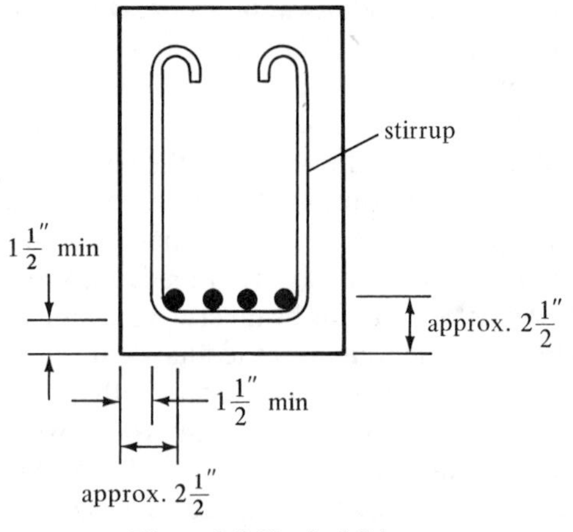

Figure 3.2 Vertical Stirrup

Minimum edge distance

Assuming #9 longitudinal bars, minimum distance from center of bars to edge of concrete

$$= \text{cover} + \text{stirrup diameter} + \tfrac{1}{2} \text{ of longitudinal bar diameter}$$

$$= 1.50 + 0.375 + \frac{1.128}{2} = 2.439'' \qquad \text{say } 2\tfrac{1}{2}''$$

The minimum cover required for concrete cast against earth, as in a footing, is 3 in. and for concrete cast not against the earth but later exposed to it, as by backfill, is 2 in. Precast and prestressed concrete or other concrete cast under plant control conditions require less cover, as described in Section 7.7 of the ACI Code.

If concrete members are exposed to very harsh surroundings, such as smoke or acid vapors, the cover should be increased above these minimums.

5. Minimum spacing of bars. The Code (7.6) states that the clear distance between parallel bars cannot be less than 1 in. nor less than the nominal bar diameter. If the bars are placed in more than one layer, those in the upper layers are required to be placed directly over the ones in the lower layers and the clear distance between the layers must not be less than 1 in.

A major purpose of these requirements is to enable the concrete to pass between the bars. The ACI Code further relates the spacing of the bars to the maximum aggregate sizes for the same purpose. In their Section 3.3.3, maximum permissible aggregate sizes are limited to the smallest of (a) one-fifth of the distance between forms, (b) one-third of the slab depth, or (c) three-fourths of the minimum clear spacing between bars.

In selecting the actual bar spacing, the designer will comply with the preceding code requirements. He will, in addition, give spacings and other dimensions in inches and fractions, not in decimals. The workers in the field are used to working with fractions and would be confused by a spacing of bars such as 3 at 1.45 in. The designer should always strive for simple spacings, as such dimensions will lead to better economy.

Each time a beam is designed, it is necessary to select the spacing and arrangement of the bars. To simplify these calculations Table A.5 (see Appendix) is given. This table shows the minimum beam widths required for different numbers of bars. The values given are based on the assumptions that $\frac{3}{8}$-in. stirrups and $1\frac{1}{2}$-in. cover are required. If 4 #11 bars are required, it can be seen from the table that a minimum beam width of 13.7 in. (say 14 in.) is required.

Example 3.1 presents the design of a rectangular beam for bending moment only.

EXAMPLE 3.1

Design the beam shown in Figure 3.3 for moment only. Compute stresses in the resulting section by using the review formulas; $f_c' = 3{,}000$ psi, $f_c = 1{,}350$ psi, $f_s = 20{,}000$ psi, and $n = 9$.

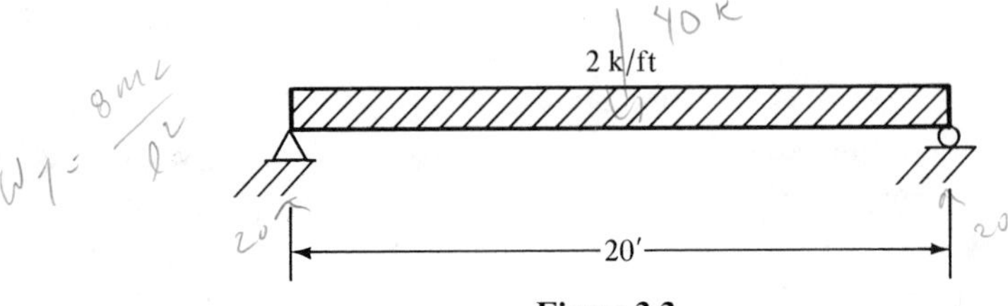

Figure 3.3

SOLUTION

Assume beam weight = 350 #/ft

$$M = \frac{(2.35)(20)^2}{8} = 117.5 \text{ ft-k}$$

$$k = \frac{f_c}{f_c + (f_s/n)} = \frac{1{,}350}{1{,}350 + (20{,}000/9)} = 0.378$$

$$j = 1 - \frac{k}{3} = 1 - \frac{0.378}{3} = 0.874$$

$$bd^2 = \frac{2M}{f_c k_j} = \frac{(2)(12)(117{,}500)}{(1{,}350)(0.378)(0.874)} = 6{,}323 \quad \left\{ \begin{array}{l} \;\; b \quad\quad d \\ 12 \times 22.95 \\ 14 \times 21.25 \\ 16 \times 19.88 \end{array} \right.$$

Try 14 × 24 beam (d = 21.5 in.)

$$\text{beam weight} = \frac{(14)(24)}{144}(150) = 350 \text{ \#/ft} \qquad \underline{\text{ok}}$$

$$A_s = \frac{M}{f_s jd} = \frac{(12)(117{,}500)}{(20{,}000)(0.874)(21.5)} = 3.75 \text{ in.}^2$$

Use 3 #10 bars (3.79 in.^{2}) (Table A.4 of Appendix).

Discussion

An effective depth of 21.25 in. was required, but to simplify the dimensions a d of 21.50 in. was selected (which when added to an edge distance of $2\frac{1}{2}$ in. gives a total depth of 24 in.). In addition, the 3 #10 bars selected have a little more area than was required. As a result of these changes, the design is no longer a perfectly balanced one, but for all practical purposes it is assumed to be.

Sketch of beam shown in Figure 3.4

Review of design

$$\rho = \frac{A_s}{bd} = \frac{3.79}{(14)(21.5)} = 0.0126$$

$$\rho n = (9)(0.0126) = 0.1134$$

$$(\rho n)^2 = 0.01286$$

$$k = \sqrt{2\rho n + (\rho n)^2} - \rho n$$

$$k = \sqrt{(2)(0.1134) + 0.01286} - 0.1134 = 0.376$$

$$j = 1 - \frac{k}{3} = 1 - \frac{0.376}{3} = 0.875$$

$$f_c = \frac{2M}{bd^2kj} = \frac{(2)(12)(117{,}500)}{(14)(21.5)^2(0.376)(0.875)} = 1{,}324 \text{ psi} < 1{,}350 \text{ psi} \quad \text{ok}$$

$$f_s = \frac{M}{A_s jd} = \frac{(12)(117{,}500)}{(3.79)(0.875)(21.5)} = 19{,}775 \text{ psi} < 20{,}000 \text{ psi} \quad \text{ok}$$

Example 3.2 illustrates the design of another simple beam by the WSD method. The beam is unsymmetrically loaded and care must be taken in calculating the maximum moment. For such cases it is usually desirable to draw the shear and moment diagrams.

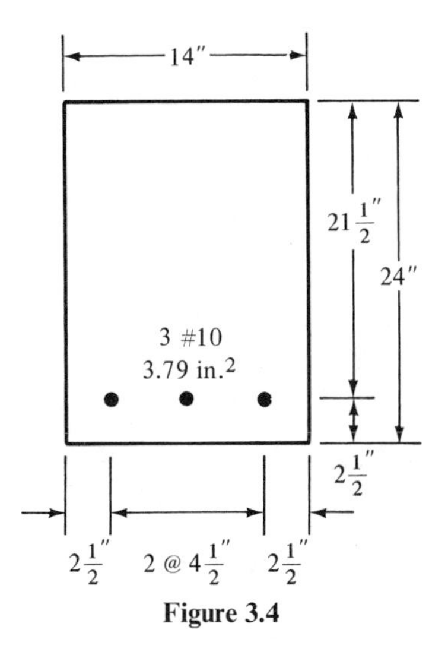

Figure 3.4

In the previous example the proportions of the beam were selected by use of the following expression:

$$bd^2 = \frac{2M}{f_c kj}$$

It will be noted that for a particular grade of concrete and a particular grade of steel, the circled part of this equation is a constant and can be taken from Table A.7 (see Appendix), where it is called K. Values of k and j are also provided in this same table.

$$K = \tfrac{1}{2} f_c kj$$

This value is used in the remaining examples of this chapter where bd^2 is needed and is written as

$$bd^2 = \frac{M}{K}$$

EXAMPLE 3.2

Design the beam shown in Figure 3.5; $f_c' = 3{,}000$ psi, $f_c = 1{,}350$ psi, $f_s = 20{,}000$ psi, and $n = 9$.

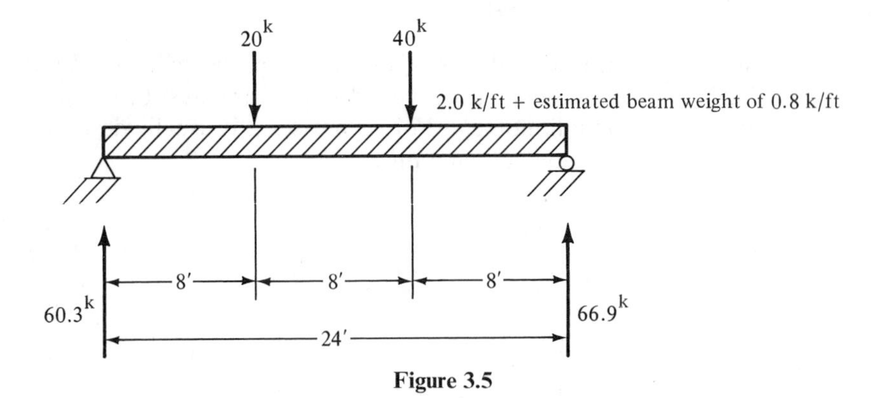

Figure 3.5

SOLUTION

$K = 223$ and $j = 0.874$ from Table A.7 (see Appendix). Assume beam weight = 800 #/ft. Drawing shear and moment diagrams (Figure 3.6),

$$bd^2 = \frac{M}{K} = \frac{(12)(450{,}000)}{223} = 24{,}215 \quad \begin{cases} b \quad d \\ 16 \times 38.9 \\ 18 \times 36.7 \\ 20 \times 34.8 \end{cases}$$

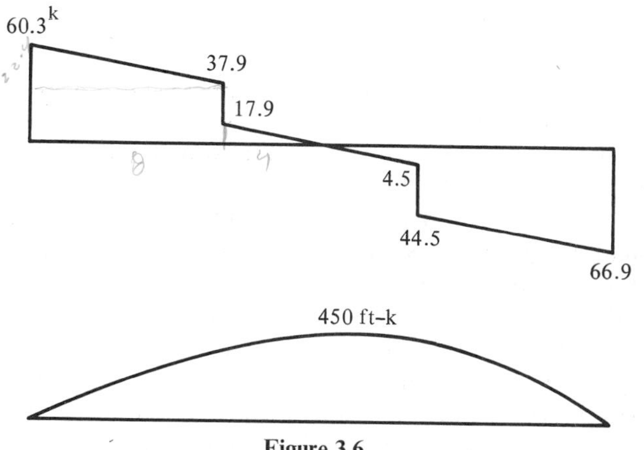

Figure 3.6

Try 20 × 38 ($d = 35.5$)

$$\text{beam weight} = \frac{(20)(38)}{144}(150) = 792\ \#/\text{ft} < 800\ \#/\text{ft} \qquad \underline{\text{ok}}$$

$$A_s = \frac{M}{f_s jd} = \frac{(12)(450{,}000)}{(20{,}000)(0.874)(35.5)} = 8.70\ \text{in.}^2$$

Use 7 #10 bars (8.86 in.^{2}).

Sketch of beam shown in Figure 3.7

$$\text{minimum edge distance} = 1.50 + 0.375 + \frac{1.27}{2} = 2.51'' \qquad \underline{\text{say } 2\tfrac{1}{2}''}$$

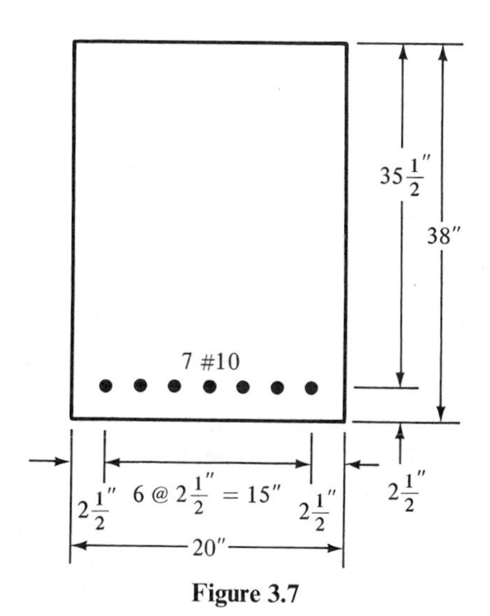

Figure 3.7

3.4 ONE-WAY SLABS

Reinforced concrete slabs are large flat plates that are supported by reinforced concrete beams, walls, or columns, by masonry walls, by structural steel beams or columns, or by the ground. If they are supported on two opposite sides only, they are referred to as *one-way slabs* since the bending is in one direction only, that is, perpendicular to the supported edges. Should the slab be supported by beams on all four edges, it is referred to as a *two-way slab* since the bending is in both directions. Actually, if a rectangular slab is supported on all four sides, but the long side is two or more times as long as the short side, the slab will, for all practical purposes, act as a one-way slab, with bending primarily occurring in the short direction. Such slabs are designed as one-way slabs. You can easily verify these bending moment ideas by supporting a sheet of paper on two opposite sides or on four sides with the support situation described. This section is concerned with one-way slabs; two-way slabs are considered in Chapter 14. It should be realized that most reinforced concrete slabs fall into the one way class.

A one-way slab is assumed to be a rectangular beam with a large ratio of width to depth. Normally, a 12-in.-wide piece of such a slab is designed as a beam, the slab being assumed to consist of a series of such beams side by side. The method of analysis used is somewhat conservative due to the lateral restraint provided by the adjacent parts of the slab. Normally a beam will tend to expand laterally somewhat as it bends, but this tendency to expand by each of the 12-in. strips is resisted by the adjacent 12-in.-wide strips which tend to expand also. In other words, Poisson's ratio is assumed to be zero. Actually, the lateral expansion tendency results in a very slight stiffening of the beam strips, which is neglected in the design procedure used here.

The 12-in.-wide beam is quite convenient when thinking of the load calculations, as loads are normally specified as so many pounds per square foot and thus the load carried per foot of the 12-in.-wide beam is the load supported per square foot by the slab. The load supported by the one-way slab including its own weight is transferred to the beams supporting the edges of the slab. Obviously, the reinforcing for flexure is placed perpendicular to these supports, that is, parallel to the long direction of the 12-in. beams. Of course, there will be some reinforcing placed in the other direction to resist shrinkage and temperature stresses.

The thickness required for a particular one-way slab depends on the bending, deflection, and shear requirements. The ACI Code (9.5.2.1) provides certain span/depth limitations for concrete flexural members where deflections are not calculated. The purpose of such limitations is to prevent deflections of such magnitudes as would interfere with the use of or cause injury to the structure. If deflections are computed for members of lesser thicknesses than listed in the table and are found to be satisfactory, it is unnecessary to abide by the thickness rules. For simply supported slabs, normal-weight concrete, and grade 60 steel, the minimum thickness given when deflections are not computed equals $l/20$, where l is the span length of the slab. *For concretes of other weights and for steels of different yield strengths, the minimum thicknesses required by the ACI Code are somewhat revised.*

The slab thickness selected is usually rounded off to the nearest $\frac{1}{4}$ in. on the high side for slabs of 6 in. or less in thickness and to the nearest $\frac{1}{2}$ in. for slabs thicker than 6 in.

As concrete hardens it shrinks. In addition, temperature changes occur that cause expansion and contraction of the concrete. When cooling occurs, the shrinkage effect and the shortening due to cooling add together. The Code (7.12) states that shrinkage and temperature reinforcement must be provided in a direction perpendicular to the main reinforcement for one-way slabs. (For two-way slabs, reinforcement is provided in both directions for bending.) The Code states that for grades 40 or 50 deformed bars, the minimum percentage of this steel is 0.002 times the gross cross-sectional area of the slab. Notice that the cross-sectional area is bt (where t is the slab thickness), not bd. Such reinforcing bars may not be spaced further apart than 5 times the slab thickness nor 18 in. When grade 60 deformed bars or welded wire fabric are used, the minimum area is $0.0018bt$. Table A.6 (see Appendix) is useful for selecting both bending and temperature bars.

Example 3.3 illustrates the design of a one-way slab. It will be noted that the Code (7.7.1.c) cover requirement for reinforcement in slabs (#11 and smaller bars) is $\frac{3}{4}$ in. clear when the slab is not exposed directly to the ground or the weather.

EXAMPLE 3.3

Design a one-way slab for the inside of a building using the span, loads, and allowable stresses given here and in Figure 3.8.

$f_c' = 3{,}000$ psi; $f_c = 1{,}350$ psi; $n = 9$; $f_s = 24{,}000$ psi (grade 60); $LL =$ 150 psf.

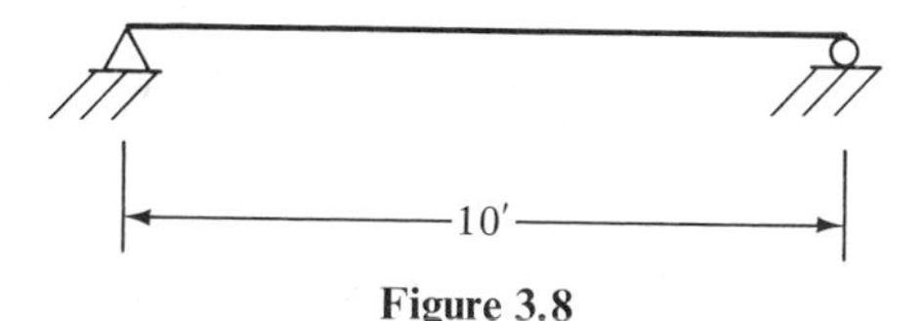

Figure 3.8

SOLUTION

Minimum t if deflection is not calculated (Code 9.5.2.1)

$$t = \frac{l}{20} = \frac{(12)(10)}{20} = 6''$$

Assume 6-in. slab (designing a 12-in.-wide strip of the slab)

$$DL = \tfrac{6}{12}(150) = 75 \text{ psf}$$
$$LL = \quad = 150$$
$$w = 225 \text{ psf} = 225 \ \#/\text{ft}$$

$$M = \frac{(0.225)(10)^2}{8} = 2.81 \text{ ft-k}$$

From the tables, $K = 201, j = 0.888$

$$bd^2 = \frac{M}{K} \qquad \text{where } b = 12''$$

$$d = \sqrt{\frac{M}{12K}} = \sqrt{\frac{(12)(2{,}810)}{(12)(201)}} = 3.74''$$

Use 6-in. slab ($d = 5$ in. with $\frac{3}{4}$-in. cover, ACl Code 7.7.1)

$$A_s = \frac{(12)(2{,}810)}{(24{,}000)(0.888)(5.00)} = 0.316 \text{ in.}^2/\text{ft}$$

Use #4 at 7 in. (0.34 in.^{2}/ft) (Table A.6 of Appendix).

Shrinkage and temperature steel

$$A_s = 0.0018bt = (0.0018)(12)(6) = 0.1296 \text{ in.}^2/\text{ft}$$

Use #3 at 10 in. (0.13 in.^{2}/ft) (Table A.6 of Appendix).

3.5 SI EXAMPLE

Example 3.4 illustrates the design of a rectangular beam using SI units.

EXAMPLE 3.4

Design the beam shown in Figure 3.9 if the weight of the concrete is 23.5 kN/m^3; $f_c' = 20.7$ MPa, $f_c = 9.3$ MPa, $f_s = 137.9$ MPa, and $n = 9$.

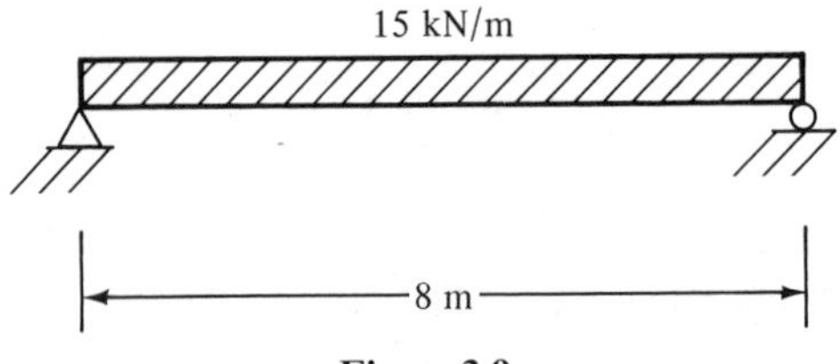

Figure 3.9

SOLUTION

Assume the beam weight is 5 kN/m.

$$M = \frac{(20)(8)^2}{8} = 160 \text{ kN}\cdot\text{m}$$

$$k = \frac{f_c}{f_c + (f_s/n)} = \frac{9.3}{9.3 + (137.9/9)} = 0.378$$

$$j = 1 - \frac{k}{3} = 1 - \frac{0.378}{3} = 0.874$$

$$K = \tfrac{1}{2} f_c k j = (\tfrac{1}{2})(9.3)(0.378)(0.874) = 1.536$$

$$bd^2 = \frac{M}{K} = \frac{(160)(10^6)}{1.536} = 104\,166\,660 \text{ mm}^3$$

b	d
300 mm ×	589 mm
350 mm ×	545 mm

Try 350 × 610 (*d* = 545 mm)

$$\text{beam weight} = \frac{(350)(610)}{10^6}(23.5) = 5.02 \text{ kN/m} > 5.00 \text{ kN/m but ok}$$

$$A_s = \frac{M}{f_s jd} = \frac{(160)(10^6)}{(137.9)(0.874)(545)} = 2\,436 \text{ mm}^2$$

Use 3 #10 bars ($A_s = 2\,457$ mm^2).

Sketch of beam shown in Figure 3.10

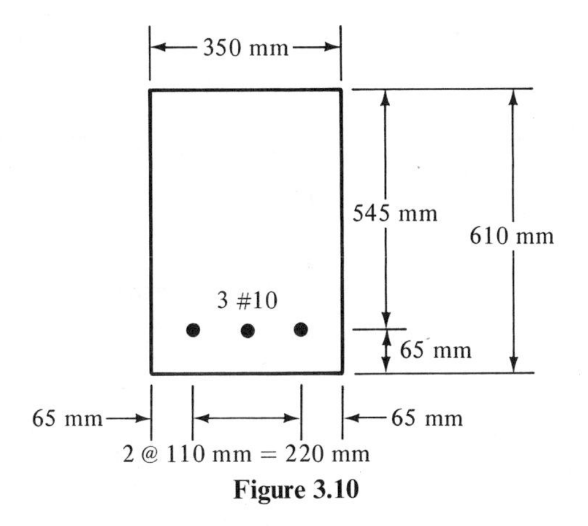

Figure 3.10

3.6 CONCLUSION OF WSD DISCUSSION

At this point the discussion of WSD is concluded and the remainder of the text is devoted to the strength design method. Only bending moment has been considered for beams and one-way slabs, and such subjects as shear, development length of bars, and doubly reinforced sections have been omitted as they will be considered in the strength design discussion. You should be aware of the wide range of flexural members that can be handled by the flexural equations presented in the preceding sections. Such members as footings, retaining walls, and many others can be designed.

PROBLEMS

3.1 to **3.8** Design and review reinforced concrete beams by the WSD method for moment only for the spans, loads, and allowable stresses given in the following figures. None of the loads shown includes the beam weights. Show a sketch of beam cross sections selected, including all dimensions.

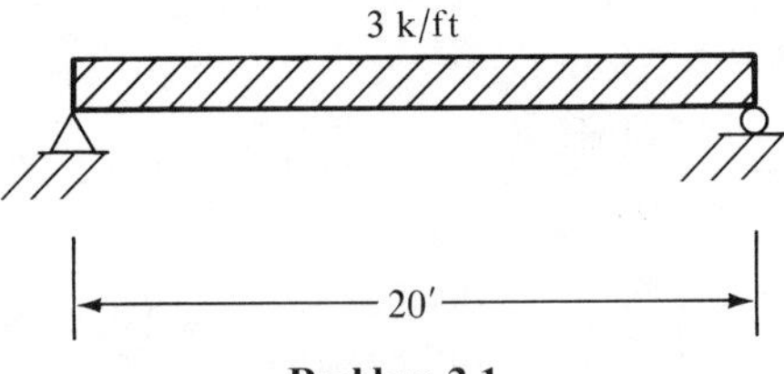

Problem 3.1

$f_c' = 3{,}000$ psi, $f_c = 1{,}350$ psi, and $f_s = 20{,}000$ psi.

3.2 Repeat Problem 3.1 using a 2.5k/ft uniform load. Select #7 bars for the reinforcing steel.

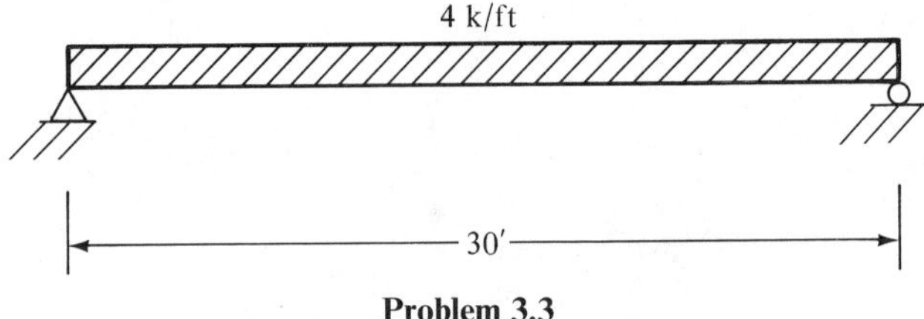

Problem 3.3

$f_c' = 3{,}000$ psi, $f_c = 1{,}350$ psi, and $f_s = 20{,}000$ psi.

3.4 Repeat Problem 3.1 if $f_c' = 4{,}000$ psi, $f_c = 1{,}800$ psi, and $f_s = 24{,}000$ psi.

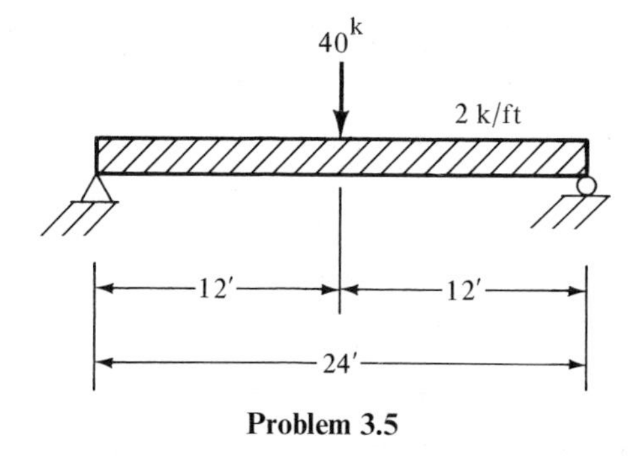

Problem 3.5

$f_c' = 2{,}500$ psi, $f_c = 1{,}125$ psi, and $f_s = 20{,}000$ psi.

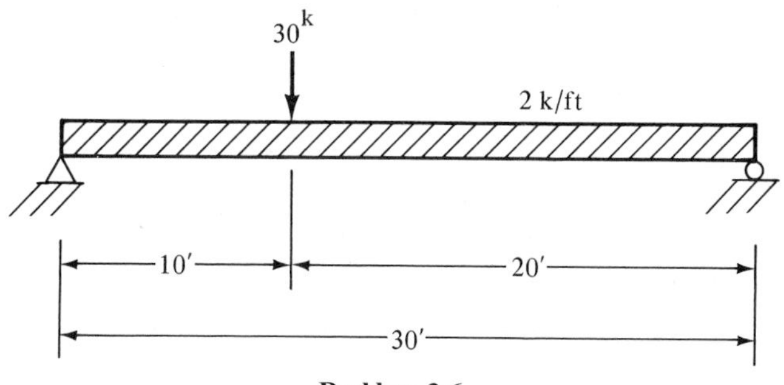

Problem 3.6

$f_c' = 4{,}000$ psi, $f_c = 1{,}800$ psi, and $f_s = 20{,}000$ psi.

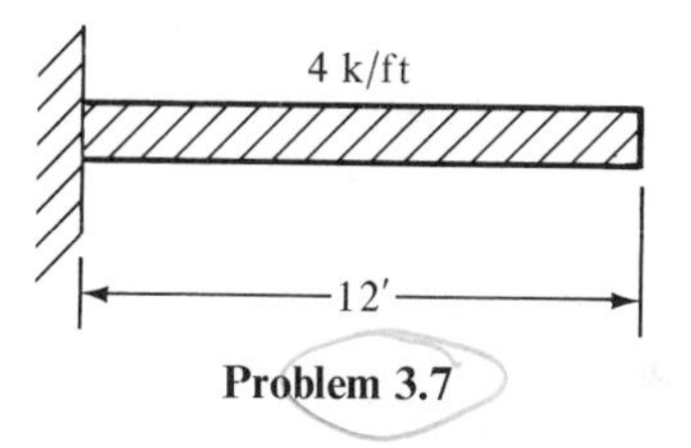

Problem 3.7

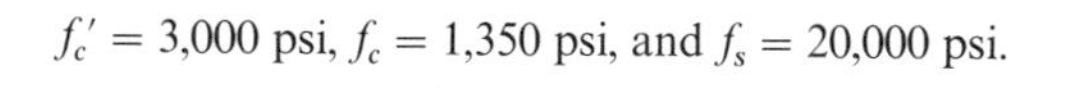
$f_c' = 3{,}000$ psi, $f_c = 1{,}350$ psi, and $f_s = 20{,}000$ psi.

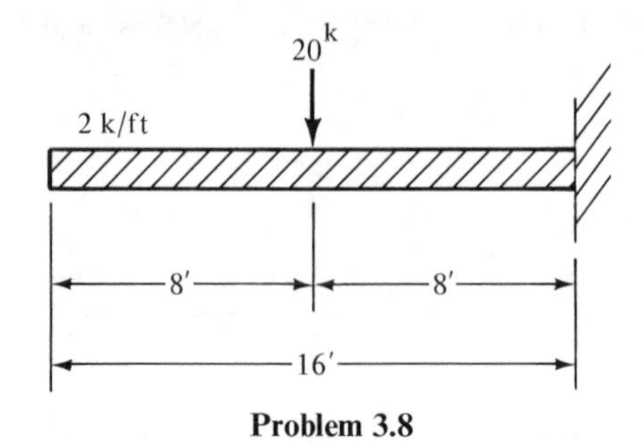

Problem 3.8

$f_c' = 3{,}000$ psi, $f_c = 1{,}350$ psi, and $f_s = 24{,}000$ psi.

3.9 to **3.10** Design the beams for maximum moments and then select reinforcing for both positive and negative moments. Loads shown do not include beam weights.

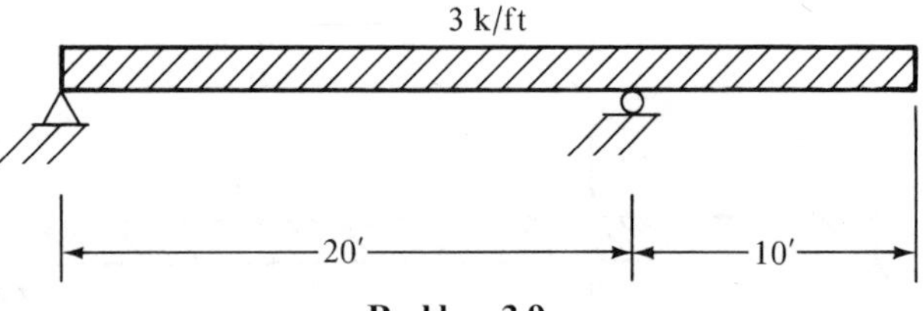

Problem 3.9

$f_c' = 3{,}000$ psi, $f_c = 1{,}350$ psi, and $f_s = 20{,}000$ psi.

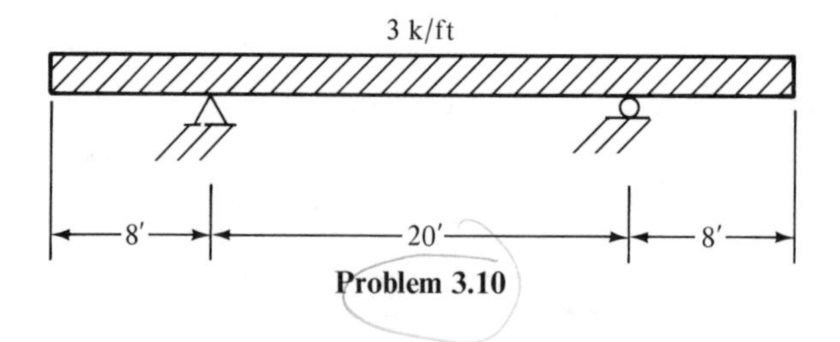

Problem 3.10

$f_c' = 4{,}000$ psi, $f_c = 1{,}800$ psi, and $f_s = 24{,}000$ psi.

3.11 Determine the depth required for a beam to support itself only for a 200-ft span. Neglect cover in weight calculations. $f_c' = 3{,}000$ psi, $f_c = 1{,}350$ psi, and $f_s =$ 20,000 psi.

3.12 Design a one-way slab for the situation shown in the accompanying illustration. $f_c' = 2{,}500$ psi, $f_c = 1{,}125$ psi, and $f_s = 20{,}000$ psi for Grade 40 steel. Use the ACI Code's thickness limitation for deflection.

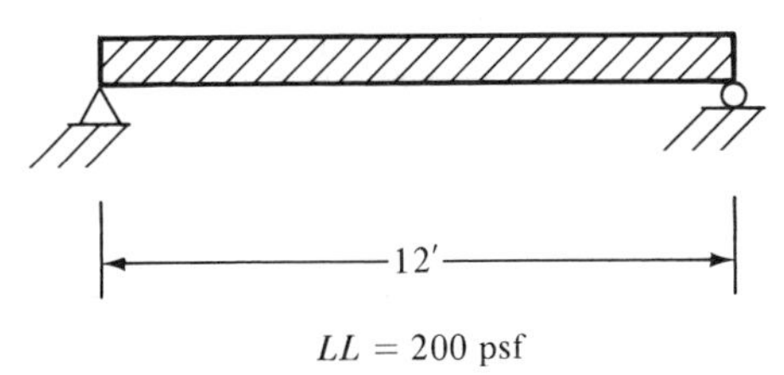

$LL = 200$ psf

Problem 3.12

3.13 Design a one-way slab for the following situation: $f_c' = 3{,}000$ psi, $f_c = 1{,}350$ psi, and $f_s = 24{,}000$ psi for Grade 60 steel. Do not use the ACI Code's thickness limitation for deflection.

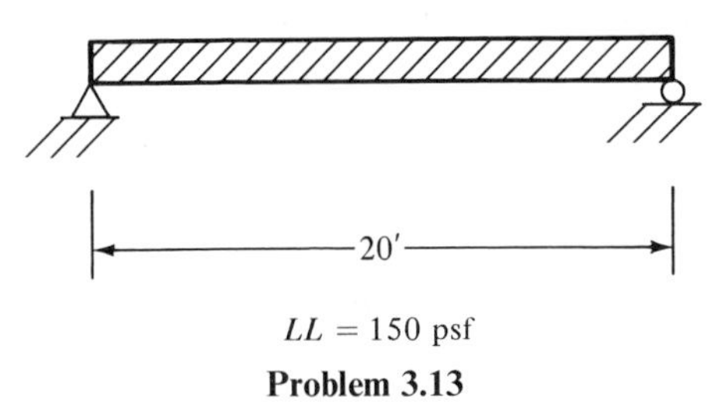

$LL = 150$ psf

Problem 3.13

3.14 Determine the stem thickness for maximum moment for the retaining wall shown in the accompanying illustration. Also determine the steel area required at the bottom and middepth of the stem. $f_c' = 3{,}000$ psi, $f_c = 1{,}350$ psi, and $f_s = 20{,}000$ psi. Assume #8 bars are to be used and assume the stem thickness is constant for the 15 ft height.

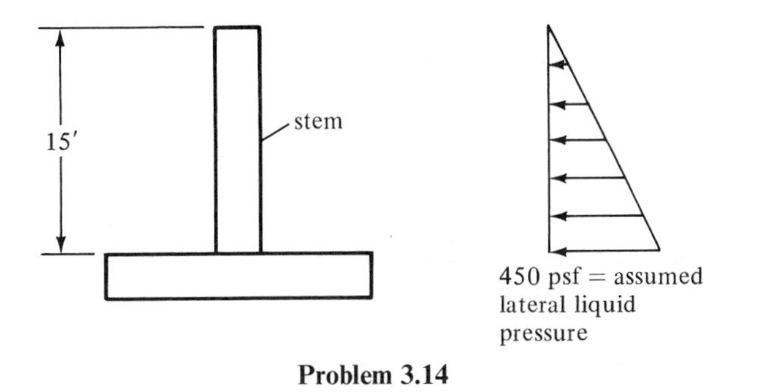

Problem 3.14

3.15 to **3.19** Design beams for the spans, loads, and allowable stresses given in the following figures. None of the loads shown include the beam weights. Assume reinforced concrete weighs 23.5 kN/m^3.

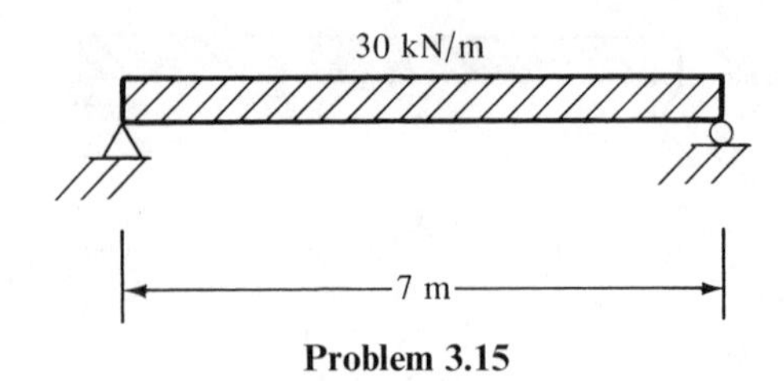

Problem 3.15

$f_c' = 20.7$ MPa, $f_c = 9.3$ MPa, $f_s = 137.9$ MPa, and $n = 9$.

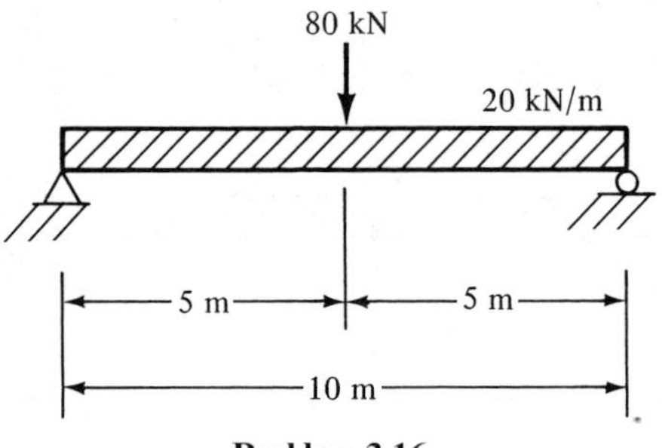

Problem 3.16

$f_c' = 17.2$ MPa, $f_c = 7.76$ MPa, $f_s = 137.9$ MPa, and $n = 10$.

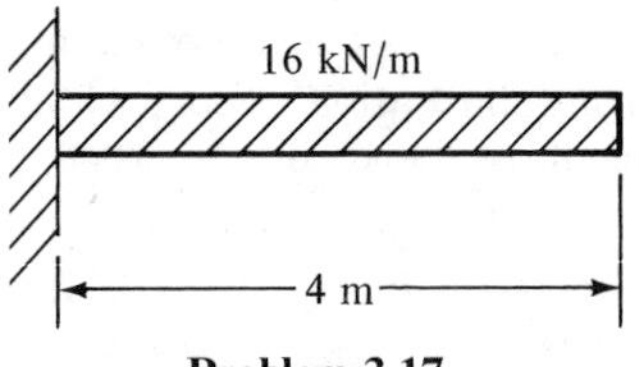

Problem 3.17

$f_c' = 20.7$ MPa, $f_c = 9.3$ MPa, $f_s = 165.5$ MPa, and $n = 9$.

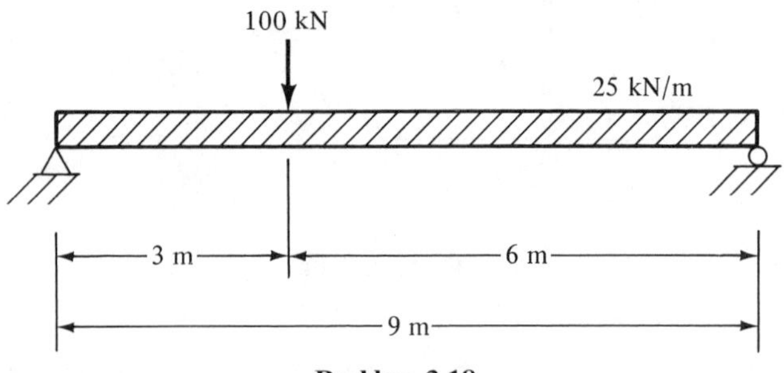

Problem 3.18

$f_c' = 27.6$ MPa, $f_c = 12.4$ MPa, $n = 8$, and $f_s = 165.5$ MPa.

3.19 Design the one-way slab shown in the accompanying figure to support a live load of 7 kN/m^2. Do not use the ACI thickness limitation for deflections and assume concrete weighs 23.5 kN/m^3.

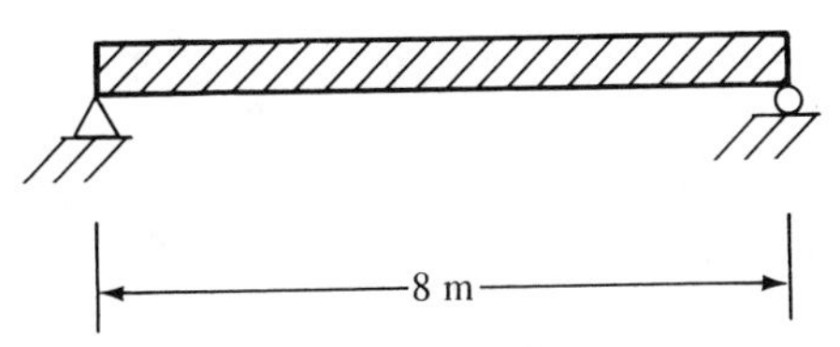

Problem 3.19

$f_c' = 20.7$ MPa, $f_c = 9.3$ MPa, $f_s = 137.9$ MPa, and $n = 9$.

CHAPTER 4

Strength Design for Beams

4.1 INTRODUCTION

In 1956 as an appendix, the ACI Code for the first time included ultimate strength design, although the concrete codes of several other countries had been based on such considerations for several decades. In 1963 the code gave ultimate strength design equal status with working stress design, and the present or 1977 Code made the method now called *strength design* the predominant method and only briefly mentioned the working stress method. In Section 8.1.2 and Appendix B of the 1977 Code, the designer is given brief permission to use the working stress method for nonprestressed sections and certain specific provisions for such designs are included.

The general method has been called ultimate strength design for the past two or three decades, but the 1977 ACI Code uses the term "strength design." The theoretical strength for a particular reinforced concrete member is a value given by the Code and is not necessarily the true ultimate strength of the member. Therefore, the more general term "strength design" is used whether beam strength, column strength, shear strength, or others are being considered.

4.2 ADVANTAGES OF STRENGTH DESIGN

Among the several advantages of the strength design method as compared to the alternate design method are the following:

1. Because of the nonlinear shapes of the stress-strain diagrams at high levels of stress are considered to result in decidedly better estimates of load-carrying ability.
2. With strength design a more consistent theory is used throughout the designs of reinforced concrete structures. For instance, with the alternate design method the transformed area or straight-line method is used for beam design and a strength design procedure is used for columns.

Water Tower Place, Tallest Reinforced Concrete Building in World, Chicago, Illinois (Courtesy Symons Corporation).

3. A more realistic factor of safety is used in strength design. With working stress design the same safety factor is used for dead and live loads, whereas such is not the case for strength design. The designer can certainly estimate the magnitudes of the dead loads that a structure will have to support more accurately than he can the live loads. For this reason the use of different safety factors in strength design for the two types of loads is a definite improvement.
4. A structure designed by the strength method will have a more uniform safety factor against collapse throughout. The strength method takes full advantage of higher-strength steels, whereas working stress design partly does so. For instance, for the alternate design method the Code (Appendix B) limits the maximum flexural allowable stress in reinforcing bars (in most cases) to 24,000 psi, but in effect much higher values may be used in strength design for higher-strength steels. The result is better economy for strength design.
5. The strength method permits more flexible designs than does the alternate design method. For instance, the percentage of steel may be varied quite a bit. As a result, large sections may be used with small percentages of steel or small sections with large percentages of steel.

Such variations are not the case in the relatively fixed alternate design method. If the same amount of steel is used in strength design for a particular beam as would be required by the alternate design method, a smaller section will result. If the same size section is used as required by working stress design, a smaller amount of steel will be required.

4.3 STRUCTURAL SAFETY

There are two methods by which the structural safety of a reinforced concrete structure might be considered. The first of these methods involves the calculations of the stresses caused by the working or service loads and their comparison with certain allowable stresses. Usually the safety factor against collapse when the working stress method is used is said to equal the smaller of f_c'/f_c or f_y/f_s.

The second approach to structural safety is the one used in strength design in which the working loads are multiplied by certain load factors that are larger than one. The resulting larger or factored loads are used for designing the structure. The values of the load factors vary depending on the type and combination of the loads.

To accurately estimate the ultimate strength of a structure, it is considered necessary to take into account the uncertainties in material strengths, dimensions, and workmanship. This is done by multiplying the theoretical ultimate strength (called the nominal strength herein) of each member by the *capacity reduction factor* ϕ, which is less than one. These values vary from 0.90 for bending down to 0.65 for plain or unreinforced concrete.

In summary, the strength design approach to safety is to select a member whose computed ultimate load capacity multiplied by its capacity reduction factor will at least equal the sum of the service loads multiplied by their respective load factors.

It is felt that member capacities obtained with the strength method are appreciably more accurate than member capacities that could be predicted with the working stress method.

4.4 LOAD FACTORS

The ACI Code (9.2) states that the required ultimate load-carrying ability of a member U provided to resist the dead load D and the live load L will at least equal

$$U = 1.4D + 1.7L \tag{4.1}$$

The load factors used for live loads logically must be larger than the ones used for dead loads because the designer can estimate the magnitudes of dead loads so much better than he can the magnitudes of live loads.

Should it be necessary to consider wind load W as well as D and L, the Code states that the structure must at least be able to support the U given at the end of this paragraph. As wind loads are of lesser duration than dead loads or than the usual gravity live loads, the ACI feels that it is only logical to reduce the value U obtained for this combination of W with D and L. This is done by multiplying the results by 0.75.

$$U = 0.75(1.4D + 1.7L + 1.7W) \tag{4.2}$$

In addition, the following condition, where L is not present, is to be considered:

$$U = 0.9D + 1.3W \tag{4.3}$$

This latter condition is included to cover cases where tension forces develop due to overturning moments. It will govern only for very tall buildings where high wind loads are present. In this expression the dead loads are reduced by 10% to take into account situations where they may have been overestimated.

In no case will the strength of the structure be less than that given by Equation 4.1. Other cases are also given in the Code (9.2) for earthquake forces E, lateral earth pressures H, and so forth.

You will notice that these load factors are not varied in proportion to the seriousness of failure. Although it may seem logical to use a higher load factor for a hospital than for a warehouse, this is not required. It is assumed, instead, that the designer will consider the seriousness of failure while he is specifying the magnitude of the service loads. Another point to remember is that the ACI load factors are minimum values and the designer is free to use larger ones if he thinks the consequences of failure so require.

4.5 CAPACITY REDUCTION FACTORS

The purpose of using capacity reduction factors is to take into consideration the uncertainties of material strengths, approximations in analysis, possible variations in dimensions of concrete sections and placement of reinforcement, and other miscellaneous workmanship items. The Code (9.3) prescribes ϕ values or capacity reduction factors for several situations. Among the values given are the following:

0.90 bending in reinforced concrete
0.85 bond, diagonal tension and anchorage
0.70 bearing on concrete
0.65 bending in plain concrete

The sizes of these factors are pretty good indications of our knowledge of the subject in question. For instance, calculated ultimate resisting moments in reinforced concrete members seem to be fairly accurate, while computed bearing capacities are more questionable.

4.6 UNDERREINFORCED AND OVERREINFORCED BEAMS

Before proceeding with the derivation of beam expressions, it is necessary to define certain terms relating to the amount of tensile steel used in a beam. These terms include *balanced steel ratio, underreinforced beams, and overreinforced beams.*

A beam that has a balanced steel ratio is one for which the steel will theoretically start to yield and the concrete reach its ultimate strain at exactly the same load. Should a beam have less reinforcement than required for a balanced ratio, it is said to be underreinforced; if more, it is said to be overreinforced.

If a beam is underreinforced and the ultimate load is approached, the steel will begin to yield even though the compression concrete is still understressed. If the load is further increased, the steel will continue to elongate, resulting in appreciable deflections and large visible cracks in the tensile concrete. As a result, the users of the structure are given notice that the load must be decreased or else the result will be considerable damage or even failure. If the load is increased further, the tension cracks will become even larger and the compression side of the concrete will become overstressed and fail.

If a beam should be overreinforced, the steel will not yield before failure. As the load is increased, deflections are not noticeable even though the compression concrete is highly stressed, and failure occurs suddenly without warning to the occupants. Rectangular beams will fail in compression when strains are about 0.003 to 0.004 for ordinary grades of concrete.

Obviously, overreinforcing is a situation to be avoided if at all possible, and the Code, by limiting the percentage of tensile steel that may be used in a beam, ensures the design of underreinforced beams and thus the ductile type of failures that provide adequate "running time."

4.7 DERIVATION OF BEAM EXPRESSIONS

Tests of reinforced concrete beams confirm that strains vary in proportion to distances from the neutral axis even on the tension sides and even near ultimate loads. Compression stresses vary approximately in a straight line until the maximum stress equals about $0.50f_c'$. This is not the case, however, after stresses go higher. When the ultimate load is reached, the strain and stress variations are approximately as shown in Figure 4.1.

The compressive stresses vary from zero at the neutral axis to a maximum value at or near the extreme fiber. The actual stress variation

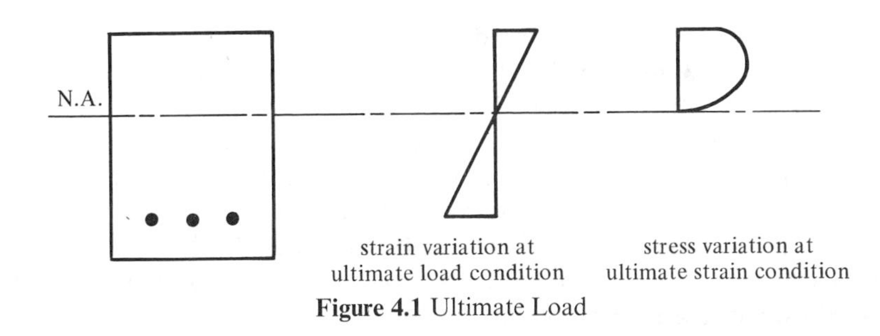

Figure 4.1 Ultimate Load

and the actual location of the neutral axis vary somewhat from beam to beam depending on such items as the magnitude and history of past loadings, shrinkage and creep of the concrete, size and spacing of tension cracks, speed of loading, and so on.

If the shape of the stress diagram was the same for every beam, it would easily be possible to derive a single rational set of expressions as was done in the preceding chapter for working stress design. Because of these stress variations, however, it is necessary to base the strength design method upon a combination of theory and test results.

Although the actual stress distribution may seem to be an important matter, any assumed shape (rectangular, parabolic, trapezoidal, etc.) can be practically used if the resulting equations compare favorably with test results. The most common shapes proposed are the rectangle, parabola, and trapezoid, with the rectangular shape used in this text as shown in Figure 4.2(c).

If the concrete is assumed to crush at a strain of about 0.003 (which is a little conservative) and the steel to yield at f_y, it is possible to make a reasonable derivation of beam formulas without knowing the exact stress distribution. However, it is necessary to know the value of the total compression and its centroid.

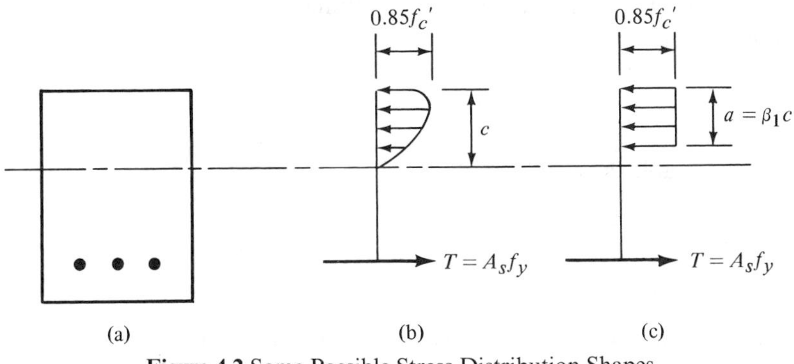

Figure 4.2 Some Possible Stress Distribution Shapes

Whitney[1] replaced the curved stress block with an equivalent rectangular block of intensity $0.85f_c'$ and depth $a = \beta_1 c$, as shown in Figure 4.2(c). The area of this rectangular block should equal that of the curved stress block and the centroids of the two blocks should coincide. Sufficient test results are available for concrete beams to provide the depths of the equivalent rectangular stress blocks. The value of a given by the Code (10.2.7) is intended to give this result. For f_c' values of 4,000 psi or less, $\beta_1 = 0.85$, and it is to be reduced continuously at a rate of 0.05 for each 1,000-psi increase in f_c' above 4,000 psi. Their value may not be less than 0.65. The values of β_1 are reduced for high-strength concretes primarily because of the less favorable shapes of their stress-strain curves (see Figure 1.1).

Based on these assumptions regarding the stress block, statics equations can easily be written for the sum of the horizontal forces and for the resisting moment produced by the internal couple. These expressions can then be solved separately for a and for the moment M_n.

A very clear statement should be made here regarding the term M_n as it otherwise can be rather confusing to the reader. M_n is defined as the theoretical resisting moment of a section. In Section 4.3 it was stated that the usable strength of a member equals its theoretical strength times the capacity reduction factor, or, in this case, ϕM_n. The usable flexural strength of a beam is defined as M_u.

$$M_u = \phi M_n$$

For writing the beam expressions, reference is made to Figure 4.3. Equating the horizontal forces C and T and solving for a,

$$0.85f_c'ab = A_s f_y$$

$$a = \frac{A_s f_y}{0.85f_c'b} = \frac{\rho f_y d}{0.85f_c'}$$

As the reinforcing steel is limited to an amount such that it will yield well before the concrete reaches its ultimate strength, the value of the ultimate moment M_n can be written as

$$M_n = T\left(d - \frac{a}{2}\right) = A_s f_y\left(d - \frac{a}{2}\right)$$

and the usable flexural strength is

$$M_u = \phi M_n = \phi A_s f_y\left(d - \frac{a}{2}\right)$$

[1] Whitney, C. S., 1942, "Plastic Theory of Reinforced Concrete Design," *Transactions ASCE*, 107, pp. 251–326.

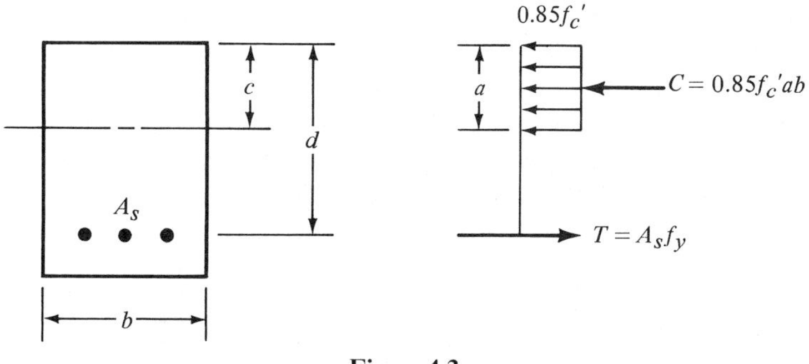

Figure 4.3

Substituting in this expression the value of a previously obtained gives an alternate form of M_u:

$$M_u = \phi A_s f_y d\left(1 - 0.59\frac{\rho f_y}{f_c'}\right)$$

4.8 MAXIMUM PERMISSIBLE STEEL PERCENTAGE

If a balanced beam (neither underreinforced nor overreinforced) is used, it will theoretically fail suddenly and without warning. Accordingly, the ACI Code (10.3.3) limits the percentage of steel used in singly reinforced

Wall Construction for Sewage Treatment Plant, Worcester, Mass. (Courtesy Symons Corporation).

concrete beams (without axial loads) to 0.75 times the percentage that would produce a balanced condition.

In this section an expression is derived for ρ_b, the percentage of steel required for a balanced design. At ultimate load for such a beam, the concrete will theoretically fail (at a strain of 0.003) and the steel will simultaneously yield. Reference is made to Figure 4.4.

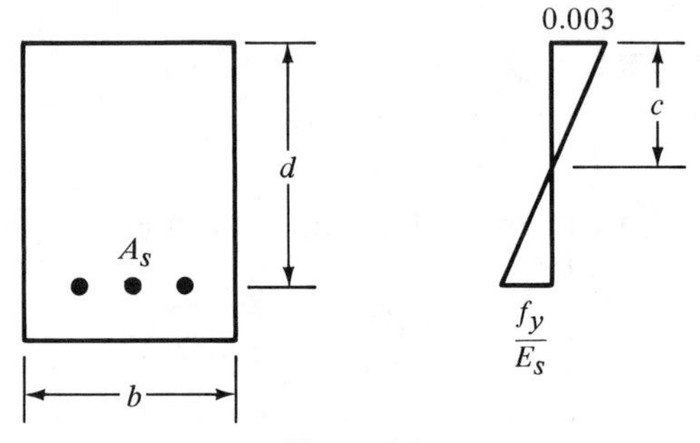

Figure 4.4

The neutral axis is located by the triangular strain relationships that follow, noting that $E_s = 29 \times 10^6$ psi for the reinforcing bars:

$$\frac{c}{d} = \frac{0.003}{0.003 + (f_y/E_s)} = \frac{0.003}{0.003 + (f_y/29 \times 10^6)}$$

$$c = \frac{87{,}000}{87{,}000 + f_y} d$$

In Section 4.7 an expression was derived for a by equating the values of C and T. This value can be converted to c by dividing it by β_1.

$$a = \frac{\rho f_y d}{0.85 f_c'}$$

$$c = \frac{a}{\beta_1} = \frac{\rho f_y d}{0.85 \beta_1 f_c'}$$

Two expressions are now available for c and they are equated to each other and solved for the percentage of steel. This is the balanced percentage ρ_b.

$$\frac{\rho f_y d}{0.85 \beta_1 f_c'} = \frac{87{,}000}{87{,}000 + f_y} d$$

$$\rho_b = \left(\frac{0.85 \beta_1 f_c'}{f_y}\right)\left(\frac{87{,}000}{87{,}000 + f_y}\right)$$

Then according to the Code, the maximum steel percentage is

$$\rho_{\max} = 0.75\rho_b$$

Values of ρ_b, $\rho_{\max}$, and so on can easily be calculated for different values of f_c' and f_y and tabulated as shown in Table A.8 (see Appendix). Included in the table are values of ρ_b, $0.75\rho_b$, and $0.50\rho_b$, the latter value being needed in certain continuous members as will be described in Chapter 12.

4.9 MINIMUM PERCENTAGE OF STEEL

A brief discussion of the modes of failure occurring for overreinforced and underreinforced beams was presented in Section 4.6. There is actually another possible mode of failure that can occur in very lightly reinforced beams. If the ultimate resisting moment of the section is less than the cracking moment, the section will fail immediately when a crack occurs. This type of failure would occur without warning, and the Code (10.5.1) provides a minimum steel percentage equal to $200/f_y$. The value was obtained by calculating the cracking moment of a plain concrete section and equating it to the strength of a reinforced concrete section of the same size and solving for the steel percentage. This value, which is not applicable to slabs of uniform thickness, applies to other sections unless the area of reinforcement provided at every section is made at least one-third greater than required by the calculations. For slabs of uniform thickness, the Code (10.5.3) states the minimum amount of reinforcing in the direction of the span shall not be less than that required for shrinkage and temperature reinforcement. The ACI Commentary (10.5.3) states that when slabs are overloaded there is a tendency for the loads to be distributed laterally, thus substantially reducing the chances of sudden failure. This explains why a reduction of the minimum reinforcing percentage below $200/f_y$ is permitted in slabs of uniform thickness. Supported slabs such as slabs on the grade are not considered to be structural slabs in this section unless they transmit vertical loads from other parts of the structure to the underlying soil.

4.10 EXAMPLE PROBLEMS

A series of problems illustrating the strength method is presented in this section. The first of these, Example 4.1, shows the calculations of the permissible ultimate capacity of a rectangular beam.

EXAMPLE 4.1

Determine the maximum permissible ultimate moment capacity of the section shown in Figure 4.5 if $f_c' = 4{,}000$ psi and $f_y = 40{,}000$ psi.

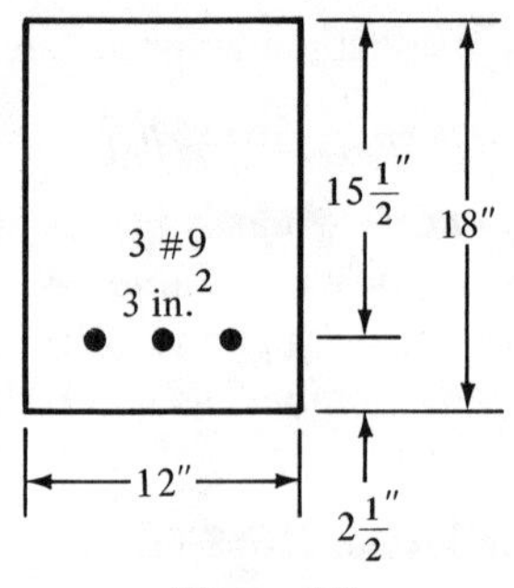

Figure 4.5

SOLUTION

$$\rho = \frac{3.00}{(12)(15.5)} = 0.0161$$

$\rho_{\max}$ from Table A.8 = 0.0371 > 0.0161 <u>ok</u>

$$\rho_{\min} = \frac{200}{f_y} = \frac{200}{40{,}000} = 0.005 < 0.0161 \qquad \underline{\text{ok}}$$

Therefore, failure is insured by tensile yielding as desired.

$$a = \frac{A_s f_y}{0.85 f_c' b} = \frac{(3.00)(40)}{(0.85)(4)(12)} = 2.94 \text{ inch}$$

$$M_u = \phi A_s f_y\left(d - \frac{a}{2}\right) = (0.90)(3.00)(40{,}000)\left(15.5 - \frac{2.94}{2}\right)$$

$$M_u = 1{,}515{,}000''\# = 126.3 \text{ ft-k}$$

Use of Graphs and Tables

In Section 4.7 the following equation was derived:

$$M_u = \phi A_s f_y d\left(1 - 0.59\frac{\rho f_y}{f_c'}\right)$$

If A_s in this equation is replaced with ρbd, the resulting expression can be solved for $M_u/\phi bd^2$ as follows:

$$M_u = \phi \rho b d f_y d\left(1 - 0.59\frac{\rho f_y}{f_c'}\right)$$

$$\frac{M_u}{\phi bd^2} = \rho f_y\left(1 - 0.59\frac{\rho f_y}{f_c'}\right)$$

For a given steel percentage ρ and for a certain concrete f_c' and certain steel f_y, the value of $M_u/\phi bd^2$ can be calculated and plotted in tables, as is illustrated in Tables A.9 through A.14 (see Appendix) or in graphs (see Graph 1 of the Appendix). It is much easier to accurately read the tables than the graphs (at least to the scale to which the graphs are shown in this text). For this reason the tables are used for the examples here. The units for $M_u/\phi bd^2$ in both the tables and the graphs are pounds per square inch.

Once $M_u/\phi bd^2$ is determined for a particular beam, the value of M_u can be calculated as illustrated in the alternate solution for Example 4.1. The same tables and graphs can be used for the design of beams as were used for analysis.

Alternate solution of Example 4.1 using Table A.10 (see Appendix)

$$\rho = 0.0161$$

$$\frac{M_u}{\phi bd^2} = 582.8 \text{ psi from table}$$

$$M_u = (0.9)(12)(15.5)^2(582.8) = 1{,}512{,}191 \text{ inch-lb} = 126.0 \text{ ft-k}$$

Another method of checking to be sure that the strength of the tensile steel does not exceed three-fourths of the compression strength of a beam at balanced conditions is presented here. Though this method does not really add to the $3/4\rho_b$ theory just presented, it provides an alternate procedure that can easily be remembered and that can be ultilized without resorting to tables or complicated equations. Furthermore, it is a useful method applicable to beams of other shapes, such as T beams, as will be illustrated in Chapter 5.

For this discussion, the beam of Example 4.1 is used and a balanced strain condition is assumed, as shown in Figure 4.6. The compression concrete is assumed to be at 0.003 and the steel at yield f_y/E_s.

From these strains the value of c can be determined:

$$c = \left(\frac{0.003}{0.003 + 0.00138}\right)(15.50) = 10.62 \text{ inch}$$

$$a = \beta_1 c = (0.85)(10.62) = 9.03 \text{ inch}$$

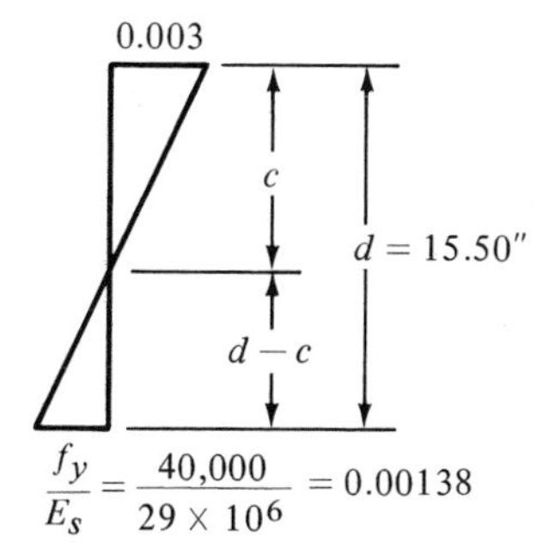

Figure 4.6

If $a = 9.03$ inch, then the area of the beam cross section in compression (A_c) at a stress of $0.85f_c'$ equals ab.

$$A_c = (9.03)(12) = 108.4 \text{ in.}^2$$

$$C = 0.85f_c'A_c = (0.85)(4)(108.4) = 368.6^{\text{k}}$$

$$\text{maximum } T = \tfrac{3}{4}C \text{ at balanced condition} = (\tfrac{3}{4})(368.4) = 276.4^{\text{k}}$$

$$\text{actual } T = A_s f_y = (3.00)(40) = 120^{\text{k}} < 276.4^{\text{k}} \qquad \underline{\text{ok}}$$

Example 4.2 illustrates the design of a beam using the maximum permissible steel percentage. This procedure, of course, provides the smallest permissible beam size for the grades of concrete and steel being used. It may yield an economical solution but probably will not. The large amount of steel is expensive and may cause such crowded conditions as to make the placing of the concrete difficult, particularly at places where beams and columns come together.

EXAMPLE 4.2

Design a rectangular beam for a 20-ft simple span to support a dead load of 2 k/ft (including the estimated beam weight) and a live load of 3 k/ft. Use ρ_{max}, $f_c' = 4{,}000$ psi, and $f_y = 40{,}000$ psi.

SOLUTION

$$w_u = 1.4D + 1.7L = (1.4)(2) + (1.7)(3) = 7.9^{\text{k}}/\text{ft}$$

$$M_u = \frac{(7.9)(20)^2}{8} = 395 \text{ ft-k}$$

from Table A.8 $\rho_{\text{max}} = 0.0371$

$$\rho_{\text{min}} = \frac{200}{f_y} = \frac{200}{40{,}000} = 0.005$$

$$M_u = \phi \rho f_y b d^2 \left(1 - 0.59\rho \frac{f_y}{f_c'}\right)$$

$$(12)(395{,}000) = (0.90)(0.0371)(40{,}000)(bd^2)\left[1 - (0.59)(0.0371)\left(\frac{40{,}000}{4{,}000}\right)\right]$$

$$bd^2 = 4543 \quad \begin{cases} 10 \times 21.31 \\ 12 \times 19.46 \\ 14 \times 18.01 \end{cases}$$

Use 12 × 24 section (d = 19.50)

$$A_s = (0.0371)(12 \times 19.50) = 8.68 \text{ in.}^2 \text{ (use 6 \#11 bars)}$$

Final section shown in Figure 4.7

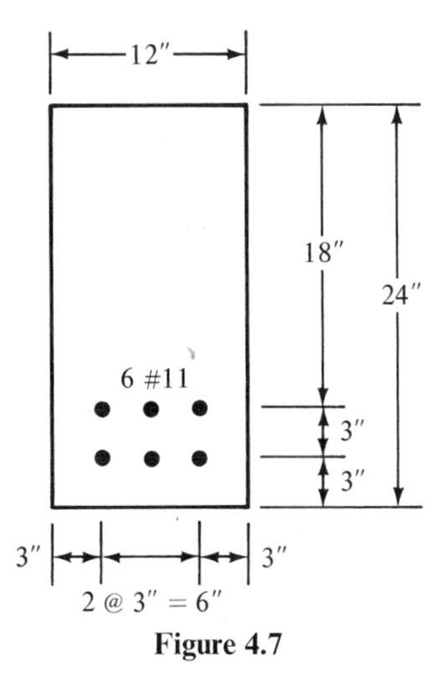

Figure 4.7

Alternate solution for bd^2 using data in appendix

$$\rho_{max} = 0.0371$$

$$\frac{M_u}{\phi bd^2} = 1159.2 \text{ psi (from Table A.10)}$$

$$bd^2 = \frac{(12)(395{,}000)}{(0.9)(1{,}159.2)} = 4{,}543$$

The beam of Example 4.2 is redesigned in Example 4.3 with a smaller steel percentage, equal to $0.18f_c'/f_y$. Through several decades of reinforced concrete experience, it has been found that if the steel percentage is kept down to about this value (which is the usual percentage obtained if the beam were designed by WSD), the beam cross section will be sufficiently large so that deflections will almost never be a problem. It is felt that from the view of both economy and deflection, the use of steel percentages in the range of $0.18f_c'/f_y$ (or perhaps $\frac{1}{2}\rho_{max} = 0.375\rho_b$) will yield very reasonable results.

If these smaller percentages of steel are used, there will be little difficulty in placing the bars and in getting the concrete between them. Of course, from the standpoint of deflection, higher percentages of steel and thus smaller beams can be used for short spans where deflections present no problem. Whatever steel percentages are used, the resulting members will have to be carefully checked for deflections for long-span beams, cantilever beams, and shallow beams and slabs.

Another reason for using a smaller percentage of steel than ρ_{max} is given in Section 8.4 of the ACI Code, where a redistribution of moments (a subject to be discussed in Chapter 12) is permitted in continuous members that have a percentage of steel less than $0.5\rho_b$. For the several reasons mentioned in these paragraphs, many designers try to keep their steel percentages somewhere in the range between $0.18f_c'/f_y$ and $0.5\rho_{max}$ where they feel they are getting the best economy.

EXAMPLE 4.3

Repeat Example 4.2 but use $\rho = 0.18f_c'/f_y$.

SOLUTION

$$\rho = \frac{(0.18)(4{,}000)}{40{,}000} = 0.018$$

$$\frac{M_u}{\phi bd^2} \text{ from Table A.10} = 643.5$$

$$bd^2 = \frac{(12)(395{,}000)}{(0.90)(643.5)} = 8{,}184 \quad \begin{cases} 14 \times 24.18 \\ 16 \times 22.62 \end{cases}$$

Use 14 × 27 beam (*d* = 24.25 in.)

$$A_s = (0.018)(14)(24.25) = 6.11 \text{ in.}^2 \text{ (use 4 \#11)}$$

Final section shown in Figure 4.8

In Example 4.4 a beam size is given for a certain bending moment and it is desired to determine the amount of steel required. The tables can

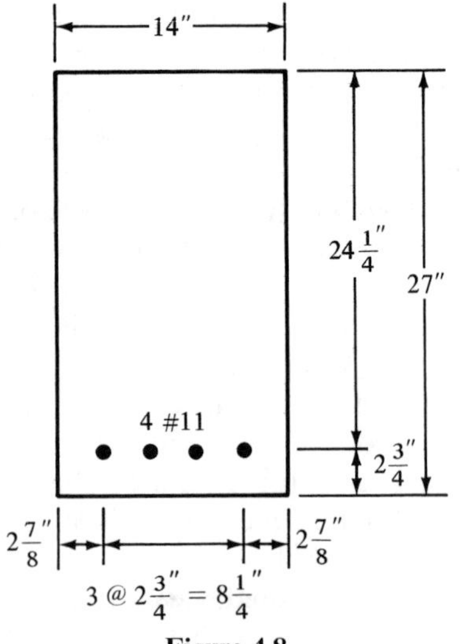

Figure 4.8

again be easily used to solve this problem. The value of $M_u/\phi bd^2$ can be calculated and used to enter the tables, from which the required percentage of steel ρ can be determined.

EXAMPLE 4.4

Determine the steel area required for the beam shown in Figure 4.9. $M_u = 100$ ft-k, $f_c' = 3{,}000$ psi, and $f_y = 40{,}000$ psi.

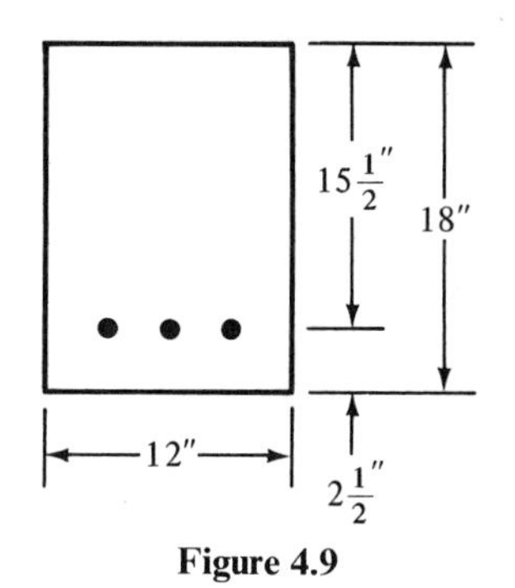

Figure 4.9

SOLUTION

$$\frac{M_u}{\phi bd^2} = \frac{(12)(100{,}000)}{(0.9)(12)(15.5)^2} = 462.5$$

$$\rho = 0.01287 \text{ (from tables)}$$

$$A_s = (0.01287)(12)(15.5) = 2.39 \text{ in.}^2$$

Example 4.5 presents the design of a one-way slab. Information about maximum span/depth ratio, temperature steel, and other related information was previously given in Section 3.4.

EXAMPLE 4.5

Design a one-way slab for a 10-ft simple span to support a live load of 150 psf. $f_c' = 4{,}000$ psi and $f_y = 40{,}000$ psi.

SOLUTION

$$\text{minimum slab } t = \left(0.4 + \frac{40{,}000}{100{,}000}\right)\left(\frac{l}{20}\right) = \frac{l}{25}$$

$$= \frac{(12)(10)}{25} = 4.80'' \qquad \underline{\text{say } 5''}$$

Assume 5 in. Slab ($d = 3\frac{7}{8}$ in.)

$$\text{slab weight} = \left(\tfrac{5}{12}\right)(150) = 62.5 \text{ psf}$$

$$w_u = (1.4)(62.5) + (1.7)(150) = 342.5 \text{ psf}$$

$$M_u = \frac{(0.3425)(10)^2}{8} = 4.28 \text{ ft-k}$$

$$\frac{M_u}{\phi b d^2} = \frac{(12)(4280)}{(0.9)(12)(3.875)^2} = 316.7$$

$$\rho = 0.00833 \text{ (from tables)}$$

$$A_s = (0.00833)(12)(3.875) = 0.387 \text{ in.}^2/\text{ft}$$

use #4 at 6″ ($A_s = 0.39$ in.^{2}/ft)

Temperature and Shrinkage Steel

$$A_s = (0.002)(12)(5) = 0.120 \text{ in.}^2/\text{ft}$$

use #3 at 11″ ($A_s = 0.12$ in.^{2}/ft)

4.11 BUNDLED BARS

Sometimes when large amounts of steel reinforcing are required in a beam or column, it is very difficult to fit all the bars in the cross section. For such situations groups of parallel bars may be bundled together. Up to four bars can be bundled provided they are enclosed by stirrups or ties. The ACI Code (7.6.6) states that bars larger than #11 should not be bundled in beams or girders. This is primarily because of crack control problems a subject discussed in Chapter 6 of this text. Typical configurations for two-, three-, and four-bar bundles are shown in Figure 4.10.

Figure 4.10 Bundled bar arrangements

When spacing limitations and cover requirements are based on bar sizes, the bundled bars may be treated as a single bar for computation purposes, the diameter of the fictitious bar to be calculated from the total equivalent area of the group. When individual bars in a bundle are cut off within the span of beams or girders, they should terminate at different points. The Code requires that there must be at least 40-bar-diameters stagger.

4.12 SI EXAMPLE

Example 4.6 illustrates the design of a beam using the SI units. It will be noted that in Tables A.9 through A.14 and in Graph 1 (see Appendix) values of $M_u/\phi bd^2$ are given in pounds per square inch. These may be converted to SI units in newtons per square millimeter by multiplying the table values by 0.006 895.

EXAMPLE 4.6

Design a rectangular beam for a 10-m simple span to support a dead load of 15 kN/m (not including beam weight) and a live load of 20 kN/m. Use $\rho = 0.5\rho_b$, $f_c' = 20.7$ MPa, $f_y = 275.8$ MPa, and assume the concrete weight is 23.5 kN/m^3.

SOLUTION

Assume beam weight is 8.5 kN/m.

$$w_u = (1.4)(23.5) + (1.7)(20) = 66.9 \text{ kN/m}$$

$$M_u = \frac{(66.9)(10)^2}{8} = 836.2 \text{ kN}\cdot\text{m}$$

$$\rho = 0.018\,6 \text{ (from Table A.8)}$$

$$M_u = \phi \rho f_y bd^2\left(1 - 0.59\rho\frac{f_y}{f_c'}\right)$$

$$(10^6)(836.2) = (0.9)(0.018\,6)(275.8)(bd^2)\left[1 - (0.59)(0.018\,6)\left(\frac{275.8}{20.7}\right)\right]$$

$$bd^2 = 212\,134\,550 \text{ mm}^3 \quad \begin{cases} 400 \times 728 \\ 450 \times 687 \\ 500 \times 651 \end{cases}$$

Use 450 × 800 (d = 695 mm) Assuming Two Rows of Steel

$$\text{beam weight} = \frac{(450)(800)}{10^6}(23.5) = 8.46 \text{ kN/m} < 8.5 \text{ kN/m} \qquad \underline{\text{ok}}$$

$$A_s = (0.018\,6)(450)(695) = 5\,817 \text{ mm}^2$$

Use 6 #11 bars = 6 036 mm^2.

Final section shown in Figure 4.11

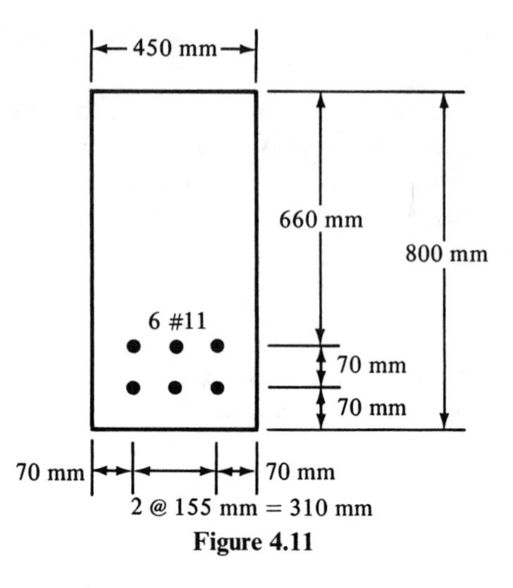

Figure 4.11

Checking bd^2 value using Appendix

$$\frac{M_u}{\phi bd^2} \text{ from Table A.9} = 635.1 \text{ psi} = (635.1)(0.006\,895)$$

$$= 4.379\,014 \text{ N/mm}^2$$

$$bd^2 = \frac{(10^6)(836.2)}{(0.9)(4.379\,014)} = 212\,173\,580 \text{ mm}^3$$

PROBLEMS

4.1 to **4.3** Determine the permissible flexural capacity of each of the beams shown in the accompanying figures if $f_y = 60{,}000$ psi and $f_c' = 4{,}000$ psi.

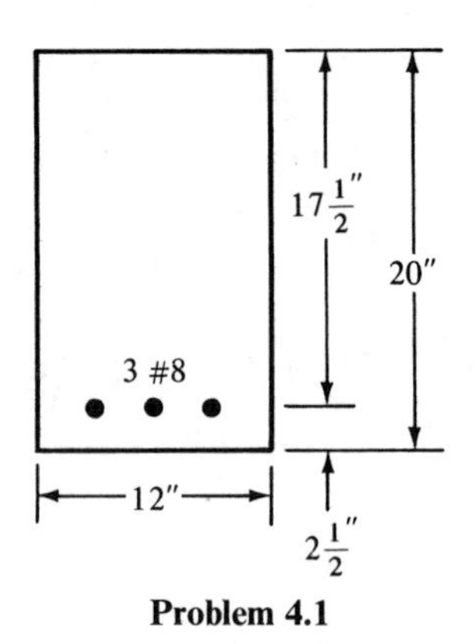

Problem 4.1

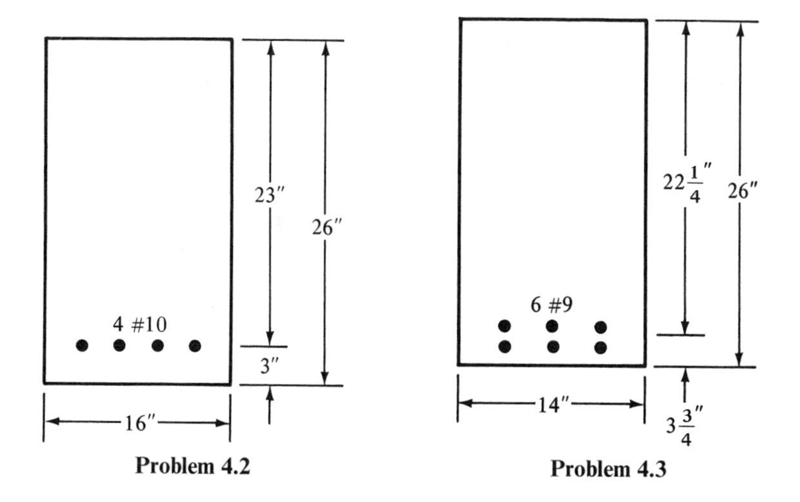

Problem 4.2 **Problem 4.3**

4.4 Repeat Problem 4.1 if $f_y = 40{,}000$ psi and $f_c' = 3{,}000$ psi.

4.5 Repeat Problem 4.2 if 4 #8 bars are used and if $f_y = 40{,}000$ psi and $f_c' = 3{,}000$ psi.

4.6 to **4.12** Design rectangular sections for the beams, loads, an ρ values shown in the accompanying illustrations. Beam weights are not included in the loads shown. Show sketches of cross sections including bar sizes, arrangement, and spacing. Assume concrete weighs 150 #/ft^3. $f_y = 60{,}000$ psi and $f_c' = 3{,}000$ psi.

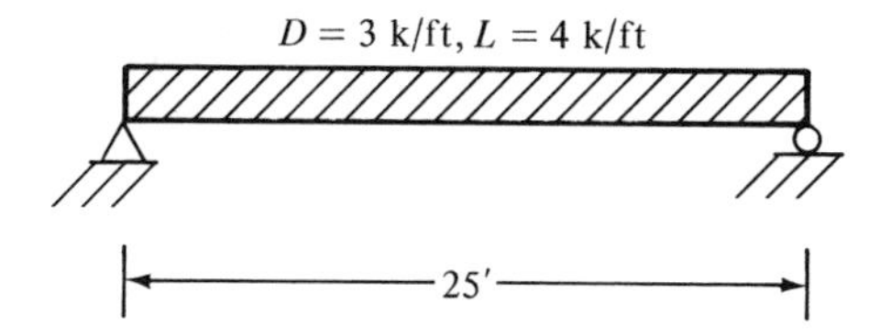

Use ρ_{max}. **Problem 4.6**

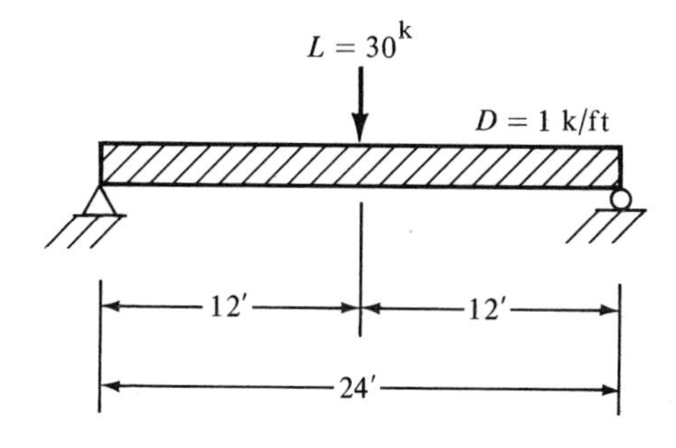

Use ρ_{max}. **Problem 4.7**

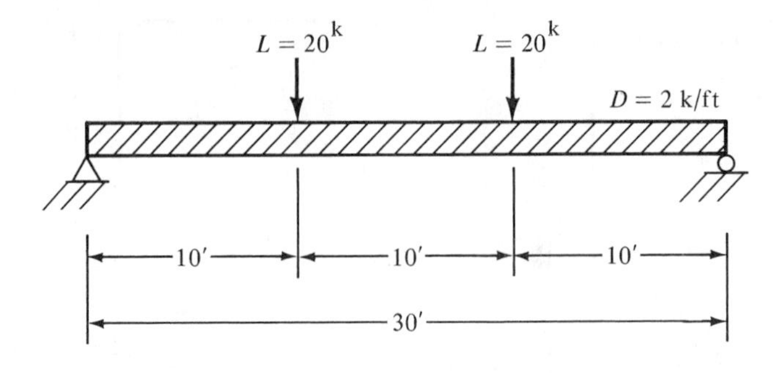

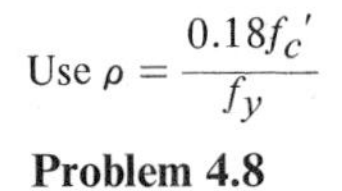

Use $\rho = \dfrac{0.18 f_c'}{f_y}$

Problem 4.8

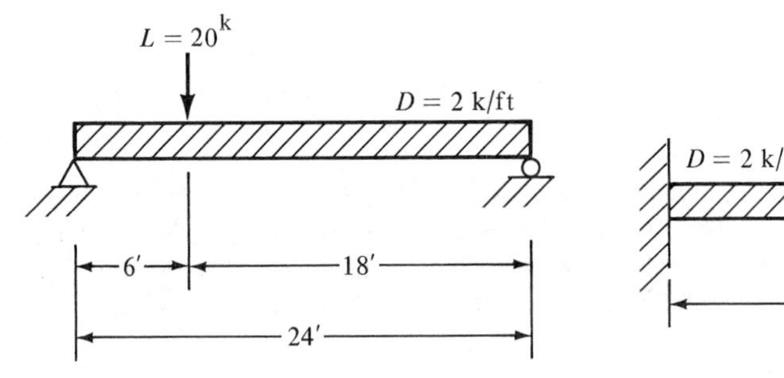

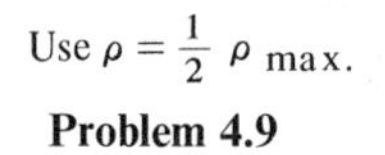

Use $\rho = \frac{1}{2}\,\rho$ max.

Problem 4.9

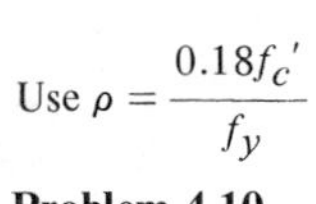

Use $\rho = \dfrac{0.18 f_c'}{f_y}$

Problem 4.10

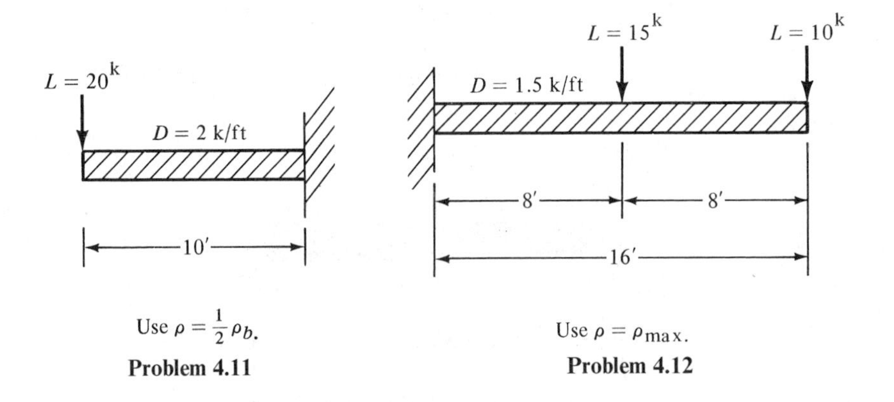

Use $\rho = \frac{1}{2}\rho_b$.

Problem 4.11

Use $\rho = \rho_{max}$.

Problem 4.12

4.13 Determine if the beam of Problem 4.2 is overreinforced or underreinforced without using tables or formulas.

4.14 to **4.15** Design rectangular sections for the beams and loads shown in the accompanying illustrations. Beam weights are not included in the given loads. $f_y =$ 60,000 psi and $f_c' = 4{,}000$ psi. Live loads are to be placed where they will cause the most severe conditions at the sections being considered. Select beam size for the largest moment and then select the steel required for maximum plus moment and show the cross section. Then select steel for maximum negative moment.

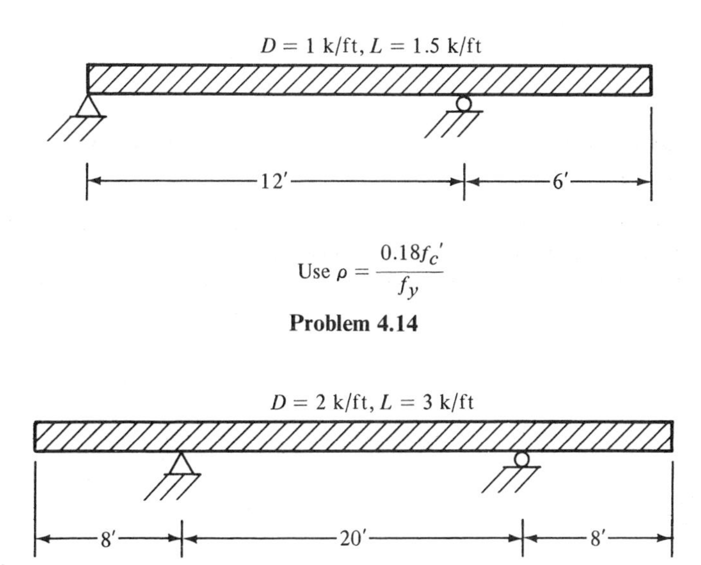

Problem 4.14

Problem 4.15

4.16 to **4.17** Design one-way slabs for the situations shown. Concrete weight = 150 #/ft^3, $f_y = 60{,}000$ psi, and $f_c' = 3{,}000$ psi. Do not use the ACI Code's minimum thickness for deflections. Steel percentages are given in the figures.

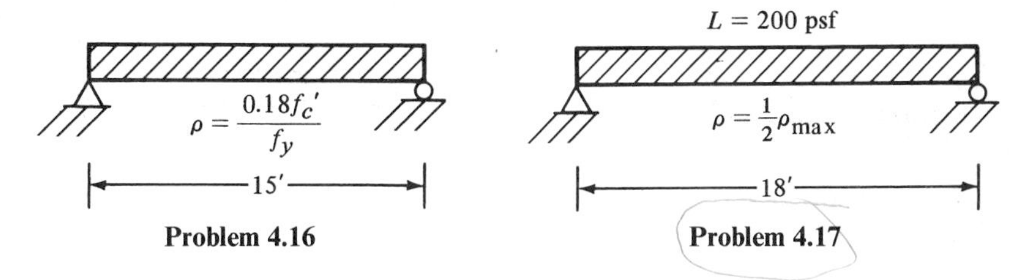

Problem 4.16 **Problem 4.17**

4.18 Using $f_c' = 4{,}000$ psi, $f_y = 40{,}000$ psi, and ρ_{max}, determine the depth of a simple beam to support itself for a 200-ft simple span. Neglect cover in calculations.

4.19 to **4.20** Determine the permissible flexural capacity of each of the beams shown if $f_y = 413.7$ MPa and $f_c' = 20.7$ MPa.

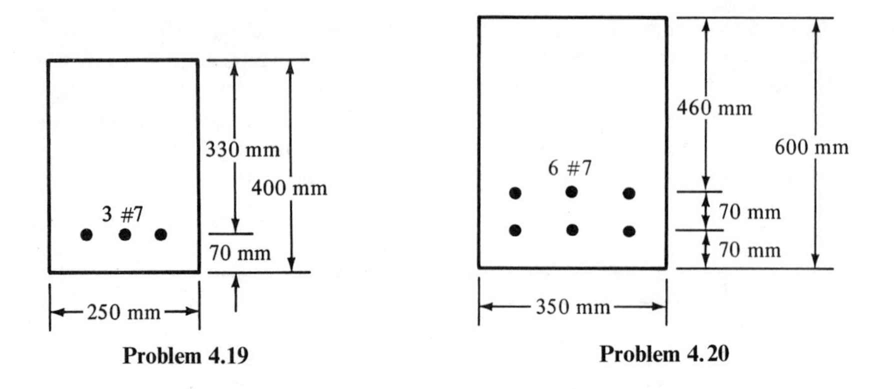

Problem 4.19

Problem 4.20

4.21 to **4.24** Design rectangular sections for the beams, loads, and ρ values shown in the accompanying illustrations. Beam weights are not included in the loads given. Show sketches of cross sections including bar sizes, arrangement, and spacing. Assume concrete weighs 23.5 kN/m^3, $f_y = 413.7$ MPa, and $f_c' = 27.6$ MPa.

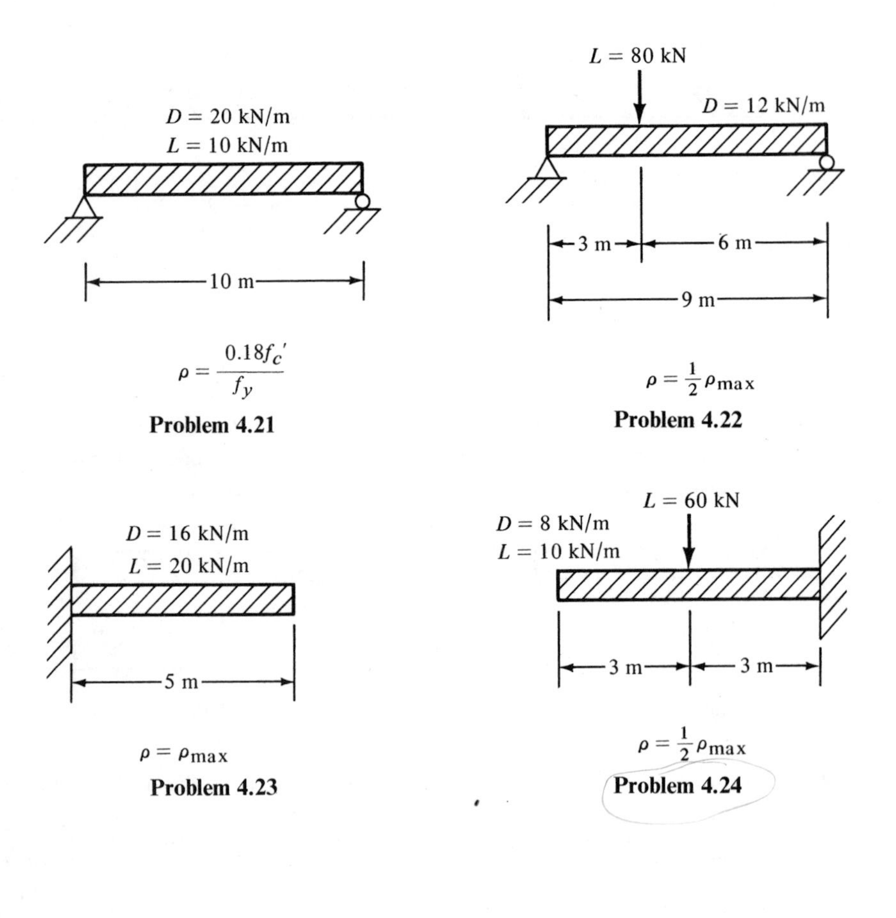

$\rho = \dfrac{0.18 f_c'}{f_y}$

Problem 4.21

$\rho = \frac{1}{2}\rho_{max}$

Problem 4.22

$\rho = \rho_{max}$

Problem 4.23

$\rho = \frac{1}{2}\rho_{max}$

Problem 4.24

CHAPTER 5

Analysis and Design of T Beams and Doubly Reinforced Beams

5.1 T BEAMS

Reinforced concrete floor systems normally consist of slabs and beams that are placed monolithically. As a result, the two parts act together to resist loads. In effect the beams have extra widths at their tops, called flanges, and the resulting T-shaped beams are called T beams. The part of a T beam below the slab is referred to as the web or stem. The stirrups in the webs extend up into the slabs, as perhaps do bent-up bars, with the result that they further make the beams and slabs act together. The crosshatched area shown in Figure 5.1 shows the effective size of a T beam.

There is a problem involved in estimating how much of the slab acts as part of the beam. Actually, the further a particular part of the slab is away from the stem, the smaller is its bending stress (because of shear strains that are not considered in the simple bending theory). Instead of prescribing a wide slab with varying stresses, the ACI Code (8.10.2) calls for a smaller width with assumed uniform stresses for design purposes. For symmetrical Ts, the Code states that the effective flange width may not exceed one-fourth of the beam span and the overhanging width on

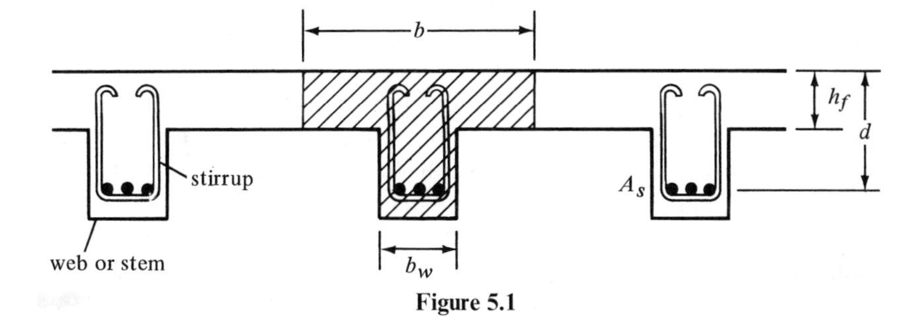

Figure 5.1

Natural History Museum, Kensington, London, England (Courtesy Cement and Concrete Association).

each side may not exceed eight times the slab thickness nor one-half the distance to the adjacent T beam. For Ts with flanges on one side only and for isolated T beams, other values are given in the Code (8.10.3 and 8.10.4).

The analysis of T beams is handled quite similarly to the method used for rectangular beams, and the tensile steel percentage is once again limited to 0.75 times the percentage required for a balanced design. It should be noticed, however, that limiting the steel percentage to a maximum of $0.75\rho_b$ very seldom presents any difficulty (at least in simple-span Ts) because the compression side of the beam is so large that compression stresses are normally quite low, and hence the designer would almost never use an amount of steel greater than $0.75\rho_b$.

The neutral axis for T beams can fall either in the flange or in the stem, depending on the proportions of the slabs and stems. If it falls in the flange, the rectangular beam formulas apply, as can be seen in Figure

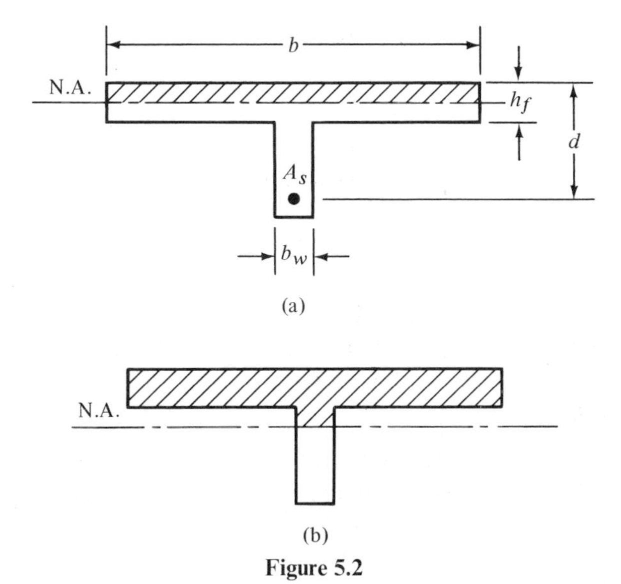

Figure 5.2

5.2(a). The concrete below the neutral axis is assumed to be cracked and its shape has no effect on the flexure calculations (other than weight). The section above the neutral axis is rectangular. If the neutral axis is below the flange, however, as shown for the beam of Figure 5.2(b), the compression concrete above the neutral axis no longer consists of a single rectangle and thus the normal rectangular beam expressions do not apply.

If the neutral axis is assumed to fall within the flange, the value of a can be computed as it was for rectangular beams:

$$a = \frac{A_s f_y}{0.85 f_c' b} = \frac{\rho f_y d}{0.85 f_c'}$$

The distance to the neutral axis c equals a/β_1 for rectangular sections. If the computed value of a is equal to or less than the flange thickness, the section for all practical purposes can be assumed to be rectangular even though the computed value of $c = a/\beta_1$ is actually greater than the flange thickness.

5.2 ANALYSIS OF T BEAMS

The calculation of the ultimate moment capacities of T beams is illustrated in Examples 5.1 and 5.2. In the first of these problems, the neutral axis falls in the flange, while for the second it is in the web. The procedure used for both examples involves the following steps:

1. The calculation of $T = A_s f_y$.
2. The calculation of the area of the concrete in compression (A_c) stressed to $0.85f_c'$.

$$C = T = 0.85f_c' A_c$$

$$A_c = \frac{T}{0.85f_c'}$$

3. The location of the center of gravity of the concrete area A_c.
4. The computation of $M_n = T$ times the lever arm from the center of gravity of the steel to the center of gravity of A_c.
5. $M_u = \phi M_n$.

For Example 5.1, where the neutral axis falls in the flange, it would be logical to apply the normal rectangular equations of Sections 4.7 and 4.8 of this book, but the author has used the couple method as a background for the solution used for Example 5.2 where the neutral axis falls in the web.

EXAMPLE 5.1

Determine the permissible ultimate moment capacity of the T beam shown in Figure 5.3. $f_c' = 3{,}000$ psi and $f_y = 50{,}000$ psi.

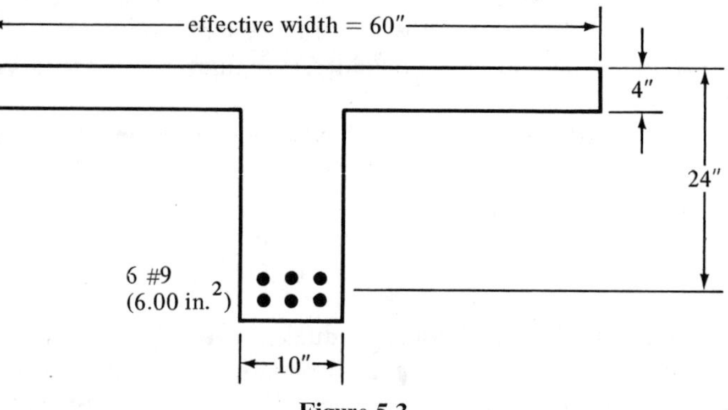

Figure 5.3

SOLUTION

Computing T

$$T = A_s f_y = (6.00)(50) = 300^{\text{k}}$$

Computing A_c

$$A_c = \frac{T}{0.85f_c'} = \frac{300}{(0.85)(3)} = 118 \text{ in.}^2 < 4 \times 60 = 240 \text{ in.}^2$$

Obviously the stress block is entirely within the flange and the rectangular formulas apply. However, using the couple method as follows:

$$a = \frac{118}{60} = 1.97''$$

$$\text{lever arm} = d - \frac{a}{2} = 24 - \frac{1.97}{2} = 23.02''$$

Calculating the Moment Capacity

$$M_n = T\left(d - \frac{a}{2}\right) = (300)(23.02) = 6906''^{k} = 576 \text{ ft-k}$$

$$M_u = \phi M_n = (0.90)(576) = \underline{518 \text{ ft-k}}$$

Checking ρ_{max} (as previously described in Section 4.10)

Assuming a balanced condition with the strain in the extreme compression fibers of the concrete at 0.003 and the steel strain equal to f_y/E_s, as shown in Figure 5.4:

$$c = \frac{0.003}{0.003 + 0.00172}(24.00) = 15.25''$$

$$a = (0.85)(15.25) = 12.96''$$

$$A_c = (60)(4) + (8.96)(10) = 329.6 \text{ in.}^2$$

$$C = (0.85)(3)(329.6) = 840.5^k$$

$$\text{maximum } T = (\tfrac{3}{4})(840.5) = 630.4^k > \text{actual } T = (6)(50) = 300^k \quad \underline{\text{ok}}$$

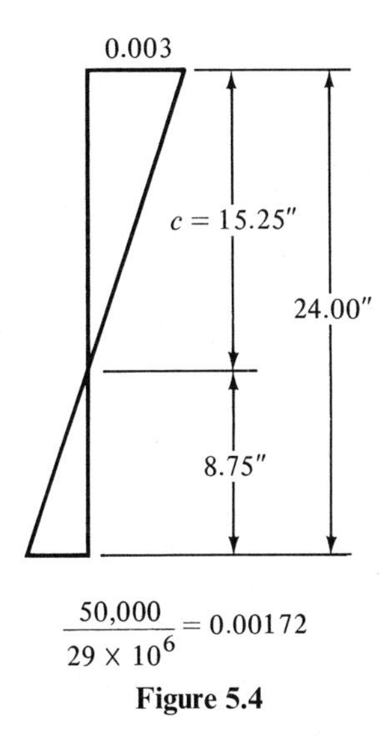

Figure 5.4

EXAMPLE 5.2

Compute the permissible ultimate moment capacity for the T beam shown in Figure 5.5 for which $f_c' = 3{,}000$ psi and $f_y = 50{,}000$ psi.

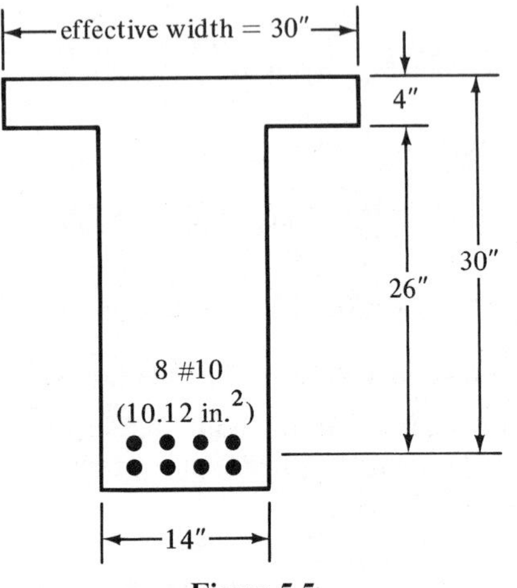

Figure 5.5

SOLUTION

Computing T

$$T = A_s f_y = (10.12)(50) = 506^{k}$$

Computing A_c

$$A_c = \frac{T}{0.85 f_c'} = \frac{506}{(0.85)(3)} = 198.4 \text{ in.}^2 > 30 \times 4 = 120 \text{ in.}^2$$

Obviously the stress block must extend below the flange to provide the necessary compression area = 198.4 − 120 = 78.4 in.^{2}, *as shown in Figure* 5.6.

Computing the distance $\bar{y}$ from the top of the flange to the center of gravity of A_c

$$\bar{y} = \frac{(120)(2) + (78.4)(6.80)}{198.4} = 3.90''$$

The lever arm distance from T to C = 30.00 − 3.90 = 26.10″

$$M_n = (506)(26.10) = 13{,}206''^{k} = 1101 \text{ ft-k}$$
$$M_u = (0.90)(1101) = 991 \text{ ft-k}$$

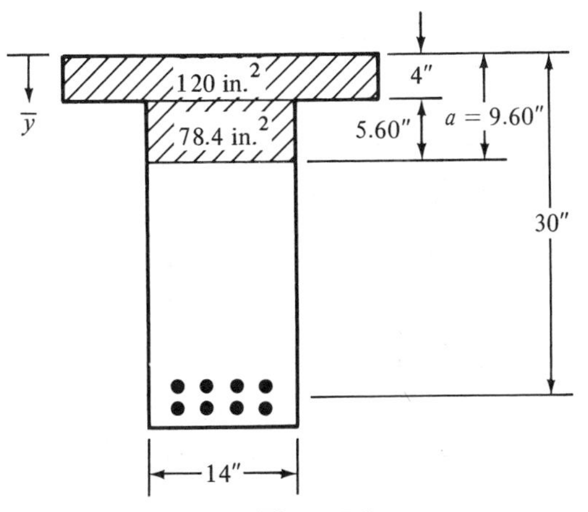

Figure 5.6

Checking ρ_{max} (Figure 5.7)

$$a = (0.85)(19.07) = 16.21''$$

$$A_c = (4)(30) + (12.20)(14) = 291 \text{ in.}^2$$

$$C = (0.85)(3)(291) = 742^k$$

$$\text{maximum } T = \tfrac{3}{4}C = (\tfrac{3}{4})(742) = 556^k > \text{actual } T = (10.12)(50) = 506^k \; \underline{\text{ok}}$$

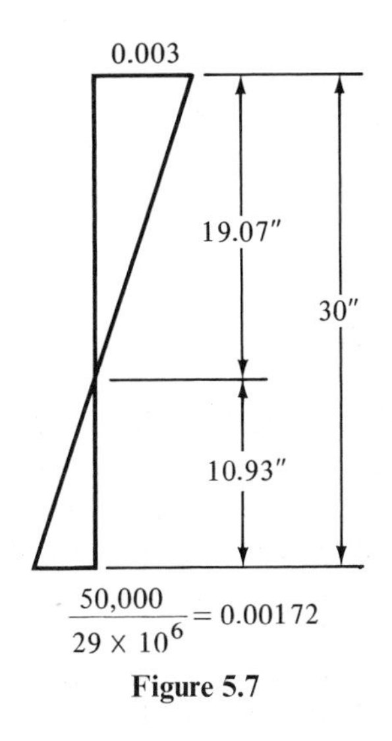

Figure 5.7

Alternate solution for Example 5.2

Breaking the T beam into the compression C_1 on the crosshatched area and the compression C_2 on the overhanging parts of the flanges,

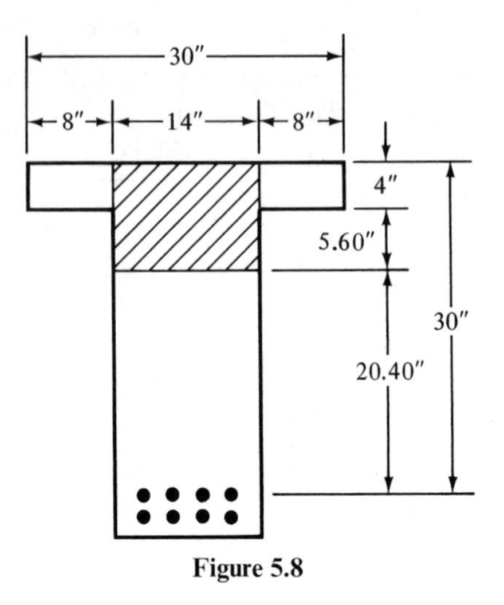

Figure 5.8

as shown in Figure 5.8:

$$C_1 = (0.85)(3)(14)(9.60) = 342.7^{\text{k}}$$
$$C_2 = (0.85)(3)(2)(8)(4) = 163.2^{\text{k}}$$

Computing M_n and M_u

$$M_n = (342.7)\left(30 - \frac{9.60}{2}\right) + (163.2)\left(30 - \frac{4}{2}\right) = 13{,}205 \text{ in.-k} = 1100 \text{ ft-k}$$

$$M_u = (0.90)(1100) = 990 \text{ ft-k}$$

5.3 DESIGN OF T BEAMS

For the design of T beams, the flange has normally already been selected in the slab design, as it is the slab. The size of the web is normally not selected on the basis of moment requirements but probably is given an area based on shear requirements; that is, a sufficient area is used so as to provide a certain minimum shear capacity as will be described in Chapter 7. It is also possible that the width of the web may be selected on the basis of the width estimated to be needed to put in the reinforcing bars. For the examples that follow (5.3 and 5.4), the values of d and b_w are given.

The flanges of most T beams are usually so large that the neutral axis probably falls within the flange and thus the rectangular beam formulas apply. Should the neutral axis fall within the web, a trial and error process is recommended for the design. In this process a lever arm from the center of gravity of the compression block to the center of gravity of the steel is estimated to equal to the larger of $0.9d$ or $d - (h_f/2)$ and from this value called z, a trial steel area is calculated ($A_s = M_n/f_y z$). Then by the process used in Example 5.2, the value of the estimated lever arm is checked. T beams are designed in Examples 5.3 and 5.4 by this process.

EXAMPLE 5.3

Design a T beam for the floor system shown in Figure 5.9 for which b_w and d are given. $M_D = 50$ ft-k, $M_L = 100$ ft-k, $f_c' = 4{,}000$ psi, $f_y = 50{,}000$ psi, and simple span = 20 ft.

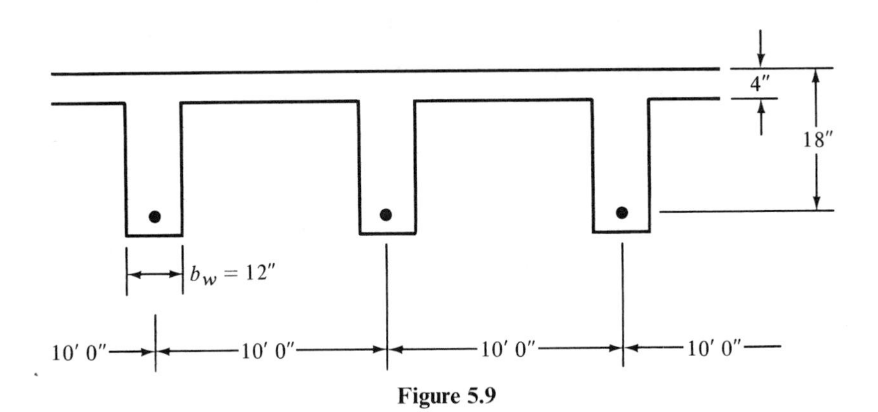

Figure 5.9

SOLUTION

Effective flange width

(a) $\frac{1}{4} \times 20 = 5'\,0'' = \underline{60''}$
(b) $12 + (2)(8)(4) = 76''$
(c) $10'\,0'' = 120''$

Moments

$$M_u = (1.4)(50) + (1.7)(100) = 240 \text{ ft-k}$$

$$M_n = \frac{240}{0.90} = 267 \text{ ft-k}$$

Assuming a lever arm z equal to the larger of $0.9d$ or $d - (h_f/2)$

$$z = (0.9)(18) = \underline{16.20''}$$

$$z = 18 - \tfrac{4}{2} = 16.00''$$

Trial steel area

$$A_s f_y z = M_n$$

$$A_s = \frac{(12)(267)}{(50)(16.20)} = 3.96 \text{ in.}^2$$

Computing values of *a* and z

$$0.85 f_c' A_c = A_s f_y$$

$$(0.85)(4)(A_c) = (3.96)(50)$$

$$A_c = 58.2 \text{ in.}^2$$

$$a = \frac{58.2}{60} = 0.97''$$

Therefore, the neutral axis is in the flange.

$$z = 18 - \frac{0.97}{2} = 17.52''$$

Calculating A_s with this revised z

$$A_s = \frac{(12)(267)}{(50)(17.52)} = 3.66 \text{ in.}^2$$

Computing values of *a* and z

$$A_c = \frac{(3.65)(50)}{(0.85)(4)} = 53.8 \text{ in.}^2$$

$$a = \frac{53.8}{60} = 0.90''$$

$$z = 18 - \frac{0.90}{2} = 17.55''$$

Calculating A_s with this revised z

$$A_s = \frac{(12)(267)}{(50)(17.55)} = 3.65 \text{ in.}^2 \quad \underline{\text{ok, close enough to previous value}}$$

Checking ρ_{max} (Figure 5.10)

$$a = (0.85)(11.44) = 9.72''$$

$$A_c = (4)(60) + (5.72)(12) = 308.6 \text{ in.}^2$$

$$C = (0.85)(4)(308.6) = 1049^{k}$$

$$\text{maximum } T = (\tfrac{3}{4})(1049) = 787^{k} > (3.65)(50) = 182.5^{k} \qquad \underline{\text{ok}}$$

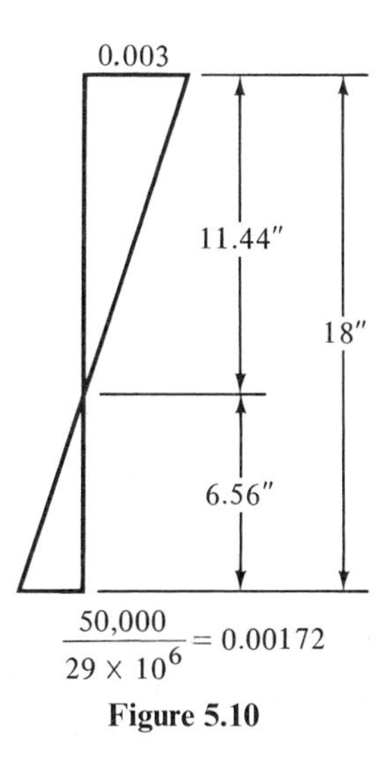

Figure 5.10

Checking ρ_{min}

Section 10.5.1 of the ACI Code states that to compute ρ to check against $\rho_{min} = 200/f_y$, the width of the stem is to be used when the stem is in tension.

$$\rho = \frac{A_s}{b_w d} = \frac{3.65}{(12)(18)} = 0.0169$$

$$\rho_{min} = \frac{200}{50,000} = 0.004 < 0.0169 \qquad \underline{\text{ok}}$$

It is possible to write an expression for the maximum amount of tensile steel permitted by the Code for a particular T beam. This can be accomplished by following exactly the procedure used for determining the maximum T value in Example 5.3. Reference is made to Figure 5.1 for the letters used.

$$c_{bal} = \frac{0.003}{0.003 + [f_y/(29 \times 10^6)]} d = \frac{87,000}{87,000 + f_y} d$$

$$a_{bal} = \beta_1 c_{bal}$$
$$C_{bal} = 0.85 f_c'[b h_f + b_w(a_{bal} - h_f)]$$
$$C_{bal} = T_{bal}$$

$$A_{s\ bal} = \frac{T_{bal}}{f_y}$$

$$A_{s\ max} = 0.75 A_{s\ bal}$$

An expression for $A_{s\,max}$ for a T beam with $f_c' = 3{,}000$ psi and $f_y = 40{,}000$ psi can be developed as follows:

$$A_{s\,max} = 0.75 A_{s\,bal} = 0.75 \frac{T_{bal}}{f_y} = 0.75 \frac{C_{bal}}{f_y}$$

$$= \frac{(0.75)(0.85 f_c')}{f_y}[bh_f + b_w(a_{bal} - h_f)]$$

$$= \frac{(0.75)(0.85 \times 3{,}000)}{40{,}000}\left[bh_f + b_w\left(0.85\frac{87{,}000}{87{,}000 + 40{,}000}d - h_f\right)\right]$$

$$A_{s\,max} = 0.0478[bh_f + b_w(0.582d - h_f)]$$

Following a similar procedure, maximum A_s values permitted by the Code for T beams can be computed for other concrete and steel grades, with the results as shown in Table 5.1. These expressions are applicable regardless of whether the neutral axis falls in the flange or in the stem.

Table 5.1 Maximum Tensile Steel Permitted in T Beams

Concrete and Steel	Formula (units in in.^{2})
$f_c' = 3{,}000$ psi $f_y = 40{,}000$ psi	$A_{s\,max} = 0.0478[bh_f + b_w(0.582d - h_f)]$
$f_c' = 4{,}000$ psi $f_y = 40{,}000$ psi	$A_{s\,max} = 0.0638[bh_f + b_w(0.582d - h_f)]$
$f_c' = 3{,}000$ psi $f_y = 50{,}000$ psi	$A_{s\,max} = 0.0382[bh_f + b_w(0.540d - h_f)]$
$f_c' = 4{,}000$ psi $f_y = 50{,}000$ psi	$A_{s\,max} = 0.0510[bh_f + b_w(0.540d - h_f)]$
$f_c' = 3{,}000$ psi $f_y = 60{,}000$ psi	$A_{s\,max} = 0.0319[bh_f + b_w(0.503d - h_f)]$
$f_c' = 4{,}000$ psi $f_y = 60{,}000$ psi	$A_{s\,max} = 0.0425[bh_f + b_w(0.503d - h_f)]$

In Section 10.3.3 of the ACI Commentary another set of expressions is presented with which ρ_{max} values can be determined for T beams. In addition ρ_{max} expressions are provided for tensilely reinforced rectangular and I beams and for rectangular sections reinforced with both tensile and compressive steel.

EXAMPLE 5.4

Design a T beam for the floor system shown in Figure 5.11 for which b_w and d are given. $M_D = 200$ ft-k, $M_L = 340$ ft-k, $f_c' = 3{,}000$ psi, $f_y = 50{,}000$ psi, and simple span = 18 ft.

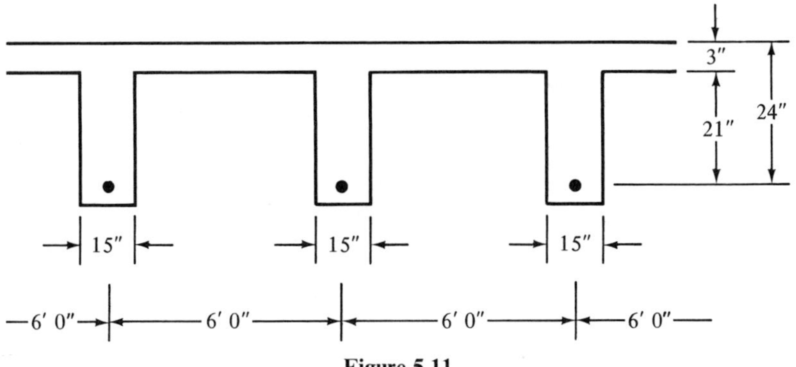

Figure 5.11

SOLUTION

Effective flange width

(a) $\frac{1}{4} \times 18' = 4'\,6'' = \underline{54''}$
(b) $15 + (2)(8)(3) = 63''$
(c) $6' - 0'' = 72''$

Moments

$$M_u = (1.4)(200) + (1.7)(340) = 858 \text{ ft-k}$$

$$M_n = \frac{858}{0.90} = 953 \text{ ft-k}$$

Assuming a lever arm z

$$z = (0.90)(24) = 21.6''$$
$$z = 24 - \tfrac{3}{2} = \underline{22.5''}$$

Trial steel area

$$A_s = \frac{(12)(953)}{(50)(22.5)} = 10.17 \text{ in.}^2$$

Checking values of *a* and *z*

$$A_c = \frac{(50)(10.17)}{(0.85)(3)} = 199.4 \text{ in.}^2$$

Stress block extends down into the web, as shown in Figure 5.12.

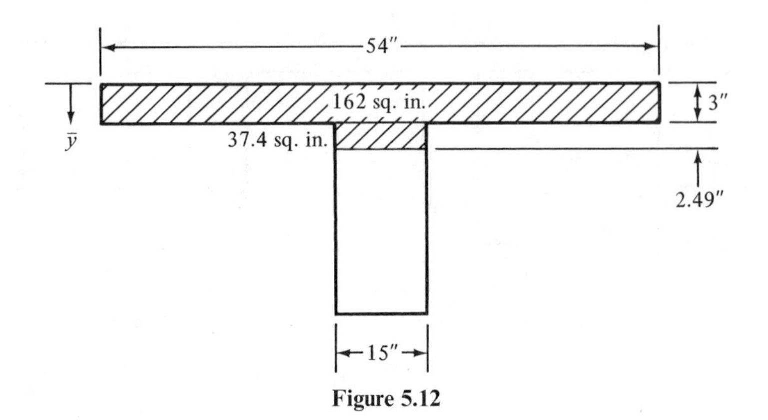

Figure 5.12

Computing the distance $\bar{y}$ from the top of the flange to the center of gravity of A_c

$$\bar{y} = \frac{(162)(1.5) + (37.4)(4.25)}{199.4} = 2.02''$$

$$z = 24 - 2.02 = 21.98''$$

$$A_s = \frac{(12)(953)}{(50)(21.98)} = 10.41 \text{ in.}^2$$ <u>close enough; use</u>

Calculating $A_{s\,\max}$

$$\begin{aligned} A_{s\,\max} &= 0.0382[bh_f + b_w(0.540d - h_f)] \\ &= 0.0382[(54)(3) + (15)(0.540 \times 24 - 3)] \\ &= 11.90 \text{ in.}^2 > 10.41 \text{ in.}^2 \end{aligned}$$ <u>ok</u>

Checking $\rho_{\min}$

$$\rho = \frac{10.41}{(15)(24)} = 0.0289$$

$$\rho_{\min} = \frac{200}{50{,}000} = 0.004 < 0.0289$$ <u>ok</u>

Construction of Locks at Smithfield, Kentucky (Courtesy Symons Corporation).

5.4 COMPRESSION STEEL

The steel that is occasionally used on the compression sides of beams is called *compression steel*, and beams with both tensile and compressive steel are referred to as *doubly reinforced beams*. Compression steel is not normally required in sections designed by the strength method because the use of the full compressive strength of the concrete decidedly decreases the need for such reinforcement, as compared to designs made with the working stress design method.

Occasionally, however, beams are limited to such small sizes by space or aesthetic requirements that compression steel is needed in addition to tensile steel. To increase the moment capacity of a beam beyond that of a tensilely reinforced beam with the maximum percentage of steel ($\frac{3}{4}\rho_b$), it is necessary to introduce another resisting couple in the beam. This is done by adding steel in both the compression and tensile sides of the beam.

Compression steel is very effective in reducing long-term deflections due to shrinkage and plastic flow. In this regard you should note the effect of compression steel on the long-term deflection expression in Section 9.5.2.5 of the Code (to be discussed in Chapter 6 of this text). Continuous compression bars are also helpful for positioning stirrups (by tying them to the compression bars) and keeping them in place during concrete placement and vibration.

Tests of doubly reinforced concrete beams have shown that even if the compression concrete crushes, the beam may very well not collapse if the compression steel is enclosed by stirrups. Once the compression concrete reaches its crushing strain, the concrete cover spalls or splits off the bars, much as in columns (see Chapter 8). If the compression bars are confined by closely spaced stirrups, the bars will not buckle until additional moment is applied. This additional moment cannot be considered in practice because beams are not practically useful after part of their concrete breaks off. (Would you like to use a building after some parts of the concrete beams have fallen on the floor?)

In doubly reinforced beams an initial assumption is made that the compression steel yields as well as the tensile steel. (The tensile steel is always assumed to yield because of the ductile requirements of the ACI Code.) If the strain at the extreme fiber of the compression concrete is assumed to equal 0.003 and the compression steel A_s' is located two-thirds of the distance from the neutral axis to the extreme concrete fiber, then the strain in the compression steel equals $\frac{2}{3} \times 0.003 = 0.002$. If this is greater than the strain in the steel at yield, as say $50{,}000/29 \times 10^6 = 0.00172$ for 50,000-psi steel, the steel has yielded. It should be noted that actually the creep and shrinkage occurring in the concrete help the compression steel to yield.

Sometimes the neutral axis is quite close to the compression steel. As a matter of fact, in some beams with low steel percentages, the neutral axis may be right at the compression steel. For such cases the addition of compression steel is probably a waste of time and money.

When compression steel is used, the ultimate resisting moment of the beam is assumed to consist of two parts: the part due to the resistance of the compression concrete and the balancing tensile reinforcing and the part due to the ultimate moment capacity of the compression steel and the balancing amount of tensile steel. This situation is illustrated in Figure 5.13. In the expressions developed here, the effect of the concrete in compression, which is replaced by the compressive steel A_s', is neglected.

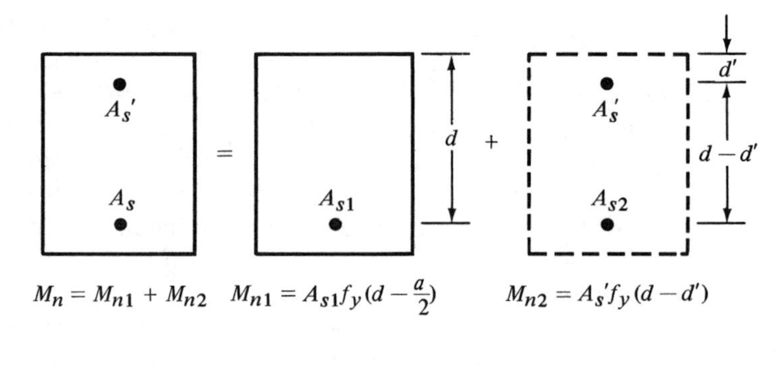

Figure 5.13

The first of the two resisting moments is illustrated in Figure 5.13(b).

$$M_{n1} = A_{s1} f_y \left(d - \frac{a}{2} \right)$$

The second resisting moment is that produced by the additional tensile and compressive steel (A_{s2} and A_s'), which is presented in Figure 5.13(c).

$$M_{n2} = A_s' f_y (d - d')$$

Up to this point it has been assumed that the compression steel has reached its yield stress. If such is the case, the values of A_{s2} and A_s' will be equal because the addition to T of $A_{s2} f_y$ must equal the addition to C of $A_s' f_y$ for equilibrium. If such is not the case, A_s' must be larger than A_{s2}, as will be described later in this section.

Combining the two values,

$$M_n = A_{s1} f_y \left(d - \frac{a}{2} \right) + A_s' f_y (d - d')$$

$$M_u = \phi M_n$$

$$M_u = \phi \left[A_{s1} f_y \left(d - \frac{a}{2} \right) + A_s' f_y (d - d') \right]$$

When the total percentage of tensile steel is equal to or less than $0.75\rho_b$ (where ρ_b is for a rectangular beam with tensile steel only), the compression steel will have little effect on the resisting moment of a doubly reinforced section because with such a tensile-yielding failure situation, the lever arm of the internal couple is not affected very much by the presence of compression steel. This is generally not the case when the tensile steel percentage is greater than $0.75\rho_b$. Nevertheless, for the problems of this chapter, the effect of the compression steel is considered regardless of the tensile steel percentage.

Examples 5.5 and 5.6 illustrate the calculations involved in determining resisting moments of doubly reinforced sections. In each of these problems, the strain in the compression steel is checked to determine if the steel has yielded. If not, as in the second of the two examples, a trial and error process is used to determine the stress. Based on the strains an estimate is made of the stress and the value of A_{s2} is computed by the following expression, where f_s' is the estimated stress in the compression steel:

$$A_{s2} f_y = A_s' f_s'$$

A new value of a is calculated, the neutral axis location is determined, a new strain in the compression steel is computed, and then f_s' is calculated and compared with the estimate. This process can be repeated until there is a reasonably close correlation between the estimated value of f_s' and its calculated value. Once the trial and error process is understood, it is possible to set up a quadratic equation that will, upon solution, yield directly the correct position of the neutral axis, thus eliminating the trial and error work. Such an equation is presented at the end of Example 5.6.

At the end of each of these two examples, the maximum permissible area of tensile reinforcing that will ensure a tensile failure is calculated. Tests of doubly reinforced beams indicate they are quite ductile and as a result *the Code* (10.3.3) *says that to insure ductile behavior in beams with compression reinforcement only the part of the tensile steel which is balanced by compression in the concrete has to be limited by the* 0.75 *factor*. The maximum tensile steel area equals the value permitted if the section were singly reinforced ($0.75\rho_b bd$) plus an area that will provide a force equal to the force produced by the compression steel. If the compression steel has yielded, this equals

$$\text{maximum permissible } A_s = 0.75\rho_b bd + A_s'$$

If the compression steel has not yielded, this expression needs to be revised. To ensure a tensile failure as required by the Code, $A_{s2} f_y$ can only equal $A_s' f_s'$ and thus A_{s2} can only equal $A_s'(f_s'/f_y)$. Thus the maximum total tensile steel is as follows when the compression steel has not yielded:

$$\text{maximum permissible } A_s = 0.75\rho_b bd + A_s' \frac{f_s'}{f_y}$$

EXAMPLE 5.5

Determine the permissible ultimate moment capacity of the beam shown in Figure 5.14. $f_c' = 3{,}000$ psi, and $f_y = 50{,}000$ psi.

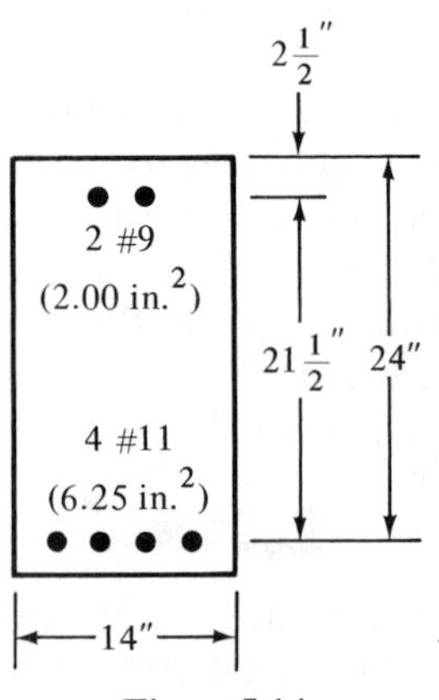

Figure 5.14

SOLUTION

Assuming the compression steel has yielded and thus $A_{s2} = A_s'$

$$A_{s2} = 2.00 \text{ in.}^2$$

$$A_{s1} = A_s - A_{s2} = 6.25 - 2.00 = 4.25 \text{ in.}^2$$

$$a = \frac{A_{s1} f_y}{0.85 f_c' b} = \frac{(4.25)(50)}{(0.85)(3)(14)} = 5.95''$$

Locating neutral axis and checking strain in compression steel

$$c = \frac{5.95}{0.85} = 7.00''$$

$$\varepsilon_s' = \frac{4.50}{7.00} \times 0.003 = 0.00193 > \frac{50{,}000}{29 \times 10^6} = 0.00172$$

Therefore, compression steel has yielded.

$$M_{n1} = A_{s1} f_y \left(d - \frac{a}{2} \right) = (4.25)(50)\left(24 - \frac{5.95}{2} \right) = 4468 \text{ in.-k} = 372.3 \text{ ft-k}$$

$$M_{n2} = A_s' f_y (d - d') = (2.00)(50)(24 - 2.5) = 2150 \text{ in.-k} = 179.2 \text{ ft-k}$$

$$M_n = M_{n1} + M_{n2} = 372.3 + 179.2 = 551.5 \text{ ft-k}$$

$$M_u = (0.90)(551.5) = 496.4 \text{ ft-k}$$

Checking the maximum amount of tensile steel permitted, noting that compression steel has yielded,

$$\begin{aligned} \text{maximum } A_s &= 0.75 \rho_b b d + A_s' \\ &= (0.75)(0.0275)(14)(24) + 2.00 \\ &= 8.93 \text{ in.}^2 > 6.25 \text{ in.}^2 \qquad \text{ok} \end{aligned}$$

Note

Should you compute the permissible resisting moment for this section with tensile steel only (6.25 in.^{2}), you will find it to be 460 ft-k. The addition of 2.00 in.^{2} of steel in the top (a 32% increase in steel) will provide an increase in permissible moment capacity equal to 36.4 ft-k (a 7.9% increase in moment). Similar calculations for other doubly reinforced beams will show that the addition of compression steel is usually not very economical.

EXAMPLE 5.6

Compute the permissible ultimate moment capacity of the section shown in Figure 5.15. $f_c' = 4{,}000$ psi and $f_y = 50{,}000$ psi.

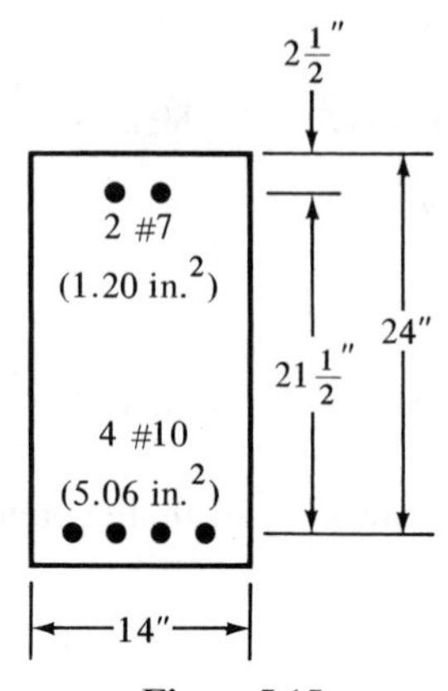

Figure 5.15

SOLUTION

Computing value of *a* (assuming compression steel yields)

$$A_{s2} = 1.20 \text{ in.}^2$$

$$A_{s1} = 5.06 - 1.20 = 3.86 \text{ in.}^2$$

$$a = \frac{(3.86)(50)}{(0.85)(4)(14)} = 4.05''$$

Locate neutral axis and compute strain in compression steel

$$c = \frac{4.05}{0.85} = 4.77''$$

$$\varepsilon_s{}' = \left(\frac{2.27}{4.77}\right)(0.003) = 0.00143 < \frac{50{,}000}{29 \times 10^6} = 0.00172$$

Therefore, compression steel has not yielded.

$$f_s' = \left(\frac{0.00143}{0.00172}\right)(50) = 41.57 \text{ ksi}$$

$$A_s'f_s' = A_{s2}f_y$$

$$(1.20)(41.57) = (A_{s2})(50)$$

$$A_{s2} = 1.00 \text{ in.}^2$$

Computing value of *a*

$$A_{s1} = 5.06 - 1.00 = 4.06 \text{ in.}^2$$

$$a = \frac{(4.06)(50)}{(0.85)(4)(14)} = 4.26''$$

Locating neutral axis and calculating strain and stress in compression steel

$$c = \frac{4.26}{0.85} = 5.01''$$

$$\varepsilon_s' = \left(\frac{2.51}{5.01}\right)(0.003) = 0.00150 < 0.00172$$

$$f_s' = \left(\frac{0.00150}{0.00172}\right)(50) = 43.60 \text{ ksi}$$

Change is not large; therefore, use $43.60^{\text{k}}/\text{in.}^2$.

$$A_s' f_s' = A_{s2} f_y$$

$$(1.20)(43.60) = (A_{s2})(50)$$

$$A_{s2} = 1.05 \text{ in.}^2$$

$$A_{s1} = 5.06 - 1.05 = 4.01 \text{ in.}^2$$

$$a = \frac{(4.01)(50)}{(0.85)(4)(14)} = 4.21''$$

$$c = \frac{4.21}{0.85} = 4.95''$$

Computing permissible ultimate resisting moment

$$M_u = \phi\left[A_{s1} f_y\left(d - \frac{a}{2}\right) + A_s' f_s'(d - d')\right]$$

$$M_u = 0.90\left[(4.01)(50)\left(24 - \frac{4.21}{2}\right) + (1.20)(43.6)(24 - 2.5)\right]$$

$$M_u = 4962''^{\text{k}} = 413.6 \text{ ft-k}$$

Checking maximum amount of tensile steel permitted, noting that compression steel has not yielded

$$\text{maximum permissible } A_s = 0.75\rho_b bd + A_s' \frac{f_s'}{f_y}$$

$$= (0.75)(0.0367)(14)(24) + (1.20)\left(\frac{43.60}{50}\right)$$

$$= 10.30 \text{ in.}^2 > 5.06 \text{ in.}^2 \qquad \underline{\text{ok}}$$

Alternate Method for Locating Neutral Axis

The expression to follow is written by equating the total compression in the compression concrete ($0.85 f_c' A_c$) and the compression steel ($A_s' f_s'$)

to the total tension in the tensile steel ($A_s f_y$). The depth to the neutral axis c is the only unknown and it can be determined by solving the following quadratic equation:

$$C + C' = T$$

$$(0.85)(4)(14)(0.85c) + (1.20)\left[0.003 \times 29{,}000\left(\frac{c - 2.5}{c}\right)\right] = 5.06 \times 50$$

$$40.46c + 104.4 - \frac{261}{c} = 253$$

$$40.46c^2 + 104.4c - 261 = 253c$$

$$40.46c^2 - 148.6c = 261$$

$$c^2 - 3.67c = 6.45$$

$$c - 1.835 = \sqrt{6.45 + (-1.835)^2} = 3.13$$

$$c = 4.96'' \text{ as compared to } 4.95''$$

5.5 DESIGN OF DOUBLY REINFORCED BEAMS

It should be remembered that sufficient tensile steel can be placed in most beams so that compression steel is not needed. But if it is needed, the design is usually quite straightforward. Examples 5.7 and 5.8 illustrate the design of doubly reinforced beams. The solutions follow the theory used for analyzing doubly reinforced sections.

EXAMPLE 5.7

Design a rectangular beam for $M_D = 250$ ft-k and $M_L = 400$ ft-k if $f_c' = 4{,}000$ psi and $f_y = 60{,}000$ psi. The maximum permissible beam dimensions are shown in Figure 5.16.

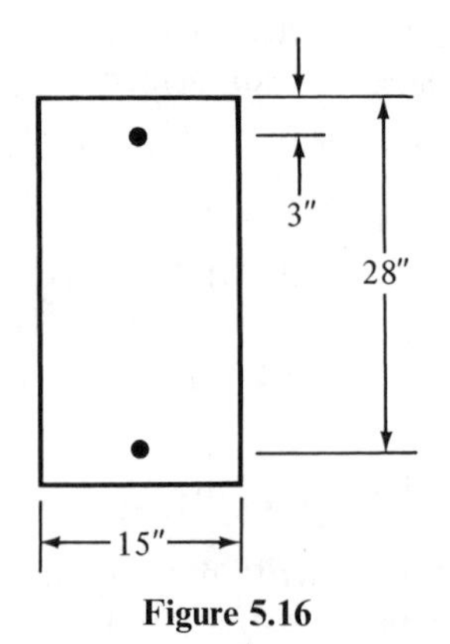

Figure 5.16

SOLUTION

$$M_u = (1.4)(250) + (1.7)(400) = 1{,}030 \text{ ft-k}$$

$$M_n = \frac{1{,}030}{0.90} = 1{,}144 \text{ ft-k}$$

$$\rho_{\max} \text{ if singly reinforced} = 0.0214$$

$$A_{s1} = (0.0214)(15)(28) = 8.99 \text{ in.}^2$$

$$\frac{M_u}{\phi b d^2} = 1040.8 \text{ (from Table A.14 in Appendix)}$$

$$M_{u1} = (1040.8)(0.90)(15)(28)^2 = 11{,}015{,}827 \text{ in.-lb}$$
$$= 918 \text{ ft-k}$$

$$M_{n1} = \frac{918}{0.90} = 1{,}020 \text{ ft-k}$$

$$M_{n2} = M_n - M_{n1} = 1{,}144 - 1{,}020 = 124 \text{ ft-k}$$

Checking to see if compression steel has yielded

$$a = \frac{(8.99)(60)}{(0.85)(4)(15)} = 10.58''$$

$$c = \frac{10.58}{0.85} = 12.45''$$

$$\varepsilon_s' = \left(\frac{9.45}{12.45}\right)(0.003) = 0.00228 > 0.00207$$

Therefore, compression steel has yielded.

$$\text{theoretical } A_s' \text{ required} = \frac{M_{n2}}{(f_y)(d - d')} = \frac{(12)(124)}{(60)(28 - 3)} = 0.99 \text{ in.}^2$$

$$A_s' f_s' = A_{s2} f_y$$

$$A_{s2} = \frac{(0.99)(60)}{60} = 0.99 \text{ in.}^2$$

$$A_s = 8.99 + 0.99 = 9.98 \text{ in.}^2$$

EXAMPLE 5.8

A beam is limited to the dimensions shown in Figure 5.17. If $M_D =$ 150 ft-k, $M_L = 220$ ft-k, $f_c' = 4{,}000$ psi, and $f_y = 60{,}000$ psi, select the reinforcing required.

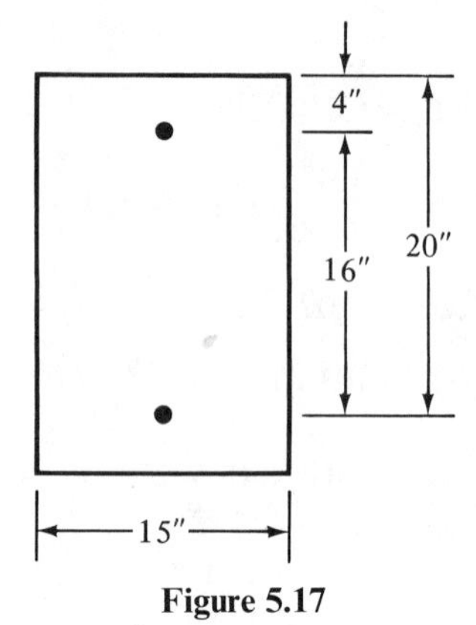

Figure 5.17

SOLUTION

$$M_u = (1.4)(150) + (1.7)(220) = 584 \text{ ft-k}$$

$$M_n = \frac{584}{0.90} = 648.9 \text{ ft-k}$$

$$\rho_{\max} \text{ if singly reinforced} = 0.0214$$

$$A_{s1} = (0.0214)(15)(20) = 6.42 \text{ in.}^2$$

Checking to see if compression steel has yielded

$$a = \frac{(6.42)(60)}{(0.85)(4)(15)} = 7.55''$$

$$c = \frac{7.55}{0.85} = 8.88''$$

$$\varepsilon_s' = \left(\frac{4.88}{8.88}\right)(0.003) = 0.00165 < \frac{60{,}000}{29 \times 10^6} = 0.00207$$

Therefore, compression steel has not yielded.

$$f_s' = \left(\frac{0.00165}{0.00207}\right)(60) = 47.83 \text{ ksi}$$

Resisting moment of singly reinforced section

$$\frac{M_u}{\phi b d^2} = 1040.8 \text{ (from table in Appendix)}$$

$$M_{u1} = (1040.8)(0.90)(15)(20)^2 = 5{,}620{,}320 \text{ in.-lb} = 468.4 \text{ ft-k}$$

$$M_{n1} = \frac{468.4}{0.90} = 520.4 \text{ ft-k}$$

Determination of steel areas

$$M_{n2} = 648.9 - 520.4 = 128.5 \text{ ft-k}$$

$$\text{theoretical } A_s' \text{ required} = \frac{M_{n2}}{(f_s')(d - d')} = \frac{(12)(128.5)}{(47.83)(20 - 4)} = 2.01 \text{ in.}^2$$

$$A_s'f_s' = A_{s2}f_y$$

$$A_{s2} = \frac{(2.01)(47.83)}{60} = 1.60 \text{ in.}^2$$

$$A_s = 6.42 + 1.60 = 8.02 \text{ in.}^2$$

PROBLEMS

5.1 to **5.6** Determine the permissible flexural capacities of the T beams shown in the accompanying illustrations if $f_y = 50{,}000$ psi and $f_c' = 4{,}000$ psi. Are the steel percentages in each case equal to or less than ρ_{max}?

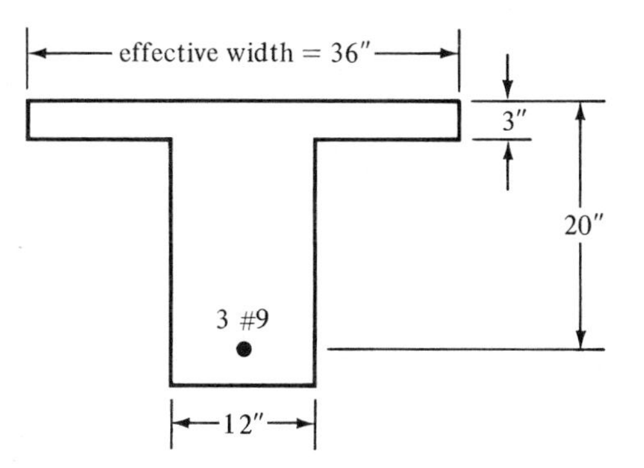

Problem 5.1

5.2 Repeat Problem 5.1 if 4 #11 bars are used.
5.3 Repeat Problem 5.1 if 8 #11 bars are used.

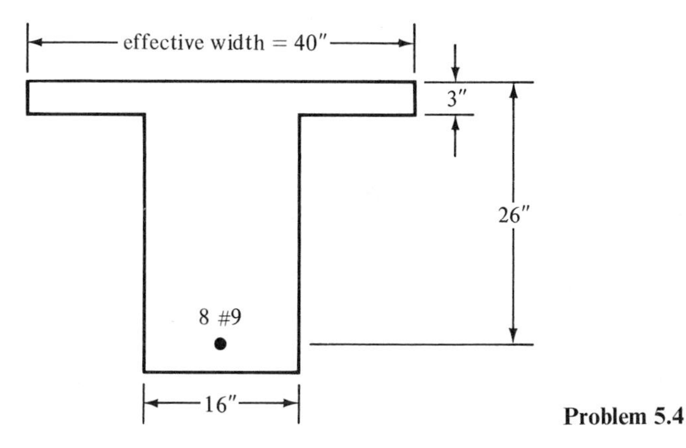

Problem 5.4

5.5 Repeat Problem 5.4 if $f_y = 60{,}000$ psi and $f_c' = 5{,}000$ psi.

5.6 Repeat Problem 5.4 if 8 #10 bars are used.

5.7 Calculate the permissible flexural capacity for one of the T beams shown if $f_c' = 3{,}000$ psi, $f_y = 40{,}000$ psi, and if the section has a 24 ft simple span. Is the steel percentage equal to or less than ρ_{max}?

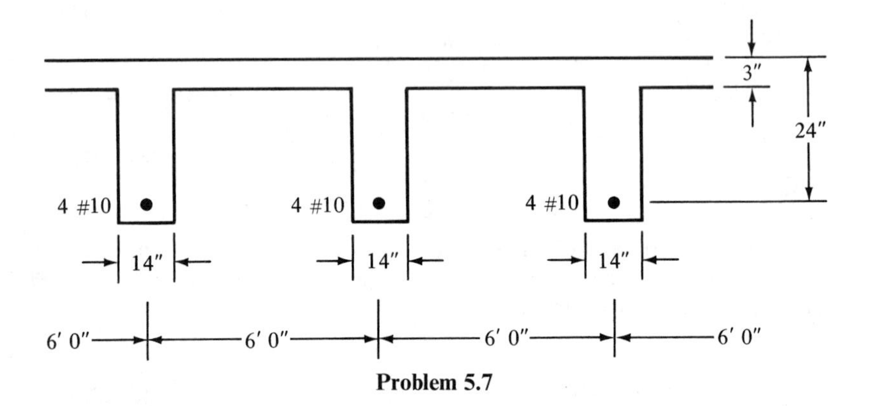

Problem 5.7

5.8 Determine the area of reinforcing steel required for the T beam shown if $f_c' = 3{,}000$ psi, $f_y = 60{,}000$ psi, and $M_u = 300$ ft-k.

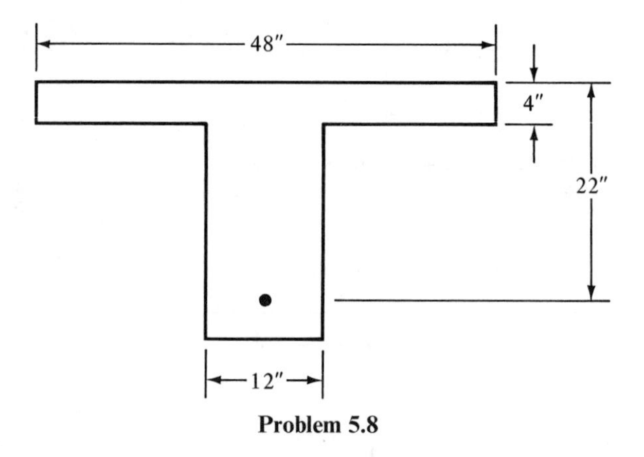

Problem 5.8

5.9 Repeat Problem 5.8 if $M_u = 500$ ft-k.

5.10 Repeat Problem 5.8 if $f_c' = 3{,}000$ psi and $f_y = 40{,}000$ psi.

5.11 Determine the amount of reinforcing steel required for each T beam in the accompanying illustration. $f_Y = 50{,}000$ psi, $f_C' = 3{,}000$ psi, simple span $= 20$ ft, $M_D = 250$ ft-k, $M_L = 325$ ft-k.

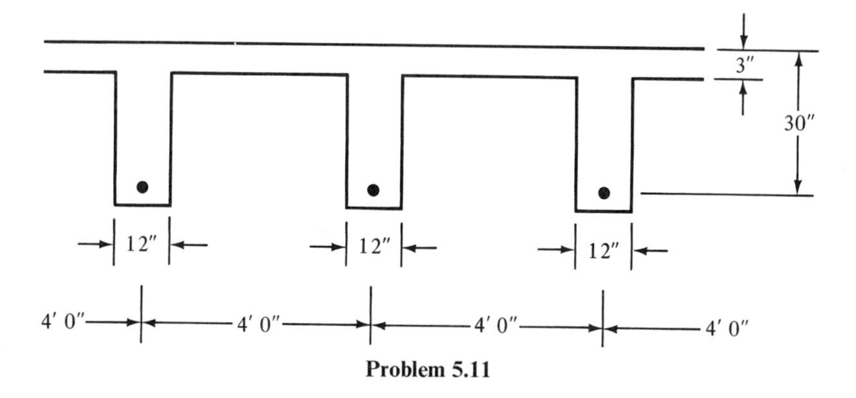

Problem 5.11

5.12 to **5.14** Compute the permissible flexural capacities of the beams shown if $f_y = 60{,}000$ psi and $f_c' = 4{,}000$ psi. Check the maximum permissible A_s in each case to ensure tensile failure.

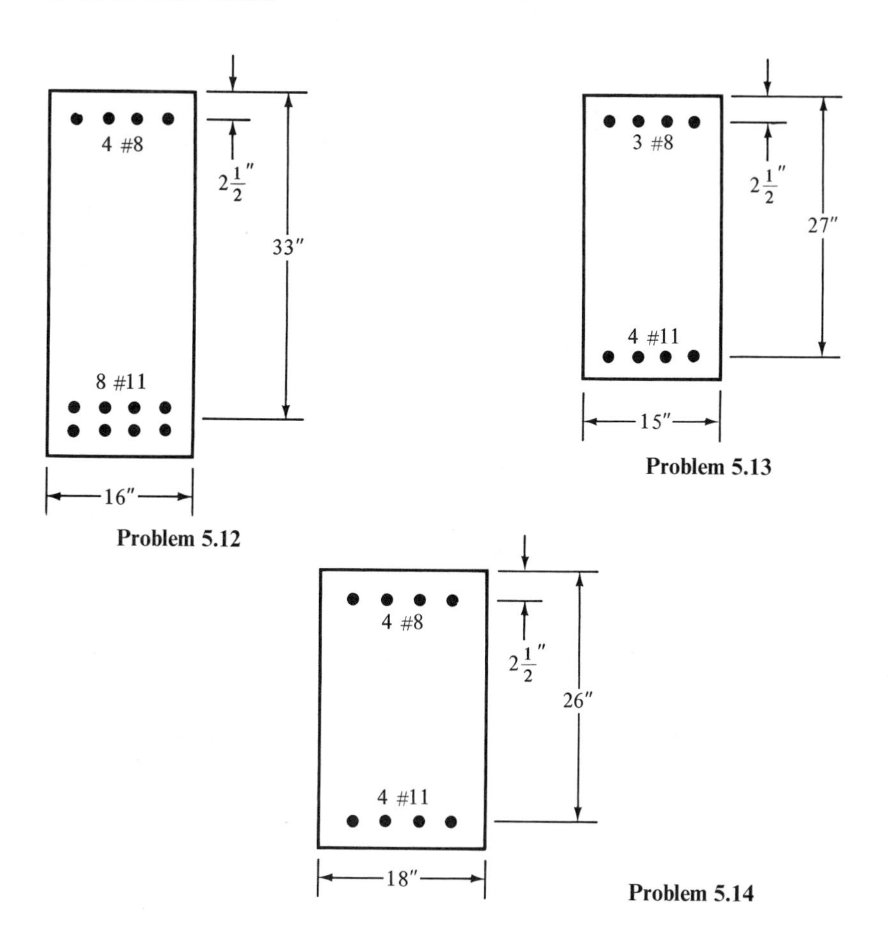

Problem 5.12

Problem 5.13

Problem 5.14

5.15 Compute the permissible ultimate capacity of the beam shown in the accompanying illustration. How much can this permissible moment be increased if 2 #9 bars are added to the top $2\frac{1}{2}$ in. from the compression face? $f_c' = 3{,}000$ psi and $f_y = 60{,}000$ psi.

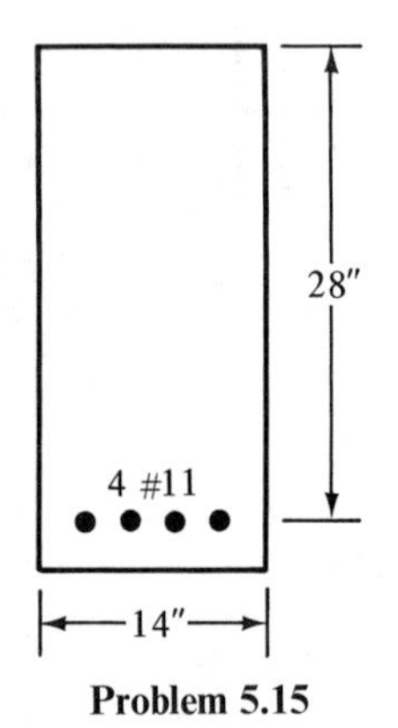

Problem 5.15

5.16 to **5.18** Determine the steel areas required for the sections shown in the accompanying illustrations. In each case the dimensions are limited to the values shown. If compression steel is required, assume it will be placed 3 in. from the compression face. $f_c' = 4{,}000$ psi and $f_y = 60{,}000$ psi.

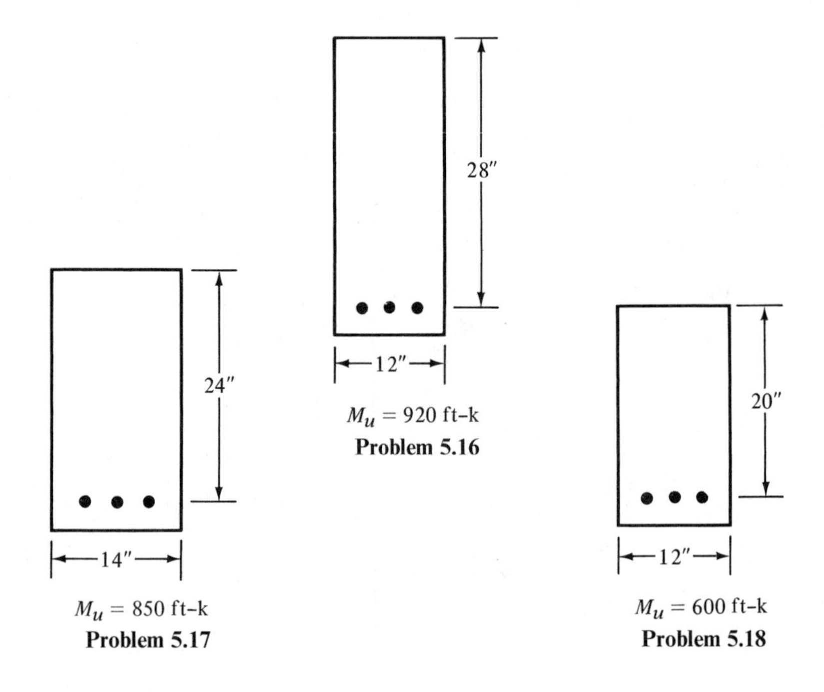

$M_u = 920$ ft–k
Problem 5.16

$M_u = 850$ ft–k
Problem 5.17

$M_u = 600$ ft–k
Problem 5.18

PROBLEMS WITH SI UNITS

5.19 to **5.20** Determine the permissible flexural capacities of the beams shown in the accompanying illustrations if $f_c' = 27.6$ MPa and $f_y = 413.7$ MPa. Are the steel percentages in each case equal to or less than ρ_{max}? $E_s = 200\,000$ MPa.

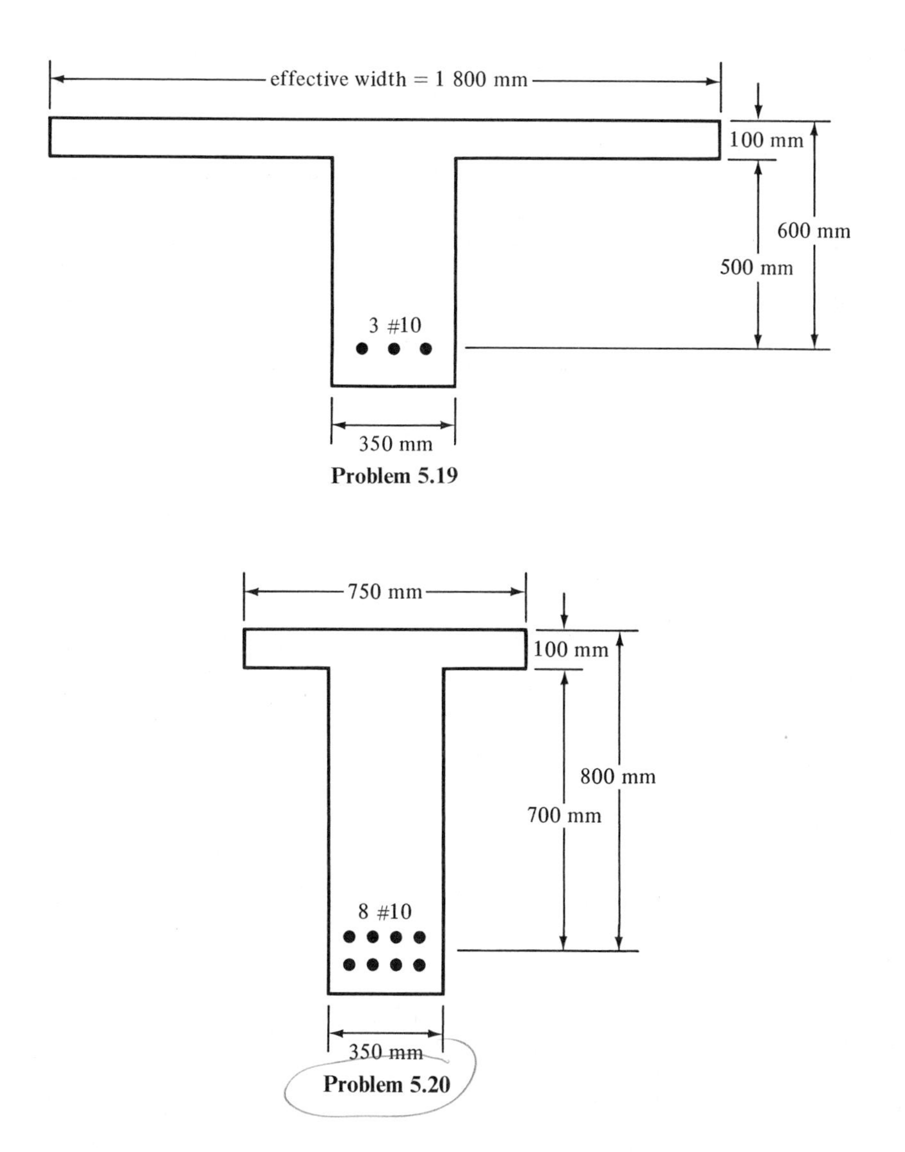

Problem 5.19

Problem 5.20

5.21 to **5.22** Determine the area of reinforcing steel required for the T beams shown if $f_c' = 20.7$ MPa and $f_y = 344.8$ MPa.

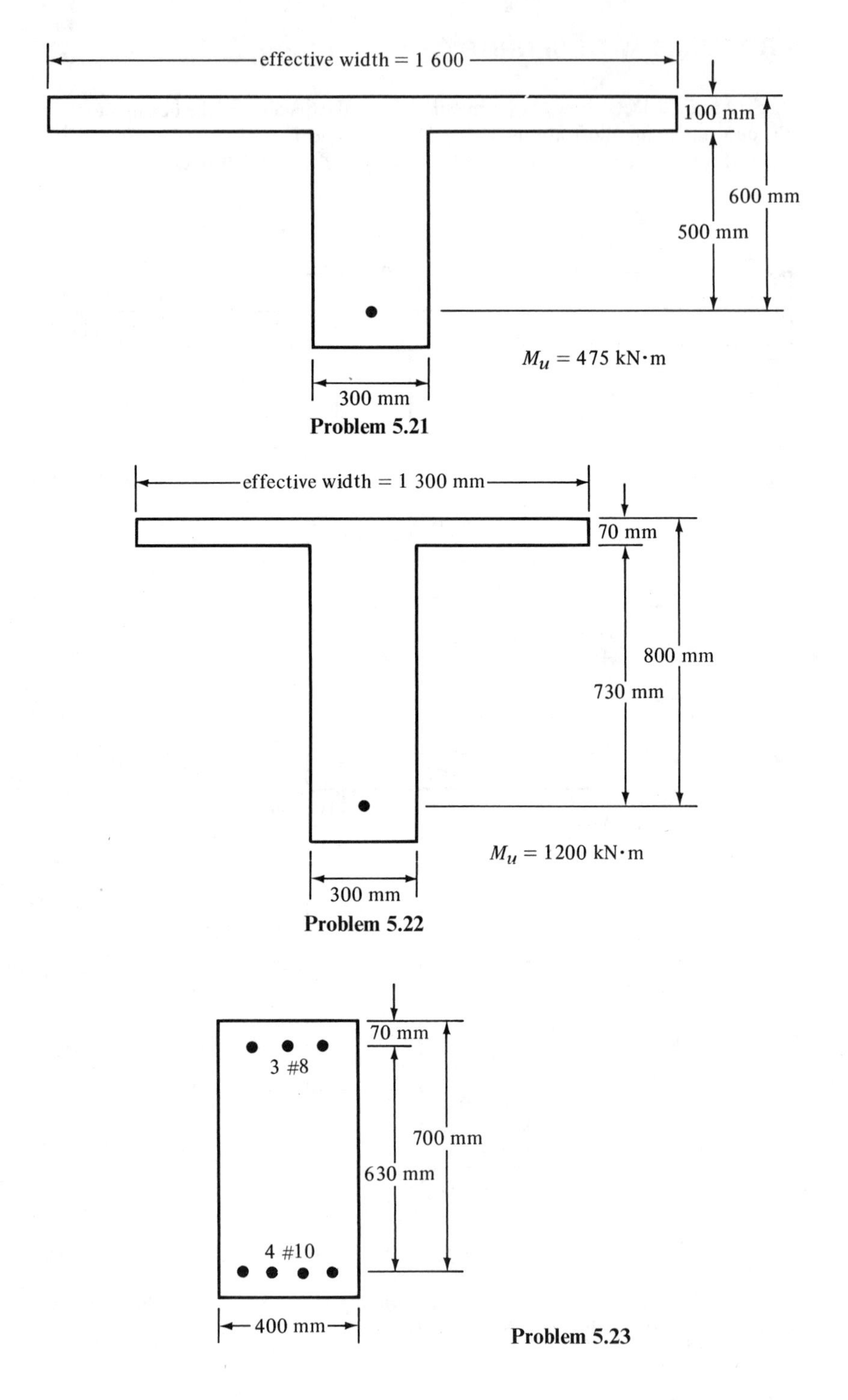

Problem 5.21

Problem 5.22

Problem 5.23

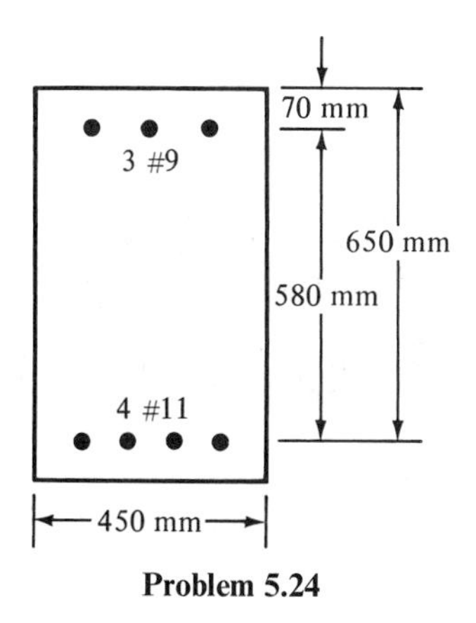

Problem 5.24

5.23 to **5.24** Compute the permissible flexural capacities of the beams shown if $f_y = 344.8$ MPa and $f_c' = 20.7$ MPa. Check the maximum permissible A_s in each case to ensure ductile failure. $E_s = 200\,000$ MPa.

5.25 to **5.26** Determine the steel areas required for the sections shown in the accompanying illustrations. In each case the dimensions are limited to the values shown. If compression steel is required, assume it will be placed 70 mm from the compression face. $f_c' = 27.6$ MPa and $f_y = 413.7$ MPa. $E_s = 200\,000$ MPa.

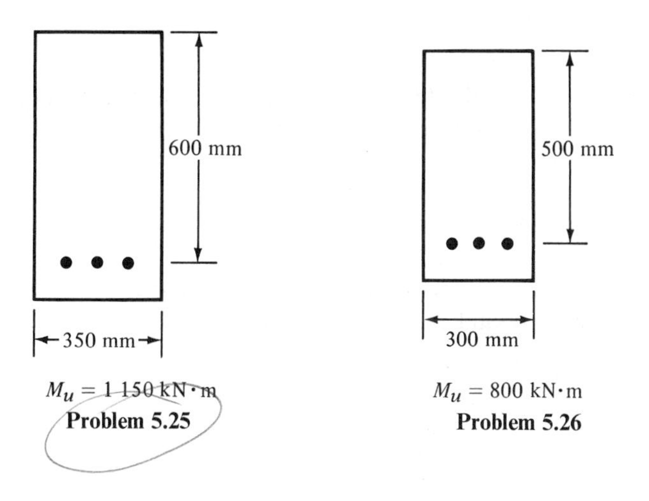

$M_u = 1\,150$ kN·m
Problem 5.25

$M_u = 800$ kN·m
Problem 5.26

CHAPTER 6

Deflections, Cracking and Development Lengths

6.1 DEFLECTIONS

The adoption of the strength design method in recent years together with the use of higher-strength concretes and steels has permitted the use of relatively slender members. As a result, deflections and deflection cracking have become more severe problems than they were a decade or two ago.

The magnitudes of deflections for concrete members can be quite important. Excessive deflections of beams and slabs may cause sagging floors, excessive vibrations, and even interference with the proper operation of supported machinery. Such deflections may damage partitions and cause poor fitting of doors and windows. In addition, they may damage a structure's appearance or frighten the occupants of the building, even though the building may be perfectly safe. Any structure used by people should be quite rigid and relatively vibration free so as to provide a sense of security.

Perhaps the most common type of deflection damage in reinforced concrete structures is the damage to light masonry partitions. They are particularly subject to injury due to concrete's long-term creep. When the floors above and below deflect the relatively rigid masonry partitions do not bend easily and they are often severely damaged. On the other hand, the more flexible gypsum board partitions are much more adaptable to such distortions.

Section 9.5.2 of the ACI Code provides a set of minimum thicknesses for beams and one-way slabs to be used, unless actual deflection calculations indicate that lesser thicknesses are permissible. These minimum thickness values should only be used for beams and slabs which are not supporting or attached to partitions or other members which are likely to be damaged by deflections. If deflections are calculated they should not exceed the values given in Table 9.5(b) of the Code. Deflections for reinforced concrete members can be calculated by the usual deflection expressions, several of which are shown in Figure 6.1.

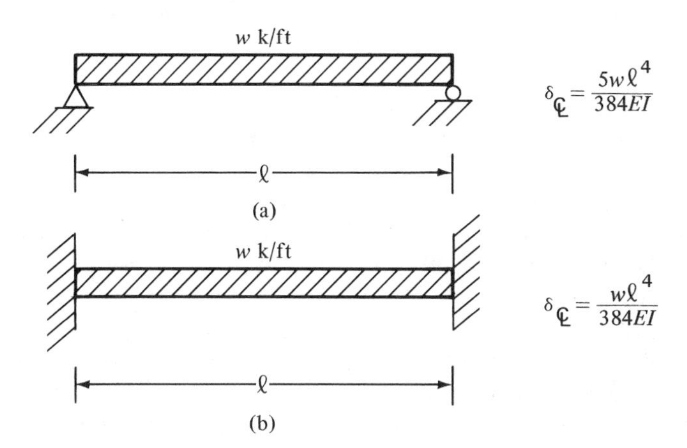

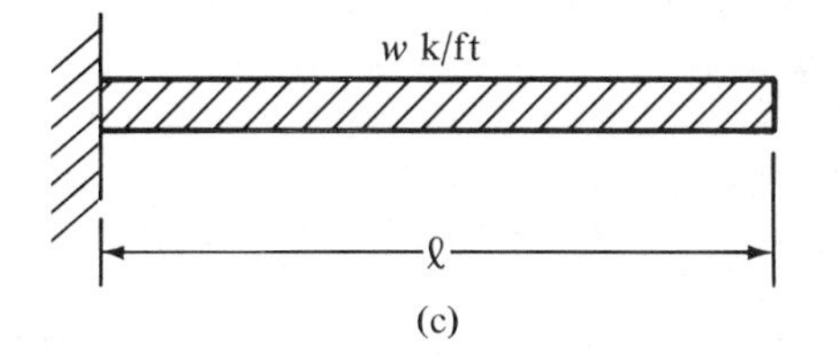

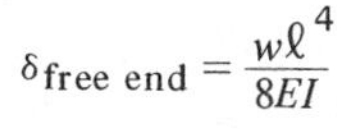

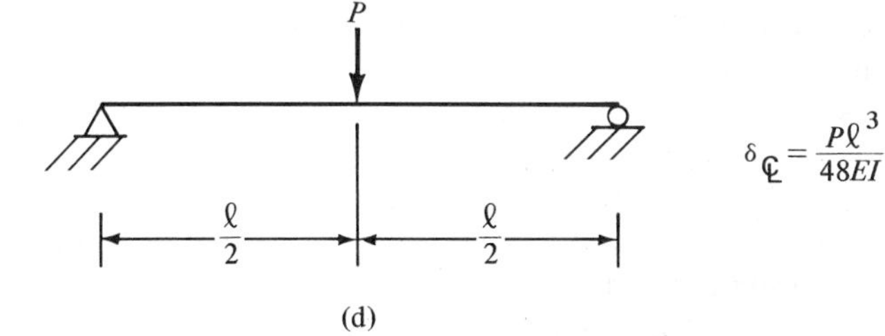

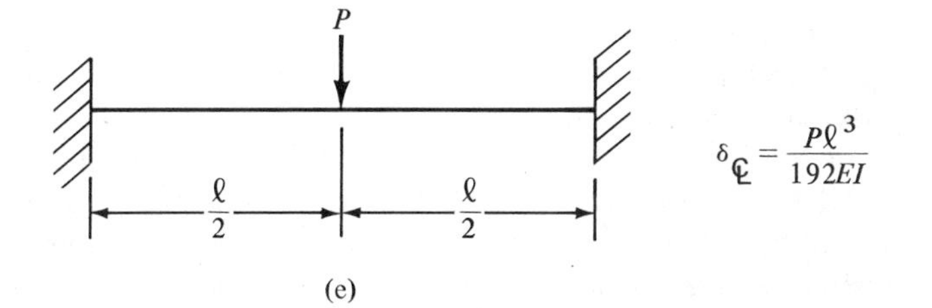

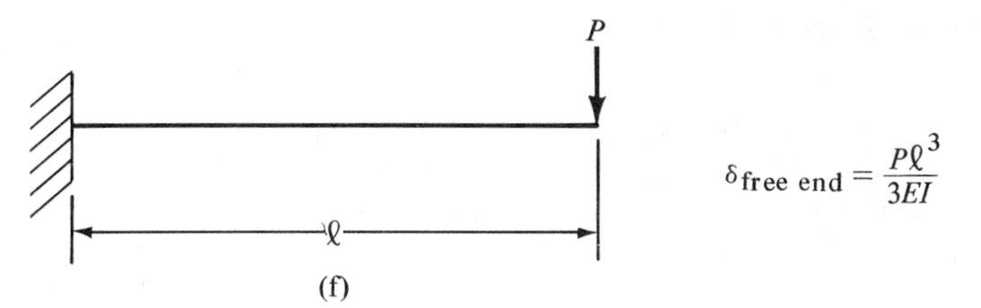

Figure 6.1 Some Deflection Expressions

A few comments should be made about the magnitudes of deflections in concrete members as determined by the expressions given in this figure. It can be seen that the ℄ deflection of a uniformly loaded simple beam [Figure 6.1(a)] is five times as large as the ℄ deflection of the same beam if its ends are fixed [Figure 6.1(b)]. Nearly all concrete beams and slabs are continuous and their deflections fall somewhere in between the two extremes mentioned here.

Because of the very large deflection variations that occur with different end restraints, it is essential that those restraints be considered if realistic deflection calculations are to be made. For most practical purposes it is sufficiently accurate to calculate the ℄ deflection of a member as though it is simply supported and to subtract from that value the deflection caused by the average of the negative moments at the member ends.

Regardless of the method used for calculating deflections, there is a problem in determining the moment of inertia to be used. The trouble lies in the amount of cracking that has occurred. If the bending moment is less than the cracking moment (that is, if the flexural stress is less than the modulus of rupture of about $7.5\sqrt{f_c'}$ for normal-weight concrete), the full uncracked section provides rigidity, and the moment of inertia for the gross section I_g is available. When larger moments are present, different-sized tension cracks occur and the position of the neutral axis varies. At cross sections where tension cracks are actually located, the moment of inertia is probably close to the transformed I, and in between cracks it is perhaps close to I_g. Furthermore, diagonal tension cracks may exist in areas of high shear, causing other variations. As a result, it is difficult to decide what I should be used.

A concrete section that is fully cracked on its tension side will have a rigidity of anywhere from one-third to three-fourths of its rigidity if it were uncracked. At different sections along the beam, the rigidity varies depending on the moment present. It is easy to see that an accurate method of calculating deflections must take these variations into account.

In Section 9.5.2.3 of the Code, a moment of inertia is given that is to be used for deflection calculations. This moment of inertia is an average value and is to be used at any point in a simple beam where the deflection is desired. It is referred to as I_e, the effective moment of inertia, and is based on an estimation of the probable amount of cracking caused by the varying moment throughout the span:[1]

$$I_e = \left(\frac{M_{cr}}{M_a}\right)^3 (I_g) + \left[1 - \left(\frac{M_{cr}}{M_a}\right)^3\right] I_{cr}$$

In this expression I_g is the gross moment of inertia (without considering the steel) of the section, M_{cr} is the cracking moment $= f_r I_g / y_t$, with $f_r =$

[1] Branson, D. E., "Instantaneous and Time-Dependent Deflections on Simple and Continuous Reinforced Concrete Beams," HPR Report No. 7, Part I, Alabama Highway Department, Bureau of Public Roads, August 1963 (1965), pp. 1–78.

$7.5\sqrt{f_c'}$ for normal-weight concrete (different for lightweight concrete, as per Section 9.5.2.3 of the Code), M_a is the maximum service-load moment occurring for the condition under consideration and I_{cr} is the transformed moment of inertia of the cracked section.

With I_e and the appropriate deflection expressions, instantaneous or immediate deflections are obtained. Long-term or sustained loads, however, cause large increases in these deflections due to shrinkage and creep. The factors affecting deflection increases include humidity, temperature, curing conditions, compression steel content, ratio of stress to strength, and the age of the concrete at the time of loading.

If concrete is loaded at an early age, its long-term deflections will be greatly increased. Excessive deflections in reinforced concrete structures can very often be traced to the early application of loads. The creep strain after about five years (after which creep is negligible) may be as high as four or five times the initial strain caused by loads applied seven to ten days after the concrete was placed, while the ratio may only be two or three for loads applied three or four months after concrete placement.

Because of the several factors mentioned in the last two paragraphs, the magnitudes of long-term deflections can only be estimated. A paper by Yu and Winter[2] forms the basis of the method recommended by the ACI Code for calculating long-term deflections. The Code (9.5.2.5) states that to determine the increase in deflection due to these causes, the part of the instantaneous deflection that is due to sustained load should be multiplied by the expression at the end of this paragraph and the result added to the instantaneous deflection. In this expression, which is applicable to both normal and lightweight concrete, A_s is the area of the tensile steel in the member while A_s' is the area of any compression steel that might be used. A glance at this expression shows that the presence of compression steel in a beam will decidedly reduce its long-term deflections.

$$2 - 1.2\frac{A_s'}{A_s} \geq 0.6$$

The full dead load of a structure can be classified as a sustained load, but the type of occupancy will determine the percentage of live load that can be called sustained. For an apartment house or for an office building, perhaps only 20% to 25% of the service live load should be considered as being sustained, while perhaps 70% to 80% of the service live load of a warehouse might fall into this category.

A study by the ACI indicates that under controlled laboratory conditions 90% of the specimens had deflections between 20% below and 30% above the values calculated as described in this section.[3]

[2] Yu, W. W., and Winter, G., 1960, "Instantaneous and Long-Time Deflections of Reinforced Concrete Beams Under Working Loads," *Journal ACI*, 57, no. 1, pp. 29–50.

[3] ACI Committee 435, 1972, "Variability of Deflections of Simply Supported Reinforced Concrete Beams," *Journal ACI*, 69, no. 1, p. 29.

Example 6.1 presents the calculation of instantaneous and long-term deflections for a uniformly loaded simple beam.

EXAMPLE 6.1

The beam of Figure 6.2 has a simple span of 20 ft and supports a dead load including its own weight of 1 klf and a live load of 0.7 klf. $f_c' = 3000$ psi.

(a) Calculate the instantaneous deflection for $D + L$.

(b) Calculate the long-term deflection for the same loads assuming that 30% of the live load is sustained.

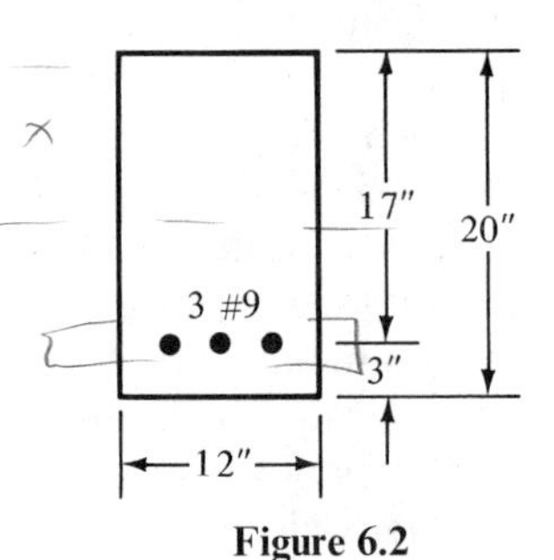

Figure 6.2

SOLUTION

(a) Instantaneous deflection:

$$I_g = (\tfrac{1}{12})(12)(20)^3 = 8{,}000 \text{ in.}^4$$

$$M_{cr} = \frac{f_r I_g}{y_t} = \frac{(7.5\sqrt{3{,}000})(8{,}000)}{10} = 328{,}633 \text{ in. lb} = 27.4 \text{ ft-k}$$

$$M_a = \frac{(1.7)(20)^2}{8} = 85 \text{ ft-k}$$

By transformed area calculations,

$$x = 6.78''$$

$$I_{cr} = 4067 \text{ in.}^4$$

$$I_e = \left(\frac{27.4}{85}\right)^3 (8{,}000) + \left[1 - \left(\frac{27.4}{85}\right)^3\right] 4067 = 4198 \text{ in.}^4$$

$$E_c = 57{,}000\sqrt{3{,}000} = 3.122 \times 10^6 \text{ psi}$$

$$\delta = \frac{5wl^4}{384E_cI_e} = \frac{(5)(\frac{1700}{12})(12 \times 20)^4}{(384)(3.12 \times 10^6)(4198)} = \underline{0.468''}$$

(b) Long-term deflection. The initial deflection due to the sustained loads is equal to $\{[1 + (0.3)(0.7)]/1.7\}(0.468) = 0.333$ in. This value is multiplied by the following value and added to the deflection determined in part (a):

$$\text{multiplier} = \left[2 - 1.2\left(\frac{0}{3.00}\right)\right] = 2.00$$

$$\text{total deflection} = 0.468 + (2)(0.333) = \underline{1.134''}$$

6.2 DEFLECTIONS FOR CONTINUOUS SPANS

For the following discussion a continuous T beam subject to both positive and negative moments is considered. As shown in Figure 6.3, the effective moment of inertia used for calculating deflections varies a great deal throughout the member. For instance, at the center of the

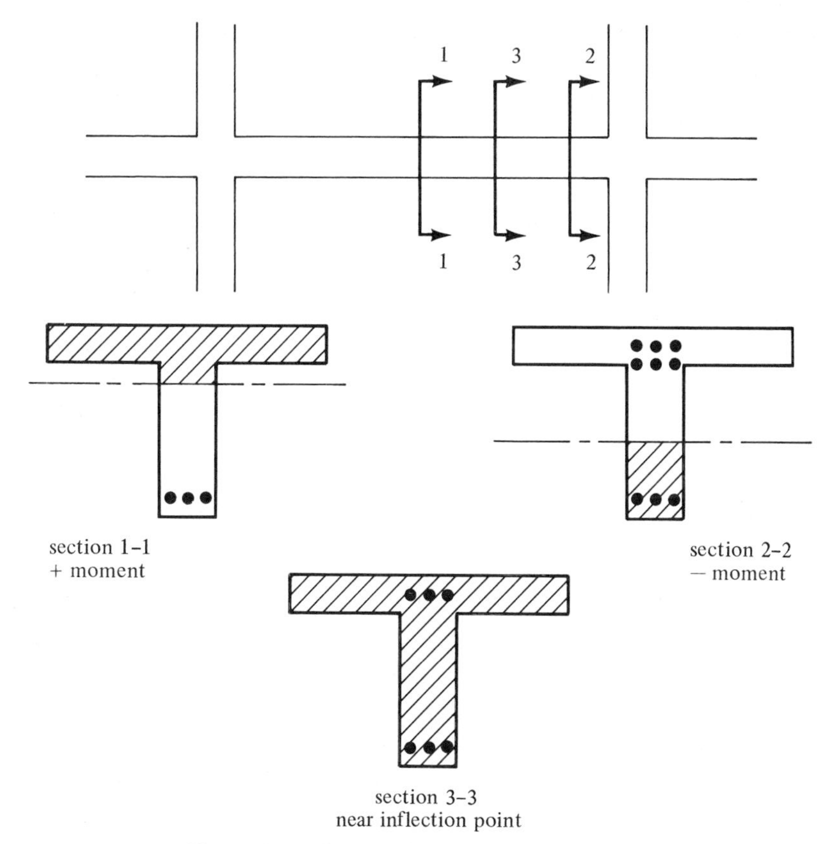

Figure 6.3 Deflections for a Continuous T Beam

span at Section 1.1 where the positive moment is largest, the web is cracked and the effective section consists of the crosshatched section plus the tensile reinforcing in the bottom of the web. At Section 2.2 in the figure, where the largest negative moment occurs, the flange is cracked and the effective section consists of the crosshatched part of the web (including any compression steel in the bottom of the web) plus the tensile bars in the top. Finally, near the points of inflection, the moments will be so low that the beam will probably be uncracked and thus the whole cross section is effective, as shown for Section 3.3 in the figure.

From the preceding discussion it is obvious that to theoretically calculate the deflection in a continuous beam, it is necessary to use a deflection procedure that takes into account the varying moment of inertia along the span. Such a procedure would be very lengthy, and it is doubtful whether the results so obtained would be within $\pm 20\%$ of the actual values. For this reason the ACI Code (9.5.2.4) permits the use of a constant moment of inertia throughout the member equal to the average of the I_e values computed at the critical positive and negative moment sections. It should also be noted that the moment multipliers at the two sections should be averaged for long-term deflections.

Example 6.2 illustrates the calculation of deflections for a continuous member. Although much of the repetitious math is omitted from the solution given herein, you can see that the calculations are still very lengthy and you will understand why approximate deflection calculations are commonly used for continuous spans.

EXAMPLE 6.2

Determine the instantaneous and long-term deflections at the midspan of the continuous T beam shown in Figure 6.4(a). The member supports a dead load including its own weight of 1.5 k/ft and a live load of 2.5 k/ft, of which 50% is assumed to be sustained. $f_c' = 3{,}000$ psi and $n = 9$. The

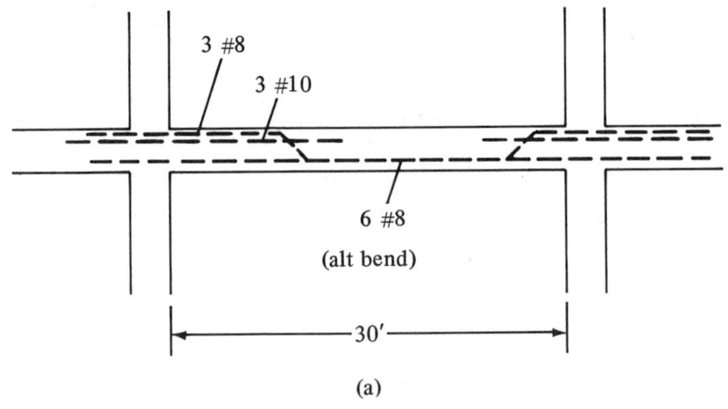

Figure 6.4(a)

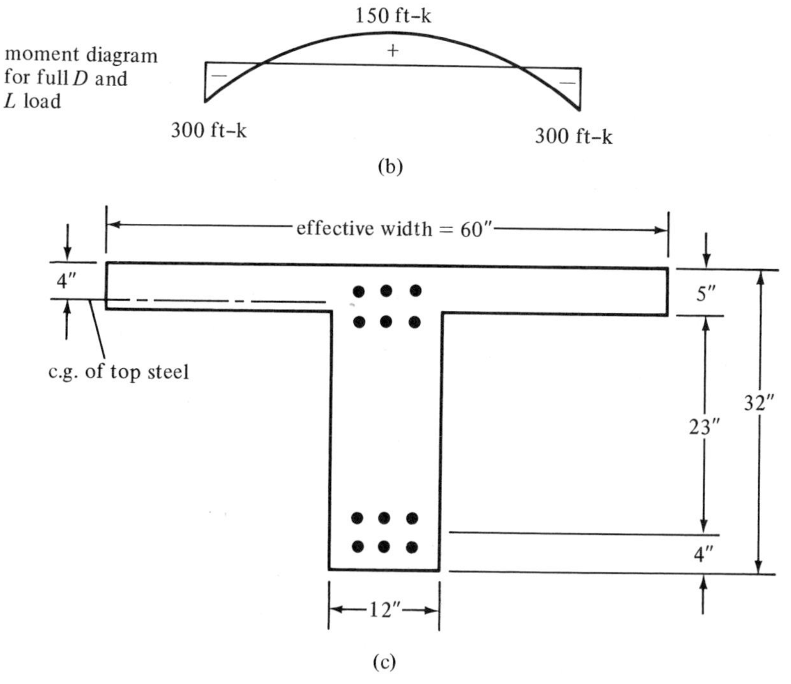

Figure 6.4(b) (c) *(cont.)*

moment diagram for full dead and live load is shown in Figure 6.4(b) while the beam cross section is shown in Figure 6.4(c).

SOLUTION

For positive moment region

1. Locating centroidal axis for uncracked section and calculating gross moment of inertia I_g and cracking moment M_{cr} for the positive moment region (Figure 6.5):

$$\bar{y} = 10.81''$$
$$I_g = 60{,}185 \text{ in.}^4$$

$$M_{cr} = \frac{(7.5)(\sqrt{3{,}000})(60{,}185)}{21.19} = 1{,}166{,}754'' \# = 97.2 \text{ ft-k}$$

2. Locating centroidal axis of cracked section and calculating transformed moment of inertia I_{cr} for the positive moment region (Figure 6.6):

$$x = 5.65''$$
$$I_{cr} = 24{,}778 \text{ in.}^4$$

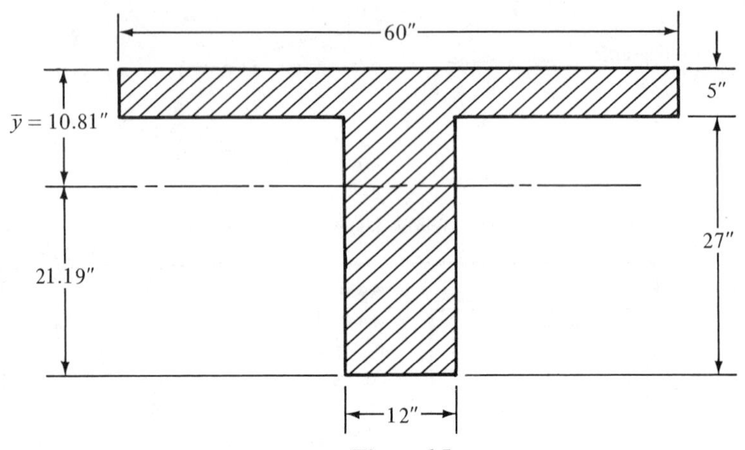

Figure 6.5

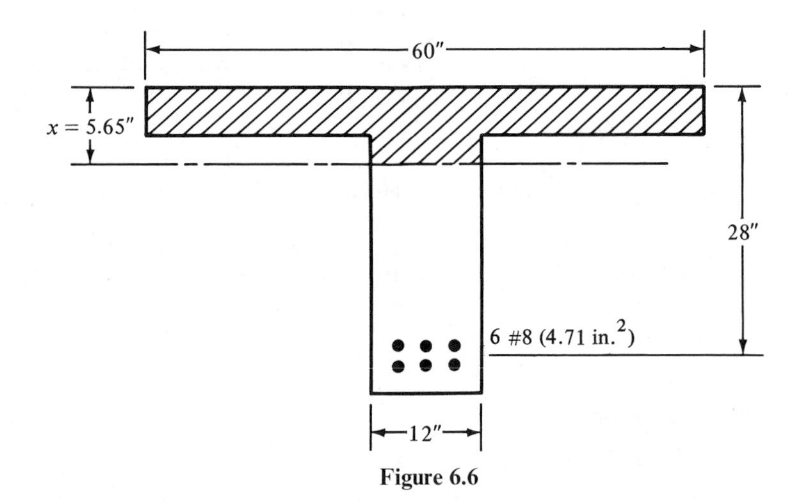

Figure 6.6

3. Calculating the effective moment of inertia in the positive moment region:

$$M_a = 150 \text{ ft-k}$$

$$I_e = \left(\frac{97.2}{150}\right)^3 (60{,}185) + \left[1 - \left(\frac{97.2}{150}\right)^3\right] 24{,}778 = 34{,}412 \text{ in.}^4$$

For negative moment region

1. Locating centroidal axis for uncracked section and calculating gross moment of inertia I_g and cracking moment M_{cr} for the negative moment region, considering only the crosshatched rectangle shown

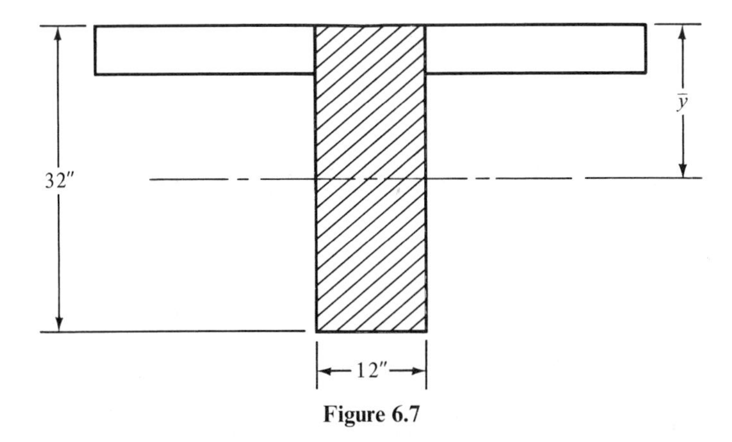

Figure 6.7

in Figure 6.7:

$$\bar{y} = \tfrac{32}{2} = 16''$$

$$I_g = (\tfrac{1}{12})(12)(32)^3 = 32{,}768 \text{ in.}^4$$

$$M_{cr} = \frac{(7.5)(\sqrt{3{,}000})(32{,}768)}{16} = 841{,}302 \text{ ft-lb} = 70.1 \text{ ft-k}$$

2. Locating centroidal axis of cracked section and calculating transformed moment of inertia I_{tr} for the negative moment region (Figure 6.8):

$$x = 10.43''$$

$$I_{cr} = 24{,}147 \text{ in.}^4$$

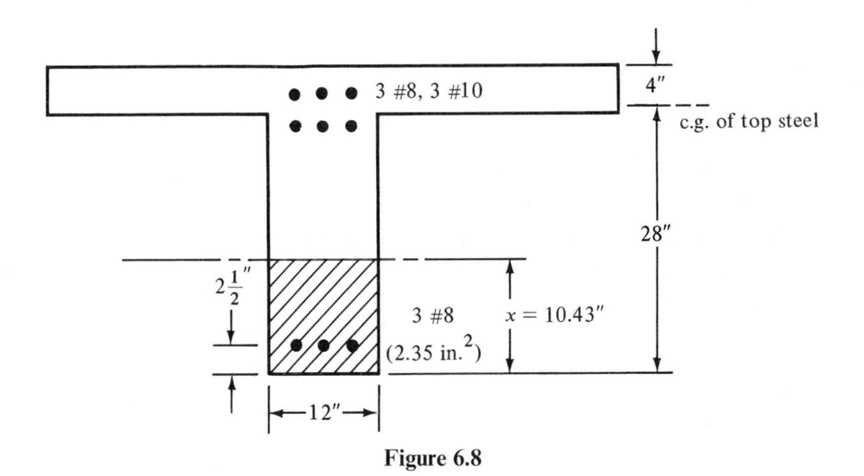

Figure 6.8

3. Calculating the effective moment of inertia in the negative moment region:

$$M_a = 300 \text{ ft-k}$$

$$I_e = \left(\frac{70.1}{300}\right)^3 (32{,}768) + \left[1 - \left(\frac{70.1}{300}\right)^3\right] 24{,}147 = 24{,}257 \text{ in.}^4$$

Instantaneous deflection

$$\text{Average } I_e = \frac{34{,}412 + 24{,}257}{2} = 29{,}334 \text{ in.}^4$$

$$E_c = 57{,}000\sqrt{3{,}000} = 3.122 \times 10^6 \text{ psi}$$

By the conjugate beam procedure, δ = moment at midspan of conjugate beam for the moment diagram of Figure 6.4(b) = (8437.5 ft^3-k)/EI:

$$\delta = \frac{(8437.5)(1{,}728)(1{,}000)}{(3.122 \times 10^6)(29{,}334)} = \underline{0.159''}$$

Long-term deflection

$$\text{initial deflection due to sustained loads} = \frac{1.5 + (0.5)(2.5)}{4.0}(0.159'') = 0.109''$$

The long-term multipliers are

$$\text{positive moment region} = 2 - 1.2\left(\frac{0}{4.71}\right) = 2.0$$

$$\text{negative moment region} = 2 - 1.2\left(\frac{2.35}{6.14}\right) = 1.54$$

$$\text{average multiplier} = \frac{2.0 + 1.54}{2} = 1.77$$

$$\text{long-term deflection} = 0.159'' + (1.77)(0.109'') = \underline{0.352''}$$

6.3 CONTROL OF CRACKING

Cracks are going to occur in reinforced concrete structures because of concrete's low tensile strength. For members with low steel stresses at service loads, the cracks may be very small and in fact may not be visible except upon careful examination. Such cracks are called microcracks and are generally initiated by bending stresses.

When steel stresses are high at service loads, particularly where high-strength steels are used, visible cracks will occur. These cracks should be limited to certain maximum sizes so that the appearance of the structure

Lake Point Tower, Chicago, Illinois (Courtesy Portland Cement Association).

is not spoiled and so that corrosion of the reinforcing does not result. The use of high-strength bars and the strength method of design have made crack control a very important item indeed. As the yield stresses of reinforcing bars in general use have increased from 40 ksi to 60 ksi and above, it has been rather natural for the designer to specify approximately the same size bars as he is accustomed to using but fewer of them. The result has been more severe cracking of members.

Although cracks cannot be eliminated, they can be limited to acceptable sizes by spreading out or distributing the reinforcement. In other words, smaller cracks will result if several small bars are used with moderate spacings rather than a few large ones with large spacings. Such a practice will usually result in satisfactory crack control even for grades 60 and 75 bars.

The maximum crack widths that are acceptable vary from approximately 0.010 in. to 0.015 in., depending on the location of the member in question, the type of structure, the surface texture of the concrete, illumination, and other factors. Somewhat smaller values may be required for members exposed to very aggressive environments such as deicing chemicals and salt water spray. Definite data are not available as to the sizes of cracks above which bar corrosion becomes particularly serious. As a matter of fact, tests seem to indicate that concrete quality, cover thickness, amount of concrete vibration, and other items may be more important than crack sizes in their effect on corrosion.

Results of laboratory tests of reinforced concrete beams to determine crack sizes vary. The sizes are greatly affected by shrinkage and other time-dependent factors. The purpose of crack control calculations is not really to limit cracks to certain rigid maximum values but rather to use reasonable bar details, as determined by field and laboratory experience, that will in effect keep cracks within a reasonable range.

In 1968 the following equation was developed for the purpose of estimating the maximum widths of cracks that will occur in the tension faces of flexural members:[4]

$$w = 0.076\beta_h f_s \sqrt[3]{d_c A}$$

where

w = the estimated cracking width in thousandth-inches

β_h = ratio of the distance to the neutral axis from the extreme tension concrete fiber to the distance from the neutral axis to the centroid of the tensile steel (values to be determined by the working stress method)

f_s = steel stress, in kips per square inch (designer is permitted to use $0.6f_y$)

d_c = the cover of the outermost bar measured from the center of the bar

A = the effective tension area of concrete around the main reinforcing (having the same centroid as the reinforcing) divided by the number of bars

For reinforced concrete beams with bars having yield stresses greater than 40,000 psi, the Code (10.6.4) in effect sets limiting values on crack sizes. This is done by requiring that member cross sections be so proportioned that the value of z computed by the expression at the end of this paragraph may not exceed 175 k/in. for members with interior exposure nor 145 k/in. for members with exterior exposure. The Code uses the Gergely-Lutz expression, with β_h equal to 1.2 to establish the limiting value of z. The limitations of z to 175 k/in. and 145 k/in. correspond,

[4] Gergely, P., and Lutz, L. A., 1968, "Maximum Crack Width in Reinforced Flexural Members," *Causes, Mechanisms and Control of Cracking in Concrete*, SP-20 (Detroit: American Concrete Institute), pp. 87–117.

respectively, to crack widths of 16 thousandth-inches and 13 thousandth-inches.

$$z = f_s \sqrt[3]{d_c A}$$

This equation does not apply to beams with extreme exposure nor to structures that are supposed to be watertight. Special consideration must be given to such situations. The value of z should be checked for both positive and negative moment bars. If the bars are not all the same size in a particular group, the number of bars should be considered to equal the total steel area divided by the area of the largest bar in the group, for purposes of determining the value of A to use in the equation.

It is felt that the Gergely-Lutz equation works reasonably well for slabs if $\beta_h = 1.35$, and the ACI Commentary (10.6.4) so recommends. In effect they are saying that to keep the same maximum crack sizes in slabs as in beams, the permissible value of z should not exceed 1.2/1.35 times the 175-k/in. and 145-k/in. values previously mentioned.

Example 6.3 illustrates the application of the crack control expression to a beam.

EXAMPLE 6.3

The beam cross section shown in Figure 6.9 was selected using $\rho = 0.18f_c'/f_y$, $f_y = 60{,}000$ psi, and $f_c' = 4{,}000$ psi. Will estimated crack sizes be within the requirements of the ACI Code if the member has exterior exposure? If not, revise the design.

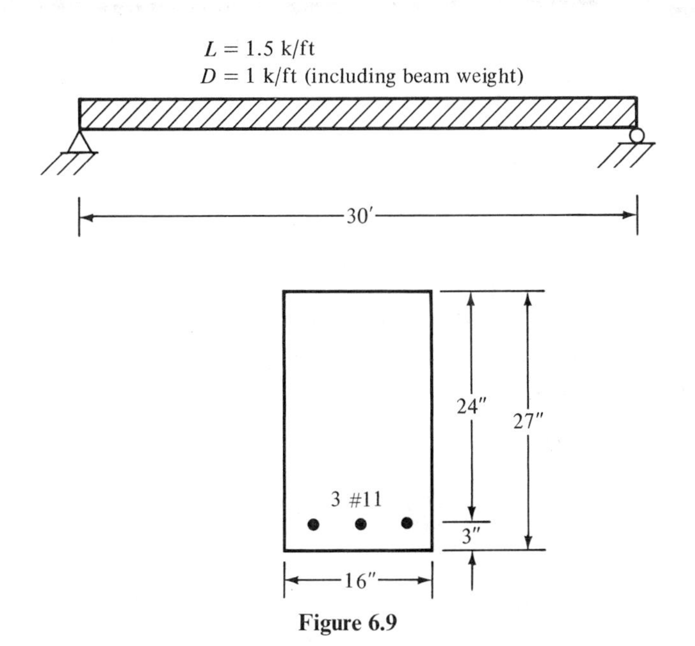

Figure 6.9

SOLUTION

$$z = (0.6 \times 60)\sqrt[3]{(3)\left(\frac{6 \times 16}{3}\right)}$$

$$= 164.8 > 145 \text{ k/in. permitted for exterior exposure} \quad \underline{\text{no good}}$$

Replace the 3 #11 bars with 5 #9 bars and try again.

$$z = (0.6 \times 60)\sqrt{(3)\left(\frac{6 \times 16}{5}\right)} = 139 \text{ ksi} < 145 \text{ ksi} \qquad \underline{\text{ok}}$$

<u>Use 5 #9</u>

6.4 CUTTING OFF OR BENDING BARS

The beams designed up to this point have been selected on the basis of maximum moments. These maximums have occurred at or near span center lines for positive moments and at the faces of supports for negative moments. At other points in the beams, the moments were less. Although it is possible to vary beam depths in some proportion to the bending moments, it is normally more economical to use prismatic sections and vary the area of the reinforcing bars in proportion to the moments. If the bending moment falls off 50% from its maximum value, 50% of the bars can theoretically be cut off or perhaps bent up or down to the other face of the beam and made continuous with the reinforcing in that other face.

As an illustration, the uniformly loaded simple beam of Figure 6.10 is considered. This beam has six bars and it is desired to cut off two bars when the moment falls off a third and two more bars when it falls off another third. For the purpose of this discussion, the maximum moment is divided into three equal parts by the horizontal lines shown. If the moment diagram is drawn to scale, a graphical method is perfectly satisfactory for finding the theoretical cutoff points.

For the parabolic moment diagram of Figure 6.10, the following expressions can be written and solved for the half-bar lengths x_1 and x_2 shown in the figure:

$$\frac{x_1^2}{(l/2)^2} = \frac{2}{6}$$

$$\frac{x_2^2}{(l/2)^2} = \frac{4}{6}$$

For different shaped moment diagrams, other mathematical expressions would have to be written or a graphical method used.

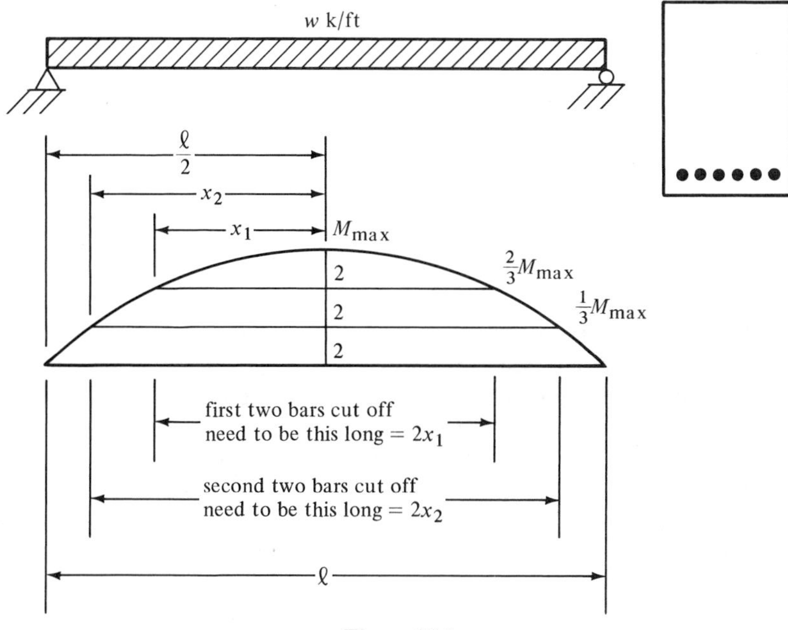

Figure 6.10

The author feels it is necessary to temporarily leave this subject and to spend some time discussing bond stresses and development lengths. In Section 6.11 some additional remarks are made about cutting off or bending bars.

6.5 BOND STRESSES AND DEVELOPMENT LENGTHS

A basic assumption made for reinforced concrete design is that there must be absolutely no slippage of the bar in relation to the surrounding concrete. In other words, the steel and the concrete should stick together or *bond* together so they will act as a unit. If there is a slipping of the steel with respect to the concrete, the result will be an immediate collapse of the beam.

In the past it was common to compute the maximum bond stresses at points in the members and to compare them with certain allowable values obtained by tests. It is the practice today, however, to look at the problem from an ultimate standpoint, where the situation is a little different. Even if the bars are completely separated from the concrete over considerable parts of their length, the ultimate strength of the beam will not be affected if the bars are so anchored at their ends that they cannot pull loose.

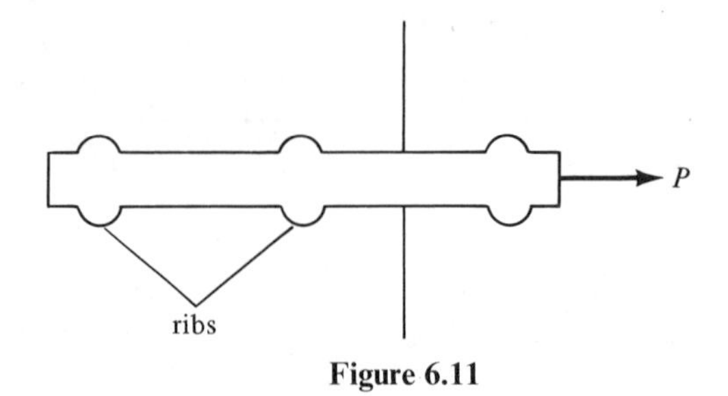

Figure 6.11

The bonding of the reinforcing bars to the concrete is due to three factors: the chemical adhesion between the two materials, the friction due to the natural roughness of the bars, and the mechanical anchorage of the closely spaced rib-shaped deformations made on the bar surfaces. The application of the force P to the bar shown in Figure 6.11 is considered in the discussion that follows.

When the force is first applied to the bar, the resistance to slipping is provided by the adhesion between the bar and the concrete. If plain bars were used, it would not take much tension in the bars to break this adhesion, particularly adjacent to a crack in the concrete. If this were to happen for a plain bar, only friction would remain to keep the bar from slipping. The introduction of deformed bars was made so that in addition to the adhesion and friction there would also be a resistance due to the bearing of the concrete on the lugs (or deformations) of the bars as well as the so-called shear-friction strength of the concrete between the lugs. Should the external shear be increased until failure occurs, it would generally occur due to a splitting of the concrete either vertically between the bar and the edge of the concrete or horizontally across the beam between the bars. When either of these types of splits run all the way to the end of a bar, the bar will slip and the beam will fail.

Splitting resistance along bars depends on quite a few factors, such as the thickness of the concrete cover, the number of bars in the vicinity, the transverse confining effect of stirrups, etc. As a result of these variables, it is impossible to make comprehensive bond tests which are good for a wide range of structures. Nevertheless, the ACI by its equations has attempted to do just this as will be described in the sections to follow.

6.6 DEVELOPMENT LENGTH REQUIRED FOR TENSION BARS

Based upon various tests with reinforcing bars, the ACI Code (12.2) specifies certain expressions for determining minimum development lengths for bars in tension. The capacity reduction factor ϕ is not included

Lab Building, Portland Cement Association, Skokie, Illinois (Courtesy Portland Cement Association).

in those expressions as allowances have already been made for understrength. It will also be noted that the required development lengths are the same for both the strength design method and the alternate or working stress design method since the equations are based on f_y in either case.

For #11 or smaller bars, the development length must not be less than the value obtained from the following expressions nor 12 in. In these expressions, l_d is the development length required, A_b is the cross sectional area of the bar in question, and d_b is the nominal diameter of the bar.

$$l_d = \frac{0.04 A_b f_y}{\sqrt{f_c'}}$$

$$l_d = 0.0004 d_b f_y$$

For #14 bars

$$l_d = \frac{0.085 f_y}{\sqrt{f_c'}}$$

For #18 bars

$$l_d = \frac{0.11 f_y}{\sqrt{f_c'}}$$

and for deformed wire

$$l_d = \frac{0.03 d_b f_y}{\sqrt{f'_c}}$$

Top bars are horizontal bars which have at least 12 in. of concrete placed beneath them. The reader has seen how air and water rise to the top of freshly placed concrete. Some of this air and water will be caught or trapped beneath top bars thus creating a zone of weakness underneath and as a result the bars are not held as tightly in place. Should top bars be used the value of l_d should be multiplied by 1.4.

Several other modifications of the calculated l_d values are to be made for various factors such as yield stresses greater than 60,000 psi, lightweight concrete, and other items. For bars with $f_y > 60{,}000$ psi l_d is to be multiplied by $2 - 60{,}000/f_y$.

When lightweight concrete is used, the bars on occasion may pull out without splitting the concrete and greater development lengths are considered necessary. If *all-lightweight concrete* is used in which both the fine and coarse aggregates are replaced with lightweight material the computed value of l_d is to be multiplied by 1.33. If *sand-lightweight concrete* is used where only the coarse aggregate is replaced with a lightweight material l_d is to be multiplied by 1.18. Linear interpolation between the two values is permissible when only partial sand replacement is used. Some relaxation of these values is permitted in the Code if the concrete mix is carefully controlled and if split cylinder tests have been made which give good results. Truthfully speaking, there is not a great deal of test data available for lightweight concrete.

If the reinforcing bars are spaced at least 6 in. on center and at least half that distance from concrete edges, the value of l_d can be multiplied by 0.8. If the bars are enclosed within a spiral which is not less than $\frac{1}{4}$ in. in diameter and which has a pitch of not more than 4 in., l_d may be multiplied by 0.75.

Finally, if more tensile steel is being used than is theoretically required, l_d may be multiplied by the ratio of the calculated steel area required divided by the actual steel area furnished.

The length of embedment required from the initial formulas given is to be multiplied by one or more of the applicable values specified for top bars, lightweight concrete, excess steel, and so on. Examples 6.4 and 6.5 show calculations for two simple cases. Table A.15 (see Appendix) can easily be used to expedite these calculations.

When bundled bars are used, the Code (12.4) states that the calculated development lengths are to be made for individual bars and are then to be increased by 20% for three-bar bundles and by 33% for four-bar bundles.

EXAMPLE 6.4

Determine the development length required for #9 bars in normal sand-gravel concrete. $f_y = 40{,}000$ psi and $f_c' = 4{,}000$ psi.

$$l_d = \frac{(0.04)(1.00)(40{,}000)}{\sqrt{4{,}000}} = \underline{25.3''}$$

$$l_d = (0.0004)(1.128)(40{,}000) = 18.0''$$

EXAMPLE 6.5

Determine the development length required for #8 top bars in sand lightweight concrete. $f_y = 50{,}000$ psi and $f_c' = 3{,}000$ psi.

$$l_d = \frac{(0.04)(0.79)(50{,}000)}{\sqrt{3{,}000}}(1.4)(1.18) = \underline{47.7''}$$

$$l_d = (0.0004)(1.00)(50{,}000)(1.4)(1.18) = 33.0''$$

6.7 CRITICAL SECTIONS FOR DEVELOPMENT LENGTH STUDY

Before the development length expressions can be applied, it is necessary to clearly understand the critical points for tensile and compressive stresses in the bars along the beam.

First of all, it is obvious that the bars will be stressed to their maximum values at those points where maximum moments occur. Thus those points must be no closer in either direction to the bar ends than the l_d values computed.

There are, however, other critical points for development lengths. As an illustration, a critical situation occurs whenever there is a tension bar whose neighboring bars have been cut off or bent over to the other face of the beam. Theoretically, if the moment is reduced by a third, one-third of the bars are cut off or bent and the remaining bars would be stressed to their yield points, and the full development lengths would be required for those bars.

This could bring up another matter in deciding the development length required for the remaining bars. The Code (12.11.3) requires that bars that are cut off or bent be extended a distance beyond their theoretical cutoff points by d or 12 bar diameters, whichever is greater. In addition, the point where the other bars are bent or cut off must also be at least a distance l_d from their points of maximum stress. Thus these two items might very well cause the remaining bars to have a stress less than f_y, thus permitting their development lengths to be reduced somewhat. A conservative approach is normally used, however, in which the remaining bars are assumed to be stressed to f_y.

6.8 DEVELOPMENT LENGTHS FOR COMPRESSION BARS

There is not a great deal of experimental information available about bond stresses and needed embedment lengths for compression steel. It is obvious, however, that embedment lengths will be smaller than those required for tension bars. For one reason, there are no tensile cracks present to encourage slipping, and, for another, there is some bearing of the end of the bars on concrete, which also helps develop the load.

The Code (12.3) states that the minimum development length provided for compression bars may not be less than the value computed from the following expression:

$$l_d = \frac{0.02 d_b f_y}{\sqrt{f_c'}} \geq 0.0003 d_b f_y$$

When bars are enclosed in spirals for any kind of concrete members, the members become decidedly stronger due to the confinement or lateral restraint of the concrete. The normal use of spirals is in spiral columns, which are discussed in Chapter 8. Should compression bars be enclosed by spirals of not less than $\frac{1}{4}$-in. diameter and with a pitch not greater than 4 in., the value of l_d may be reduced by a fourth. Also, if more compression steel is used than theoretically required, l_d may be multiplied by (A_s required)/(A_s furnished). In no case can the development length be less than 8 in.

6.9 HOOKS

Hooks may be used to anchor tension bars where sufficient space is not available to run them straight for full development. Actually, hooks are not quite as effective as equivalent lengths of straight bars; but they are very useful where space is limited. It should be noted that hooks on compression bars are considered to be useless for development length purposes (ACI Code 12.5.3). Figure 6.12 shows the standard 180° and 90° hooks specified by the Code (7.1). Either the 180° hook with an extension of 4 bar diameters at the free end or a 90° hook with an extension of 12 bar diameters at the free end may be used. The radii and diameters shown are measured on the insides of the bends.

The Code (12.5.1) states that standard hooks may be assumed to develop a tensile resistance equal to f_h, which is given by the following expression:

$$f_h = \xi \sqrt{f_c'}$$

For this expression ξ values are given in Table 6.1 (which is Table 12.5.1 of the ACI Code).

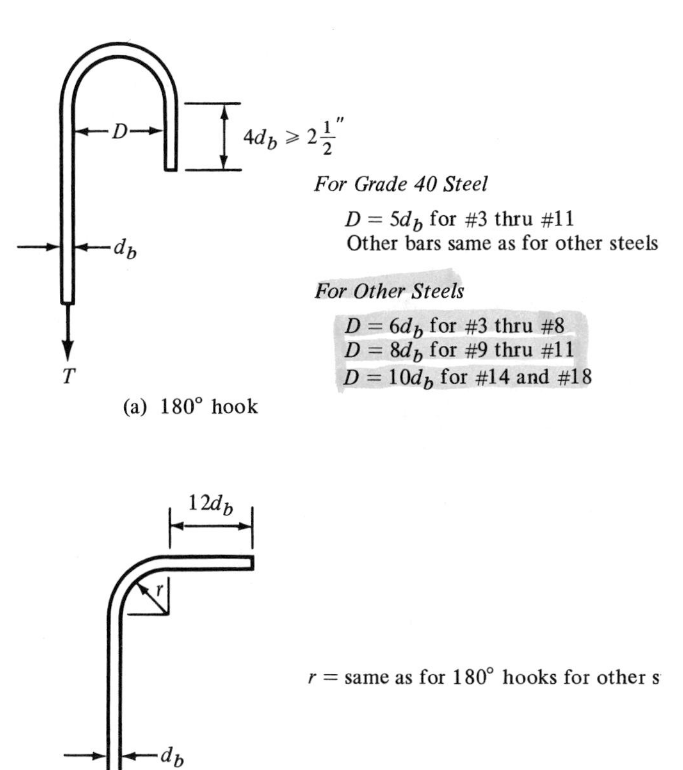

(a) 180° hook

(b) 90° hook

Figure 6.12

Table 6.1 ξ Values

Bar Size	$f_y = 60$ ksi Top Bars	$f_y = 60$ ksi Other Bars	$f_y = 40$ ksi Bars All
#3 to #5	540	540	360
#6	450	540	360
#7 to #9	360	540	360
#10	360	480	360
#11	360	420	360
#14	330	330	330
#18	220	220	220

Should the bars be enclosed by spirals or ties perpendicular to the plane of the hook, the values of ξ may be increased by 30%. Example 6.6 illustrates the calculations needed for the determination of the required embedment length with and without hooks for a cantilever beam.

EXAMPLE 6.6

Determine the embedment length required for the bars of the beam shown in Figure 6.13.

(a) If the bars are straight.
(b) If a 180° hook is used.
(c) If a 90° hook is used.

The 6 #9 bars shown are considered to be top bars. $f_c' = 4{,}000$ psi and $f_y = 40{,}000$ psi.

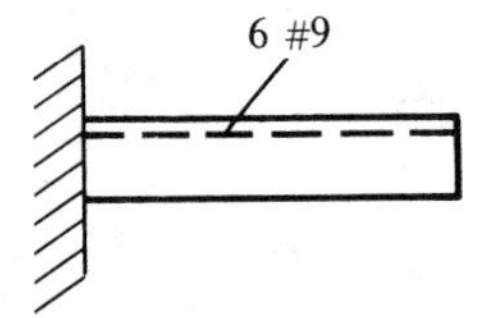

Figure 6.13

SOLUTION

(a) If bars are straight:

$$l_d = 35'' \text{ (from Table A.15)}$$

(b) If 180° hooks are used:

$$f_h = \xi\sqrt{f_c'} = 360\sqrt{4{,}000} = 22{,}768 \text{ psi}$$

$$\text{remaining stress to be developed} = 40{,}000 - 22{,}768 = 17{,}232 \text{ psi}$$

$$\text{extra embedment besides hook} = \left(\frac{17{,}232}{40{,}000}\right)(35) = 15''$$

See Figure 6.14.

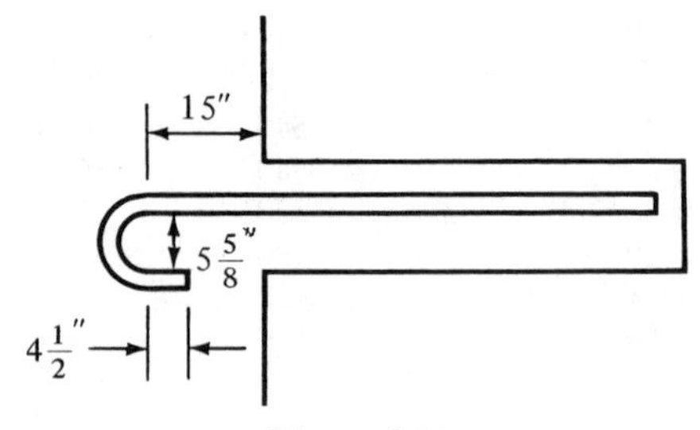

Figure 6.14

(c) If 90° hooks are used. (See Figure 6.15.)

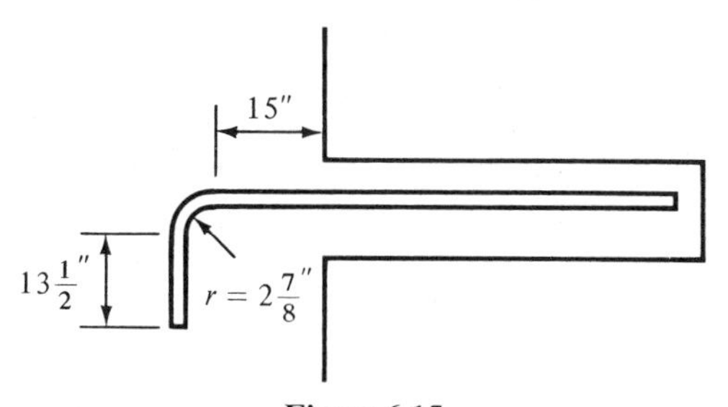

Figure 6.15

6.10 DETAILED DEVELOPMENT LENGTH REQUIREMENTS FOR POSITIVE AND NEGATIVE REINFORCEMENT

In the preceding several sections a general introduction to the subject of development length has been presented. In addition to this information it is necessary to understand the detailed requirements given specifically by the Code for positive and negative reinforcement. After studying this information the reader will be rather happy to read Section 6.11 of this chapter where simplified design office practices for determining bar lengths are discussed.

Positive Moment Reinforcement

Section 12.12 of the Code provides several detailed requirements for the lengths of positive moment reinforcement. These are briefly summarized in the paragraphs to follow:

1. At least one-third of the positive steel in simple beams and one-fourth of the positive steel in continuous members must extend uninterrupted along the same face of the member at least 6 inches into the support (12.12.1).

2. The positive reinforcement required in the preceding paragraph must, *if the member is part of a primary lateral load resisting system*, be extended into the support a sufficient distance to develop the yield stress of the bars at the face of the support. This requirement is included by the Code (12.12.2) to assure a ductile response to severe overstress as might occur with moment reversal during an earthquake or explosion. As a result of this requirement it is necessary to have bottom bars lapped at interior supports and to use additional embedment lengths and hooks at exterior supports.

3. Section 12.12.3 of the Code says that at simple supports and at points of inflection the positive moment tension bars must have their diameters

limited to certain maximum sizes. The purpose of the limitation is to keep bond stresses within reason at these points of low moments and large shears. It has not been shown that long anchorage lengths are fully effective in developing bars in a short distance between a P.I. and a point of maximum bar stress a condition which might occur in heavily loaded short beams with large bottom bars. It is specified that l_d as computed by the requirements mentioned in the preceding sections of this chapter may not exceed the following:

$$\frac{M_n}{V_u} + l_a$$

In this expression M_n is the computed theoretical flexural strength of the member if all reinforcing in that part of the beam is assumed stressed to f_y and V_u is the maximum applied shear at the section. At a support l_a is equal to the sum of the embedment length beyond the ℄ of the support and the equivalent embedment length of any furnished hooks or mechanical anchorage. At a point of inflection l_a is equal to the larger of the effective depth of the member or $12d_b$. When the ends of the reinforcement are confined by a compression reaction such as where there is a column below but not when a beam frames into a girder the value $1.3M_n/V_u$ is allowed. The values described here are summarized in Figure 6.16 and

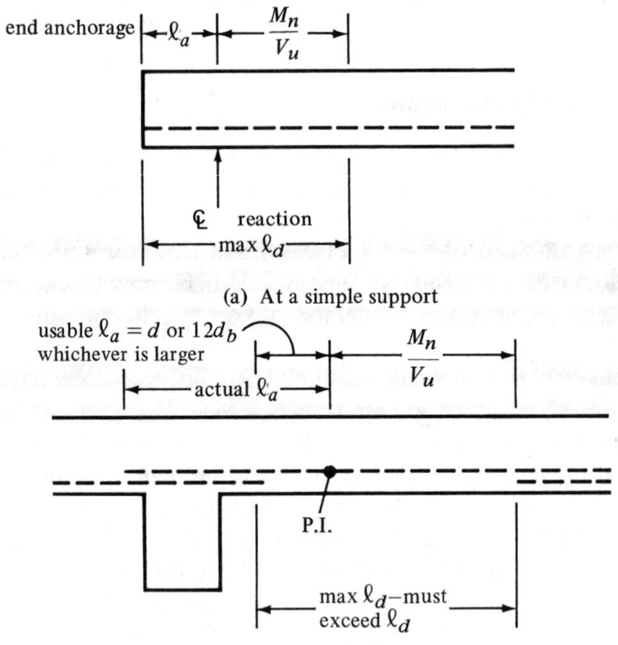

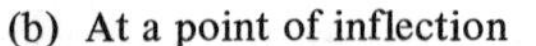

Figure 6.16 Development Length Requirements for Positive Moment Reinforcing

a brief numerical illustration for a simple end supported beam is provided in Example 6.7.

EXAMPLE 6.7

At the simple support shown in Figure 6.17 two #9 bars have been extended from the maximum moment area and into the support. Are the bar sizes satisfactory if $f_y = 60$ ksi, $f_c' = 3$ ksi, $b = 12$ in., $d = 24$ in., $V_u = 65^k$, if normal sand-gravel concrete is used and if the reaction is compressive?

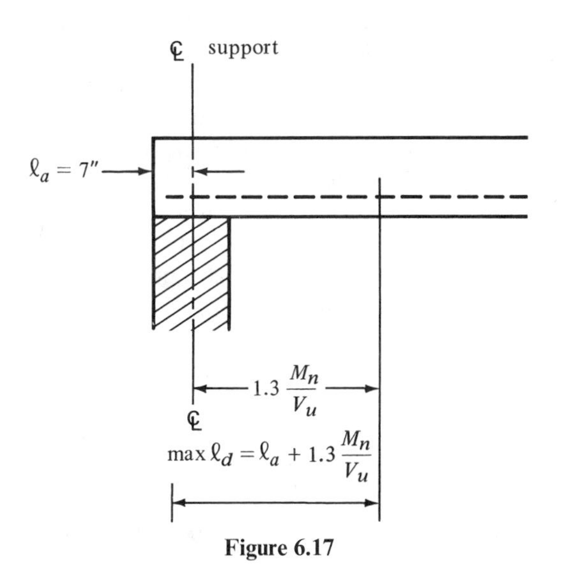

Figure 6.17

SOLUTION

$$a = \frac{A_s f_y}{0.85 f_c' b} = \frac{(2.00)(60)}{(0.85)(3)(12)} = 3.92''$$

$$M_n = A_s f_y\left(d - \frac{a}{2}\right) = (2.00)(60)\left(24 - \frac{3.92}{2}\right) = 2644.8''^{k}$$

l_d needed from Table 15 in Appendix = 44″

$$\text{maximum permissible } l_d = \frac{M_n}{V_u} + l_a$$

$$= (1.3)\left(\frac{2644.8}{65}\right) + 7 = 59.90'' > 44'' \qquad \text{ok}$$

Note: If this condition had not been satisfied the permissible value of l_d could have been increased by using smaller bars or by increasing the end anchorage l_a as by the use of hooks.

Negative Moment Reinforcement

Section 12.13 of the Code says that the tension reinforcement required by negative moments must be properly anchored into or through the supporting member. Section 12.13.1 of the Code says that the tension in these negative bars shall be developed on each side of the section in question by embedment length or end anchorage or a combination thereof. Hooks may be used as these are tension bars.

Section 12.11.3 of the Code says that the reinforcement must extend beyond the point where it is no longer required for moment by a distance equal to the effective depth of the member or 12 bar diameters whichever is larger. Finally Section 12.13.3 of the Code says that at least one-third of the total reinforcement provided for negative moment at the support must have an embedment length beyond the point of inflection no less than the effective depth of the member, 12 bar diameters or one-sixteenth of the clear span of the member whichever is greatest.

Example 6.8 which follows presents a detailed consideration of the lengths required by the Code for a set of straight negative moment bars at an interior support of a continuous beam.

EXAMPLE 6.8

At the first interior support of the continuous beam shown in Fig. 6.18 six #8 straight bars (4.71 in.^{2}) are used to resist a moment M_u of 390 ft-k for which the calculated A_s required is 4.24 in.^{2} If $f_y = 60$ ksi, $f_c' = 3$ ksi and normal sand gravel concrete is used determine the length of the bars as required by the ACI Code. The dimensions of the beam are $b_w = 14$ in. and $d = 24$ in.

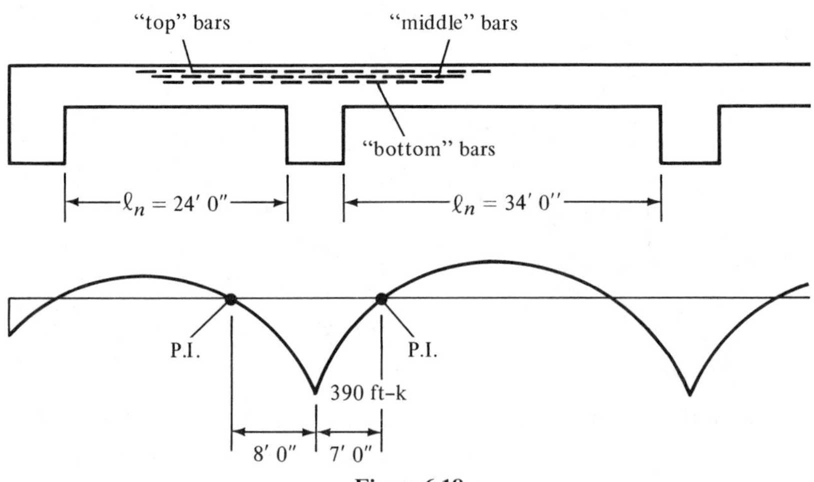

Figure 6.18

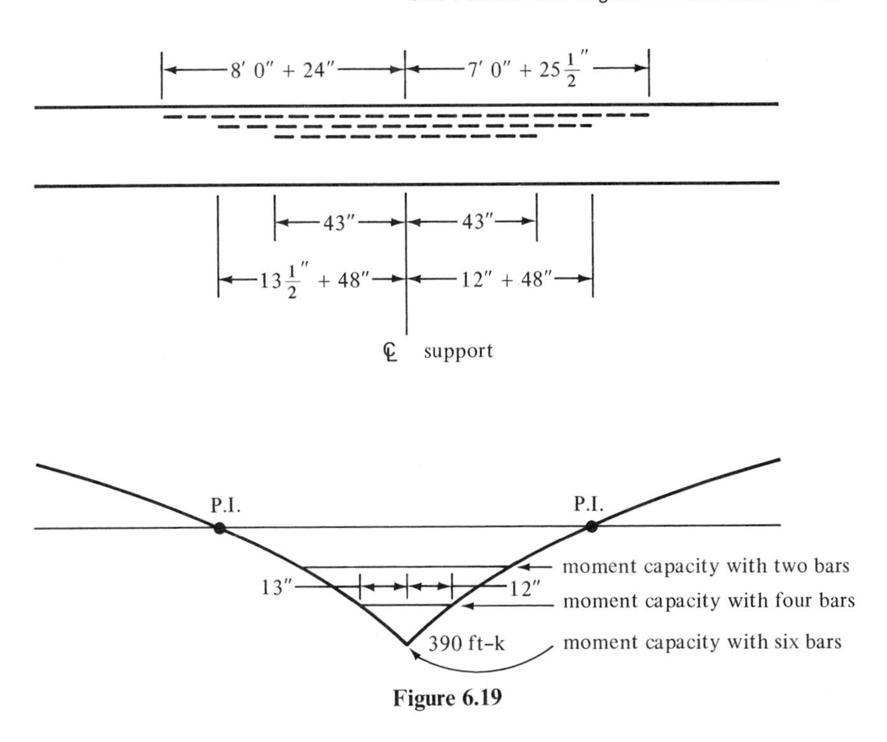

Figure 6.19

Note: The six negative tension bars are actually placed in one layer in the top of the beam but they are shown for illustration purposes in Figs. 6.18 and 6.19 as though they are arranged in 3 layers of two bars each. In the solution that follows the "bottom" two bars are cut off first, at the point where the moment plus the required development length is furnished; the "middle" two are cut when the moment plus the needed development length permits and; the "top" two are cut off at the required distance beyond the point of inflection (P.I.).

SOLUTION

1. Basic development lengths required (noting these are top bars)

$$l_d = 48 \text{ in. from Table A.15 in Appendix}$$

2. Section 12.13.3 of the Code requires one-third of the bars to extend beyond P.I. ($1/3 \times 6 = 2$ bars) for a distance equal to the largest of the following values (applies to "top" bars in figure).
   (a) d of web = 24″
   (b) $12d_b = (12)(1) = 12''$
   (c) $(\frac{1}{16})(24 \times 12)$ for 24 ft span = 18″
   $(\frac{1}{16})(34 \times 12)$ for 34 ft span = $25\frac{1}{2}''$
   Use 24″ for short span and $25\frac{1}{2}''$ for long span

3. Section 12.11.3 of the Code requires that the bars shall extend beyond the point where they are no longer required by moment for a distance equal to the greater of:
   (a) $d = 24''$
   (b) $12d_b = (12)(1) = 12''$
4. Section 12.2.4(a) of the Code says if bars are spaced at least 6 in. on center and at least 3 in. from the side face of the member l_d can be multiplied by 0.8. (it is assumed that such is not the case here).
5. Section 12.3.3 says required l_d can be multiplied by $A_{s\,\text{reqd}}/A_{s\,\text{furn}}$. This only applies to the first bars cut off ("bottom" bars here).

$$\text{Reduced } l_d = (48)\left(\frac{4.24}{4.71}\right) = 43''$$

6. In Fig. 6.19 these values are applied to the bars in question and the results given.

6.11 CUTTING OFF OR BENDING BARS CONTINUED

At this point the author would like to make a few concluding remarks about the material presented in Section 6.4. Considerable information has been presented in the last several sections which affect the points where reinforcing bars can be cut off or bent. A brief summary of these requirements is included in this section as they affect this discussion.

The structural designer is seldom involved with a fixed moment diagram—the loads move and the moment diagram changes. Therefore, the Code (12.11.3) says that reinforcing bars should be continued a distance of 12 bar diameters or the effective depth of the member, whichever is greater, (except at the supports of simple spans and the free ends of cantilevers) beyond their theoretical cutoff points.

As previously mentioned, the bars must be embedded a distance l_d from their point of maximum stress.

Next, the Code (12.12.1) says that at least one third of the positive steel in simple spans and one fourth of the positive steel in continuous spans must be continued along the same face of the beam at least 6 in. into the support.

Somewhat similar rules are provided by the Code (12.13.3) for negative steel. At least one third of the negative steel provided at a support must be extended beyond its point of inflection a distance equal to one sixteenth of the clear span, 12 bar diameters or the effective depth of the member, whichever is greater. Other negative bars must be extended beyond their theoretical point of cutoff by the effective depth, 12 bar diameters and at least l_d from the face of the support.

Trying to go through these various calculations for cutoff or bend points for all of the bars in even a modest sized structure can be a very

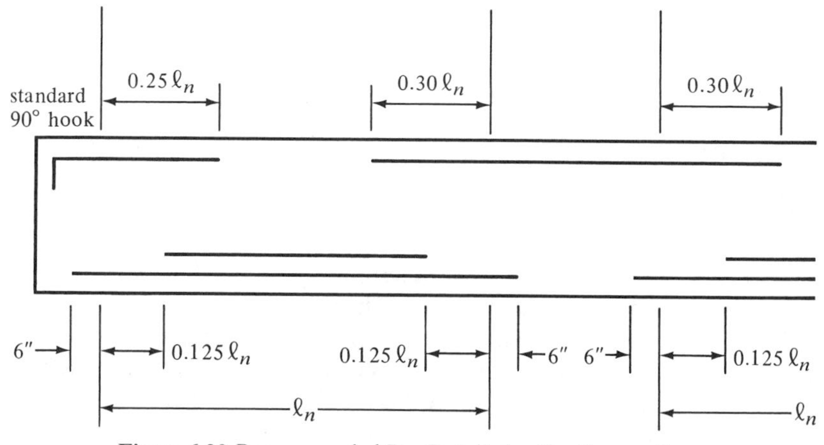

Figure 6.20 Recommended Bar Details for Continuous Beams

large job. Therefore, the average designer or perhaps the structural draftsman will cut off or bend bars by certain rules of thumb, which have been developed to meet the code rules described here. In Figure 6.20 a sample set of such rules are given. In the *CRSI Handbook*[5] such rules are provided for several different types of structural members, such as solid one-way slabs, one-way concrete joists, two-way slabs, and so forth.

6.12 BAR SPLICES IN FLEXURAL MEMBERS

Field splices of reinforcing bars are often necessary because of limitations of bar lengths available, requirements at construction joints, and changes from larger bars to smaller bars. Although steel fabricators normally stock reinforcing bars in 60-ft lengths, it is often convenient to work in the field with bars of shorter lengths, thus necessitating the use of rather frequent splices.

The most common method of splicing #11 or smaller bars is simply to lap the bars one over the other. Lapped bars may be either spaced from each other or placed in contact, with the contact splices being much preferred since the bars can be wired together. Such bars also hold their positions better during the placing of the concrete. Although lapped splices are easy to make, the complicated nature of the resulting stress transfer and the local cracks that frequently occur in the vicinity of the bar ends are disadvantageous. Obviously, bond stresses play an important part in transferring the forces from one bar to another. Thus the required splice lengths are closely related to development lengths.

[5] Concrete Reinforcing Steel Institute (Chicago, 1975).

Splicing of bars can also be accomplished by welding or by mechanical devices. Welded splices, from the view of stress transfer, are the best splices but they may be expensive and may cause metallurgical problems. The ACI Code (12.15.3.3) states that welded splices must be accomplished by butting the bars and welding them together so that the connection will be able to develop at least 125% of the specified yield strength of the bars. Splices not meeting this strength requirement can be used at points where the bars are not stressed to their maximum tensile stresses.

A special kind of welding called thermite welding, which is a kind of fusion welding, can be used for splicing. The bars to be welded are lined up with a gap between them, molds placed around the bars, and the cavity filled with an exothermic powder which is ignited and burns with sufficient heat to melt the steel. The molten steel flows into the gap between the bars and into a cavity in the mold around the bar ends. This type of welding has been successfully used for connecting #14 and #18 bars and also is not as sensitive to the chemical makeup of the bars.

Mechanical connectors usually consist of some type of sleeve splice which fits over the ends of the bars to be joined and into which a metallic grout filler is placed to interlock the grooves inside the sleeve with the bar deformations. From the standpoint of stress transfer good mechanical connectors are next best to welded splices. They do have the disadvantage that some slippage may occur in the connections and as a result there may be some concrete cracks in the area of the splices.

Before the specific provisions of the ACI Code are introduced a few general comments are presented which should explain the background for these provisions. These remarks are taken from a paper by George F. Leyh of the CRSI.[6]

1. Splicing of reinforcement can never reproduce exactly the same effect as continuous reinforcing.
2. The goal of the splice provisions is to require a ductile situation where the reinforcing will yield before the splices fail. Splice failures occur suddenly without warning and with dangerous results.
3. Lap splices fail by splitting of the concrete along the bars. If some type of closed reinforcing is wrapped around the main reinforcing (such as ties and spirals described for columns in Chapter 8) the chances of splitting are reduced and smaller splice lengths are needed.
4. When stresses in reinforcement are reduced at splice locations the chances of splice failure are correspondingly reduced. For this reason the Code requirements are less restrictive where stresses are low.

The minimum laps required for tension splices are specified in Section 12.16.1 of the ACI Code. In that section tension splices are classified as

[6] Proceedings of the PCA-ACI Teleconference on ACI 318-71 Building Code Requirements. 1972. Skokie, Illinois: Portland Cement Association (p 14-1).

being Class A, B or C (depending on percentage of bars spliced and maximum tensile stresses) and the required splice lengths (l_s) are given as some multiple of the development length l_d (determined for the full f_y of the bars). The values which may not be less than 12 in. are $1.0l_d$, $1.3l_d$ and $1.7l_d$, respectively.

Splices should be located away from points of maximum tensile stress. Furthermore, all of the bars should not be spliced at the same locations—that is the splices should be staggered. Should two bars of different diameters be lap spliced the lap length used should equal the value required for the smaller bar or the development length required for the larger bar whichever is greater.

The length of lap splices for bundled bars must be equal to the required lap lengths for individual bars of the same size, but increased by 20% for 3 bar bundles and 33% for 4 bar bundles (ACI Code 12.4). Furthermore, individual splices within the bundles are not permitted to overlap each other.

Compression bars may be spliced by lapping, by end bearing and by welding or mechanical devices. The Code (12.17.1) says that the minimum splice length of such bars should be the development length l_d but may not be less than $0.0005f_y d_b$ for bars with f_y of 60,000 psi or less than $(0.0009f_y - 24)d_b$ for steels with higher f_y values nor 12 in. Should the concrete strengths be less than 3,000 psi it is necessary to increase the computed laps by one-third. Reduced values are given in the Code for cases where the bars are enclosed by ties or spirals.

The Code (12.15.2) with one exception prohibits the use of lap splices for #14 or #18 bars. When column bars of those sizes are in compression it is permissible to connect them to footings by means of dowels of smaller sizes with lap splices as described in Section 15.8.6 of the Code.

PROBLEMS

6.1 to **6.7** Calculate the instantaneous and long-term deflections for the problems shown. Use $f_y = 60{,}000$ psi, $f_c' = 4{,}000$ psi, and assume the D values shown include beam weights and that 30% of the live loads are sustained.

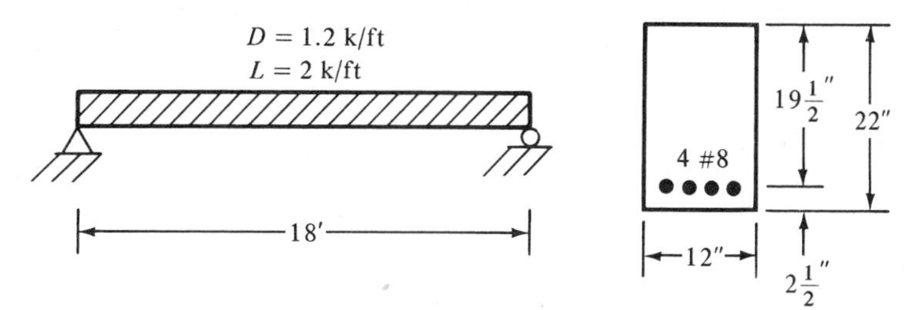

Problem 6.1

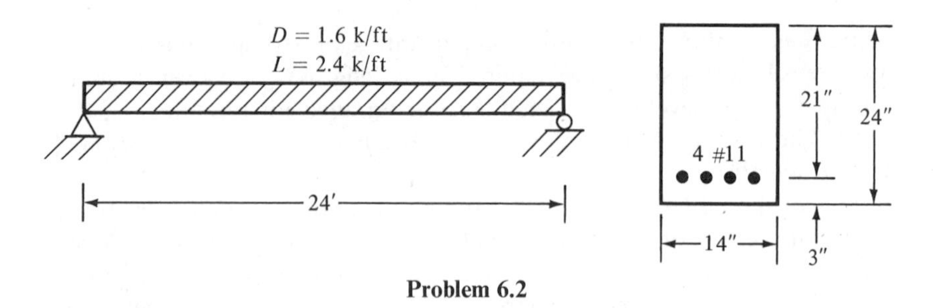

Problem 6.2

6.3 Repeat Problem 6.1 if 2 #9 bars are placed $2\frac{1}{2}$ in. from the top of the section.

6.4 Repeat Problem 6.2 if a 20^k concentrated live load is added at the ℄ of the span.

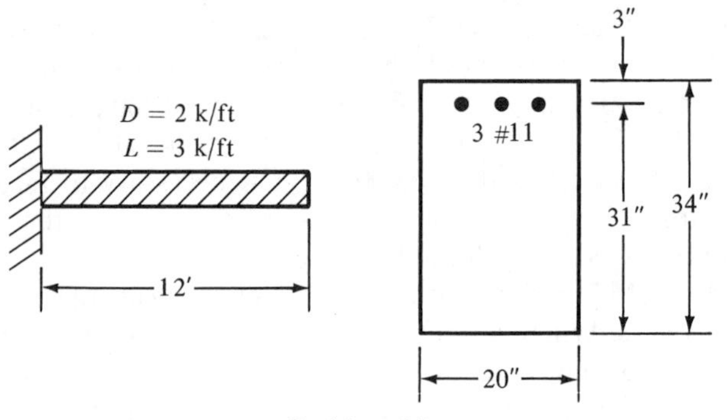

Problem 6.5

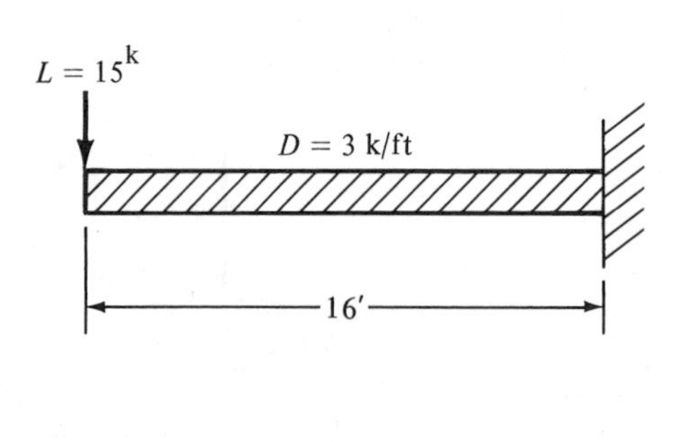

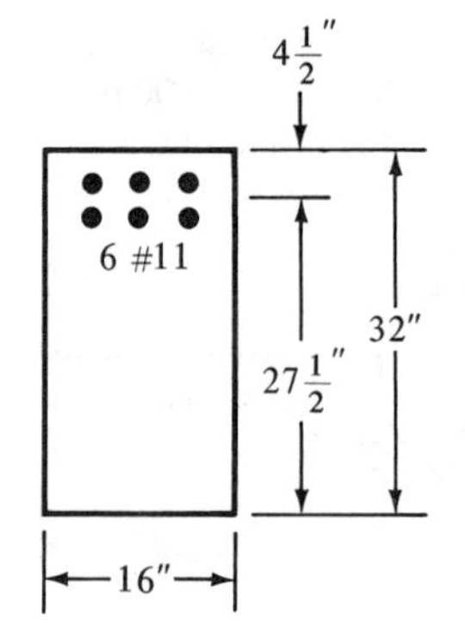

Problem 6.6

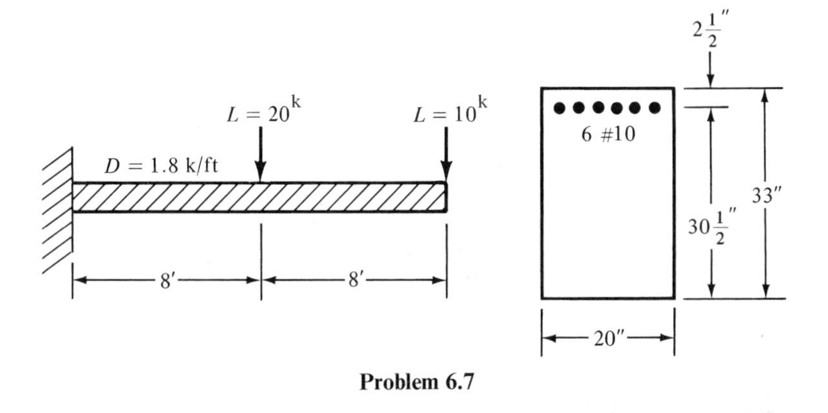

Problem 6.7

6.8 Select a rectangular beam section for the span and loads shown in the accompanying illustration. Use ρ_{max}, #10 bars, $f_c' = 4{,}000$ psi, and $f_y = 60{,}000$ psi. Will the estimated crack sizes be within the requirements of the ACI Code if the member has interior exposure? If not change the bar sizes.

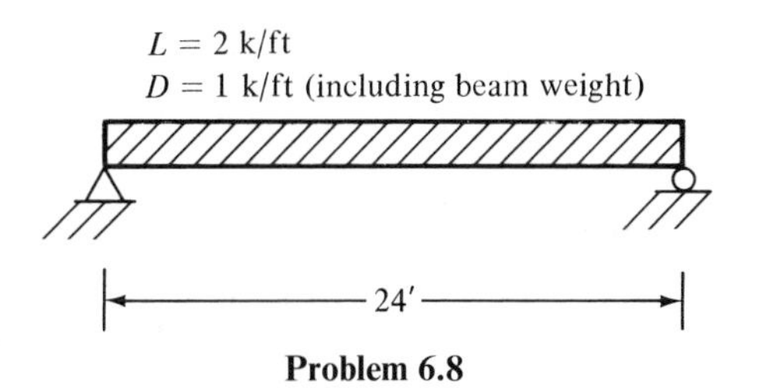

Problem 6.8

6.9 Check the ACI crack control provisions for the beam shown assuming exterior exposure and $f_y = 50{,}000$ psi.

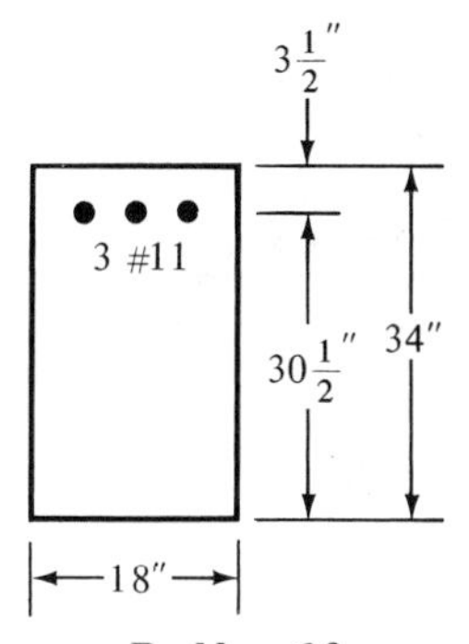

Problem 6.9

6.10 (a) Determine the development length required for the #8 bars of Problem 6.1, assuming normal sand-gravel concrete.

(b) Determine the development length required for the #11 bars of Problem 6.2 if sand lightweight concrete is used.

6.11 Determine the development length required for the #9 compression bars of Problem 6.3.

6.12 (a) Determine the embedment length required for the #10 bars of Problem 6.7 if normal sand-gravel concrete is used and the bars are straight.

(b) Repeat part (a) if 180° hooks are used.

(c) Repeat part (b) if 90° hooks are used.

PROBLEMS WITH SI UNITS

6.13 to **6.15** Calculate the instantaneous and long-term deflections for the problems shown. Use $f'_c = 20.7$ MPa, $f_y = 344.8$ MPa, $f_r = 2.832$ MPa and $n = 9$. Assume the D values shown include the beam weights and that 20% of the live loads are sustained. $E_s = 200\,000$ MPa. $E_c = 21\,650$ MPa.

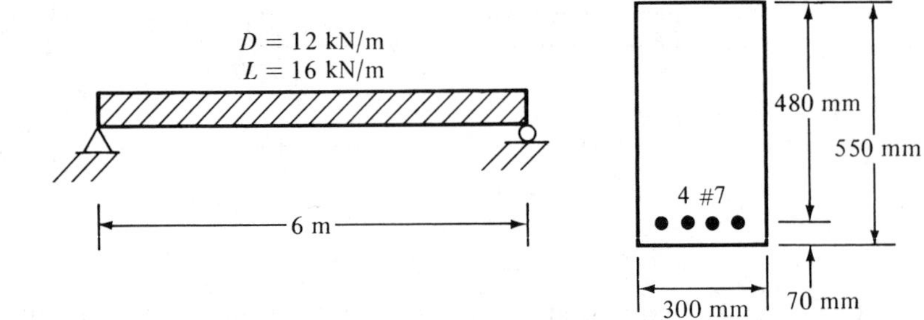

Problem 6.13

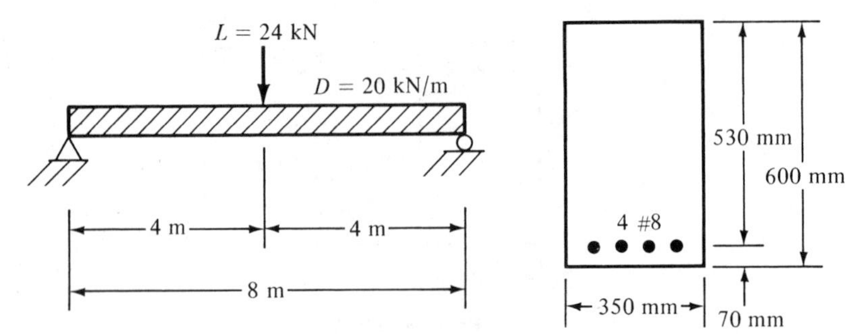

Problem 6.14

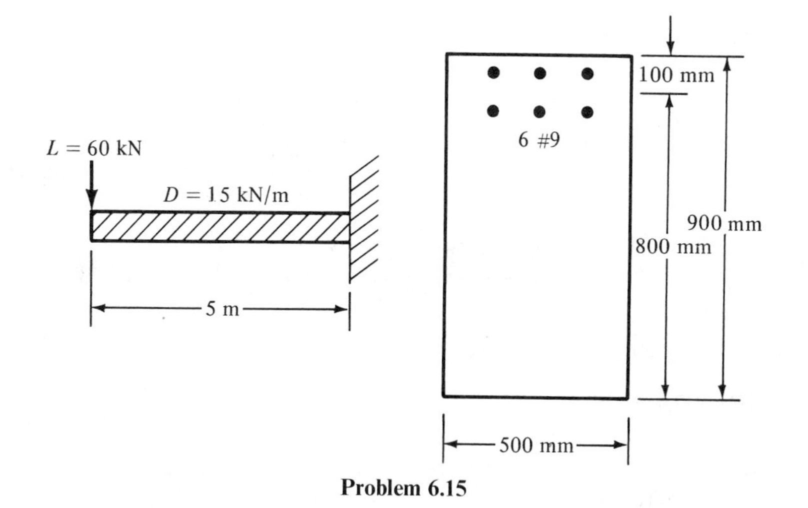

Problem 6.15

6.16 (a) Determine the embedment length required for the #9 bars of Problem 6–15 if normal sand-gravel concrete is used and the bars are straight.

(b) Repeat part (a) if 180° hooks are used.

CHAPTER 7

Shear and Diagonal Tension

7.1 INTRODUCTION

Although no one has ever been able to accurately determine the resistance of concrete to pure shearing stress, the matter is not very important because pure shearing stress is probably never encountered in concrete structures. Furthermore, according to engineering mechanics, if pure shear is produced in a member, a principal tensile stress of equal magnitude will be produced on another plane. As the tensile strength of concrete is less than its shearing strength, the concrete will fail in tension before its shearing strength is reached.

You have previously learned that in elastic homogeneous beams, where stresses are proportional to strains, two kinds of stresses occur (bending and shear) and they can be calculated with the following expressions:

$$f = \frac{Mc}{I}$$

$$v = \frac{VQ}{Ib}$$

An element of a beam not located at an extreme fiber or at the neutral axis is subject to both bending and shear stresses. These stresses combine into inclined compressive and tensile stresses, called principal stresses, which can be determined from the following expression:

$$f_p = \frac{f}{2} \pm \sqrt{\left(\frac{f}{2}\right)^2 + v^2}$$

The direction of the principal stresses can be determined with the formula to follow, in which α is the inclination of the stress to the beam's axis:

$$\tan 2\alpha = 2\frac{v}{f}$$

Obviously, at different positions along the beam, the relative magnitudes of v and f change and thus the directions of the principal stresses change. It can be seen from the preceding equation that at the neutral axis the principal stresses will be located at a 45° angle with the horizontal.

You understand by this time that tension stresses in concrete are a serious matter. Diagonal principal tensile stresses, called *diagonal tension*, occur at different places and angles in concrete beams and they must be carefully considered. If they reach certain values, additional reinforcing, called *web reinforcing*, must be supplied.

The discussion presented up to this point relating to diagonal tension applies rather well to plain concrete beams. If, however, reinforced concrete beams are being considered, the situation is quite different because the longitudinal bending tension stresses are resisted quite satisfactorily by the longitudinal reinforcing. These bars, however, do not provide resistance to the diagonal tension stresses.

A great deal of research has been done into the subject of shear and diagonal tension for nonhomogeneous reinforced concrete beams, and many theories have been developed. Despite all this work and all the resulting theories, no one has been able to clearly explain the failure mechanism involved. As a result, design procedures are based primarily on test data.

If V_u is divided by the effective beam area $b_w d$ the result is what is called an average shearing stress. This stress is not equal to the diagonal tension stress but merely serves as an indicator of its magnitude. Should this indicator exceed a certain value, shear or web reinforcing is considered necessary. In the ACI Code the basic shear equations are presented in terms of shear forces and not shear stresses. In other words the average shear stresses described in this paragraph are multiplied by the effective beam areas to obtain total shear forces.

For this discussion V_n is considered to be the nominal or theoretical shear strength of a member. This strength is provided by the concrete and by the shear reinforcement

$$V_n = V_c + V_s$$

The permissible shear strength of a member ϕV_n is equal to ϕV_c plus ϕV_s which must at least equal the factored shear force to be taken, V_u.

$$V_u = \phi V_c + \phi V_s$$

The shear strength provided by the concrete V_c is considered to equal an average shear stress (normally $2\sqrt{f_c'}$) times the effective cross sectional area of the member $b_w d$ where b_w is the width of a rectangular beam or of the web of a T beam or an I beam.

$$V_c = 2\sqrt{f_c'}\,b_w d$$

Beam tests have shown some interesting facts about the occurrence of cracks at different average shear stress values. For instance, where large moments occur for which appropriate longitudinal steel has been selected, diagonal cracks begin to occur at shear stresses as low as $1.9\sqrt{f_c'}$. On the other hand, in regions where large amounts of shear and little bending moment exists, cracks begin at average shear stress values of about $3.5\sqrt{f_c'}$.[1]

As a result of this information the Code (11.3.1.1) suggests that conservatively V_c (the shear force which the concrete can resist without web reinforcing) can go as high as $2\sqrt{f_c'}b_w d$. As an alternative the following shear force (from Section 11.3.2.1 of the Code) may be used which takes into account the effects of the longitudinal reinforcing and the moment and shear quantities. This value must be calculated separately for each point being considered in the beam.

$$V_c = \left(1.9\sqrt{f_c'} + 2500\rho_w \frac{V_u d}{M_u}\right) b_w d \leq 3.5\sqrt{f_c'} b_w d$$

In this expression $\rho_w = A_s/b_w d$ and M_u is the factored moment occurring simultaneously with V_u the factored shear at the section considered.

7.2 SHEAR CRACKING OF REINFORCED CONCRETE BEAMS

Inclined cracks can develop in the webs of reinforced concrete beams either as extensions of flexural cracks or occasionally as independent cracks. The first of these two types is the *flexure shear crack*, an example of which is shown in Figure 7.1. These are the ordinary types of shear cracks found in both prestressed and nonprestressed beams. They run at angles of about 45° with the beam axis and probably start at the top of a flexure crack. The approximately vertical flexure cracks shown are not dangerous unless there is a critical combination of shear stress and flexure stress occurring at the top of one of the flexure cracks.

Occasionally an inclined crack will develop independently in a beam, even though no flexure cracks are in that locality. Such cracks, which are called *web shear cracks*, will sometimes occur in the webs of prestressed sections, particularly those with large flanges and thin webs. They also sometimes occur near the points of inflection of continuous beams. Figure 7.2 illustrates these types of cracks.

[1] ACI-ASCE Committee 326, 1962, "Shear and Diagonal Tension," part 2, *Journal ACI*, 59, p. 277.

Massive Reinforced Concrete Members (Courtesy Bethlehem Steel Corporation).

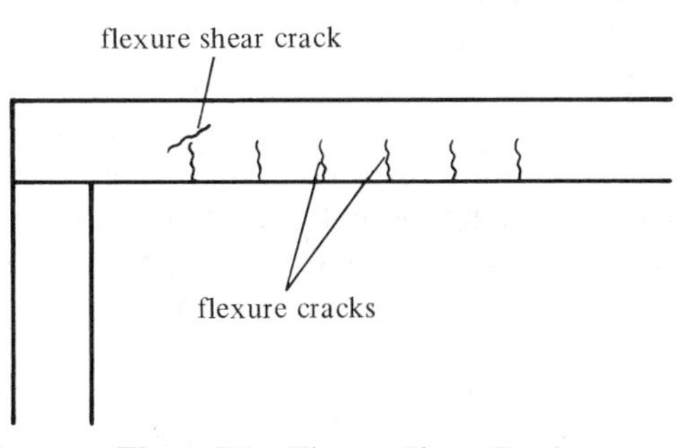

Figure 7.1 Flexure Shear Crack

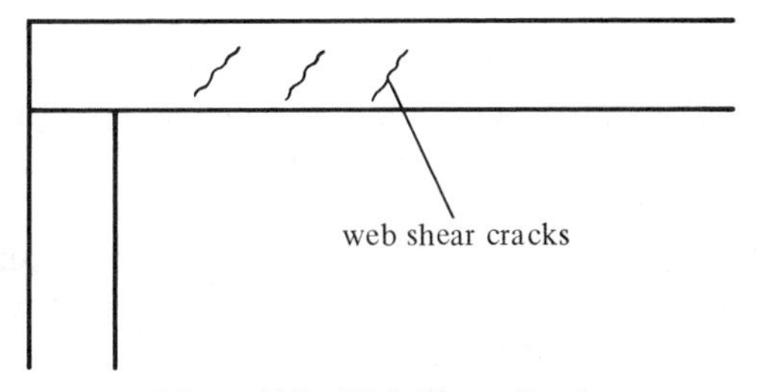

Figure 7.2 Web Shear Cracks

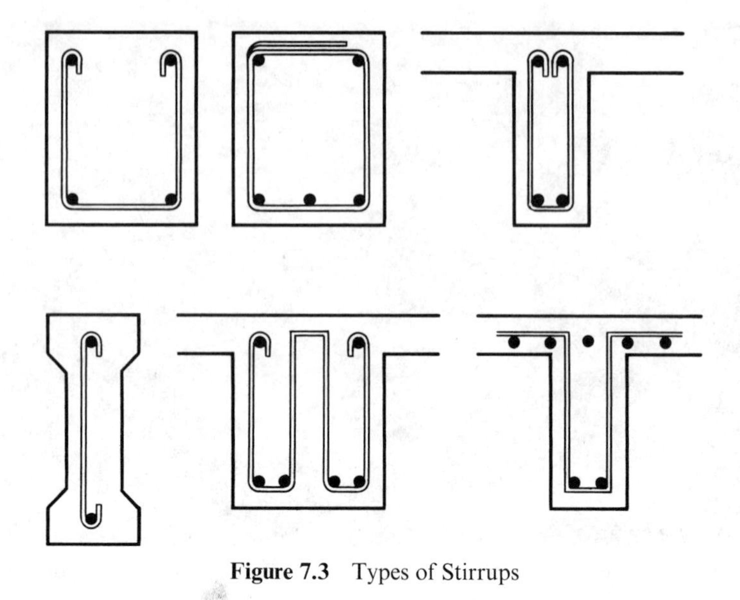

Figure 7.3 Types of Stirrups

7.3 WEB REINFORCEMENT

When the factored shear V_u is high, it shows that serious cracks are going to occur unless some type of additional reinforcing is provided. This reinforcing is called web reinforcing and usually takes the form of stirrups that enclose the longitudinal reinforcing along the faces of the beam. The most common stirrups are ⊔ shaped but they can be ⊔⊓⊔ shaped or perhaps have only a single vertical prong, as shown in Figure 7.3.

The actual behavior of web reinforcement is not really understood, although several theories have been presented through the years. One theory, which has been widely used almost since the turn of the century, is the so-called truss analogy, wherein a reinforced concrete beam with shear reinforcing is said to behave much as a statically determinate parallel chord truss with pinned joints. The flexural compression concrete is thought of as the top chord of the truss while the tensile reinforcing is said to be the bottom chord. The truss web is made up of stirrups acting as vertical tension members and pieces of concrete between the approximately 45° diagonal tension cracks acting as diagonal compression members.[2] Such a "truss" is shown in Figure 7.4.

The Code requires web reinforcement for all major beams. In Section 11.5.5.3, a minimum area of web reinforcing is required for all concrete flexure members except slabs and footings, certain concrete floor joists,

[2] Mörsch, E., 1908, "Der Eisenbeton-Seine Theorie Und Anwendung," (Stuttgart: Wittwer).

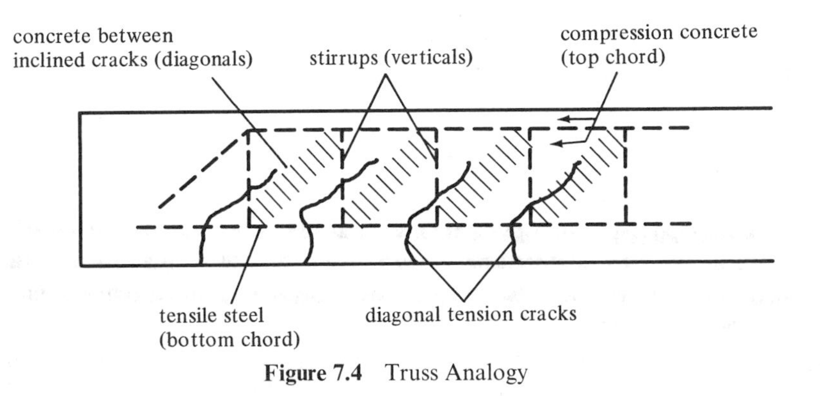

Figure 7.4 Truss Analogy

and in beams whose total depths are not more than 10 in. nor more than $2\frac{1}{2}$ times their flange thicknesses or one-half their web widths, whichever is greater.

Actually, inclined or diagonal stirrups lined up approximately with the principal stress directions are more efficient in carrying the shears and preventing or delaying the formation of diagonal cracks. Such stirrups, however, are not usually considered to be very practical in the United States because of the great amount of labor required for positioning them. Actually, they can be rather practical for precast concrete beams where the bars and stirrups are preassembled into cages before being used and where the same beams are duplicated many times.

Bent-up bars (usually at 45° angles) are another satisfactory type of web reinforcing (see Figure 7.5). Although bent-up bars are very commonly used in flexural members in the United States, the average designer seldom considers the fact that they can resist some of the diagonal tension. Two reasons for not counting their contribution to diagonal tension resistance are that there are only a few bent-up bars in a beam and they may not be conveniently located for use as web reinforcement.

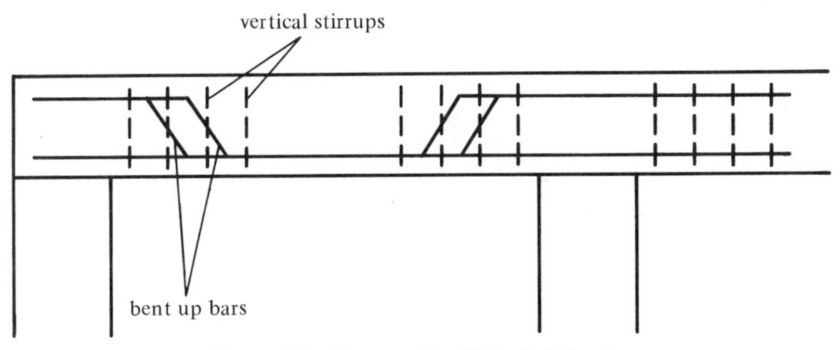

Figure 7.5 Bent-up Bar Web Reinforcing

Welded wire fabric with wires perpendicular to the axis of a beam is another satisfactory type of web reinforcing.

7.4 SPACING OF STIRRUPS

The purpose of stirrups is to minimize the size of diagonal tension cracks or to carry diagonal tension from one side of such a crack to the other. Truthfully, these stirrups do not really resist much tension until after the crack begins to form.

Tests made on reinforced concrete beams show that a beam cannot fail by widening of these diagonal cracks until the stirrups going across the cracks have been stressed to their yield stresses. If a crack is assumed to occur at 45°, the horizontal projection of the crack running through the beam will be d and the number of stirrups used in that distance will be d/s, where s is the center-to-center spacing of the stirrups (see Figure 7.6).

In Figure 7.7 the section at the crack is shown. Each stirrup is assumed to carry the tensile force halfway on each side to the next stirrup. The tensile stress on the crack equals the average shearing stress $V_u/b_w d$ minus the permissible shearing stress in the concrete $\phi V_c/b_w d$. The vertical com-

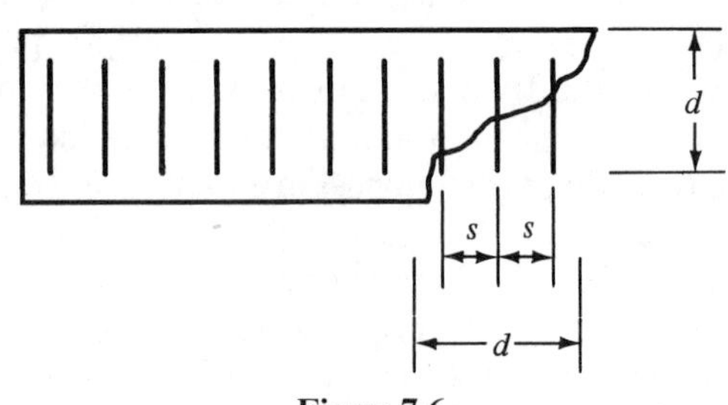

Figure 7.6

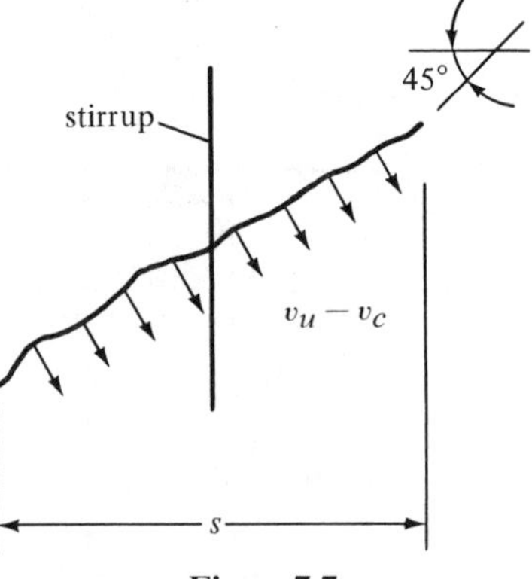

Figure 7.7

ponent of this tensile stress is assumed to be carried by the stirrups. The horizontal component has already been covered in the design of the longitudinal reinforcement.

$$\text{Total diagonal force on section} = \left(\frac{V_u}{b_w d} - \frac{\phi V_c}{b_w d}\right)(1.414 s b_w)$$

$$\text{Vertical component of force} = (0.707)\left(\frac{V_u}{b_w d} - \frac{\phi V_c}{b_w d}\right)(1.414 s b_w)$$

$$= \left(\frac{V_u}{d} - \frac{\phi V_c}{d}\right)s = \left(\frac{V_u - \phi V_c}{d}\right)s$$

Letting $V_s = \dfrac{V_u - \phi V_c}{\phi}$ the previous expression becomes

$$= \frac{\phi V_s}{d} s$$

This vertical component must be carried by the stirrup, whose strength equals its cross sectional area (A_v) times its yield stress times ϕ.

$$\phi A_v f_y = \frac{\phi V_s}{d} s$$

$$V_s = \frac{A_v f_y d}{s}$$

Or, as frequently used when the stirrup size is assumed and the theoretical spacing is required,

$$s = \frac{A_v f_y d}{V_s}$$

Going through a similar derivation, the following expression can be determined for the required area for inclined stirrups, in which α is the angle between the stirrups and the longitudinal axis of the member

$$V_s = \frac{A_v f_y(\sin\alpha + \cos\alpha)}{s}$$

7.5 ACI CODE REQUIREMENTS

In this section a detailed list of the Code requirements controlling the design of web reinforcing is presented, even though some of these items have been previously mentioned in this chapter.

Placement of Reinforcing Bars for Hemispherical Dome of Nuclear Power Plant Reinforced Concrete Containment Structure (Courtesy Bethlehem Steel Corporation).

1. When the factored shear V_u exceeds the permissible shear ϕV_c the Code (11.3.1.1) requires the use of web reinforcing. The value of V_c is normally taken as $2\sqrt{f_c'}b_w d$ but the Code (11.3.2.1) permits the use of the following less conservative value

$$V_c = \left(1.9\sqrt{f_c'} + 2500\rho_w \frac{V_u d}{M_u}\right) b_w d \leq 3.5\sqrt{f_c'}b_w d$$

In this expression M_u is the moment occurring simultaneously with V_u at the section in question. The value of $V_u d/M_u$ must not be taken greater than 1.0 in calculating V_c says the Code.

2. In Section 11.5.5 the Code provides a minimum amount of web reinforcing for all prestressed and nonprestressed concrete flexural members (other than the exceptions previously mentioned) for any sections of the beam where V_u is greater than $\phi V_c/2$. The expression to follow provides the minimum area (which is to be used as long as the factored torsional moment T_u does not exceed $\phi(0.5\sqrt{f_c'}\sum x^2 y)$, the computation of which is described in Chapter 13).

$$A_v = 50 \frac{b_w s}{f_y}$$

Though you may not feel the use of such minimum shear reinforcing is necessary, studies of earthquake damage in recent years have shown very large amounts of shear damage occurring in reinforced concrete structures, and it is felt that the use of this minimum value will greatly improve the resistance of such structures to seismic forces. Actually, a good many designers feel that the minimum area of web reinforcing should be used throughout beams and not just where V_u is greater than $\phi V_c/2$.

3. The Code does not permit a diagonal 45° crack to run a vertical distance of more than $d/2$ without being intercepted by a stirrup. For this reason the maximum spacing of vertical stirrups (11.5.4.1) is limited to the lesser of $d/2$ or 24 in. as long as V_s is not greater than $4\sqrt{f_c'}b_w d$. Should V_s exceed $4\sqrt{f_c'}b_w d$, the Code states that every potential crack must intercept two stirrups while it runs a vertical distance equal to $d/2$. In other words, the maximum center-to-center spacing of stirrups is reduced to $d/4$ (11.5.4.3). Under no circumstances may V_s be allowed to exceed $8\sqrt{f_c'}b_w d$ (Code 11.5.6.8).

4. Section 12.14 of the Code provides requirements about dimensions, embedment lengths, and so forth. For stirrups to develop their design strengths, they must be adequately anchored. Stirrups may be crossed by diagonal tension cracks at various points along their depths. As these cracks may cross very close to the tension or compression edges of the members, the stirrups must be able to develop their yield strengths along the full extent of their lengths. It can then be seen why they should be bent around longitudinal bars of greater diameters than their own and extended beyond by adequate development lengths.

Web reinforcing is required to be carried as close to the tension and compression faces of the beam as possible. The ends of stirrups must be anchored by one of the following means: (a) by standard hooks with an embedment of $0.5l_d$, the embedment distance being measured from the middepth of the member ($d/2$) and the beginning of the hook, (b) by embedment above or below middepth on the compression side by l_d but not less than 24 bar diameters or (c) by bending around the longitudinal reinforcing through at least 180°. Such bends can only be considered effective where the stirrup makes an angle of at least 45° with the longitudinal bars.

5. When a beam reaction causes compression in the end of a member in the same direction as the external shear, the shearing strength of that part of the member is increased. Tests of such reinforced concrete members have found that in general as long as a gradually varying shear is present (as with a uniformly loaded member), the first crack occurs a distance d from the face of the support. It is therefore permissible, according to the Code (11.1.3), to decrease somewhat the calculated shearing stress for a distance d from the face of the support. This is done by using a V_u in that

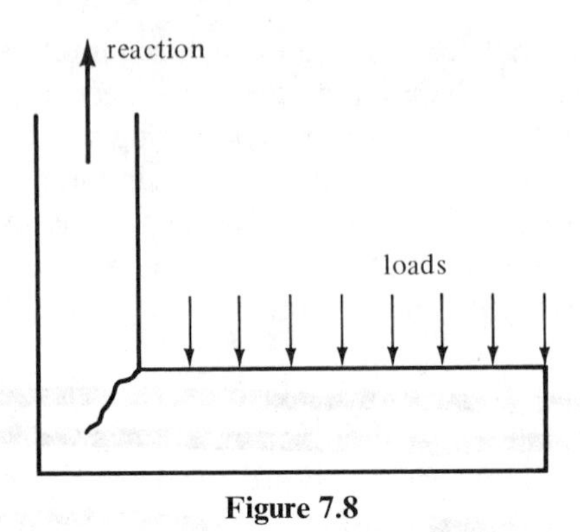

Figure 7.8

range equal to the calculated V_u at a distance d from the face of the support. Should a concentrated load be applied in this region, no such shear reduction is permissible.

Should the reaction tend to produce tension in this zone, no shear stress reduction is permitted because tests have shown that cracking may occur at the face of the support or even inside it. One such example is shown in Figure 7.8, while another case occurs in the design of a retaining wall footing in Section 11.10 of this text.

6. Various tests of reinforced concrete beams of normal proportions with sufficient web reinforcing have shown that shearing forces have no significant effect on the flexural capacities of the beams. Experiments with deep beams, however, show that large shears will often keep those members from developing their full flexural capacities. As a result, the Code requirements given in the preceding paragraphs are not applicable to beams whose clear spans divided by their effective depths are less than five and where the loads are applied at the tops or compression faces.

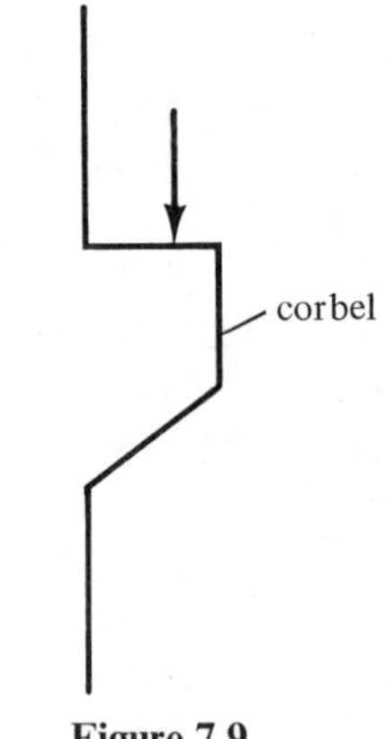

Figure 7.9

Members falling into this class include deep beams, short cantilevers, and corbels. Corbels are brackets that project from the sides of columns and are used to support beams and girders, as shown in Figure 7.9. They are quite commonly used in precast construction. Special web reinforcing provisions are made for such members in Section 11.8 of the Code.

7.6 WEB REINFORCING DESIGN EXAMPLES

Shear and diagonal tension calculations are illustrated by Examples 7.1 through 7.5. Maximum vertical stirrup spacings have been previously given, whereas no comment has been made about minimum spacings. Stirrups must be spaced far enough apart to permit the aggregate to pass through, and, in addition, they must be reasonably few in number so as to keep within reason the amount of labor involved in placing them. Accordingly, minimum spacings of 3 or 4 in. are normally used. Usually #3 stirrups are assumed, and if the calculated theoretical spacings are less than 3 or 4 in., larger-diameter stirrups can be used. Another alternative is to use ⊔⊓⊔ stirrups instead of ⊔ stirrups. Different diameter stirrups should not be used in the same beam, as confusion will result.

As is illustrated in Examples 7.2, 7.4 and 7.5, it is quite convenient to draw the V_u diagram and carefully label it with values of such items as ϕV_c, $\phi V_c/2$, V_u at a distance d from the face of the support and to show the dimensions involved.

Some designers place their first stirrup a distance d from the face of the support, although the more common procedure is to place it one-half of the end-calculated spacing requirement from the face. This latter procedure is followed for the example problems here.

In these example problems the author has attempted to select stirrup spacings which correspond closely to the theoretical spacings required with the result that some fractional dimensions are used. From a practical view, however, stirrups are often spaced with center to center dimensions which are multiples of 3 or 4 in., to simplify the field work. Though this procedure may require an additional stirrup or two total costs should be less because of reduced labor costs. A common field procedure is to place chalk marks at 2 ft intervals on the forms and to place the stirrups by eye in between the marks.

EXAMPLE 7.1

Determine the minimum cross section required for a rectangular beam from a shear standpoint so that no web reinforcing is required by the ACI Code if $V_u = 38^k$ and $f_c' = 4{,}000$ psi. Use the conservative value of $V_c = 2\sqrt{f_c'}b_w d$.

SOLUTION

Shear strength provided by concrete

$$\phi V_c = (0.85)(2\sqrt{4000}b_w d) = 107.52 b_w d$$

But the ACI Code 11.5.5.1 states that a minimum area of shear reinforcement is to be provided if V_u exceeds $\frac{1}{2}\phi V_c$

$$38{,}000 = (\tfrac{1}{2})(107.52 b_w d)$$
$$b_w d = 706.8 \text{ in.}$$

Use 22″ × 35″ beam ($d = 32.5''$)

EXAMPLE 7.2

Select #3 ⊔ stirrups for the beam shown in Figure 7.10 for which $D = 4^k/\text{ft}$ and $L = 6^k/\text{ft}$. $f_c' = 4{,}000$ psi, $f_y = 60{,}000$ psi, and $V_c = 2\sqrt{f_c'}b_w d$.

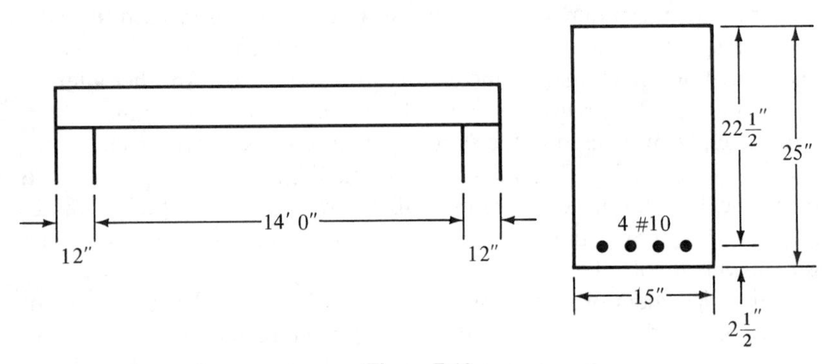

Figure 7.10

SOLUTION

$$V_u \text{ @ left end} = 7(1.4 \times 4 + 1.7 \times 6) = 110.6^k = 110{,}600\,\#$$

V_u @ a distance d from face of support

$$= \left(\frac{84 - 22.5}{84}\right)(110{,}600) = 80{,}975\,\#$$

$$V_c = 2\sqrt{f_c'}b_w d = (2)(\sqrt{4000})(15)(22.5) = 42{,}691\,\#$$

These values are shown in Figure 7.11.

$$V_u = \phi V_c + \phi V_s$$
$$\phi V_s = V_u - \phi V_c = 80{,}975 - (0.85)(42{,}961) = 44{,}688\,\#$$

$$V_s = \frac{44{,}688}{0.85} = 52{,}574\,\#$$

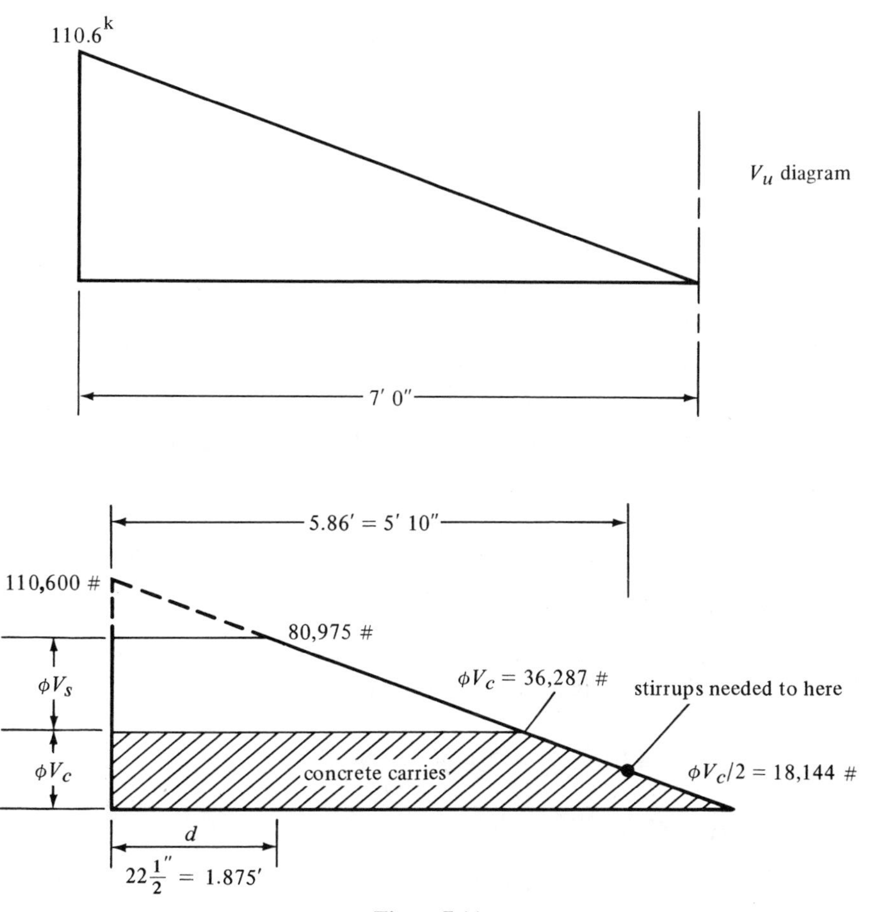

Figure 7.11

Maximum spacing of stirrups $= d/2 = 11.25''$ since V_s is $< 4\sqrt{f'_c}b_w d =$ 85,382#. Maximum theoretical spacing @ left end

$$s = \frac{A_v f_y d}{V_s} = \frac{(2)(0.11)(60{,}000)(22.5)}{52{,}574} = 5.65''$$

Maximum spacing to provide minimum A_v of stirrups

$$s = \frac{A_v f_y}{50 b_w} = \frac{(2)(0.11)(60{,}000)}{(50)(15)} = 17.6''$$

For convenience, theoretical spacings are calculated at different points along the span and are listed in the accompanying table.

Distance from Face of Support	V_u	$V_s = \frac{V_u - \phi V_c}{\phi}$	Theoretical $s = \frac{A_v f_y d}{V_s}$
0 to $d = 1{,}875'$	80,975#	52,574#	5.65″
2′	79,000	50,251	5.91″
3′	63,200	31,662	9.38″
4′	47,400	13,074	Maximum 11.25″

Spacings selected

1 @ 2″ = 0′ 2″
6 @ 5″ = 2′ 6″
2 @ 8″ = 1′ 4″
2 @ 11″ = 1′ 10″
5′ 10″ symmetric about ℄

EXAMPLE 7.3

Compute the value of V_c at a distance 3 ft from the left end of the beam of Example 7.2 and Figure 7.10 by using the ACI value for

$$V_c = \left(1.9\sqrt{f_c'} + 2500\rho_w \frac{V_u d}{M_u}\right) b_w d \not> 3.5\sqrt{f_c'} b_w d$$

SOLUTION

$$w_u = (1.4)(4) + (1.7)(6) = 15.8^k/\text{ft}$$

$$V_u \text{ at } 3' = (7)(15.8) - (3)(15.8) = 63.2^k$$

$$M_u \text{ at } 3' = (110.6)(3) - (15.8)(3)(1.5) = 260.7 \text{ ft-k}$$

$$\rho_w = \frac{5.06}{(15)(22.5)} = 0.0150$$

$$\frac{V_u d}{M_u} = \frac{(63.2)(22.5)}{(12)(260.7)} = 0.455 < 1.0 \qquad \underline{\text{ok}}$$

$$V_c = [1.9\sqrt{4000} + (2500)(0.0150)(0.455)](15)(22.5)$$

$$= \underline{46{,}315\#} < 3.5\sqrt{4000}(15)(22.5) = 74{,}709\#$$

EXAMPLE 7.4

Select #3 ⊔ stirrups for the beam of Example 7.2 assuming the live load is placed to produce maximum shear at beam ℄ although this is not required by the Code.

SOLUTION

$$\text{maximum } V_u \text{ at left end} = (7)(1.4 \times 4 + 1.7 \times 6) = 110.6^k = 110{,}600\#$$

For maximum V_u at ℄, the live load is placed as shown in Figure 7.12.

$$V_u \text{ at ℄} = 57.05 - (7)(1.4 \times 4) = 17.85^k = 17{,}850\#$$
$$V_c = 2\sqrt{4000}(15)(22.5) = 42{,}691\#$$

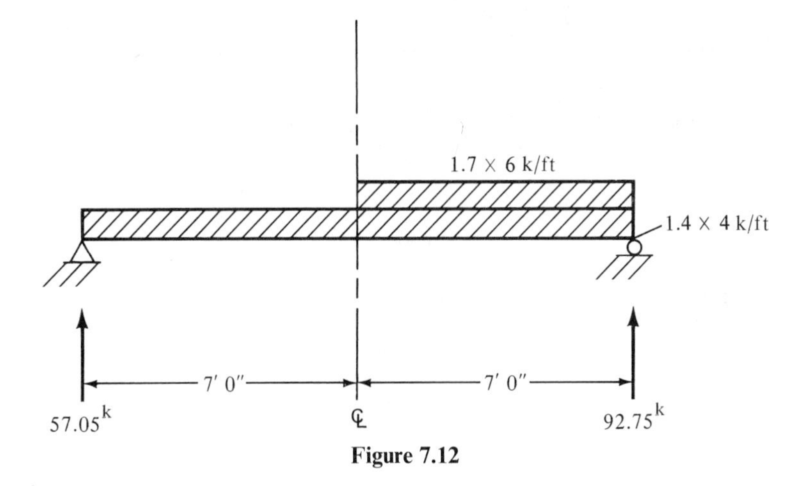

Figure 7.12

V_u at a distance d from face of support = 85,756# as determined by proportions from Figure 7.13

$$V_u = \phi V_c + \phi V_s$$
$$\phi V_s = V_u - \phi V_c = 85{,}756 - (0.85)(42{,}691) = 49{,}470\# \text{ @ left end}$$
$$V_s = \frac{49{,}470}{0.85} = 58{,}200\#$$

The limited spacings are the same as in Example 7.2. The theoretical spacings are given in the accompanying table.

Distance from Face of Support	V_u	$V_s = \dfrac{V_u - \phi V_c}{\phi}$	Theoretical Spacing Required $s = \dfrac{A_v f_y d}{V_s}$
0 to 1.88′	85,756#	58,199#	5.10″
2′	84,100	56,251	5.28″
3′	70,850	40,662	7.30″
4′	57,600	25,074	Maximum 11.25″

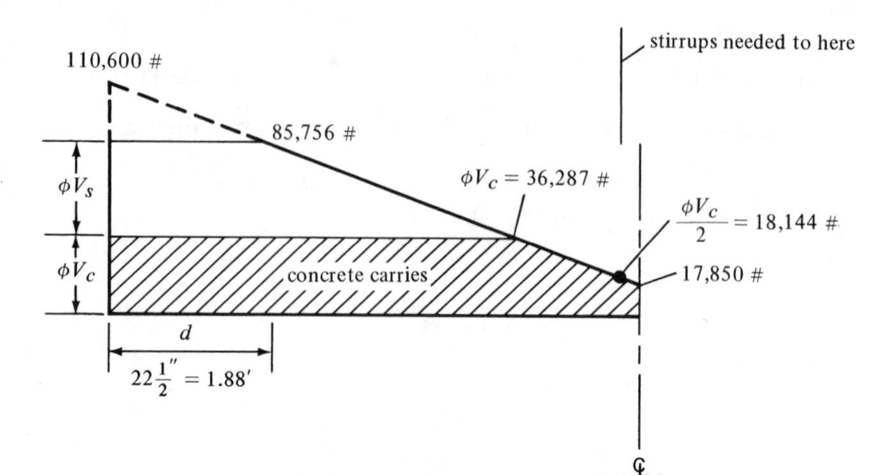

Figure 7.13

Spacings selected

$$\begin{aligned} 1 \ @ \ \ 3'' &= 0'\ 3'' \\ 5 \ @ \ \ 5'' &= 2'\ 1'' \\ 3 \ @ \ \ 7'' &= 1'\ 9'' \\ 3 \ @ \ 10'' &= \underline{2'\ 6''} \\ & \ 6'\ 7'' \end{aligned}$$

symmetric about ℄

Example 7.5 presents a few items not considered in the previous examples. First a T-beam is involved and secondly V_s for part of the beam is between $4\sqrt{f_c'}b_wd$ and $8\sqrt{f_c'}b_wd$ so that the maximum stirrup spacing in that range is $d/4$.

EXAMPLE 7.5

Select spacings for #3 ⊔ stirrups for a T-beam with $b_w = 10$ in. and $d = 20$ in. for the V_u diagram shown in Figure 7.14. $f_y = 60{,}000$ psi. $f_c' = 3{,}000$ psi.

SOLUTION

V_u at a distance d from face of support

$$= 44{,}000 + \left(\frac{72 - 20}{72}\right)(68{,}000 - 44{,}000) = 61{,}333\,\#$$

$$\phi V_c = (0.85)(2\sqrt{3000})(10)(20) = 18{,}623\,\#$$

$$\frac{\phi V_c}{2} = \frac{18{,}623}{2} = 9311\,\#$$

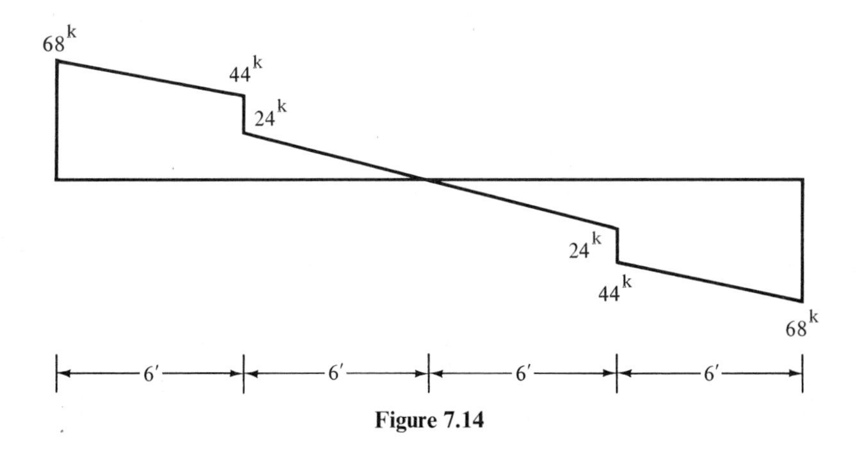

Figure 7.14

Stirrups are needed for a distance

$$= 72 + \left(\frac{24{,}000 - 9311}{24{,}000}\right)(72) = 116''$$

ϕV_s @ left end $= V_u - \phi V_c = 61{,}333 - 18{,}623 = 42{,}710\#$ which is larger than $\phi 4\sqrt{f_c'}b_w d = (0.85)(4\sqrt{3000})(10)(20) = 37{,}245\#$, but less than $\phi 8\sqrt{f_c'}b_w d$. Therefore the maximum spacing of stirrups in that range is $d/4 = 5''$. The point where $V_u - \phi V_c$ is equal to 37,245 is 36″ from the left end as shown in Figure 7.15.

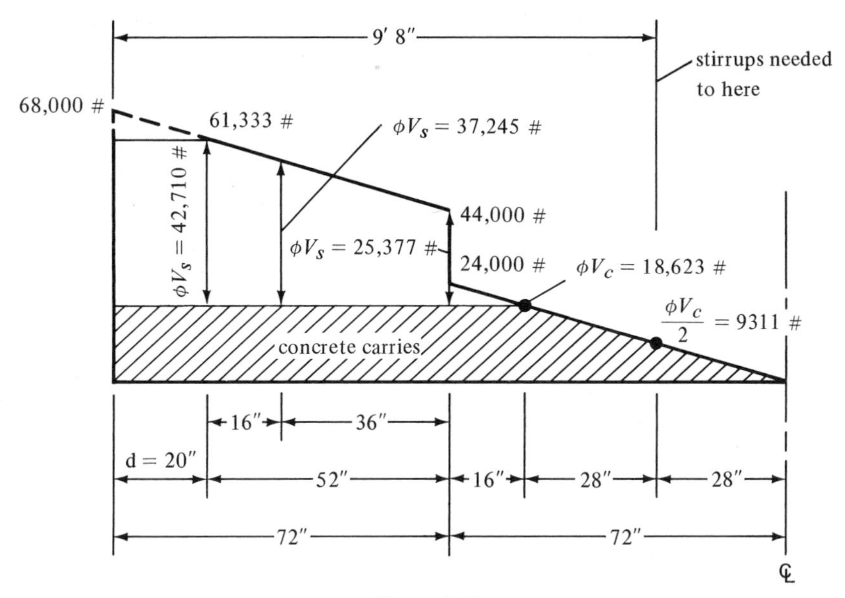

Figure 7.15

Theoretical spacing of #3 stirrups where $V_u - \phi V_c$ is 42,710# is as follows

$$V_s = \frac{42{,}710}{0.85} = 50247\#$$

$$s = \frac{A_v f_y d}{V_s} = \frac{(2)(0.11)(60{,}000)(20)}{50{,}247} = 5.25''$$

Maximum spacing permitted by ACI 11.5.5.3

$$s = \frac{A_v f_y}{50 b_w} = \frac{(2)(0.11)(60{,}000)}{(50)(10)} = 26.4''$$

After $V_u - \phi V_c$ falls to $\phi 4\sqrt{f_c'} b_w d = (0.85)(4\sqrt{3000})(10)(20) = 37{,}245\#$ the Code permits a maximum spacing of $d/2 = 10''$ and the theoretical spacing is

$$s = \frac{(2)(0.11)(60{,}000)(20)}{(37{,}245/0.85)} = 6.02''$$

The theoretical spacing where $V_u - \phi V_c = 44{,}000 - 18{,}623 = 25{,}377\#$ is

$$s = \frac{(2)(0.11)(60{,}000)(20)}{(25{,}377/0.85)} = 8.84''$$

A summary of the results of the preceding calculations is shown in Figure 7.16 where the solid dark line represents the maximum stirrup spacings permitted by the Code while the dotted line represents the calculated theoretical spacings required for $V_u - \phi V_c$. From this information the author selected the following spacings.

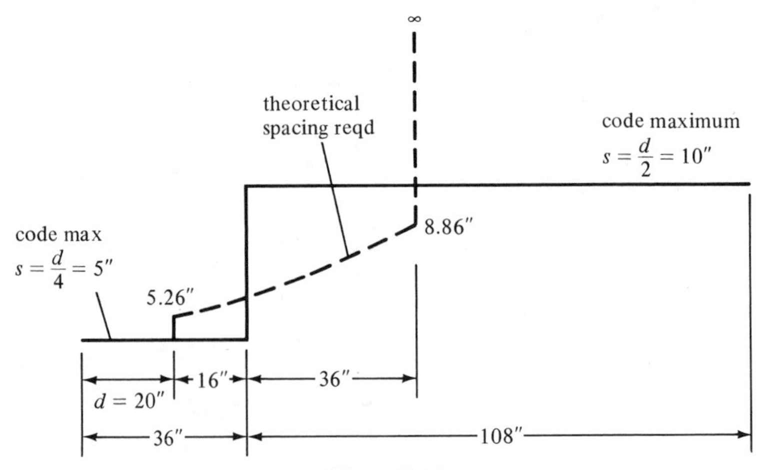

Figure 7.16

$$\begin{aligned} 1 \text{ at } 2 &= 2 \\ 15 \text{ at } 5'' &= 75'' \\ 4 \text{ at } 10'' &= \underline{40''} \\ & \quad 9'9 \end{aligned}$$

symmetrical about ℄

7.7 INTRODUCTORY TORSION COMMENTS

Until recent years the safety factors required by design codes for ordinary reinforced concrete members were sufficiently large as to permit the designer to neglect torsion in all but the most extreme cases. This is not altogether true today, and torsion needs to be considered more frequently than in the past. It may be quite significant for curved beams, spiral staircases, beams that have large loads applied laterally off center, and even in the spandrel beams running between exterior columns of buildings. These beams support the edges of floor slabs and their supporting beams and thus are laterally loaded on one side.

If factored torsional moments are less than $\phi(0.5\sqrt{f_c'}\sum x^2 y)$, they will not appreciably reduce either the shear or flexural strengths of reinforced concrete members. If they are higher than $\phi(0.5\sqrt{f_c'}\sum x^2 y)$, however, it is considered necessary to consider their interaction with shear and flexural stresses. For such cases the stirrups will be made closed in form and will probably have to be increased in size and spaced closer together. In addition, more longitudinal steel will have to be added. Chapter 13 of this textbook is devoted to this subject.

PROBLEMS

7.1 to **7.7** For the beams and loads given, select stirrup spacings. $f_c' = 3{,}000$ psi and $f_y = 60{,}000$ psi. The dead loads shown include beam weights. Do not consider movement of live loads unless specifically requested.

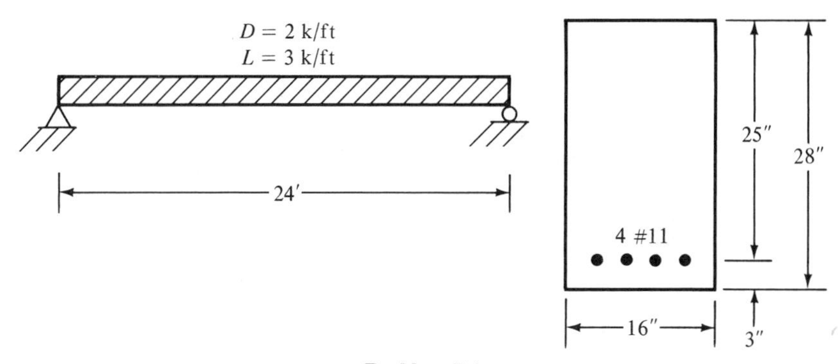

Problem 7.1

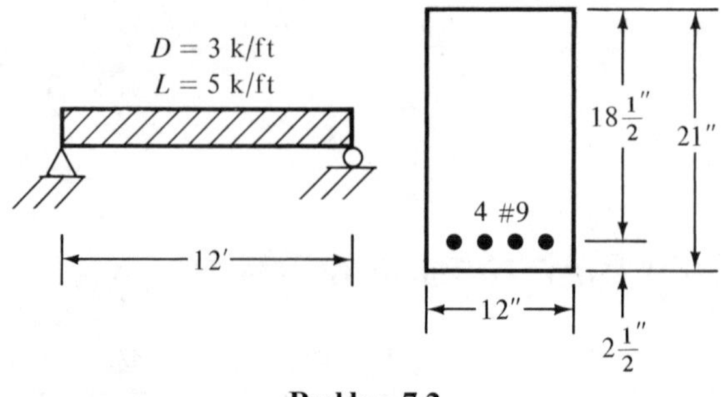

Problem 7.2

7.3 Repeat Problem 7.2 if live load positions are considered to cause maximum end shear and maximum ℄ shear.

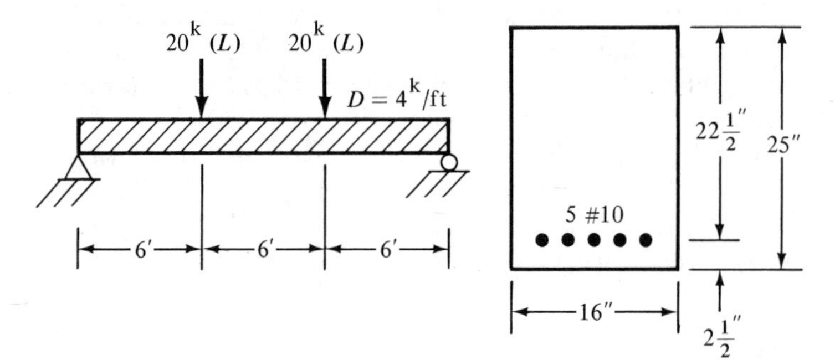

Problem 7.4

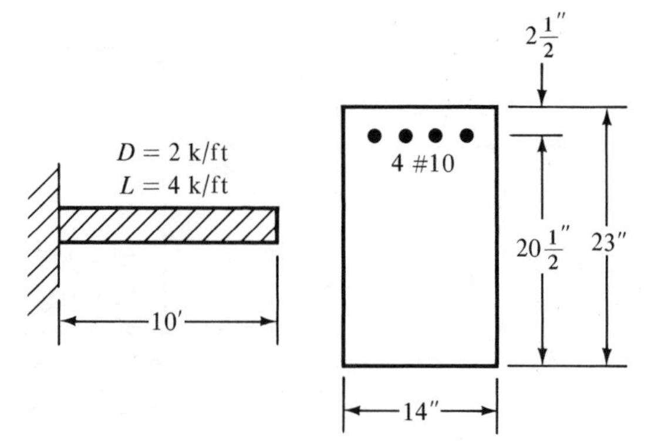

Problem 7.5

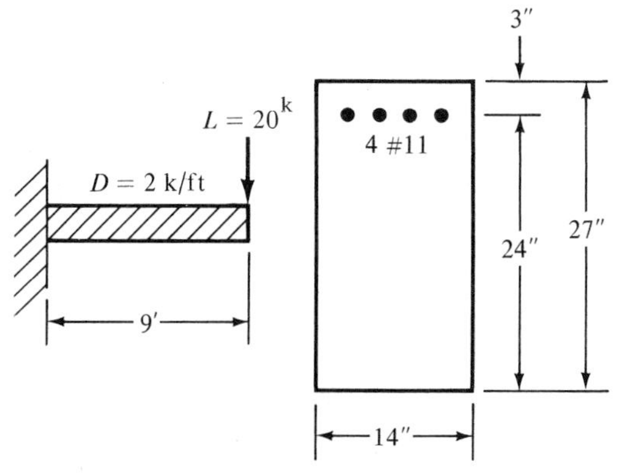

Problem 7.6

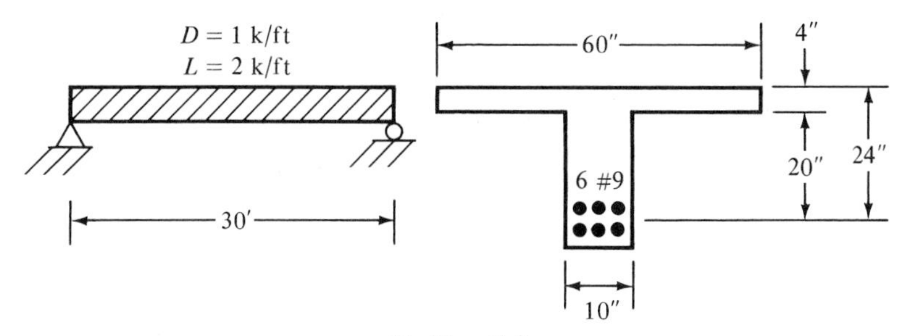

Problem 7.7

PROBLEMS WITH SI UNITS

7.8 to **7.11** For the beams and loads given, select stirrup spacings. $f_c' = 20.7$ MPa and $f_y = 275.8$ MPa. The dead loads shown include beam weights. Do not consider movement of live loads. Shear stress carried by concrete, V_c, shall not exceed $0.166\sqrt{f_c'}b_w d$.

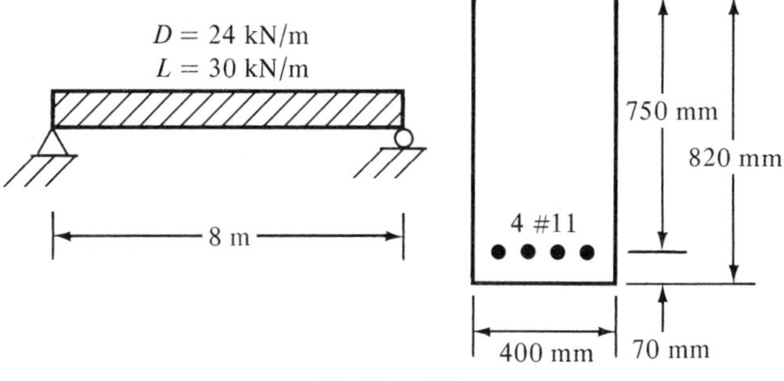

Problem 7.8

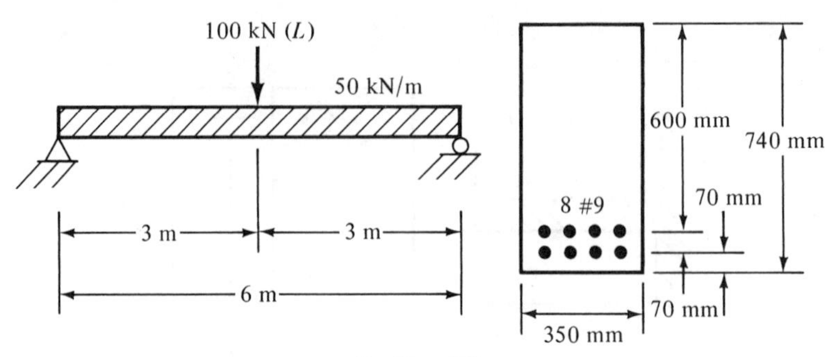

Problem 7.9

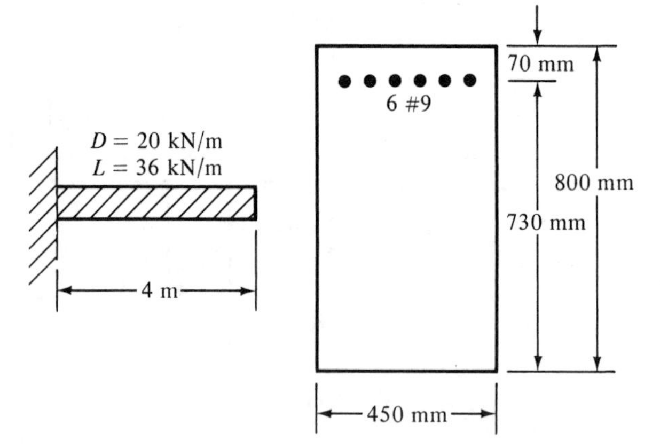

Problem 7.10

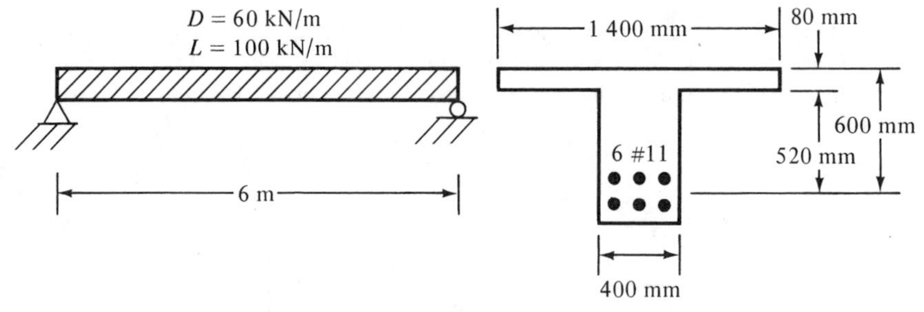

Problem 7.11

CHAPTER 8

Introduction to Columns

8.1 INTRODUCTION

This chapter is devoted to the analysis and design of axially loaded columns and to the introductory analysis of columns subjected to axial load and bending. Only short columns are considered in this chapter; Chapter 9 is devoted to the design of both short and long compression members subject to axial load and bending. The terms "column" and "compression member" are used interchangeably throughout the text.

A *short column* is one that is in little danger of buckling due to its slenderness. The load which it will support is controlled only by the

Water Tower Place, Chicago, Illinois (Courtesy Symons Corporation).

dimensions of the cross section and the strength of the materials involved. When the length of a column becomes rather large as compared to its least lateral dimension, a buckling failure is possible and it is said to be a *long column.* The ACI Column Committee has estimated that about 90% of all braced columns and about 40% of all unbraced columns should be classified as short columns.[1]

For members such as building columns, which normally carry mostly axial compressive loads, it is economical to have the large percentage of the load carried by the concrete. Nevertheless, no matter how close a column is assumed to being an axially loaded member, some reinforcing will be needed because there is always some bending present due to loads not being perfectly centered, due to movement of loads, due to members not being perfectly straight, due to types of connections, and so on. As a result, there will be some tension and reinforcing bars will be required to supply the needed tensile strength.

Section B.6.1 of Appendix B of the Code says that for working stress design the combined flexural and axial load capacities of compression members is to be taken as 40% of the values computed by the strength design procedure. Slenderness is to be handled by the same method used in strength design.

8.2 TYPES OF COLUMNS

Plain concrete columns cannot support much load and are not used as columns although they can be used for pedestals if their height is not greater than three times their least lateral dimension (Code 2.1 and 15.11). The addition of longitudinal bars will greatly increase their load-carrying capacity. Further large strength increases may be made by providing lateral restraint for these longitudinal bars. Under compressive loads columns tend not only to shorten lengthwise but also to expand laterally due to the Poisson effect. The capacity of such members can be greatly increased by providing lateral restraint in the form of closed ties or helical spirals wrapped around the longitudinal reinforcing.

Reinforced concrete columns are referred to as *tied* or *spiral* columns depending on the method used for laterally bracing or holding the bars in place. If the column has a series of closed ties, as shown in Figure 8.1(a), it is referred to as a *tied column*. These ties are very effective in increasing the column strength. They prevent the longitudinal bars from being displaced during construction and they resist the tendency of the same bars to buckle outwards under load, which would cause the outer concrete cover to break off. Tied columns are ordinarily square or rectangular but they can be octagonal, round, L-shaped, and so forth.

[1] American Concrete Institute, 1972, *Notes on ACI 318-71 Building Code with Design Applications.* Skokie, Illinois, (page 10–2).

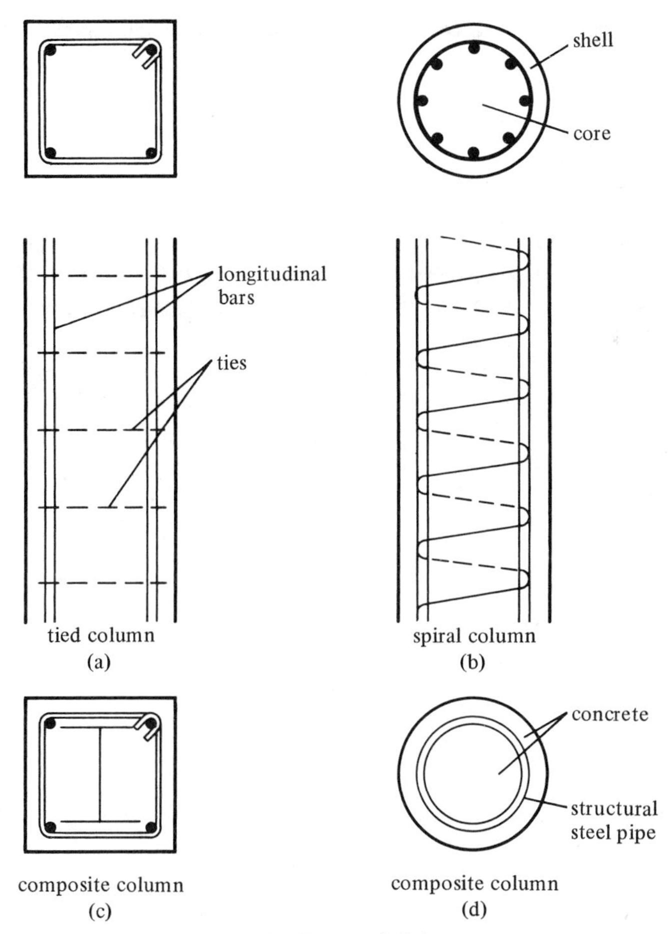

Figure 8.1 Types of Columns

If a continuous helical spiral made from bars or heavy wire is wrapped around the longitudinal bars, as shown in Figure 8.1(b), the column is referred to as a *spiral column.* Spirals are even more effective than ties in increasing a column's strength. The closely spaced spirals do a better job of holding the longitudinal bars in place, and they also confine the concrete inside and greatly increase its resistance to axial compression. As the concrete inside the spiral tends to spread out laterally under the compressive load, the spiral that restrains it is put into hoop tension, and the column will not fail until the spiral yields or breaks, permitting the bursting of the concrete inside. Spiral columns are normally round but they also can be made into rectangular, octagonal, or other shapes.

Composite columns, illustrated in Figures 8.1(c) and (d), are those concrete columns that are reinforced longitudinally by structural steel shapes, which may or may not be surrounded by structural steel bars, or they may consist of structural steel tubing filled with concrete (commonly called "lally columns").

8.3 AXIAL LOAD CAPACITY OF COLUMNS

In actual practice there are no perfect axially loaded columns, but a discussion of such members provides an excellent starting point for explaining the theory involved in designing real columns with their eccentric loads. Several basic ideas can be explained with the aid of purely axially loaded columns, and the strengths obtained provide upper theoretical limits that can be clearly verified with actual tests.

It has been known for several decades that the stresses in the concrete and the reinforcing bars of a column supporting a long-term load cannot be calculated with any degree of accuracy. You might think that such stresses could be determined by multiplying the strains by the appropriate moduli of elasticity. But this idea does not work too well practically because the modulus of elasticity of the concrete is changing during loading due to creep and shrinkage. So it can be seen that the parts of the load carried by the concrete and by the steel vary with the magnitude and duration of the loads. For instance, the larger the percentage of dead loads and the longer they are applied, the greater the creep in the concrete and the larger the percentage of load carried by the reinforcement.

Though stresses cannot be predicted in columns in the elastic range with any degree of accuracy, several decades of testing have shown that the ultimate strength of columns can be estimated very well. Furthermore, it has been shown that the proportion of live and dead loads, the length of loading, and other such items have little effect on the ultimate strength. It does not even matter whether the concrete or the steel approaches its ultimate strength first. If one of the two materials is stressed close to its ultimate strength, its large deformations will cause the stress to increase quicker in the other material.

For these reasons only the ultimate strength of columns is considered here. At failure the theoretical ultimate strength or nominal strength of a short axially loaded column is quite accurately determined by the expression that follows, in which A_g is the gross concrete area and A_{st} is the total area of longitudinal reinforcement including bars and steel shapes:

$$P_n = 0.85f_c'(A_g - A_{st}) + f_yA_{st}$$

8.4 FAILURE OF TIED AND SPIRAL COLUMNS

Should a short tied column be loaded until it fails, parts of the shell or covering concrete will spall off and, unless the ties are quite closely spaced, the longitudinal bars will buckle almost immediately as their lateral support (the covering concrete) is gone. Such failures may often be quite sudden, and apparently they have occurred rather frequently in structures subjected to earthquake loadings.

When spiral columns are loaded to failure, the situation is quite different. The covering concrete or shell will spall off but the core will continue to stand, and if the spiral is closely spaced, the core will be able to resist an appreciable amount of additional load beyond the load that causes spalling. The closely spaced loops of the spiral together with the longitudinal bars form a cage that very effectively confines the concrete. As a result, the spalling off of the shell of a spiral column provides a warning that failure is going to occur if the load is further increased.

American practice is to neglect any excess capacity after the shell spalls off since it is felt that once the spalling occurs the column will no longer be useful—at least from the viewpoint of the occupants of the building. For this reason the spiral is designed so that it is just a little stronger than the shell that is assumed to spall off. The spalling gives a warning of impending failure, and then the column will take a little more load before it fails. Designing the spiral so that it is just a little stronger than the shell does not increase the column's ultimate strength much but it does result in a more gradual or ductile failure.

The strength of the shell is given by the following expression, where A_c is the area of the core, which is considered to have a diameter that extends from out to out of the spiral:

$$\text{shell strength} = 0.85f_c'(A_g - A_c)$$

It can be shown, by considering the estimated hoop tension that is produced in spirals due to the lateral pressure from the core and by tests, that spiral steel is at least twice as effective in increasing the ultimate column capacity as is longitudinal steel.[2,3] Therefore, the strength of the spiral can be computed by the following expression, in which ρ_s is the percentage of spiral steel:

$$\text{spiral strength} = 2\rho_s A_c f_y$$

Equating these expressions and solving for the required percentage of spiral steel:

$$0.85f_c'(A_g - A_c) = 2\rho_s A_c f_y$$

$$\rho_s = 0.425\frac{(A_g - A_c)f_c'}{A_c f_y} = 0.425\frac{[(A_g/A_c) - 1]f_c'}{f_y}$$

[2] Park, A., and Paulay, T., 1975, *Reinforced Concrete Structures* (New York: Wiley), pp. 25, 119–121.

[3] Considere, A., 1902, "Compressive Resistance of Concrete Steel and Hooped Concrete, Part I," *Engineering Record*, December 20, pp. 581–583; "Part II," December 27, pp. 605–606.

To make the spiral a little stronger than the spalled concrete, the Code (10.9.3) specifies the minimum spiral percentage to be as follows:

$$\rho_s = 0.45\frac{[(A_g/A_c) - 1]f_c'}{f_y}$$

Once the required percentage of spiral steel is determined, the spiral may be selected with the expression to follow, in which ρ_s is written in terms of the volume of the steel in one loop:

$$\rho_s = \frac{\text{volume of spiral in one loop}}{\text{volume of concrete core for a pitch } s}$$

$$= \frac{V_{\text{spiral}}}{V_{\text{core}}}$$

$$= \frac{a_s\pi(D_c - d_b)}{(\pi D_c^2/4)s} = \frac{4a_s(D_c - d_b)}{sD_c^2}$$

In this expression D_c is the diameter of the core out to out of the spiral, a_s is the cross-sectional area of the spiral bar, and d_b is the diameter of the spiral bar. Here reference is made to Figure 8.2. The designer can assume

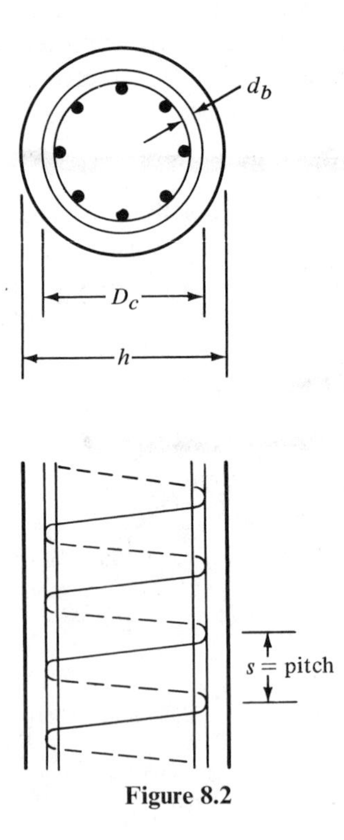

Figure 8.2

a spiral diameter and solve for the pitch required. If the results do not seem reasonable, he can try another diameter. The pitch used must be within the limitations listed in the next section of this chapter. Actually, Table A.16 (see Appendix), which is based on this expression, permits the designer to select spirals directly.

8.5 CODE REQUIREMENTS FOR CAST-IN-PLACE COLUMNS

The ACI Code specifies quite a few limitations on the dimensions, reinforcing, lateral restraint, and other items pertaining to concrete columns. Some of the most important limitations are listed in the paragraphs to follow.

1. The percentage of longitudinal reinforcement (ACI Code 10.9.1) shall not be less than 1% nor greater than 8% of the gross cross-sectional area. It is felt that if the amount of steel is less than 1%, there will be a distinct possibility of a sudden nonductile failure, as might occur in a plain concrete column. Actually the Code does permit the use of less than 1% steel if the column has been made larger than is necessary to carry the loads because of architectural or other reasons. In no circumstances, however, will they permit the steel area to be less than 0.005 times the given area of the concrete actually provided. The maximum percentage of steel is given to prevent too much crowding of the reinforcing bars. Even at 8% it is rather difficult to place the bars and get the concrete down into the forms around them. Usually the percentage of reinforcement should not exceed 4% when the bars are to be lap spliced.

2. If a rectangular bar arrangement is used, a minimum of four bars is permitted, while a minimum of six is specified for circular arrangements (ACI Code 10.9.2).

3. When tied columns are used, the ties shall not be less than #3, provided the longitudinal bars are #10 or smaller. The minimum size is #4 for longitudinal bars larger than #10 and for bundled bars. The center-to-center spacing of ties shall not be more than 16 times the diameter of the longitudinal bars, 48 times the diameter of the ties, nor the least lateral dimension of the column. The ties must be arranged so that every corner and alternate longitudinal bar have lateral support provided by the corner of a tie having an included angle not greater than 135°. No bars can be located a greater distance than 6 in. clear on either side from such a laterally supported bar. These requirements are given by the ACI Code in its Section 7.10.5. Figure 8.3 shows tie arrangements for several column cross sections.

4, The Code (7.10.4) states that the clear spacing of spirals may not be less than 1 in. nor greater than 3 in. Should splices be necessary in spirals, they are to be provided by welding or by lapping the spiral bars or wires

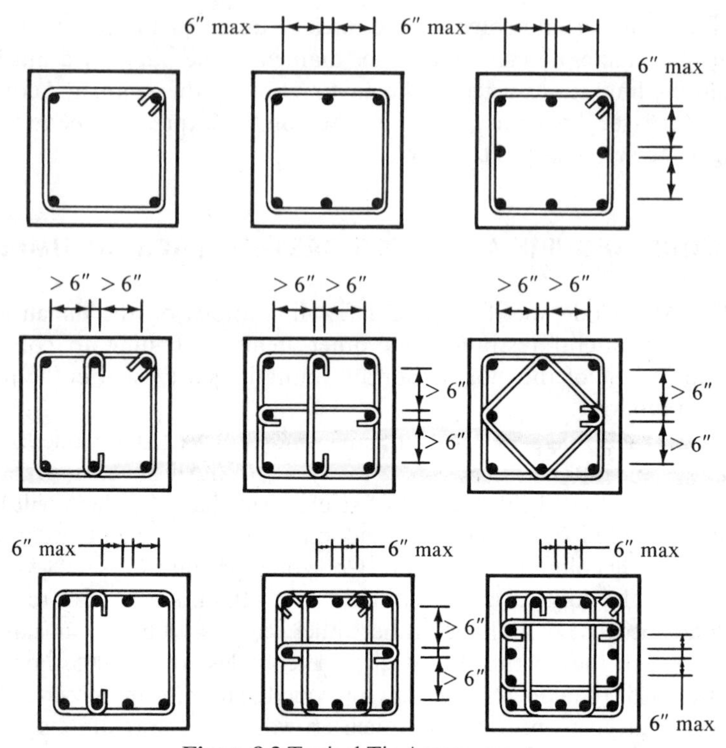

Figure 8.3 Typical Tie Arrangements

by the larger of 48 diameters or 12 in. Special spacer bars are used to hold the spirals in place and at the desired pitch until the concrete hardens. These spacers consist of vertical bars with small hooks.

8.6 SAFETY PROVISIONS FOR COLUMNS

The values of ϕ to be used for columns as specified in Section 9.3.2 of the Code are well below those used for flexure and shear (0.90 and 0.85, respectively). A value of 0.70 is specified for tied columns and 0.75 for spiral columns because of their greater toughness.

The failure of a column is generally a more severe matter than is the failure of a beam because a column generally supports a larger part of a structure than does a beam. In other words, if a column fails in a building, a larger part of the building will fall down than if a beam fails. This is particularly true for a lower level column in a multistorey building. As a result, lower ϕ values are desirable for columns.

There are other reasons for using lower ϕ values in columns. As an example, it is more difficult to do as good a job in placing the concrete for a column than it is for a beam. The reader can readily see the difficulty of getting concrete down into narrow column forms and between the longitudinal and lateral reinforcing. As a result, the quality of the resulting concrete columns is probably not as good as that of beams and slabs.

The failure strength of a beam is normally dependent on the yield stress of the tensile steel—a property which is quite accurately controlled in the steel mills. On the other hand, the failure strength of a column is closely related to the concrete's ultimate strength a value which is quite variable. The length factors of columns are another variable which drastically affect their strength and thus makes the use of lower ϕ factors necessary.

It seems impossible for a column to be perfectly axially loaded. Even if loads could be perfectly centered at one time they would not stay in place. Furthermore, columns may be initially crooked or have other flaws with the result that lateral bending will be produced. Wind and other lateral loads cause columns to bend, and the columns in rigid frame buildings are subjected to moments even when the frame is supporting gravity loads alone.

Despite the facts mentioned in the preceding paragraph, there are many situations where there are no calculated moments for the columns of a structure. For many years the Code specified that such columns had to be designed for certain minimum moments even though no calculated moments were present. This was accomplished by requiring designers to assume certain minimum eccentricities for their column loads. These minimum values were 1 in. or $0.05h$ for spiral columns and 1 in. or $0.10h$ for tied columns (the term h representing the outside diameter of round columns or the total depth of square or rectangular columns). A moment equal to the axial load times the minimum eccentricity was used for design.

In today's Code minimum eccentricities are not specified but the same objective is accomplished by requiring that theoretical axial load capacities be multiplied by 0.80 for tied columns and 0.85 for spiral columns. Thus, as shown in Section 10.3.5 of the Code the axial load capacity of columns may not be greater than the following values:

For spiral columns

$$P_u = \phi P_n = \phi 0.85[0.85f_c'(A_g - A_{st}) + f_y A_{st}]$$

For tied columns

$$P_u = \phi P_n = \phi 0.80[0.85f_c'(A_g - A_{st}) + f_y A_{st}]$$

When calculated moments are available they must be used for designs but the sizes obtained may not be less than would result from application of the preceding "axial load" equations. In addition, slenderness effects must be considered as described in Chapter 9.

Royal Towers, Baltimore, Maryland (Courtesy Simpson Timber Company).

8.7 COMMENTS ON ECONOMICAL COLUMN DESIGN

Reinforcing bars are quite expensive and thus the percentage of longitudinal reinforcing used in reinforced concrete columns is a very major factor in their economy. This means that under normal circumstances the smallest possible percentage of steel should be used. This can be accomplished by using larger column sizes and/or higher strength concretes.

Higher strength concretes can be more economically used in columns than in beams. Under ordinary loads, only 30 to 40% of a beam cross section is in compression while the remaining 60 to 70% is in tension and thus assumed to be cracked. This means that if a high strength concrete is used for a beam, 60 to 70% of it is wasted. For the usual column, however, the situation is quite different because a much larger percentage of its cross section is in compression. As a result, it is quite economical to use high strength concretes for columns. Although some designers have used concretes with ultimate strengths as high as 9,000 psi for column design with apparent economy,[4] the use of 4,000 and 5,000 psi columns is the normal rule when higher strengths are specified for columns.

[4] Schmidt, W. and Hoffman, E. S., "9000 psi Concrete—Why? Why Not?," *Civil Engineering*, May 1975, pp. 52–55.

In general, tied columns are more economical than spiral columns particularly if square or rectangular columns are to be used. Of course, spiral columns, high strength concretes, and high percentages of steel save floor space.

8.8 DESIGN OF AXIALLY LOADED COLUMNS

As a brief introduction to columns, the design of two axially loaded short columns is presented in this section. Moment and length effects are completely neglected. Example 8.1 presents the design of an axially loaded square tied column while Example 8.2 illustrates the design of a similarly loaded round spiral column. Table A.17 of the Appendix provides several properties for circular columns which are particularly useful for designing spiral columns.

EXAMPLE 8.1

Design a square tied column to support an axial dead load D of 130^k and an axial live load L of 180^k. Assume 2% longitudinal steel is desired. $f_c' = 4{,}000$ psi, $f_y = 60{,}000$ psi.

SOLUTION

$$P_u = (1.4)(130) + (1.7)(180) = 488^k$$

Selecting column dimensions

$$P_u = \phi 0.80[0.85 f_c'(A_g - A_{st}) + f_y A_{st}]$$

$$488 = (0.70)(0.80)[(0.85)(4)(A_g - 0.02A_g) + (60)(0.02A_g)]$$

$$A_g = 192.3 \text{ in.}^2 \qquad \underline{\text{Use } 14 \times 14 (A_g = 196 \text{ in.}^2)}$$

Selecting longitudinal bars

Substituting into column equation with known A_g and solving for A_{st}

$$488 = (0.70)(0.80)[(0.85)(4)(196 - A_{st}) + 60A_{st}]$$

$$A_{st} = 3.62 \text{ in.}^2 \qquad \underline{\text{Use 6 \#7 Bars } (3.61 \text{ in.}^2)}$$

Design of ties (assuming #3 bars)

Spacing: (a) $48 \times \frac{3}{8} = 18''$

(b) $16 \times \frac{7}{8} = 14'' \leftarrow$

(c) Least dim. $= 14'' \leftarrow$ Use #3 Ties at 14″

Sketch of column cross section shown in Figure 8.4

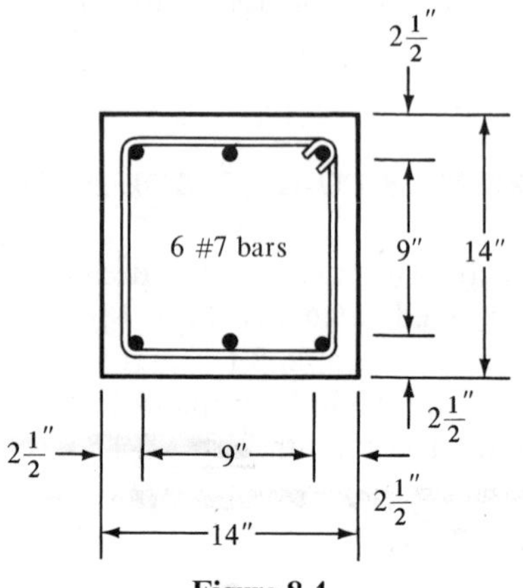

Figure 8.4

EXAMPLE 8.2

Design a round spiral column to support an axial dead load D of 180^k and an axial live load L of 300^k. Assume 2% longitudinal steel is desired. $f_c' = 4{,}000$ psi, $f_y = 60{,}000$ psi.

SOLUTION

$$P_u = (1.4)(180) + (1.7)(300) = 762^k$$

Selecting column dimensions and bar sizes

$$P_u = \phi 0.85[0.85 f_c'(A_g - A_{st}) + f_y A_{st}]$$

$$762 = (0.75)(0.85)[(0.85)(4)(A_g - 0.02A_g) + (60)(0.02A_g)]$$

$$A_g = 263.7 \text{ in.}^2 \qquad \underline{\text{Use } 18'' \text{ diameter column } (255 \text{ in.}^2)}$$

$$762 = (0.75)(0.85)[(0.85)(4)(255 - A_{st}) + 60A_{st}]$$

$$A_{st} = 5.80 \text{ in.}^2 \qquad \underline{\text{Use 6 \#9 Bars } (6.00 \text{ in.}^2)}$$

Sketch of column cross section shown in Figure 8.5

Design of spiral

$$A_c = \frac{(\pi)(15)^2}{4} = 177 \text{ in.}^2$$

$$\text{minimum } \rho_s = (0.45)\left(\frac{A_g}{A_c} - 1\right)\frac{f_c'}{f_y} = (0.45)\left(\frac{255}{177} - 1\right)\left(\frac{4}{60}\right) = 0.0132$$

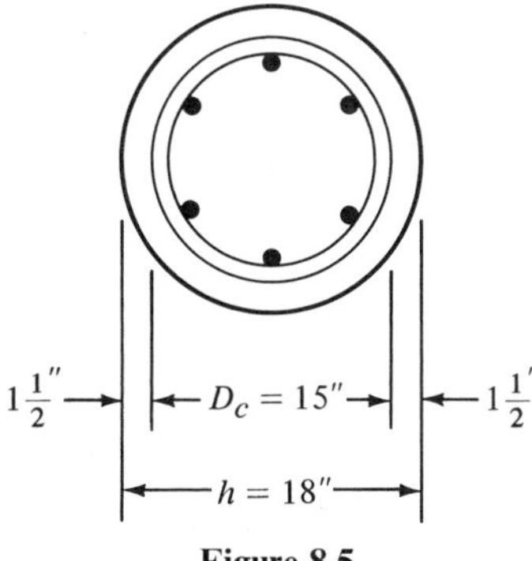

Figure 8.5

Assume a #3 spiral.

$$\rho_s = \frac{4a_s(D_c - d_b)}{sD_c^2}$$

$$0.0132 = \frac{(4)(0.11)(15 - 0.375)}{(s)(15)^2}$$

$$s = 2.17''$$ say 2″

(Checked with Table A.16 in Appendix.)

8.9 AXIAL LOAD AND BENDING

Although some discussion was presented in the last section about axially loaded columns, all columns are subjected to some bending as well as axial forces and they must be designed to resist both. As a result, columns will bend, and the moment will tend to produce compression on one side of the column and tension on the other. Depending on the relative magnitudes of the moments and axial loads, there are theoretically several ways in which the sections might fail. These are as follows:

1. Large moment with no appreciable axial load. The failure would occur as in a beam.
2. Large moment with small axial load. Failure is initiated by the yielding of the bars on the tensile side of the column.
3. A balanced loading situation such that the moment and axial load are of such proportions that the concrete fails on the compression side by crushing at the same time as the tensile steel yields on the other side.
4. Large axial load and small moment, with failure occurring by the crushing of the concrete on the compression side while the tensile steel has not reached its yield point.

5. Large axial load and small moment such that the entire cross section is in compression. Failure occurs by the crushing of the concrete with all the reinforcing bars in compression.
6. Large axial load with negligible moment. Failure occurs by the crushing of the concrete with all bars in the member having reached their yield stress in compression.

Should numerical values be determined for these failure modes for a particular column and plotted as a diagram, the result would be a so-called interaction diagram, such as the one shown in Figure 9.1 in the next chapter. Sample calculations for problems of these types are presented in the remaining sections of this chapter.

8.10 COLUMNS WITH LARGE MOMENTS AND SMALL AXIAL LOADS

Example 8.3 shows that it is possible, with the usual statics equations, to determine the ultimate load P_n at which a column will fail. For this example the load P_n is assumed to be located some eccentric distance e from the x-axis of the column. You might be puzzled about this distance and wonder what it represents. You might say, "All I have for my column is a moment and axial load but no e." You should realize that e is merely the distance the load would have to be off center to produce the moment involved. In other words, $P_n e = M_n$. If you have a load P_n and a moment M_n, you can calculate the value of e as follows, remembering to use consistent units:

$$e = \frac{M_n}{P_n}$$

The eccentricity e is measured from the *plastic centroid* of the column. The plastic centroid represents the location of the resultant force produced by the steel and the concrete. For locating the plastic centroid, all concrete is assumed to be stressed in compression to $0.85f_c'$ and all steel to f_y in compression. For symmetrical sections the plastic centroid obviously coincides with the centroid of the column cross section, while for nonsymmetrical sections it can be located by taking moments.

In Example 8.3 the eccentricity is quite large, so that the load is outside of the column cross section. For such a case a large moment is present, and it is logically assumed that failure is initiated by the yielding of the tensile steel on the far side of the column. A pair of simultaneous equations can be written involving the two unknowns—the ultimate load P_n and the depth of the stress block a.

In the solution the four forces involved—P_n, T (the force in the tensile bars), C_s' (the force in the compression bars), and C_c (the compression in the concrete)—are shown in Fig. 8.6(b). It is assumed in this first example

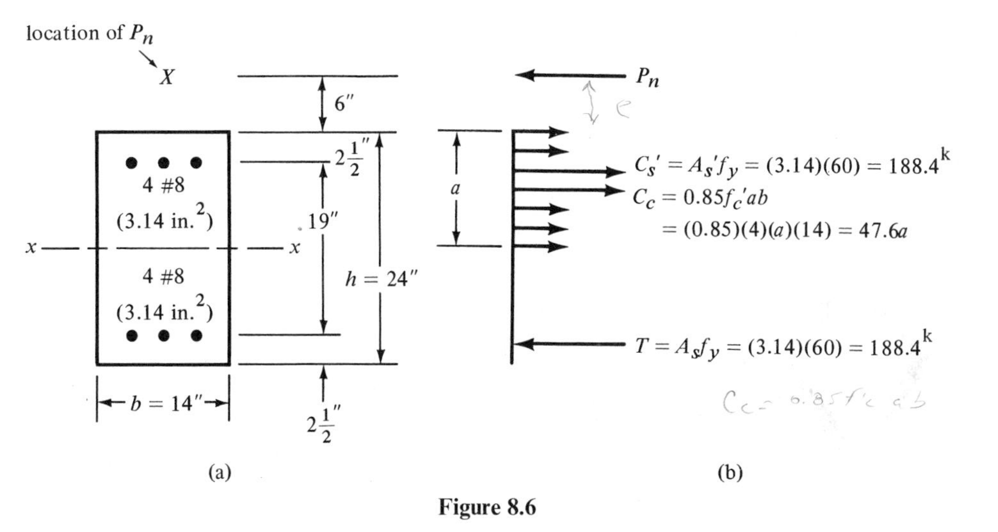

Figure 8.6

that both the compression steel and the tensile steel have reached their yield stresses. Thus both T and C_s' are known, leaving two unknowns, C_c and P_n. The value of C_c equals $0.85f_c'ab$ with a being the only unknown.

The four forces are equated to zero, giving one equation, and then moments are taken about the centroid of the tensile steel, giving another equation involving the two unknowns. These are solved simultaneously and P_n is obtained. Finally, the strains in the compression and tensile steel are checked to be sure that yielding has occurred.

EXAMPLE 8.3

Determine P_n for the column shown in Figure 8.6(a). $f_c' = 4{,}000$ psi and $f_y = 60{,}000$ psi.

SOLUTION

Assuming all bars have yielded and equating the horizontal forces shown in Figure 8.6(b) to zero

$$-P_n + C_s' + C_c - T = 0$$

$$-P_n + 188.4 + 47.6a - 188.4 = 0$$

$$P_n = 47.6a$$

Taking moments about the tensile steel

$$-(P_n)(27.5) + (188.4)(19) + (47.6a)\left(21.5 - \frac{a}{2}\right) = 0$$

$$-27.5P_n + 3579.6 + 1023.4a - 23.8a^2 = 0$$

Substituting into this equation the value $P_n = 47.6a$

$$-1309a + 3579.6 + 1023.4a - 23.8a^2 = 0$$

$$a = 7.65''$$

$$P_n = (47.6)(7.65) = 364.1^{\text{k}}$$

Checking to see if bars have yielded (Figure 8.7)

$$c = \frac{7.65}{0.85} = 9.00''$$

$$\frac{60{,}000}{29 \times 10^6} = 0.00207$$

$$\varepsilon_s' = \frac{6.50}{9.00} \times 0.003 = 0.00217 > 0.00207$$

$$\varepsilon_s = \frac{12.50}{9.00} \times 0.003 = 0.00417 > 0.00207$$

Therefore, both steels have yielded, as assumed.

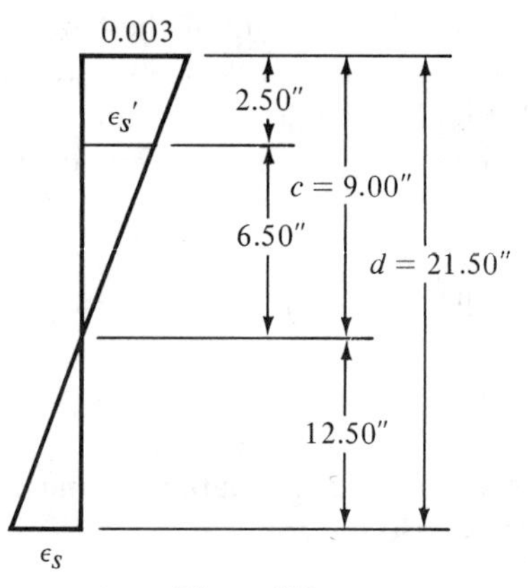

Figure 8.7

8.11 COLUMNS WITH LARGE AXIAL LOADS AND SMALL MOMENTS

In Example 8.4 the eccentricity is relatively small, and it is logical to assume that the tensile steel has not yielded. An estimate is made for the depth of the stress block a. As the eccentricity is small and the load P_n

located within the cross section, it is logical to assume that failure is initiated by the crushing of the concrete. A large part of the column is in compression and a value of a is assumed which is a little greater than half the column depth.

From the assumed value of a, the strains in the steels can be calculated, as can their stresses and the values of T and C_s' (for the assumed value of a). Moments can be taken about the centroid of the tensile steel and the resulting equation solved for P_n. Then as a check the four forces are equated to zero, leaving a as the unknown. The resulting equation is solved for a and compared with the assumed value. If the discrepancy is large, another value of a is assumed and the process repeated.

EXAMPLE 8.4

Determine P_n for the column shown in Figure 8.8(a). $f_c' = 4{,}000$ psi and $f_y = 50{,}000$ psi.

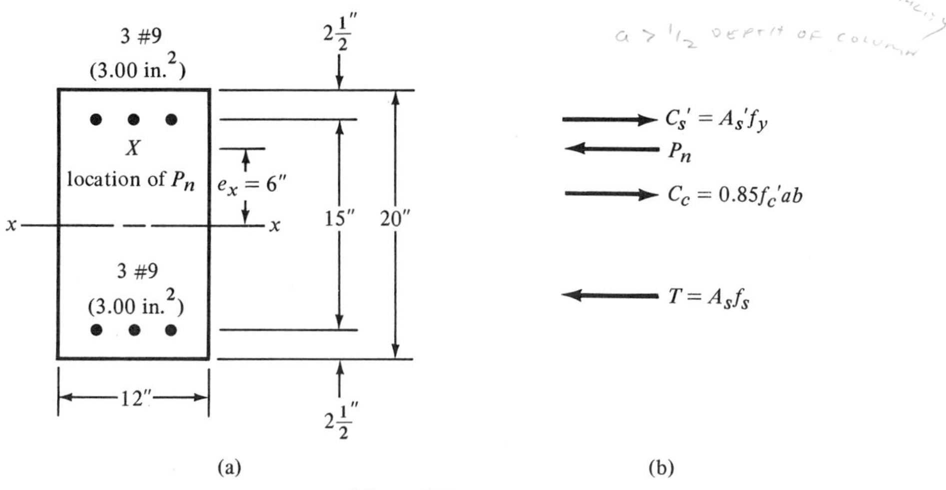

Figure 8.8

SOLUTION

Assume a = 12.00 in.

$$c = \frac{12.00}{0.85} = 14.12''$$

Working with strain diagram and calculating steel stress f_s (Figure 8.9)

$$\varepsilon_s' = \left(\frac{11.62}{14.12}\right)(0.003) = 0.00247 > \frac{50{,}000}{29 \times 10^6} = 0.00172$$

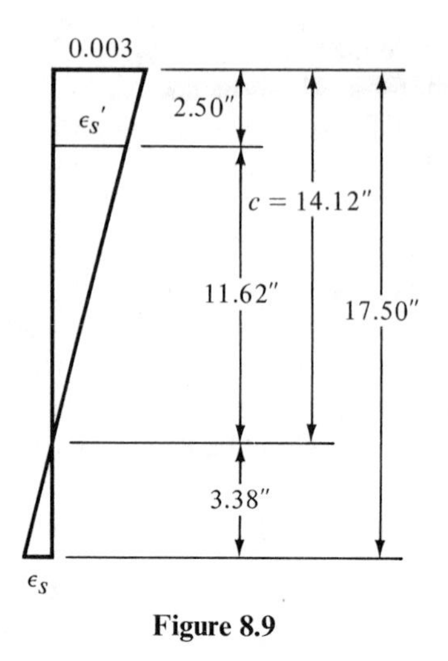

Figure 8.9

Therefore, the compression steel has yielded.

$$\varepsilon_s = \left(\frac{3.38}{14.12}\right)(0.003) = 0.000718 < 0.00172$$

$$f_s = (0.000718)(29 \times 10^6) = 20{,}822 \text{ psi.}$$

Taking moments about centroid of tensile steel

$$(P_n)(13.5) - (0.85 \times 4 \times 12 \times 12)(17.5 - \tfrac{12}{2}) - (3)(50)(15) = 0$$

$$P_n = 583.7^{\text{k}}$$

Checking assumed value of *a* by summing axial forces

$$P_n + T - C_s' - C_c = 0$$

$$583.7 + (3)(20.822) - (3)(50) - (0.85)(4)(a)(12) = 0$$

$$a = 12.16'' \qquad \underline{\text{ok}}$$

8.12 BALANCED LOADING

A column normally fails by either tension or compression. In between the two extremes lies the so-called balanced load condition where the failure may be of either type. In Chapter 4 the term "balanced section" was used in referring to a section whose compression concrete strain reached 0.003 at the same time as the tensile steel reached its yield strain at f_y/E_s. In a beam this situation theoretically occurred when the steel percentage equaled ρ_b.

For columns the definition of balanced loading is the same as it was for beams—that is, a column that has a strain of 0.003 on its compression side at the same time that its tensile steel on the other side has a strain of f_y/E_s. Though it is easily possible to prevent a balanced condition in beams by limiting the maximum steel percentage to $0.75p_b$, such is not the case for columns. For columns it is not possible to prevent sudden compression failures nor balanced failures. For every column there is a balanced loading situation where an ultimate load P_{bn} placed at an eccentricity e_b will produce a moment M_{bn}, at which time the balanced strains will simultaneously be reached.

It is not difficult to obtain the value of the balanced load P_{bn} for a particular column. For a numerical example, reference is made to the column of Figure 8.8, which was previously considered in Example 8.4. This column cross section is repeated in Figure 8.10 together with a balanced strain situation. It is assumed that $f_c' = 4{,}000$ psi and $f_y = 50{,}000$ psi.

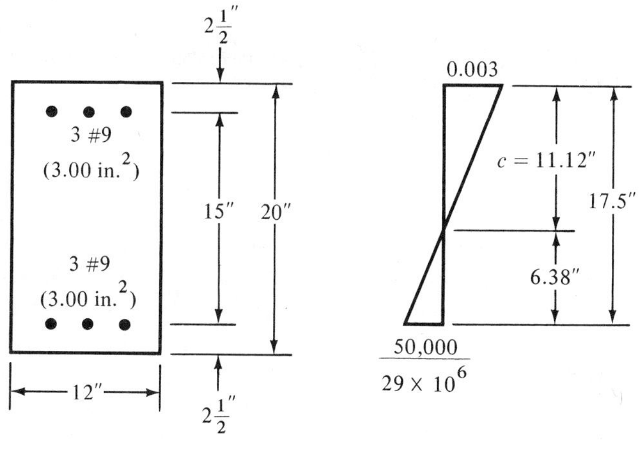

Figure 8.10

Assuming a balanced strain situation as shown in the figure, the values of c and a can be determined as follows:

$$c = \left(\frac{0.003}{0.003 + 0.00172}\right)(17.5) = 11.12''$$

$$a = (11.12)(0.85) = 9.45''$$

Similarly, the strain in the compression steel is

$$\varepsilon_s' = \left(\frac{11.12 - 2.50}{11.12}\right)(0.003) = 0.00233 > 0.00172$$

Therefore compression steel has yielded

Then the value of P_{bn} can be determined as follows:

$$P_{bn} = A_s'f_y - A_sf_y + 0.85f_c'ab$$

$$P_{bn} = (3.00)(50) - (3.00)(50) + (0.85)(4)(9.45)(12) = 385.6^k$$

The value of e_b can be determined by taking moments about the plastic centroid, where, for equilibrium, the moment of the balanced load $P_b'e_b$ must be balanced by the moment of the forces C_c, C_s', and T.

$$-P_{bn}e_b + (C_c)\left(10 - \frac{9.45}{2}\right) + (C_s')(7.5) + (T)(7.5) = 0$$

$$-(385.6)(e_b) + (0.85)(4)(12)(9.45)\left(10 - \frac{9.45}{2}\right)$$

$$+ (3)(50)(7.5) + (3)(50)(7.5) = 0$$

$$e_b = 11.11''$$

$$M_{bn} = (385.6)(11.11) = 4284''^k = 357 \text{ ft-k}$$

It will be noted that if P_n is greater than P_{bn} (or if $e < e_b$), the ultimate capacity of a column is controlled by compression.

8.13 SUMMARY

Several loading situations that can occur in compression members have been presented in the last few sections. These included axial load only, axial load and bending with a compression failure, axial load and bending with yielding of the tensile steel, and a balanced loading situation. Actually, another situation can occur when the eccentricity is so small that the entire cross section is in compression. For such cases it is necessary to determine the strain in the steels, determine their stresses if less than f_y, and then write an expression for P_n as follows:

$$P_n = 0.85f_c'bh + A_sf_s + A_s'f_s'$$

For each of the several cases considered, it is possible to write an equation in terms of f_c', f_y, a, b, and so on, which can be solved directly for P_n. Although these equations would expedite the P_n solutions, space is not taken to derive them here because graphs provide a much quicker approach. Such graphs are presented and used extensively in Chapter 9.

It will be noted that in this chapter P_n values were obtained only for rectangular tied columns. The same theory could be used for round columns but the mathematics would be somewhat complicated because of the circular layout of the bars, and the calculations of distances would be rather tedious. Several approximate methods have been developed that

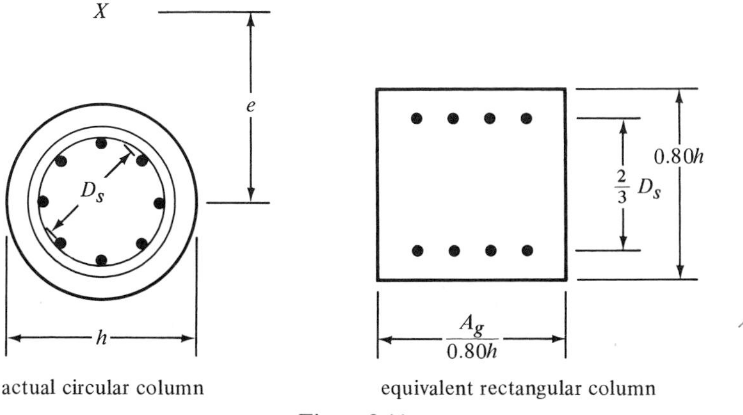

Figure 8.11

greatly simplify the mathematics. Perhaps the best known of these is the one proposed by Whitney, in which equivalent rectangular columns are used to replace the circular ones.[5] This method gives results that correspond quite closely with test results.

In Whitney's method the area of the equivalent column is made equal to the area of the actual circular column and its depth in the direction of bending is 0.80 times the outside diameter of the real column. One-half the steel is assumed to be placed on one side of the equivalent column and one-half on the other. The distance between these two areas of steel is assumed to equal two-thirds of the diameter (D_s) of a circle passing through the center of the bars in the real column. These values are illustrated in Figure 8.11. Once the equivalent column is established, the calculations for P_n and other values are made as for rectangular columns.

PROBLEMS

8.1 to **8.6** Design columns for axial load only. Include the design of ties or spirals and a sketch of the cross sections selected, including bar arrangements. All columns are assumed to be short and are not exposed to the weather.

8.1 Square tied column; $P_D = 200^k$, $P_L = 300^k$, $f_c' = 4{,}000$ psi, and $f_y = 60{,}000$ psi. Assume $\rho_g = 2\%$.

8.2 Repeat Problem 8.1 if ρ_g is to be 5%.

8.3 Round spiral column; $P_D = 160^k$, $P_L = 250^k$, $f_c' = 3{,}500$ psi, and $f_y = 50{,}000$ psi. Assume $\rho_g = 3\%$.

8.4 Round spiral column; $P_D = 200^k$, $P_L = 300^k$, $f_c' = 5{,}000$ psi, $f_y = 60{,}000$ psi, and $\rho_g = 4\%$.

[5] Whitney, Charles S., 1942, "Plastic Theory of Reinforced Concrete Design," *Transactions ASCE*, 107, pp. 251–326.

8.5 Smallest possible square tied column; $P_D = 240^k$, $P_L = 300^k$, $f_c' = 3{,}000$ psi, and $f_y = 40{,}000$ psi.

8.6 Design a rectangular tied column with the long side equal to two times the length of the short side. $P_D = 240^k$, $P_L = 360^k$, $f_c' = 4{,}000$ psi, $f_y = 60{,}000$ psi, and $\rho_g = 2\%$.

8.7 to **8.9** Using statics equations determine the value of P_n for each of the short columns shown.

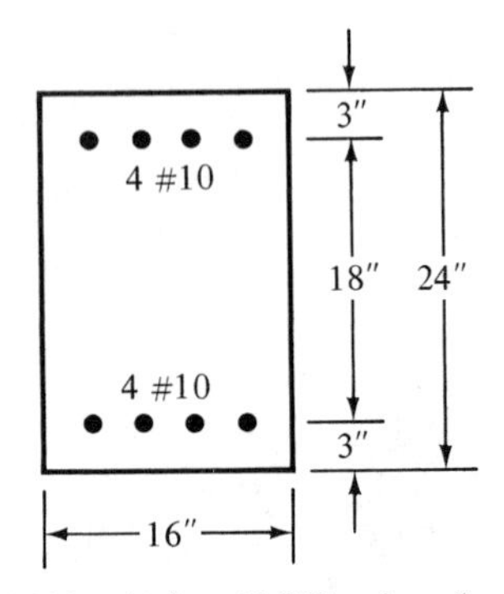

Problem 8.7 $f_c' = 4{,}000$ psi, $f_y = 40{,}000$ psi, and e_x from x-axis = 15 in.

8.8 Repeat Problem 8.7 if $e_x = 7$ in.

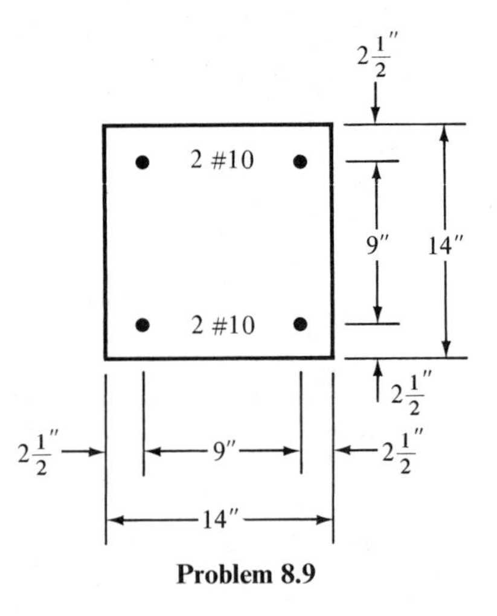

Problem 8.9

$f_c' = 3{,}000$ psi, $f_y = 50{,}000$ psi, and $e_y = 6$ in.

8.10 Assuming a balanced loading condition, determine P_{bn}, e_b, and M_{bn} for the column of Problem 8.7.

8.11 Using Whitney's equivalent rectangular column method, determine P_n for the circular column shown. $f_c' = 4{,}000$ psi, $f_y = 60{,}000$ psi, and $e = 8$ in.

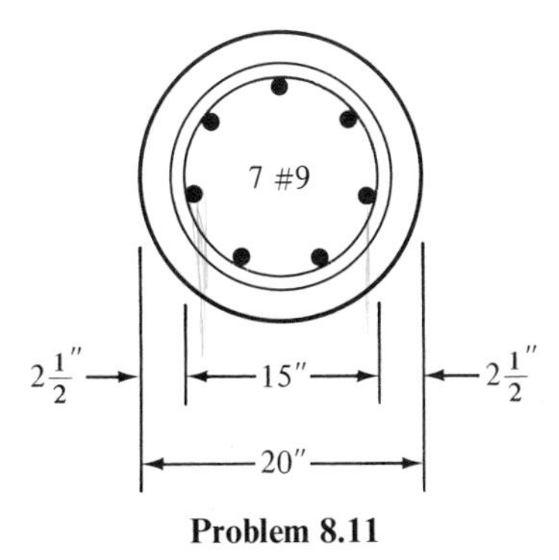

Problem 8.11

8.12 Repeat Problem 8.11 if $e = 4$ in. and 6#10 bars are used.

PROBLEMS WITH SI UNITS

8.13 to **8.14** Design columns for axial load only for the conditions described. Include the design of ties or spirals and a sketch of the cross sections selected, including bar arrangements. All columns are assumed to be short.

8.13 Square tied column; $P_D = 600$ kN, $P_L = 800$ kN, $f_c' = 20.7$ MPa, and $f_y = 344.8$ MPa. Assume $\rho = 0.02$.

8.14 Smallest possible square tied column; $P_D = 700$ kN, $P_L = 900$ kN, $f_c' = 27.6$ MPa, and $f_y = 413.7$ MPa.

8.15 Round spiral column; $P_D = 500$ kN, $P_L = 650$ kN, $f_c' = 17.2$ MPa, $f_y = 275.8$ MPa, and $\rho = 0.03$.

8.16 to **8.18** Using statics equations determine the value of P_n for each of the short columns shown. $E = 200\,000$ MPa.

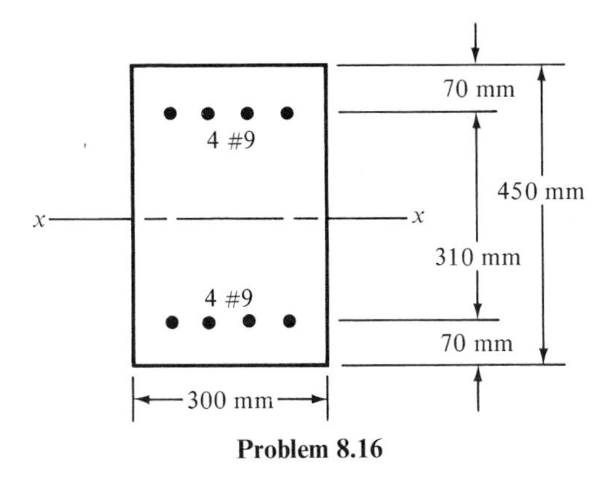

Problem 8.16

$f_c' = 27.6$ MPa, $f_y = 275.8$ MPa, and e_x from x-axis = 350 mm.

8.17 Repeat Problem 8.16 if $e_x = 150$ mm.

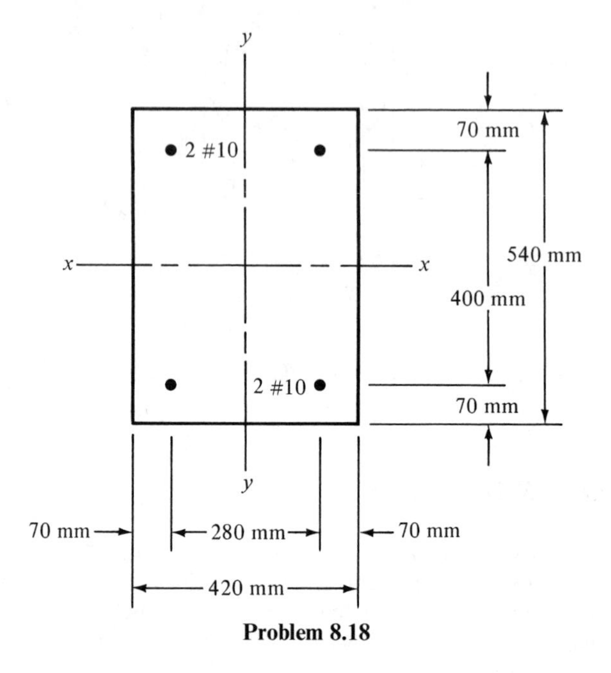

Problem 8.18

$f_c' = 27.6$ MPa, $f_y = 413.7$ MPa, and $e_y = 200$ mm.

CHAPTER 9

Design of Members Subject to Axial Load and Bending

9.1 COLUMN INTERACTION DIAGRAMS

The analysis and design of a large number of columns with the equations described in Chapter 8 would be a very time-consuming job. As a result, several organizations such as the ACI and the CRSI have developed tables or graphs with which the calculations can be greatly expedited. Graphs 2 through 9 presented in the Appendix of this text for symmetrical tied and spiral columns are examples of such design aids.

A diagram that shows the axial load capacity of a column plotted versus its moment is called a *column interaction diagram*. An example of such a diagram is shown in Figure 9.1. These diagrams are very useful for studying the strengths of columns with varying proportions of axial loads and moments. Any combination of loading that falls inside the curve is satisfactory, while any combination falling outside the curve represents failure.

An interaction diagram can be constructed for a given concrete column by making use of the equations considered in Chapter 8. Such a curve is drawn for a column as the load changes from one of a pure axial nature through varying combinations of axial loads and moments and on to pure bending. For practical use in analysis and design, an interaction curve should be constructed in nondimensional form. This can be done by dividing P_u by $\phi f_c'bh$ and by dividing M_u by $\phi f_c'bh^2$ for rectangular columns ($\phi f_c'h^2$ and $\phi f_c'h^3$, respectively, for round columns).

If a column is loaded to failure with an axial load only, the failure will occur at point A on the diagram. Moving out from point A on the curve, the axial load capacity decreases as the proportion of bending moment increases. At the very bottom of the curve, point C represents the bending strength of the member if it is subjected to moment only with no axial load present. In between the extreme points A and C, the column fails due to a combination of axial load and bending. Point B is called the balanced

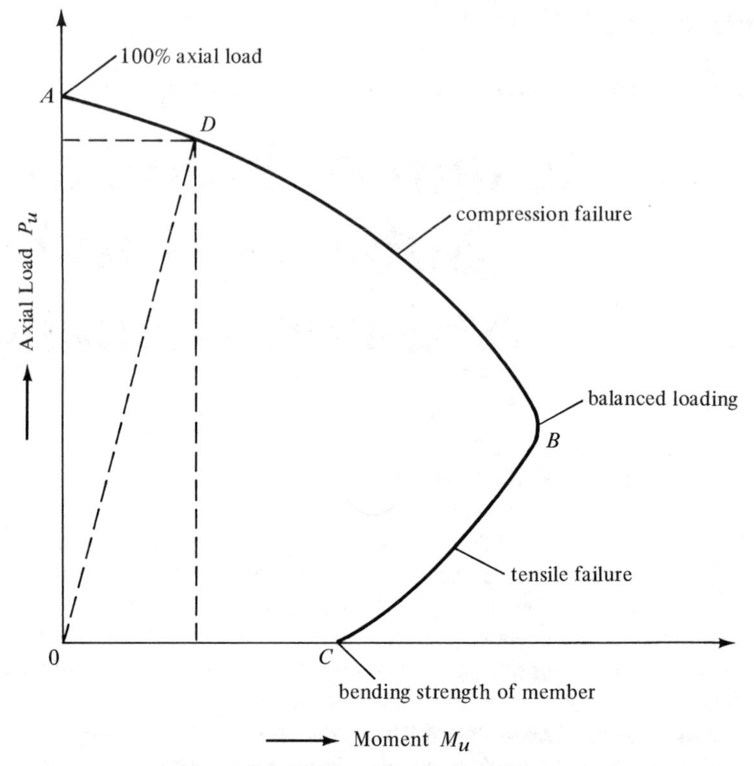

Figure 9.1 Column Interaction Diagram

point and it represents the balanced loading case where theoretically a compression failure and tensile yielding occur simultaneously.

Refer to point D on the curve. The horizontal and vertical dotted lines to this point indicate a particular combination of axial load and moment at which the column will fail. Should a radial line be drawn from point 0 to the interaction curve at any point (as to D in this case), it will represent a constant eccentricity of load, that is, a constant ratio of moment to axial load.

You may be somewhat puzzled by the shape of the lower part of the curve. From A to B on the curve the moment capacity of a section increases as the axial load decreases, but just the opposite occurs from B to C. A little thought on this point however shows the result is quite logical after all. The part of the curve from B to C represents the range of tensile failures. Any axial compressive load in that range tends to reduce the stresses in the tensile bars, with the result that a larger moment can be resisted.

At this time a further comment should be made about the ϕ values. The ACI Code specifies values of 0.70 and 0.75 for tied and spiral columns, respectively. Should a column have quite a large moment and a very small

Massive Reinforced Concrete Columns (Courtesy Bethlehem Steel Corporation).

axial loading, so that it falls on the lower part of the curve between points B and C, the use of these ϕ values will be unreasonable. For instance, for a member in pure bending (point C on the curve), the required ϕ is 0.90, but if the same member has a very small axial load added, the ϕ would immediately fall to 0.70 or 0.75. Therefore, the Code (9.3.2c) states that for members with f_y not exceeding 60,000 psi with symmetric reinforcing and with $(h - d' - d_s)/h$ not less than 0.7, the value of ϕ may be increased linearly from 0.7 or 0.75 to 0.90 as ϕP_n decreases from $0.10f_c'A_g$ to zero.

Finally, it will be noted that the graphs shown in the Appendix are for 4,000-psi concretes and 60,000-psi steels. Nevertheless, the results obtained from using them for other concretes and steels are quite satisfactory unless concretes much stronger than 4,000 psi are used. It will be remembered that β_1 decreases from 0.85 for higher-strength concretes and the particular curves given were based on 0.85 values for determining the properties of the compression stress block. Of course, graphs can be obtained for these other strength materials.

The charts (2-9) which are presented in the appendix of this text are a very small sample of the column design aids which are available to the

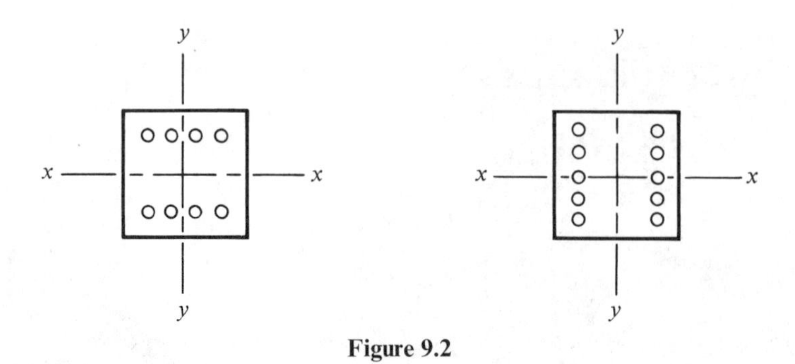

Figure 9.2

designer. These diagrams were developed using the basic statics equations for eccentrically loaded columns which were presented in Chapter 8. *It is extremely important to understand that this particular group of charts is applicable only to columns with reinforcing bars located in two faces as shown in Figure 9.2.*

If the bars are placed in all four faces of a column the designer needs a set of charts developed on that basis. Such charts are normally constructed assuming that the steel has been uniformly distributed around the column in a thin tubular shape having a cross sectional area equal to the cross sectional area of the bars. Should the columns be rectangular rather than square the total steel is usually assumed to be placed one quarter in each face for constructing the curves.

For the numerical examples presented in this chapter an approximate procedure is used when steel is placed in all four faces of a column. If bending occurs about the x axis the bars darkened in Figure 9.3(a) are the only ones considered in using the appendix charts. For bending about the y axis the bars darkened in part (b) of the figure are the only ones considered. *In actual practice it is necessary to use a set of charts or tables developed on the basis of the correct bar arrangements.* The previously

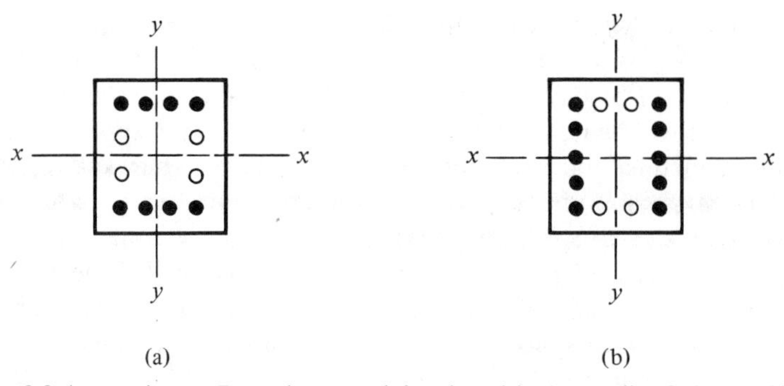

Figure 9.3 Approximate Procedure used herein with Appendix Column Charts. (a) For ρ_x only darkened bars considered; (b) for ρ_y only darkened bars considered.

mentioned *CRSI Handbook* and the *ACI Design Handbook* are sources of such information.

9.2 EXAMPLE REVIEW PROBLEMS WITH CHARTS

Examples 9.1 and 9.2 illustrate the determination of P_n values for two columns using the interaction curves given in the Appendix. To enter the curves for analysis, it is necessary to compute the values e/h and $\rho\mu$, where ρ is the percentage of steel obtained by dividing the steel area by the gross area of the column and where $\mu = f_y/(0.85f_c')$.

Entering the curves with these values, it is possible to pick the value $(P_n e)/(f_c' bh^2)$, which is also equal to $(P_u e)/(\phi f_c' bh^2)$ or to $K'(e/h)$. Once this value is obtained, the expression may be solved for P_n or P_u, as desired.

EXAMPLE 9.1

Using the graphs in the Appendix, determine the value of P_n for the tied column shown in Figure 9.4. $f_c' = 4{,}000$ psi and $f_y = 60{,}000$ psi.

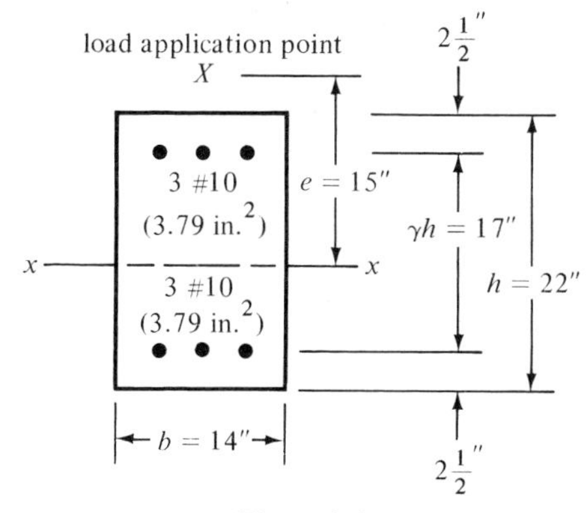

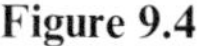

Figure 9.4

SOLUTION

$$\gamma = \tfrac{17}{22} = 0.77$$

Therefore, we must interpolate between Graphs 3 and 4.

$$\frac{e}{h} = \frac{15}{22} = 0.68$$

$$\rho = \frac{(2)(3.79)}{(14)(22)} = 0.0246$$

$$\mu = \frac{f_y}{0.85f_c'} = \frac{60}{(0.85)(4)} = 17.65$$

$$\rho\mu = (0.0246)(17.65) = 0.434$$

From graphs (interpolating between $\gamma = 0.80$ and 0.70)

γ	0.80	0.77	0.70
$\frac{P_n e}{f_c' b h^2}$	0.247	0.242	0.229

$$P_n = \frac{(0.242)(4)(14)(22)^2}{15} = 437.3$$

EXAMPLE 9.2

Using the graphs in the Appendix, determine the value of P_n for the spiral column shown in Figure 9.5. $f_c' = 3{,}500$ psi and $f_y = 50{,}000$ psi.

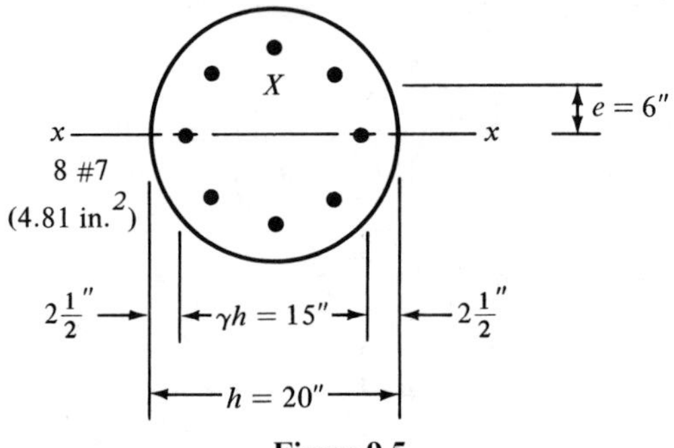

Figure 9.5

SOLUTION

$$\gamma = \tfrac{15}{20} = 0.75$$

Therefore, we must interpolate between Graphs 7 and 8.

$$\frac{e}{h} = \frac{6}{20} = 0.30$$

$$\rho = \frac{4.81}{314} = 0.0153$$

$$\mu = \frac{f_y}{0.85 f_c'} = \frac{50}{(0.85)(3.5)} = 16.81$$

$$\rho\mu = (0.0153)(16.81) = 0.257$$

From graphs

γ	0.80	0.75	0.70
$\frac{P_n e}{f_c' h^3}$	0.103	0.100	0.097

$$P_n = \frac{(0.100)(3.5)(20)^3}{6} = \underline{466.7^{\text{k}}}$$

Another comment should be made concerning the use of these graphs, although it does not affect the solution of the preceding examples. For tensile failures below the balanced points on the curves, a reduction in the axial load would reduce the moment capacity. As a result, the designer may have to check two loading situations for columns in this range. Not only would he have to check the column for the largest P_n and M_n values but also he would have to check for the smallest axial load value that could occur at the same time as the maximum M_n.

9.3 EXAMPLE DESIGN PROBLEMS WITH CHARTS

Example 9.3 illustrates the application of the interaction curves to the design of a member subject to axial load and bending about one axis. The procedure used is very close to the one used for the analysis of such members except that instead of determining P_n from the curves, the percentage of steel required is determined. This means that to enter the curves, the values of e/h and $(P_ne)/f_c'bh^2$ are calculated and the value $\rho\mu$ selected from the curves.

There are several methods available for estimating column sizes, but a trial and error method is about as good as any. The designer estimates the column size and then determines the steel percentage required for that column from the interaction curves. If he feels the ρ determined is not reasonable, he can quickly try another column size and redetermine the ρ required.

EXAMPLE 9.3

The 12 × 20 tied column of Figure 9.6 is to be used to support the following: $P_D = 100^k$, $P_L = 140^k$, $M_D = 60$ ft-k, and $M_L = 80$ ft-k. If $f_c' = 4{,}000$ psi and $f_y = 60{,}000$ psi, select the reinforcing bars to be used by referring to the interaction curves.

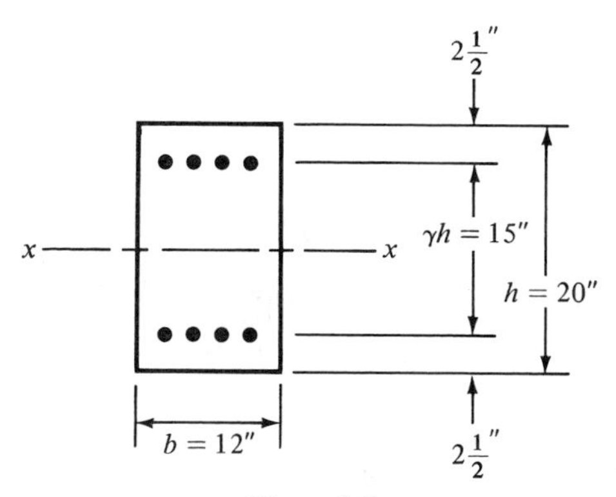

Figure 9.6

SOLUTION

$$P_u = (1.4)(100) + (1.7)(140) = 378^k$$

$$P_n = \frac{378}{0.70} = 540^k$$

$$M_u = (1.4)(60) + (1.7)(80) = 220 \text{ ft-k}$$

$$M_n = \frac{220}{0.70} = 314.3 \text{ ft-k}$$

$$e = \frac{(12)(314.3)}{540} = 6.98''$$

$$\frac{e}{h} = \frac{6.98}{20} = 0.349$$

$$\frac{P_n e}{f_c' b h^2} = \frac{(540)(6.98)}{(4)(12)(20)^2} = 0.196$$

$$\gamma = \frac{15}{20} = 0.75$$

From graphs

γ	0.80	0.75	0.70
$\rho\mu$	0.38	0.405	0.43

$$\mu = \frac{f_y}{0.85 f_c'} = \frac{60}{(0.85)(4)} = 17.65$$

$$\rho\mu = 0.405$$

$$\rho = \frac{0.405}{17.65} = 0.0229$$

$$A_s = (0.0229)(12 \times 20) = 5.50 \text{ in.}^2$$

use 6 #9 = 6.00 in.^{2}

9.4 SLENDERNESS EFFECTS

Section 10.10.1 of the Code states that the design of a compression member should desirably be based on a theoretical analysis of the structure that takes into account the effects of axial loads, moments, deflections, duration of loads, varying member sizes, end conditions, and so on. If such a theoretical procedure is not used, the Code (10.11.5) provides an

Clemson University Library, Clemson, S.C.
(Courtesy Clemson University Communications Center).

approximate method for determining slenderness effects. This method, which is based on the factors just mentioned for an "exact" analysis, results in a moment magnifier δ, which is to be multiplied by the larger moment (later called M_2) at the end of the column and that value δM_2 used in design. If bending occurs about both axes, δ is to be computed separately for each direction and the values obtained multiplied by the respective M_2 values.

Several items involved in the calculation of δ are discussed in the next several paragraphs. These include unsupported column lengths, effective length factors, radii of gyration, and the ACI Code requirements.

Unsupported Lengths

The length used for calculating the slenderness ratio of a column l_u is its unsupported length. This length is considered to be equal to the clear distance between slabs, beams, or other members that provide lateral support to the column. If haunches or capitals are present, the clear distance is measured from the bottom of the capitals or haunches.

Effective Length Factors

To calculate the slenderness ratio of a particular column, it is necessary to estimate its effective length. This is the distance between points of zero moment in the column. For this initial discussion it is assumed that no sidesway or joint translation is possible. Sidesway or joint translation means that one or both ends of a column can move laterally with respect to each other.

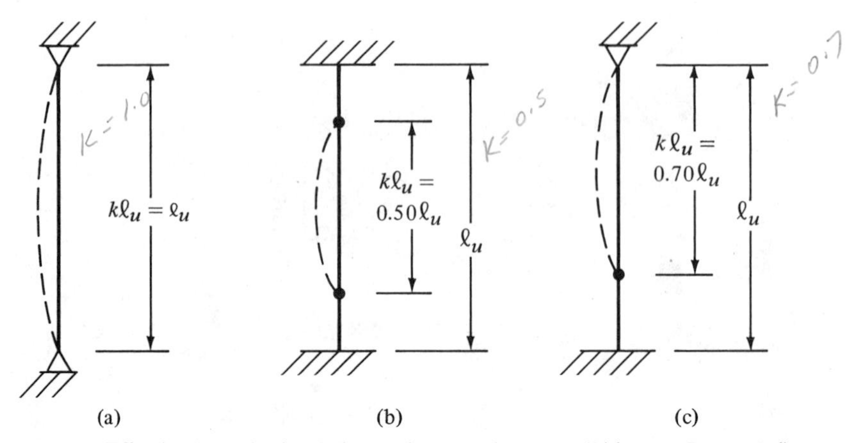

Figure 9.7 Effective Lengths for Columns in Braced Frames (Sidesway Prevented)

If there were such a thing as a perfectly pinned end column, its effective length would be its unsupported length, as shown in Figure 9.7(a). The *effective length factor k* is the number which must be multiplied by the column's unsupported length to obtain its effective length. For a perfectly pinned end column, $k = 1.0$.

Columns with different end conditions have entirely different effective lengths. For instance, if there were such a thing as a perfectly fixed end column, its points of inflection (or points of zero moment) would occur at its one-fourth points, and its effective length would be $l_u/2$, as shown in Figure 9.7(b). As a result, its k value would equal 0.50.

Obviously, the smaller the effective length of a particular column, the smaller its danger of buckling and the greater its load-carrying capacity. In Figure 9.7(c) is shown a column with one end fixed and one end pinned. The k factor for this column is theoretically 0.70.

Reinforced concrete columns serve as parts of frames, and these frames are sometimes *braced* and sometimes *unbraced*. A braced frame is one for which sidesway or joint translation is prevented by means of bracing, shear walls, or lateral support from adjoining structures. An unbraced frame does not have any of these types of bracing supplied and must depend on the stiffness of its own members to prevent lateral buckling. For braced frames k values can never be greater than 1.0 but for unbraced frames the k values will always be greater than 1.0 because of sidesway.

An example of an unbraced column is shown in Figure 9.8(a). The base of this particular column is assumed to be fixed while its upper end is assumed to be completely free to both rotate and translate. The elastic curve of such a column will take the shape of the elastic curve of a pinned end column of twice its length. Its effective length will therefore equal $2l_u$, as shown in the figure. In Figure 9.8(b) another unbraced column case is illustrated.

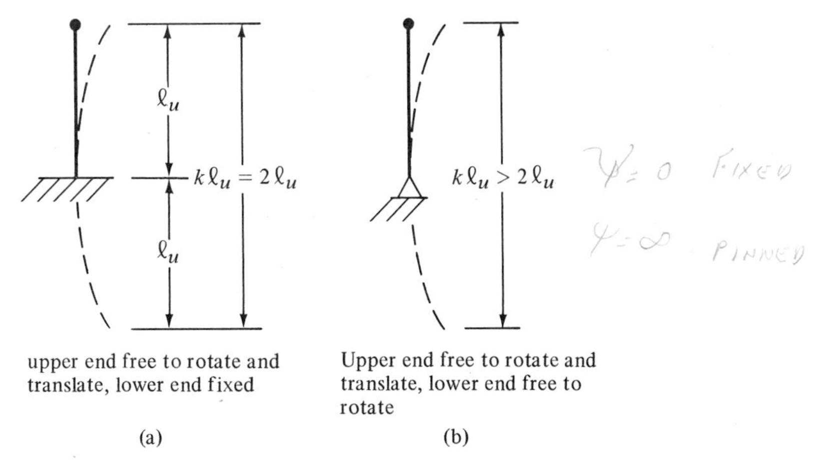

upper end free to rotate and translate, lower end fixed

(a)

Upper end free to rotate and translate, lower end free to rotate

(b)

Figure 9.8 Columns for Unbraced Frames

Alignment Charts

A method often used for estimating effective lengths involves the use of the alignment charts shown in Figures 9.9 and 9.10.[1,2] The first of these charts is applicable to braced frames where the bracing members or shear walls supposedly provide a stiffness EI estimated to be at least equal to six times the stiffness of all the columns on that story. Should such bracing not be provided, the columns are assumed to be free to translate and the second chart for unbraced frames is assumed to be applicable.

To use the alignment charts for a particular column, ψ factors are computed at each end of the column. The ψ factor at one end of the column equals the sum of the stiffnesses ($\sum EI$) of the columns meeting at that joint, including the column in question, divided by the sum of all the stiffnesses of the beams meeting at the joint. Should one end of the column be pinned, ψ is theoretically equal to ∞, and if fixed, $\psi = 0$. As a perfectly fixed end is practically impossible to have, ψ is usually taken as 1.0 instead of 0 for assumed fixed ends.

One of the two ψ values is called ψ_A and the other ψ_B. After these values are computed, the effective length factor k is obtained by placing a straightedge between ψ_A and ψ_B. The point where the straightedge crosses the middle monograph is k.

It can be seen that the ψ factors used to enter the alignment charts and thus the resulting effective length factors are dependent on the relative stiffnesses of the compression and flexural members. This brings up the

[1] Johnston, B. G., editor, 1976, *Guide to Stability Design for Metal Structures*, 3rd ed. (New York: Wiley), p. 420.

[2] Julian, O. G., and Lawrence, L. S., 1959, "Notes on J and L Nomograms for Determination of Effective Lengths," unpublished.

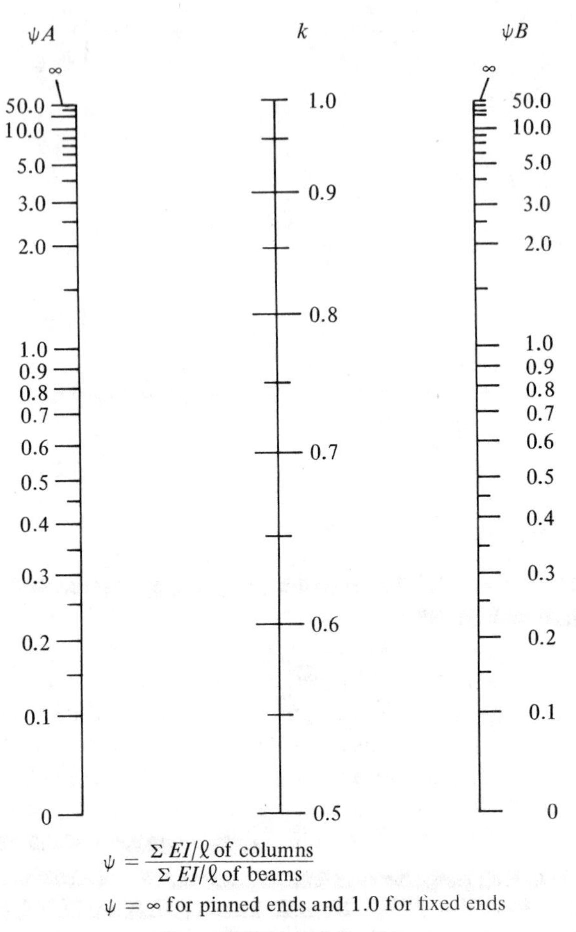

$$\psi = \frac{\Sigma\, EI/\ell \text{ of columns}}{\Sigma\, EI/\ell \text{ of beams}}$$

$\psi = \infty$ for pinned ends and 1.0 for fixed ends

Figure 9.9 Braced Frames

question as to what moments of inertia should be used in determining the ψ values. One acceptable practice (according to Section 10.11.2 of the ACI Commentary) is to calculate the rigidity of the flexural members on the basis of the transformed moments of inertia and the rigidity of the compression members with Equation 9.3 given in the next section of this chapter using $\beta_d = 0$. Another acceptable practice is to use gross moments of inertia for the columns and 50% of gross moments of inertia for the flexural members.

Radii of Gyration

The radius of gyration is equal to 0.25 times its diameter and 0.289 times the dimension of a rectangular column in the direction being considered. The ACI Code (10.11.3) permits the approximate value 0.30 to be used in

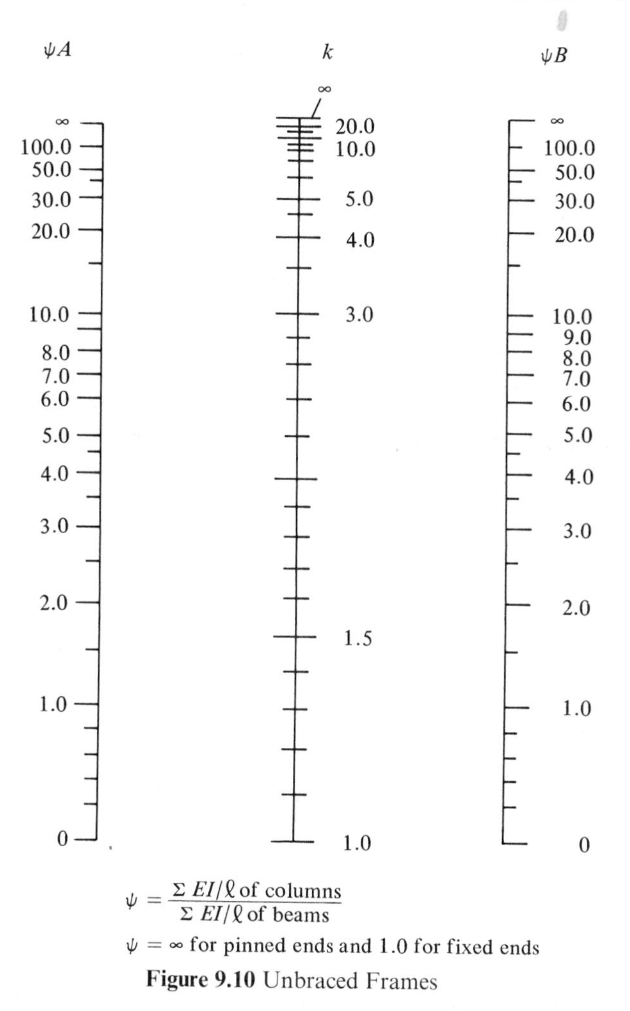

$$\psi = \frac{\Sigma\, EI/\ell \text{ of columns}}{\Sigma\, EI/\ell \text{ of beams}}$$

$\psi = \infty$ for pinned ends and 1.0 for fixed ends

Figure 9.10 Unbraced Frames

place of 0.289 and this is done herein. For other sections the values of r can be computed from the properties of the gross section.

Comments on Slenderness

The Code (10.11.2.1) states that the effective length factor is to be taken as 1.0 for compression members in frames braced against sidesway unless a theoretical analysis shows that a lesser value can be used. Should the member be in a frame not braced against sidesway, the Code states that the value of k is to be larger than 1.0 and is to be determined with proper consideration given to the effects of cracking and reinforcing on the column stiffness. ACI-ASCE Committee 441 suggests that it is not realistic to assume that k will be less than 1.2 for such columns, and therefore it seems logical to make preliminary designs with k equal to or larger than that value.

It has been found that as long as slenderness ratios do not exceed certain values, the reduction in column strengths due to slenderness will be less than 5%. It is the intent of the ACI Code to permit the design of columns in this range as short columns. If $(kl_u)/r \leq 22$ for columns in unbraced frames, or is less than $34 - 12(M_1/M_2)$ for columns in braced frames, slenderness effects may be neglected (ACI Code 10.11.4).

With reference to the ratio given for braced frames, the moments at the ends of a column are referred to as M_1 and M_2. M_2 is the larger of the two moments and is always considered to have a positive sign. The smaller moment M_1 is considered to have a positive sign if the member is bent in single curvature and negative if bent in double curvature. You can see that if the moments cause single curvature, the column will be in a worse situation as to the danger of buckling than if they cause double curvature (where the moments tend to oppose each other). Should a column be subjected to transverse loads so that the largest calculated moments occur somewhere other than at the column ends, that moment should be used as M_2. For this situation $C_m = 1.0$, according to the Code (10.11.5.3).

In computing the moment magnifier, a moment coefficient C_m is used. This coefficient is greatly affected by the magnitudes and directions of M_1 and M_2.

$$C_m = 0.6 + 0.4\frac{M_1}{M_2} \geq 0.4 \tag{9.1}$$

Should there be no calculated moments at the ends of the columns, the values of M_1/M_2 and C_m are to be taken as 1.0. If the factored column moments are zero or close to it the value of M_2 for slender columns should be based on the minimum eccentricity described in Section 10.11.5.4 of the Code. It is not necessary to apply these eccentricities to both axes simultaneously.

9.5 MOMENT MAGNIFIER

The calculation of the so-called moment magnifier δ involves several different items. These include the following:

1. $E_c = 57{,}000\sqrt{f_c'}$.
2. I_g = gross inertia of the column cross section about the centroidal axis being considered.
3. $E_s = 29 \times 10^6$ psi.
4. I_{se} = moment of inertia of reinforcing about the centroidal axis of the section.

5. β_d = ratio of maximum design dead load moment to the maximum design total load moment. This value is always assumed to have a positive value.

Once these values are determined, EI can be determined as follows:

$$EI = \frac{(E_c I_g/5) + E_s I_{se}}{1 + \beta_d} \tag{9.2}$$

or more conservatively as

$$EI = \frac{E_c I_g/2.5}{1 + \beta_d} \tag{9.3}$$

The Euler buckling load enters into the calculation and equals

$$P_c = \frac{\pi^2 EI}{(kl_u)^2} \tag{9.4}$$

Finally, the moment magnifier δ can be computed as follows:

$$\delta = \frac{C_m}{1 - (P_u/\phi P_c)} \geq 1.0 \tag{9.5}$$

Example 9.4 illustrates the design of a long column using this approximate procedure from the ACI Code.

It will be noted that if $(kl_u)/r$ is larger than 100, an exact analysis must always be considered (Code 10.11.4).

EXAMPLE 9.4

For the column shown in Figure 9.11, f_y = 60,000 psi, f_c' = 4,000 psi, $P_D = 60^k$, $P_L = 100^k$, M_D = 120 ft-k, and M_L = 140 ft-k. The frame is braced against side sway, the column is bent in single curvature about the x-axis, and the moments are equal at each end of the member. Is the

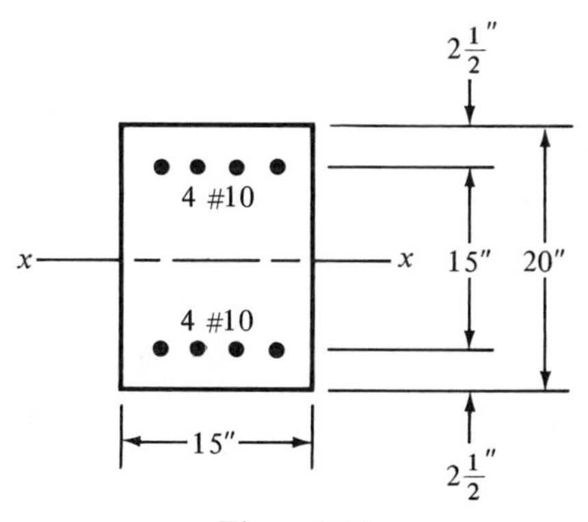

Figure 9.11

member satisfactory if:

(a) $l_u = 10$ feet?
(b) $l_u = 20$ feet?

If not, calculate δ and determine the steel area required.

SOLUTION

$$P_u = (1.4)(60) + (1.7)(100) = 254^{\text{k}}$$
$$M_u = (1.4)(120) + (1.7)(140) = 406 \text{ ft-k}$$
$$e = \frac{(12)(406)}{254} = 19.18''$$
$$\frac{e}{h} = \frac{19.18}{20} = 0.959$$
$$\rho = \frac{10.12}{300} = 0.0337$$
$$\mu = \frac{60}{(0.85)(4)} = 17.65$$
$$\rho\mu = (0.0337)(17.65) = 0.595$$

From the charts by interpolation, $P_u = 255^{\text{k}}$ with no consideration given to slenderness.

(a) $l_u = 10$ ft

$$r = (0.30)(20) = 6.00''$$

$k = 1.0$ since frame is braced against side sway

$$\frac{kl_u}{r} = \frac{(1)(12 \times 10)}{6} = 20$$

Noting that M_1/M_2 is positive for single curvature,

$$34 - 12(1) = 22 > 20$$

Slenderness effect can be neglected and member is satisfactory.

(b) $l_u = 20' \, 0''$

$$\frac{kl_u}{r} = \frac{(1)(12 \times 20)}{6} = 40$$

$$34 - (12)(1) = 22 < 40$$

Slenderness must be considered.

$$C_m = 0.60 + 0.4(1.0) = 1.0$$
$$E_c = 57{,}000\sqrt{4{,}000} = 3{,}605 \text{ ksi}$$
$$I_g = (\tfrac{1}{12})(15)(20)^3 = 10{,}000 \text{ in.}^4$$

$$\beta_d = \frac{(1.4)(120)}{(1.4)(120) + (1.7)(140)} = 0.414$$

$$I_{se} = (2)(5.06)(7.5)^2 = 569 \text{ in.}^4$$

$$EI = \frac{(E_c I_g/5) + E_s I_{se}}{1 + \beta_d} = \frac{[(3{,}605)(10{,}000)/5] + (29 \times 10^3)(569)}{1 + 0.414}$$

$$EI = 16{,}768{,}741 k\text{-in.}^2$$

$$P_c = \frac{\pi^2 EI}{(kl_u)^2} = \frac{(\pi)^2(16{,}768{,}741)}{(12 \times 20)^2} = 2873$$

$$\delta = \frac{C_m}{1 - (P_u/\phi P_c)} = \frac{1.0}{1 - [254/(0.7)(2873)]} = 1.14$$

Determine steel area required with a moment of δM_n

$$P_n = \frac{254}{0.70} = 362.9^{\text{k}}$$

$$M_u = (1.4)(120) + (1.7)(140) = 406 \text{ ft-k}$$

$$M_n = \frac{406}{0.7} = 580 \text{ ft-k}$$

$$\delta M_n = (1.14)(580) = 661.2 \text{ ft-k}$$

Entering design curves

$$\gamma = \tfrac{15}{20} = 0.75$$

$$e = \frac{(12)(661.2)}{362.9} = 21.86''$$

$$\frac{e}{h} = \frac{21.86}{20} = 1.093$$

$$\frac{P_n e}{f_c' b h^2} = \frac{(362.9)(21.86)}{(4)(15)(20)^2} = 0.331$$

$$\mu = \frac{60}{(0.85)(4)} = 17.65$$

$$\rho\mu \text{ from charts by interpolation} = 0.735$$

$$\rho = \frac{0.735}{17.65} = 0.0416$$

$$A_s = (0.0416)(15)(20) = 12.48 \text{ in.}^2$$

use 8 #11 bars

Example 9.5 is presented to show the trial and error calculations involved in the design of a long column. The column size is roughly estimated as well as δ. With the graphs and an assumed percentage of steel, a better column size is predicted, the value of δ calculated, and the required percentage of steel determined. The design can then be refined a little more by repeating the calculations again if so desired.

EXAMPLE 9.5

Design a 24-ft-long square tied column bent in single curvature in a braced frame with ψ values assumed to be equal to 1.0 at each end. $P_D = 180^k$. $P_L = 200^k$, $M_D = 100$ ft-k, $M_L = 130$ ft-k, $f_y = 60{,}000$ psi, and $f_c' = 4{,}000$ psi. Use a ρ of about 0.02.

SOLUTION

$$P_u = (1.4)(180) + (1.7)(200) = 592^k$$

$$M_u = (1.4)(100) + (1.7)(130) = 361 \text{ ft-k}$$

$$e = \frac{(12)(361)}{592} = 7.32''$$

Trial size

Assume an 18-in. × 18-in. column.

$$r = (0.30)(18) = 5.4''$$

$$k = 0.77 \text{ (from alignment chart of Figure 9.9)}$$

$$\frac{kl_u}{r} = \frac{(0.77)(12 \times 24)}{5.4} = 41.07 > 34 - 12 = 22 \text{ Therefore, it is a long column.}$$

$$\mu = \frac{60}{(0.85)(4)} = 17.65$$

$$\rho\mu = (0.02)(17.65) = 0.353$$

The moment will be magnified as it is a long column. Assume $\delta = 1.20$ and multiply by M_u or by e.

$$\delta\frac{e}{h} = (1.20)\left(\frac{7.32}{18}\right) = 0.488$$

For use of the graphs assume

$$\gamma = \frac{18 - (2)(2.5)}{18} = 0.722$$

By interpolation from left side of Graphs 3 and 4,

$$\frac{P_u}{\phi f_c' bh} = 0.43$$

From which

$$bh = h^2 = \frac{592}{(0.7)(4)(0.43)} = 22.17''$$

Try 22-in. × 22-in. column

$$\frac{kl_u}{r} = \frac{(0.77)(12 \times 24)}{(0.3)(22)} = 33.6 > 22 \quad \underline{\text{Therefore, it is a long column.}}$$

$$C_m = 0.60 + (0.4)(1.0) = 1.0$$

$$E_c = 57{,}000\sqrt{4{,}000} = 3{,}605{,}000 \text{ psi} = 3{,}605 \text{ ksi}$$

$$I_g = (\tfrac{1}{12})(22)(22)^3 = 19{,}521 \text{ in.}^4$$

$$\beta_d = \frac{(1.4)(100)}{(1.4)(100) + (1.7)(130)} = 0.388$$

As the steel area is not known yet, the conservative value of EI (which does not involve the reinforcing) from the ACI Code (Section 10.11.5.2) is used.

$$EI = \frac{E_c I_g/2.5}{1 + \beta_d} = \frac{[(3{,}605)(19{,}521)/2.5]}{1 + 0.388} = 20{,}280{,}463 \text{ k-in.}^2$$

$$P_c = \frac{(\pi)^2(20{,}280{,}463)}{(0.77 \times 12 \times 24)^2} = 4070^{\text{k}}$$

$$\delta = \frac{1.0}{1 - [592/(0.70)(4070)]} = 1.26$$

Return to graphs with $P_u = 592^{\text{k}}$ and $\delta M_u = (1.26)(361) = 454.9'^{\text{k}}$

$$e = \frac{(12)(454.9)}{592} = 9.22''$$

$$\frac{e}{h} = \frac{9.22}{22} = 0.419$$

$$\frac{P_u e}{\phi f_c' b h^2} = \frac{(592)(9.22)}{(0.70)(4)(22)(22)^2} = 0.183$$

$$\gamma = \frac{22 - (2)(2.5)}{22} = 0.772$$

By interpolation between Graphs 3 and 4, $\rho\mu = 0.288$.

$$\rho = \frac{0.288}{17.65} = 0.0163$$

$$A_s = (0.0163)(22 \times 22) = 7.89 \text{ in.}^2 \qquad \underline{\text{say 8 \#9}}$$

Note: The designer could now repeat this design using the more precise value of EI, which in turn will slightly change δ and thus e/h and so on, but space is not taken here for such calculations.

The ACI Commentary 10.11.7 provides a far simpler method (called the "Modified R Method") for handling slender columns. As the results obtained are of equivalent accuracy with those given by the very tedious method just described it is quite popular with the design profession. To determine the capacity of a particular column the value of a strength reduction factor R is calculated with a simple equation and is multiplied by the short column capacity. This method is not considered to be suitable for columns which are pinned at both ends because the results are unconservative.

9.6. BIAXIAL BENDING

Many columns are subjected to biaxial bending, that is, bending about both axes. Corner columns in buildings where beams and girders frame into the columns from both directions are the most common cases but there are others, such as where columns are cast monolithically as part of frames in both directions or where columns are supporting heavy spandrel beams.

Circular columns have polar symmetry and thus the same ultimate capacity in all directions. The design process is the same, therefore, regardless of the directions of the moments. If there is bending about both the x- and y-axes, the biaxial moment can be computed by combining the two moments or their eccentricities as follows:

$$M_u = \sqrt{(M_{ux})^2 + (M_{uy})^2} \tag{9.6}$$

or

$$e = \sqrt{(e_x)^2 + (e_y)^2} \tag{9.7}$$

For shapes other than circular ones, it is necessary to consider the three-dimensional interaction effects.

You might quite logically think that you could determine P_n for a biaxially loaded column by using static equations, as was done in Examples 8.3 and 8.4 of Chapter 8. Such a procedure will lead to the correct answer but the mathematics involved is so complicated due to the shape of the compression side of the column that the method is not a practical one. Nevertheless, a few comments are made about this type of solution and reference is made to Figure 9.12.

An assumed location is selected for the neutral axis and the appropriate strain triangles drawn as shown in the figure. The usual equations are written with $C_c = 0.85f_c'$ times the shaded area A_c and with each bar having a force equal to its cross-sectional area times its stress. The solution

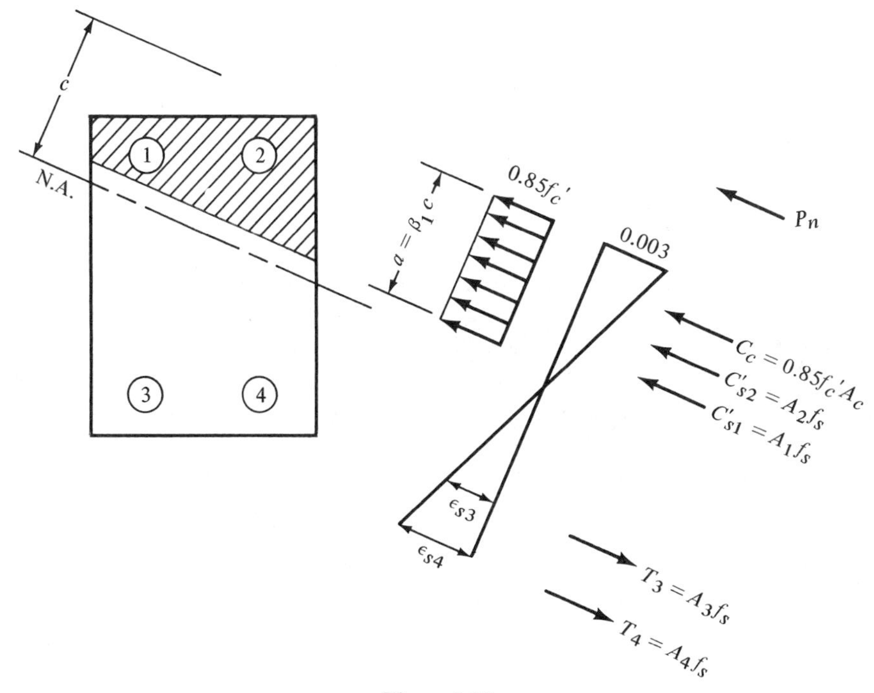

Figure 9.12

of the equation yields the load that would establish that neutral axis—but the designer usually starts with certain loads and eccentricities and he does not know the neutral axis location and, furthermore, the neutral axis is probably not even perpendicular to the resultant $e = \sqrt{(e_x)^2 + (e_y)^2}$.

For column shapes other than circular ones, it is desirable to consider three-dimensional interaction curves such as the one shown in Figure 9.13. In this figure the curve labeled M_{nxo} represents the interaction curve if bending occurs about the x-axis only, while the one labeled M_{nyo} is the one if bending occurs about the y-axis only.

In this figure, for a constant P_n, the cross-hatched plane shown represents the contour of M_n for bending about any axis.

The Code does not provide empirical equations for the design of columns subject to biaxial bending. There is available, however, for such cases a so-called reciprocal interaction equation developed by Professor Boris Bresler of the University of California at Berkeley that provides satisfactory results.[3] In this equation, which follows, P_{no} is the pure axial load capacity of the section normally taken as $0.85f_c'A_g + A_sf_y$, P_{nx} is

[3] Bresler, B., 1960, "Design Criteria for Reinforced Concrete Columns under Axial Load and Biaxial Bending," *Journal ACI*, 57, p. 481.

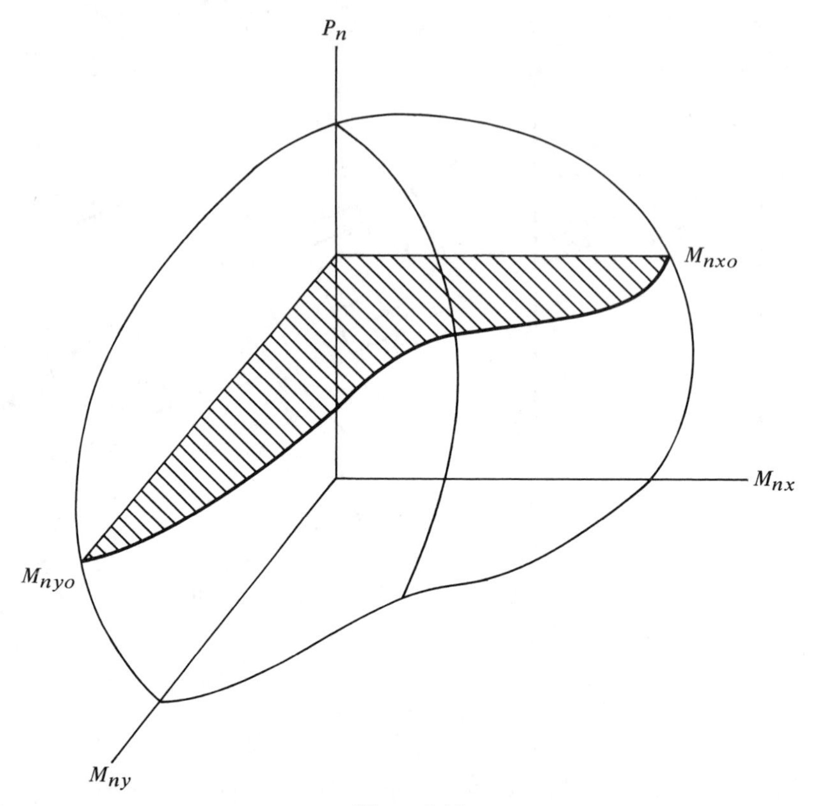

Figure 9.13

the axial load capacity when the load is placed at an eccentricity e_x with $e_y = 0$, while P_{ny} is the axial load capacity at e_y with $e_x = 0$.

$$\frac{P_n}{P_{nx}} + \frac{P_n}{P_{ny}} - \frac{P_n}{P_{no}} = 1.0 \tag{9.8}$$

The Bresler equation works rather well as long as P_n is at least as large as $0.10P_{no}$. Should P_n be less than $0.10P_{no}$, it is satisfactory to neglect the axial force completely and design the section as a member subject to biaxial bending only. This procedure is a little on the conservative side. For this lower part of the interaction curve, it will be remembered that a little axial load increases the moment capacity of the section. The Bresler equation does not apply to axial tension loads. Professor Bresler found that the ultimate loads predicted by his equation for the conditions described do not vary from test results by more than 10%.

Example 9.6 illustrates the use of the reciprocal theorem for the analysis of a column subjected to biaxial bending. The procedure for calculating P_{nx} and P_{ny} is the same as the one used for the prior examples of this chapter. Note, however, that in calculating the percentage of steel for the

x-axis only, the bars in the top and bottom rows are considered (six bars). Similarly, in calculating ρ for the y-axis only, the bars in the left column and the right column are considered (four bars).

EXAMPLE 9.6

Determine the permissible load capacity P_u of the short tied column shown in Figure 9.14, which is subjected to biaxial bending. $f_c' = 3{,}500$ psi, $f_y = 60{,}000$ psi, $e_x = 16$ in., and $e_y = 8$ in.

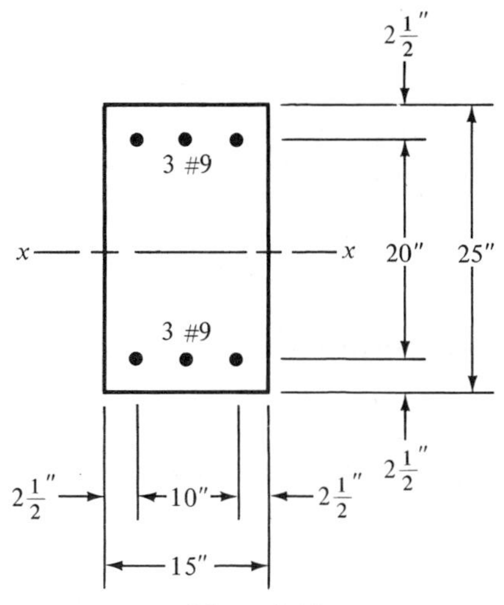

Figure 9.14

SOLUTION

For bending about x axis

$$\gamma = \tfrac{20}{25} = 0.80$$

$$\frac{e}{h} = \frac{16}{25} = 0.64$$

$$\rho = \frac{(6)}{(15)(25)} = 0.016$$

$$\mu = \frac{f_y}{0.85 f_c'} = \frac{60}{(0.85)(3.5)} = 20.17$$

$$\rho\mu = (0.016)(20.17) = 0.323$$

$$\frac{P_{nx} e_x}{f_c' b h^2} \text{ from chart} = 0.212$$

$$P_{nx} = \frac{(3.5)(15)(25)^2(0.212)}{16} = 434.8^{k}$$

For bending about *y*-axis

$$\gamma = \tfrac{10}{15} = 0.667$$

$$\frac{e}{h} = \frac{8}{15} = 0.533$$

$$\rho = \frac{4}{(15)(25)} = 0.0107$$

$$\mu = \frac{60}{(0.85)(3.5)} = 20.17$$

$$\rho\mu = (0.0107)(20.17) = 0.216$$

$$\frac{P_{ny}e_y}{f_c'bh^2} \text{ from chart} = 0.160 \text{ (after interpolation)}$$

$$P_{ny} = \frac{(3.5)(25)(15)^2(0.160)}{8} = 393.8^{\text{k}}$$

Determining axial load capacity of section

$$P_{no} = (0.85)(15 \times 25)(3.5) + (60)(6) = 1475.6^{\text{k}}$$

Using the Bresler expression to determine P_n

$$\frac{P_n}{P_{nx}} + \frac{P_n}{P_{ny}} - \frac{P_n}{P_{no}} = 1.0$$

$$\frac{P_n}{434.8} + \frac{P_n}{393.8} - \frac{P_n}{1475.6} = 1.0$$

$$3.39P_n + 3.75P_n - P_n = 1475.6$$

$$\underline{P_n = 240.3^{\text{k}}}$$

If the moments in the weak direction (y-axis here) are rather small as compared to bending in the strong direction (x-axis), it is rather common to neglect the smaller moment. This practice is probably reasonable as long as e_y is less than about 20% of e_x since the Bresler expression will show little reduction for P_n. For the example just solved, an e_y equal to 50% of e_x caused the axial load capacity to be reduced by 44.7%.

Examples 9.7 and 9.8 illustrate the design of columns subject to biaxial bending. The Bresler expression, which is of little use in the proportioning of such members, is used to check the capacities of the sections selected by some other procedure. Exact theoretical designs of columns subject to biaxial bending are very complicated and, as a result, are seldom made in design offices. They are either proportioned by approximate methods or with "canned" computer programs.

During the past few decades several approximate methods have been introduced for the design of columns with biaxial moments. For instance, there are available quite a few design charts with which satisfactory designs may be made. The problems are reduced to very simple calculations in which coefficients are taken from the charts and used to magnify the moments about a single axis. Designs are then made with the regular uniaxial design charts.[4,5,6]

Not only are charts used but also some designers use rules of thumb for making initial designs. One very simple method (it's not too good) involves the following steps: (1) the selection of the reinforcement required in the x direction considering P_n and M_{nx}, (2) the selection of the reinforcement required in the y direction considering P_n and M_{ny}, and (3) the determination of the total column steel area required by adding the areas obtained in steps (1) and (2). This method on occasions will result in large design errors on the unsafe side because the strength of the concrete is counted twice, once for the x direction and once for the y direction.[7]

Another approximate procedure that works fairly well for design office calculations is used for the last two examples of this chapter (Examples 9.7 and 9.8). If this simple method is applied to square columns, the values of both M_{nx} and M_{ny} are assumed to act about both the x-axis and the y-axis (i.e., $M_x = M_y = M_{nx} + M_{ny}$). The steel is selected about one of the axes and is spread around the column, and the Bresler expression is used to check the ultimate load capacity of the eccentrically loaded column.

Should a rectangular section be used where the y-axis is the weaker direction, it would seem logical to calculate $M_y = M_{nx} + M_{ny}$ and to use that moment to select the steel required about the y-axis and spread the computed steel area over the whole column cross section. Although such a procedure will produce safe designs, the resulting columns may be rather uneconomical because they will often be much too strong about the strong axis. A fairly satisfactory approximation is to calculate $M_y = M_{nx} + M_{ny}$ and multiply it by b/h, and with that moment, design the column about the weaker axis.[8]

Example 9.7 illustrates the design of a short square column subject to biaxial bending, while Example 9.8 illustrates the design of a short rectangular column subject to biaxial bending. The approximate method described in the last two paragraphs is used for both problems and the Bresler expression is used for checking the results. If either of these

[4] Parme, A. L.; Nieves, J. M.; and Gouwens, A., 1966, "Capacity of Reinforced Rectangular Columns Subject to Biaxial Bending," *Journal ACI*, 63, no. 11, pp. 911–923.

[5] Weber, D. C., 1966, "Ultimate Strength Design Charts for Columns with Biaxial Bending," *Journal ACI*, 63, no. 11, pp. 1205–1230.

[6] Row, D. G., and Paulay, T., 1973, "Biaxial Flexure and Axial Load Interaction in Short Reinforced Concrete Columns," *Bulletin of New Zealand Society for Earthquake Engineering*, 6, no. 2, pp. 110–121.

[7] Park, R., and Paulay, T., 1975, *Reinforced Concrete Structures* (New York: Wiley), pp. 158–159.

[8] Fintel, M., ed., 1974, *Handbook of Concrete Engineering* (New York: Van Nostrand), pp. 27–28.

columns had been a long column, it would have been necessary to magnify the design moments for slenderness effects regardless of the design method used.

EXAMPLE 9.7

Select the reinforcing needed for the short square tied column shown in Figure 9.15 for the following: $P_D = 100^k$, $P_L = 140^k$, $M_{DX} = 50$ ft-k, $M_{LX} = 70$ ft-k, $M_{DY} = 40$ ft-k, and $M_{LX} = 60$ ft-k. $f_C' = 4{,}000$ psi and $f_Y = 60{,}000$ psi.

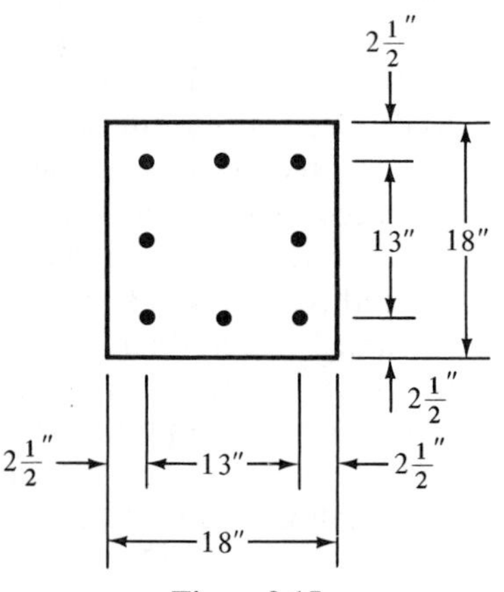

Figure 9.15

SOLUTION

Computing design values

$$P_u = (1.4)(1{,}100) + (1.7)(140) = 378^k$$

$$\frac{P_u}{f_c' A_g} = \frac{378}{(4)(324)} = 0.292 > 0.10 \qquad \text{(See page 205)}$$

Therefore, $\phi = 0.70$.

$$P_n = \frac{378}{0.70} = 540^k$$

$$M_{ux} = (1.4)(50) + (1.7)(70) = 189 \text{ ft-k}$$

$$M_{nx} = \frac{189}{0.70} = 270 \text{ ft-k}$$

$$M_{uy} = (1.4)(40) + (1.7)(60) = 158 \text{ ft-k}$$

$$M_{ny} = \frac{158}{0.70} = 225.7 \text{ ft-k}$$

As a result of biaxial bending, the design moment about the x- or y-axis is assumed to equal $M_{nx} + M_{ny} = 270 + 225.7 = 495.7$ ft-k

Determining steel required

$$e_x = e_y = \frac{(12)(495.7)}{540} = 11.02''$$

$$\gamma = \frac{13}{18} = 0.722$$

$$\frac{e}{h} = \frac{11.02}{18} = 0.612$$

$$\frac{P_n e}{f_c' b h^2} = \frac{(540)(11.02)}{(4)(18)(18)^2} = 0.255$$

$$\mu = \frac{60}{(0.85)(4)} = 17.65$$

$$\rho\mu \text{ (by interpolation)} = 0.55$$

$$\rho_t = \frac{0.55}{17.65} = 0.0312$$

$$A_{s\,\text{total}} = (0.0312)(18)^2 = 10.11 \text{ in.}^2$$

Use 8 #11 (12.50 in.^{2}), as shown in Figure 9.15.

A review of the column with the Bresler expression gives a nominal $P_n = 575.2^k > 540^k$, which is satisfactory.

EXAMPLE 9.8

Select the reinforcing required for the short rectangular tied column shown in Figure 9.16 for the following: $P_n = 600^k$, $M_{nx} = 300$ ft-k, and $M_{ny} = 250$ ft-k. $f_c' = 4{,}000$ psi and $f_y = 60{,}000$ psi.

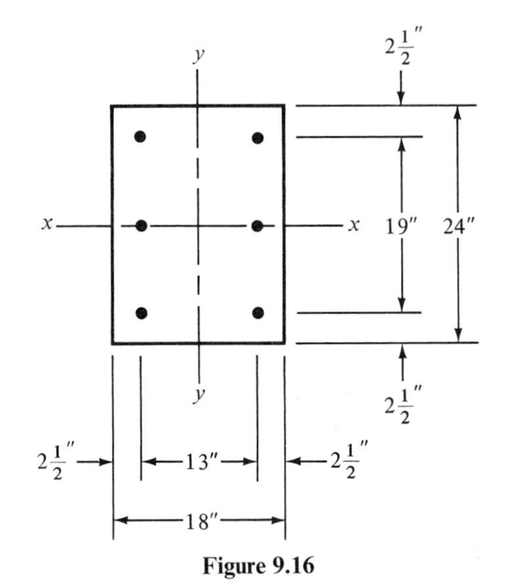

Figure 9.16

SOLUTION

Designing the column about the minor axis for $P_n = 600^k$ and for $M_y = (b/h)/(M_{nx} + M_{ny})$,

$$M_y = \tfrac{18}{24}(300 + 250) = 412.5 \text{ ft-k}$$

$$e_y = \frac{(12)(412.5)}{600} = 8.25''$$

$$\gamma = \tfrac{13}{18} = 0.722$$

$$\frac{e}{h} = \frac{8.25}{18} = 0.458$$

$$\frac{P_n e}{f_c' b h^2} = \frac{(600)(8.25)}{(4)(24)(18)^2} = 0.159$$

$$\mu = \frac{60}{(0.85)(4)} = 17.65$$

$$\rho\mu \text{ (by interpolation)} = 0.186$$

$$\rho_t = \frac{0.186}{17.65} = 0.0105$$

$$A_{s\,\text{total}} = (0.0105)(18)(24) = 4.54 \text{ in.}^2$$

Use 6 #8 (4.71 in.^{2}), as shown in Figure 9.16.

A review of the column with the Bresler expression gives a permissible $P_n = 622^k > 600^k$, which is satisfactory.

When a beam is subjected to biaxial bending the following approximate interaction equation may be used for design purposes

$$\frac{M_x}{M_{ux}} + \frac{M_y}{M_{uy}} \leq 1.0$$

In this expression M_x and M_y are the design moments while M_{ux} is the permissible moment capacity of the section if bending occurs about the x axis only and M_{uy} is the permissible moment capacity if bending occurs about the y axis only. This same expression may be satisfactorily used for axially loaded members if the design axial load is about 15% or less of the axial load capacity of the section. For a detailed discussion of this subject you are referred to the *Handbook of Concrete Engineering.*[9]

[9] Fintel, M., ed. 1974, New York, Van Nostrand, pp. 34–36.

PROBLEMS

9.1 to **9.7** Using the interaction curves in the Appendix, determine P_n values for the short columns shown if $f_c' = 4{,}000$ psi and $f_y = 60{,}000$ psi.

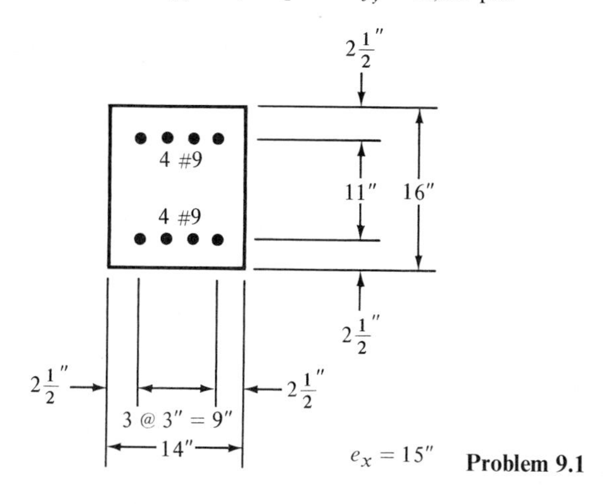

Problem 9.1

9.2 Repeat Problem 9.1 if $e_x = 6$ in.

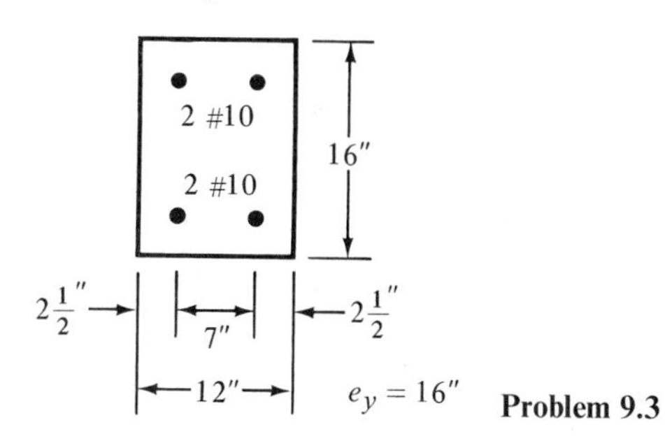

Problem 9.3

9.4 Repeat Problem 9.3 if $e_y = 8$ in.

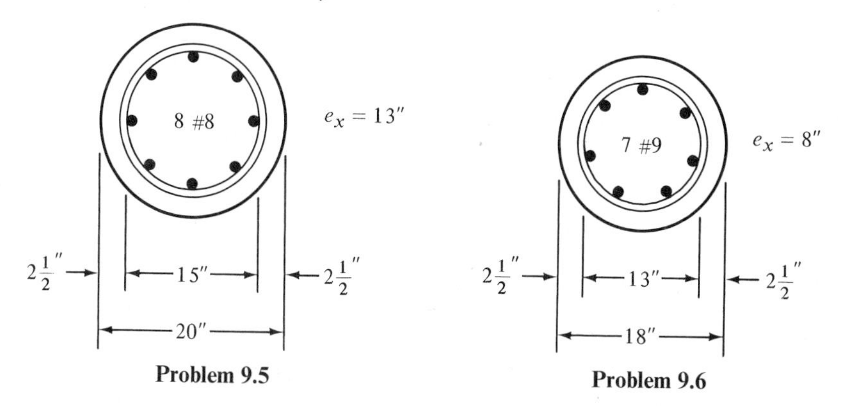

Problem 9.5

Problem 9.6

9.7 Repeat Problem 9.6 if $e_x = 8$ in. and $e_y = 6$ in.

9.8 The column of Problem 9.1 is 18 ft long, is bent in single curvature, and is a part of a braced frame and has a k factor of 1.0. Is it satisfactory to support a P_n of 200^k and an M_{nx} of 150 ft-k? Assume $\beta_d = 0.40$.

9.9 The column of Problem 9.1 is 15 ft long, is a part of an unbraced frame, and has a k factor of 1.3. Can it safely support a P_n of 250^k and an M_{ny} of 125 ft-k? Assume $\beta_d = 0.45$ and $e_x = 0$.

9.10 to **9.16** Select reinforcing for the short columns shown. $f_c' = 3{,}000$ psi and $f_y = 50{,}000$ psi.

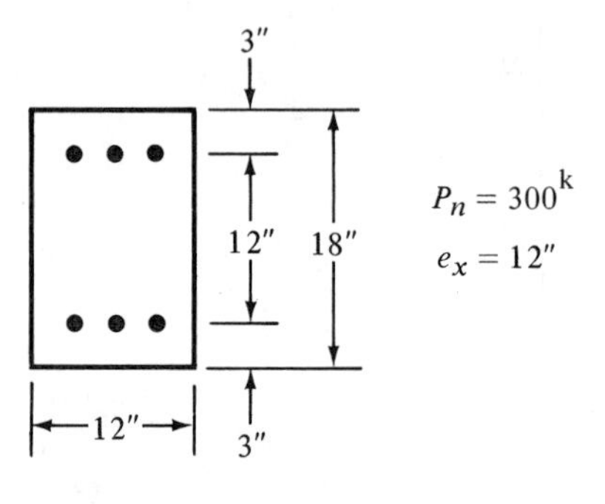

Problem 9.10

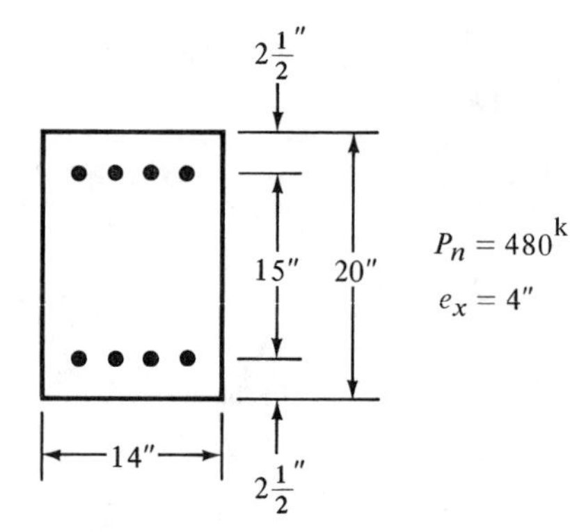

Problem 9.11

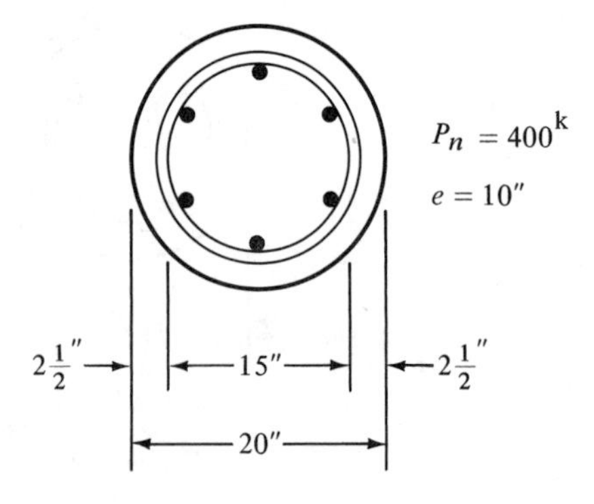

Problem 9.12

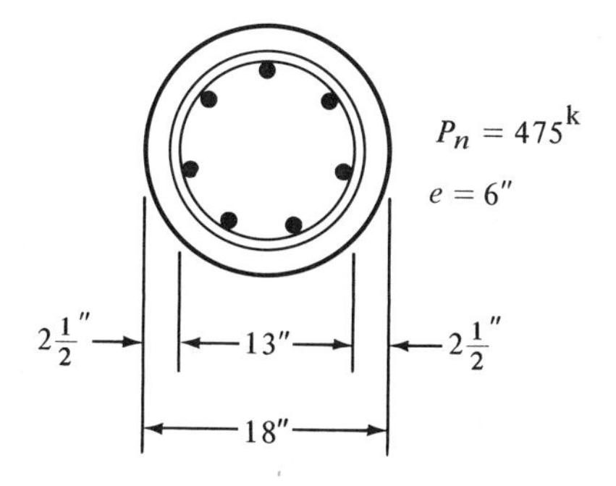

Problem 9.13

9.14 Repeat Problem 9.13 if $e = 4$ in.

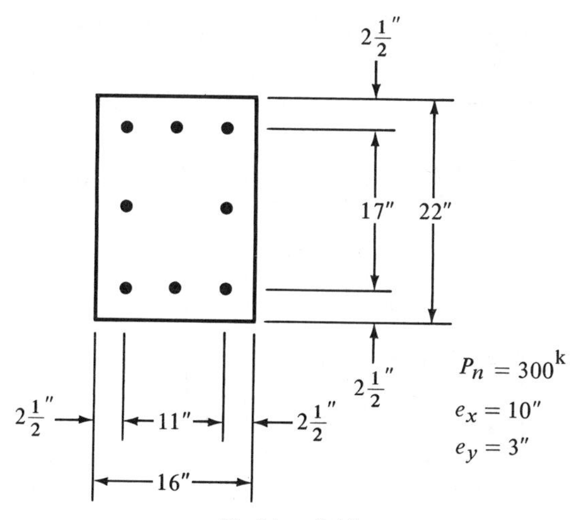

Problem 9.15

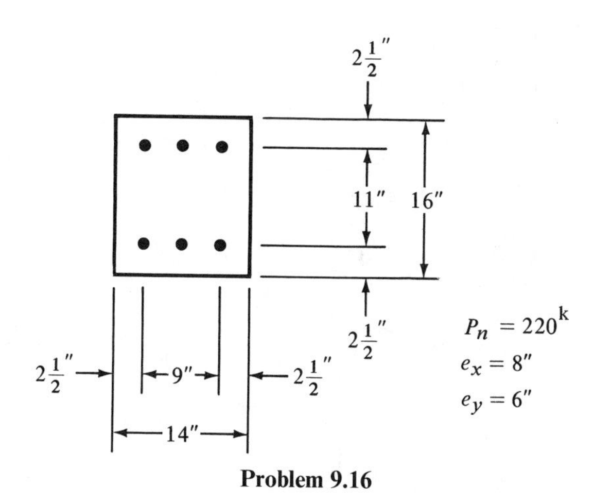

Problem 9.16

PROBLEMS WITH SI UNITS

9.17 to **9.19** Using the interaction curves in the Appendix, determine P_n values for the short columns shown if $f_c' = 20.7$ MPa and $f_y = 344.8$ MPa.

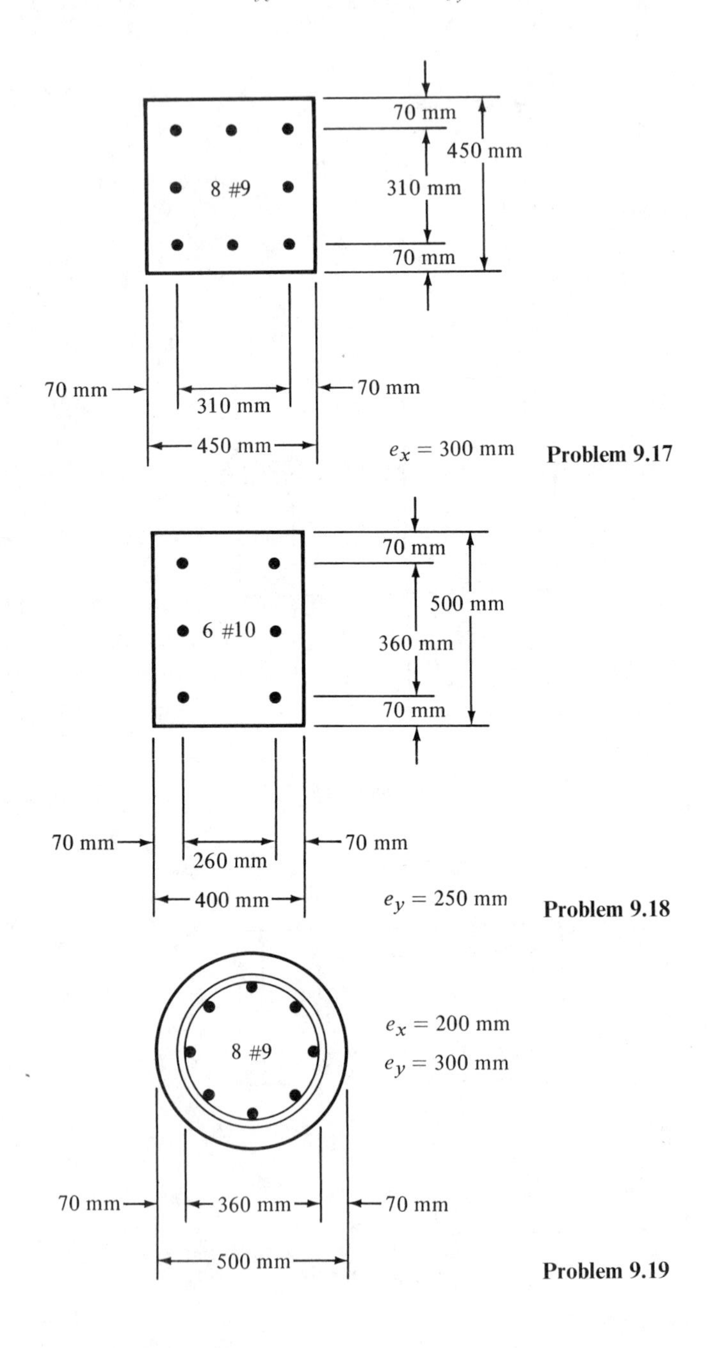

Problem 9.17

Problem 9.18

Problem 9.19

9.20 to **9.22** Select reinforcing for the short columns shown if $f_c' = 27.6$ MPa and $f_y = 413.7$ MPa.

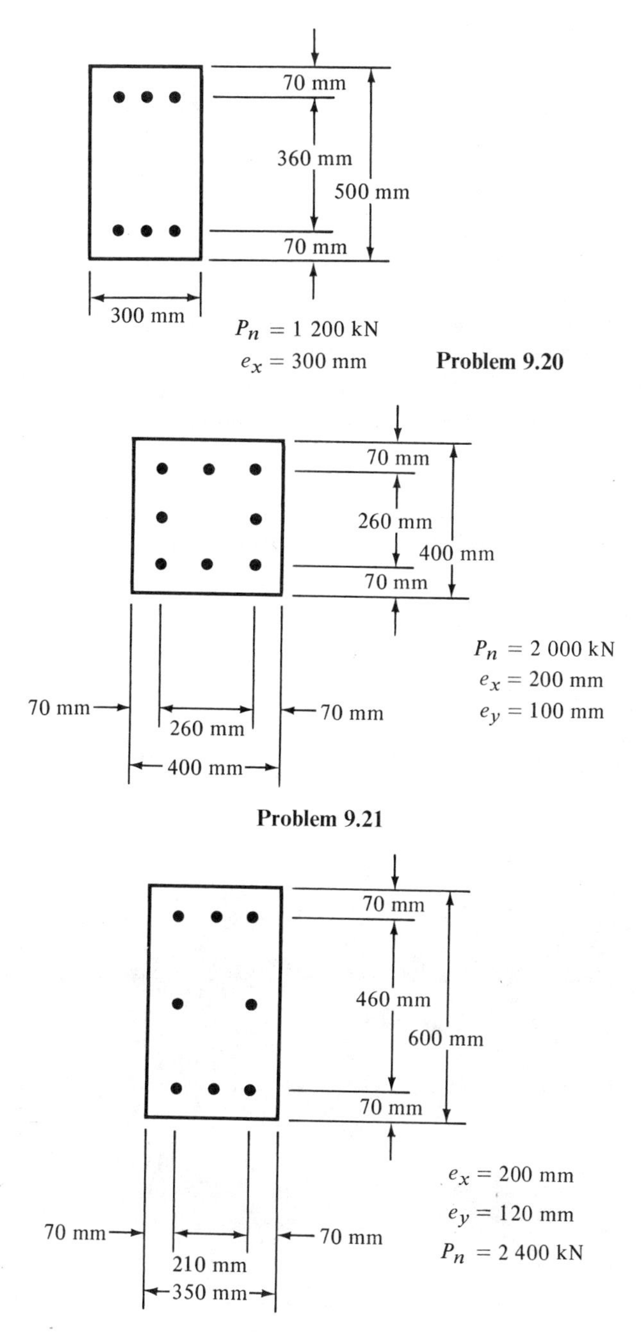

Problem 9.20

Problem 9.21

Problem 9.22

CHAPTER 10

Footings

10.1 INTRODUCTION

Footings are those structural members used to support columns and walls and transmit their load to the underlying soils. Reinforced concrete is a material admirably suited for footings and is used as such for both reinforced concrete and structural steel buildings, bridges, towers, and other structures.

The permissible pressure on a soil beneath a footing is normally a few tons per square foot. The compressive stresses in the walls and columns of an ordinary structure may run as high as a few hundred tons per square foot. It is therefore necessary to spread these loads over sufficient soil areas to permit the soil to support the loads safely.

Construction of Foundations for Louisiana Super Dome and Convention Center (Courtesy United States Steel).

Not only is it desired to transfer the superstructure loads to the soil beneath in a manner that will prevent excessive or uneven settlements and rotations but it is also necessary to provide sufficient resistance to sliding and overturning.

To accomplish these objectives it is necessary to transmit the supported loads to a soil of sufficient strength and there to spread it out over an area such that the unit pressure is within a reasonable range. If it is not possible to dig a short distance and find a satisfactory soil, it will be necessary to use piles or caissons to do the job. These latter subjects are not considered to be within the scope of this text.

10.2 TYPES OF FOOTINGS

Among the several types of reinforced concrete footings in common use are the wall, isolated, combined, raft, and pile cap types. These are briefly introduced in this section; the remainder of the chapter is used to provide more detailed information about the simpler types in this group.

1. *A wall footing* [Figure 10.1(a)] is simply an enlargement of the bottom of a wall that will sufficiently distribute the load to the foundation soil. Wall footings are normally used around the perimeter of a building and perhaps for some of the interior walls.

2. An *isolated* or *single-column footing* [Figure 10.1(b)] is used to support the load of a single column. These are the most commonly used footings, particularly where the loads are relatively light and the columns are not closely spaced.

3. *Combined footings* are used to support two or more column loads [Figure 10.1(c)]. A combined footing might be economical where two or more heavily loaded columns are so spaced that normally designed single-column footings would run into each other. Single-column footings are usually square or rectangular and, when used for columns located right at property lines, would extend across those lines. A footing for such a column combined with one for an interior column can be designed to fit within the property lines.

4. A *raft or mat foundation* [Figure 10.1(d)] is a continuous reinforced concrete slab over a large area used to support many columns and walls. This kind of foundation is used where soil strength is low or where column loads are large but where piles or caissons are not used. For such cases isolated footings would be so large that it is more economical to use a continuous raft or mat under the entire area. This type of footing is particularly useful in reducing differential settlements throughout the building.

5. *Pile caps* [Figure 10.1(e)] are slabs of concrete used to distribute column loads to groups of piles.

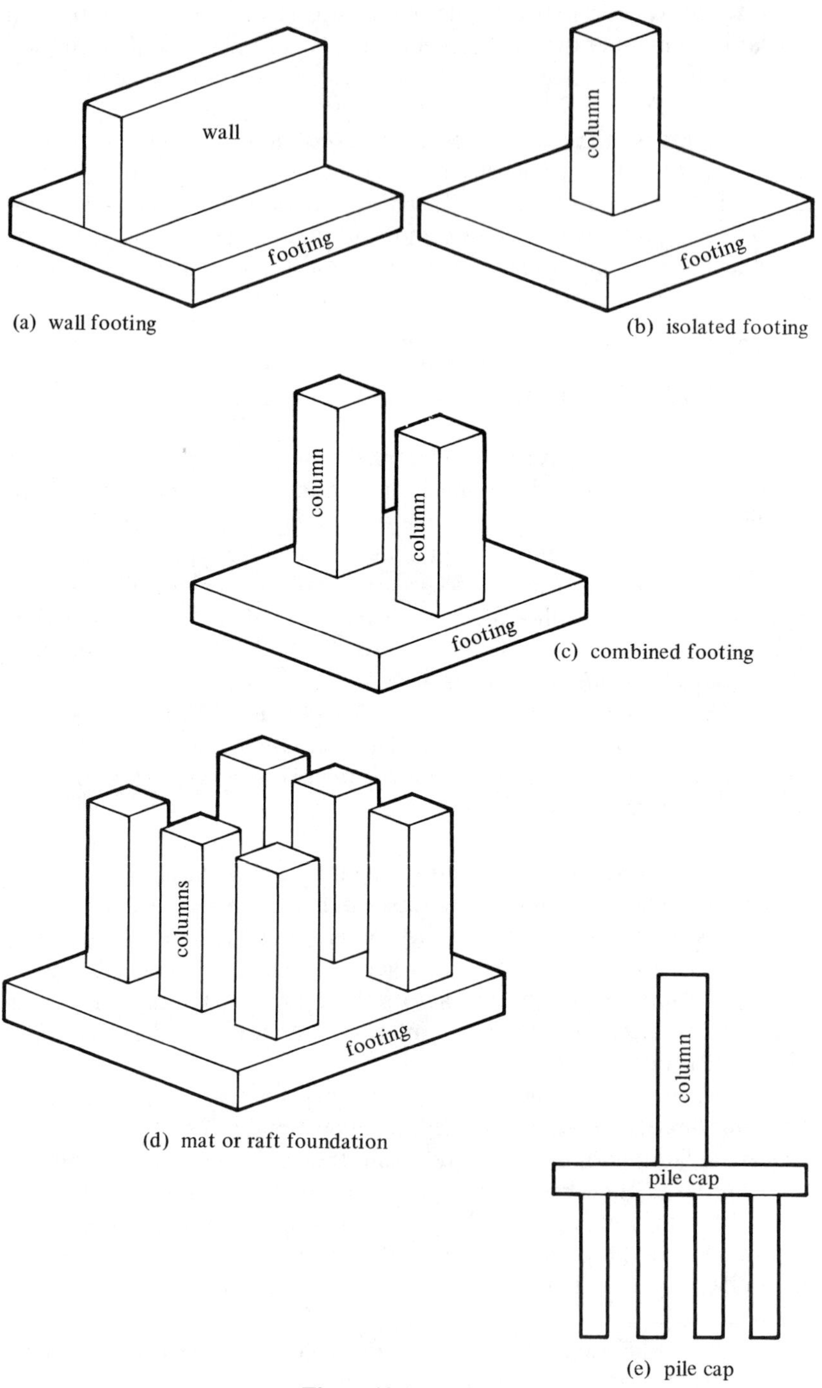

(a) wall footing

(b) isolated footing

(c) combined footing

(d) mat or raft foundation

(e) pile cap

Figure 10.1 Footings

10.3 ACTUAL SOIL PRESSURES

The soil pressure at the surface of contact between a footing and the soil is assumed to be uniformly distributed as long as the load above is applied at the center of gravity of the footing. This assumption is made even though many tests have shown that soil pressures are unevenly distributed due to variations in soil properties, footing rigidity, and other factors. A uniform-pressure assumption, however, usually provides a conservative design since the calculated shears and moments are usually larger than those that actually occur.

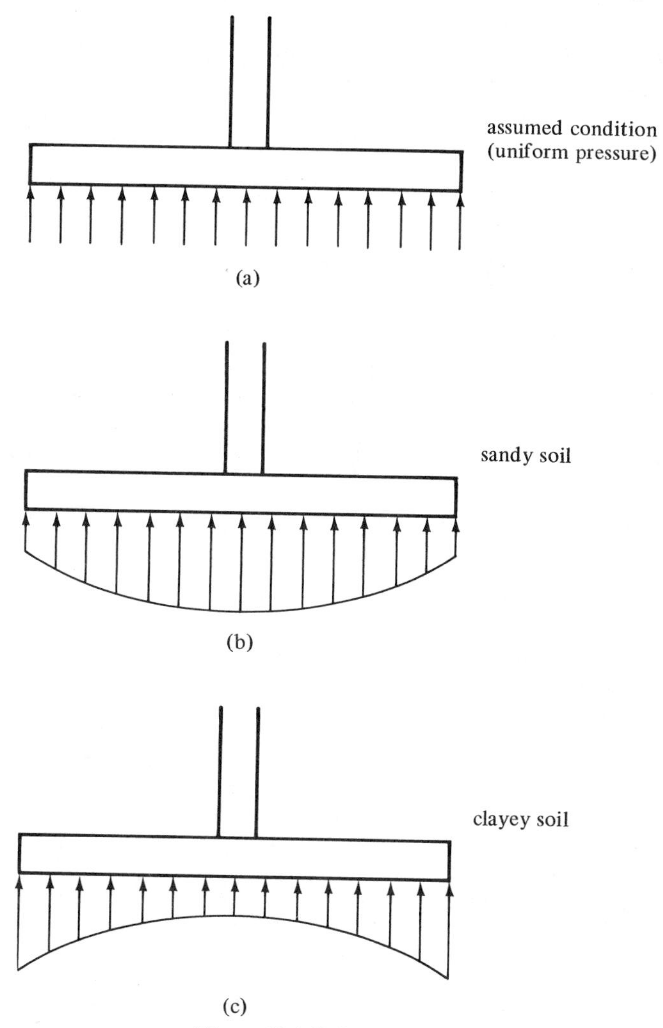

Figure 10.2 Soil Conditions

As an example of the variation of soil pressures, footings on sand and clay soils are considered. When footings are supported by sandy soils, the pressures are larger under the center of the footing and smaller near the edge [Figure 10.2(b)]. The sand at the edges of the footing does not have a great deal of lateral support and tends to move from underneath the footing edges, with the result that more of the load is carried near the center of the footing.

Just the opposite situation is true for footings supported by clayey soils. The clay under the edges of the footing sticks to or has a cohesion with the surrounding clay soil. As a result, more of the load is carried at the edge of the footing than near the middle. [See Figure 10.2(c).]

The designer should clearly understand that the assumption of uniform soil pressure underneath footings is made for reasons of simplifying calculations and may very well have to be revised for some soil conditions.

Should the load be eccentrically applied to a footing with respect to the center of gravity of the footing, the soil pressure is assumed to vary uniformly in proportion to the moment, as illustrated in Section 10.11.

10.4 ALLOWABLE SOIL PRESSURES

The allowable soil pressures to be used for designing the footings for a particular structure are desirably obtained by using the services of a soils engineer. He will determine safe values from the principles of soil mechanics on the basis of test borings, load tests and other experimental investigations.

As such investigations may often not be feasible, most building codes provide certain approximate allowable bearing pressures that can be used for the types of soils and soil conditions occurring in that locality. Table 10.1 shows a set of allowable values that are typical of such building

Table 10.1 Maximum Allowable Soil Pressures

Class of Material	Maximum Allowable Soil Pressure (kip/ft^2)
Rock	20% of ultimate crushing strength
Compact coarse sand, compact fine sand, hard clay, or sand clay	8
Medium stiff clay or sandy clay	6
Compact inorganic sand and silt mixtures	4
Loose sand	3
Soft sandy clay or clay	2
Loose inorganic sand-silt mixtures	1
Loose organic sand and silt mixtures, muck, or bay mud	0

codes. These particular values were taken from the Uniform Building Code (Section 2905, 1970).

Section 15.2.4 of the ACI Code states that the required area of a footing is to be determined by dividing the anticipated total load, including the footing weight, by the allowable soil pressure. It will be noted that this total load is the unfactored load, and yet the design of footings described in this chapter is based on strength design, where the loads are multiplied by the appropriate load factors. It is obvious that an ultimate load cannot be divided by an allowable soil pressure to determine the bearing area required.

The designer can handle this problem in two ways. He can determine the bearing area required by summing up the actual or unfactored dead and live loads and dividing them by the allowable soil pressure. Once this area is determined and the dimensions selected, an ultimate soil pressure can be computed by dividing the factored or ultimate load by the area provided. The remainder of the footing can then be designed by the strength method using this ultimate soil pressure. This simple procedure is used for the footing examples here.

The 1971 ACI Commentary (15.2) provided an alternate method for determining the footing area required that will give exactly the same answers as the procedure just described. By this latter method the allowable soil pressure is increased to an ultimate value by multiplying it by a ratio equal to that used for increasing the service loads. For instance, the ratio for D and L loads would be

$$\text{ratio} = \frac{1.4D + 1.7L}{D + L}$$

Or for $D + L + W$,

$$\text{ratio} = \frac{0.75(1.4D + 1.7L + 1.7W)}{D + L + W}$$

The resulting ultimate soil pressure can be divided into the ultimate column load to determine the area required.

10.5 DESIGN OF WALL FOOTINGS

The theory used for designing beams is applicable to the design of footings with only a few modifications. The upward soil pressure under the wall footing of Figure 10.3 tends to bend the footing into the deformed shape shown. The footing will be designed for the shears and moments involved.

It appears that the maximum moment in this footing occurs under the middle of the wall, but tests have shown that this is not correct because of the rigidity of such walls. If the walls are of reinforced concrete with

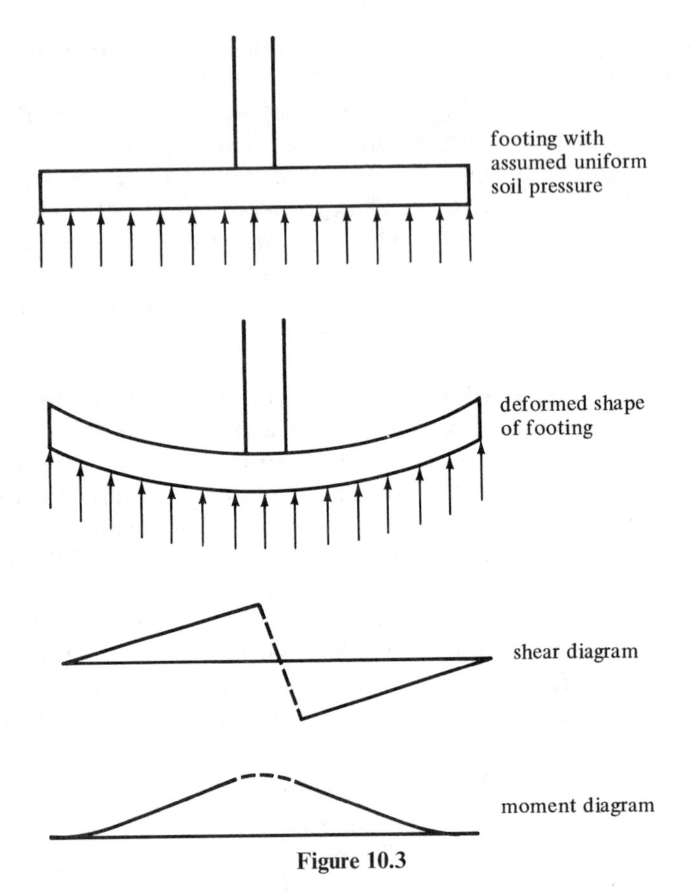

Figure 10.3

their considerable rigidity, it is considered satisfactory to compute the moments at the faces of the walls (ACI Code 15.4.2). Should a footing be supporting a masonry wall with its greater flexibility, the Code states that the moment should be taken at a section halfway from the face of the wall to its center.

To compute the bending moments and shears in a footing, it is necessary to compute only the net upward pressure q_u caused by the factored wall loads above. In other words, the weight of the footing and soil on top of the footing can be neglected. These items cause an upward pressure equal to their downward weights and they cancel each other for purposes of computing shears and moments. In a similar manner, it is obvious that there are no moments or shears existing in a book lying flat on a table.

Should a wall footing be loaded until it fails in shear, the failure will not occur on a vertical plane at the wall face but rather at an angle of approximately 45° with the wall, as shown in Figure 10.4. Apparently the diagonal tension, which one would expect to cause cracks in between the

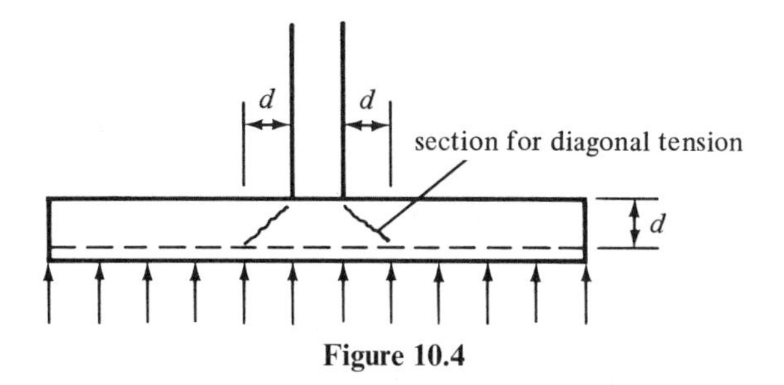

Figure 10.4

two diagonal lines, is opposed by the squeezing or compression caused by the downward wall load and the upward soil pressure. Outside this zone the compression effect is negligible in its effect on diagonal tension. Therefore, shear may be calculated at a distance d from the face of the wall (ACI Code 11.11.1.1) due to the loads located outside the section.

The use of stirrups in footings is considered to be impractical and uneconomical. For this reason the effective depth of wall footings is selected such that V_u is limited to the permissible shear ϕV_c which the concrete can carry without web reinforcing that is $\phi 2\sqrt{f_c'}bd$. The following expression is used to select the depths of wall footings:

$$d = \frac{V_u}{(\phi)(2\sqrt{f_c'})(b)}$$

The design of wall footings is conveniently handled by using 12-in. widths of the wall, as shown in Figure 10.5. Such a practice is followed for the design of a wall footing in Example 10.1. It will be noted that Section 15.7 of the Code states that the depth of a footing above the bottom reinforcing bars may be no less than 6 inches for footings on soils and 12 in. for those on piles.

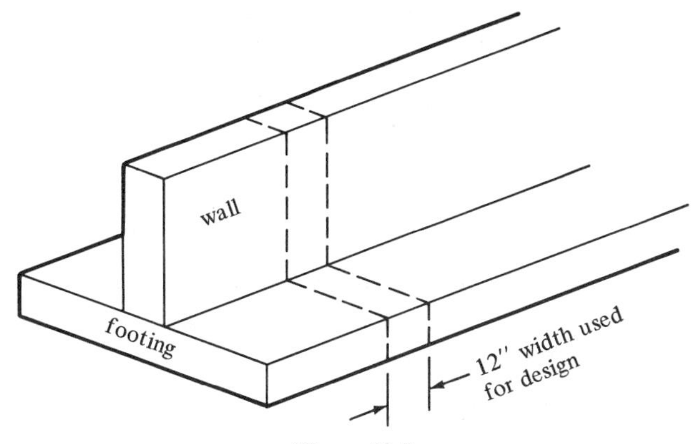

Figure 10.5

EXAMPLE 10.1

Design a wall footing to support a 12-in.-wide reinforced concrete wall with a dead load $D = 20$ k/ft and a live load $L = 15$ k/ft. The bottom of the footing is 4 ft below the final grade, the soil weighs 100 #/ft^3, the allowable soil pressure q_a is 4 ksf, $f_y = 40$ ksi, and $f_c' = 4$ ksi.

SOLUTION

Assume a 12-in.-thick footing ($d = 8.5$ in.). The cover is determined by referring to the Code (7.7.1), which says that for concrete cast against and permanently exposed to the earth, a minimum of 3 in. clear distance outside any reinforcing is required.

The footing weight is $(\frac{12}{12})(150) = 150$ psf and the soil fill on top of the footing is $(\frac{36}{12})(100) = 300$ psf. So 450 psf of the allowable soil pressure q_a is used to support the footing itself and the soil fill on top. The remaining soil pressure is available to support the wall loads. It is called q_e, the effective soil pressure.

$$q_e = 4{,}000 - (\tfrac{12}{12})(150) - (\tfrac{36}{12})(100) = 3{,}550 \text{ psf}$$

$$\text{width of footing required} = \frac{35}{3.55} = 9.86' \qquad \underline{\text{say } 10'0''}$$

Bearing pressure for strength design

$$q_u = \frac{(1.4)(20) + (1.7)(15)}{10.00} = 5.35 \text{ ksf}$$

Depth required for shear (at a distance *d* from face of wall)

$$V_u = \left(\frac{10.00}{2} - \frac{6}{12} - \frac{8.5}{12}\right)(5.35) = 20.29^{\text{k}}$$

$$d = \frac{20{,}290}{(0.85)(2\sqrt{4{,}000})(12)} = 15.73'' + 3.5'' = 19.23'' > 12''$$

$\underline{\text{try again}}$

Assume 18-in. footing ($d = 14.5$ in.)

$$q_e = 4{,}000 - (\tfrac{18}{12})(150) - (\tfrac{30}{12})(100) = 3{,}525 \text{ psf}$$

$$\text{width required} = \frac{35}{3.525} = 9.93' \qquad \underline{\text{say } 10'0''}$$

Bearing pressure for strength design

$$q_u = \frac{(1.4)(20) + (1.7)(15)}{10.00} = 5.35 \text{ ksf}$$

Depth required for shear

$$V_u = \left(\frac{10.00}{2} - \frac{6}{12} - \frac{14.5}{12}\right)(5.35) = 17.61^k$$

$$d = \frac{17{,}610}{(0.85)(2\sqrt{4{,}000})(12)} = 13.65'' + 3.5'' = 17.15''$$

use 18″ total depth

Steel area (using d = 14.5 in.)

Taking moments at face of wall,

$$\text{cantilever length} = \frac{10.00}{2} - \frac{6}{12} = 4.50'$$

$$M_u = (4.50)(5.35)(2.25) = 54.17 \text{ ft-k}$$

$$\frac{M_u}{\phi b d^2} = \frac{(12)(54{,}170)}{(0.9)(12)(14.5)^2} = 286$$

From Table A.10 (see Appendix), $\rho = 0.00749$.

$$A_s = (0.00749)(12)(14.5) = 1.30 \text{ in.}^2$$

use #8 at 7″

Development length

required development length = 20″

actual development length assuming bars are cut off 3″ from edge of footing $= \frac{10'0''}{2} - 6'' - 3''$

= 4′ 3″ from face of wall at section of maximum moment > 20″ ok

Longitudinal temperature and shrinkage steel

$$A_s = (0.002)(12)(18) = 0.432 \text{ in.}^2$$

use #6 at 12″

10.6 DESIGN OF SQUARE ISOLATED FOOTINGS

Single-column footings usually provide the most economical column foundations. Such footings are generally square in plan but they can also be rectangular or even circular or octagonal. Rectangular footings are

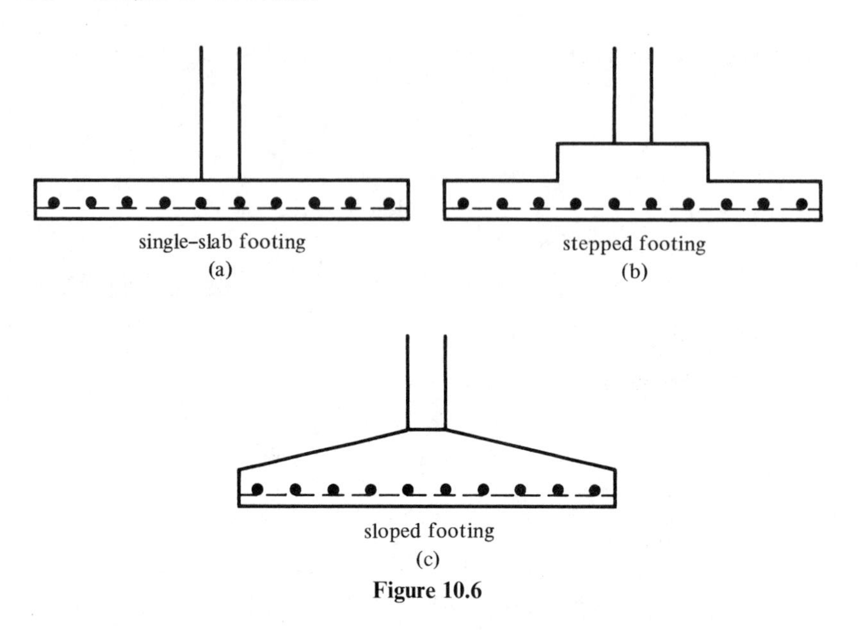

Figure 10.6

used where such shapes are dictated by the available space or where the cross sections of the columns are very pronounced rectangles.

Most footings consist of single slabs, such as the one shown in Figure 10.6(a), but if calculated thicknesses are greater than 3 or 4 ft, it may be economical to use stepped footings, that is, ones with pedestals or caps, as illustrated in Figure 10.6(b). The shears and moments in a footing are obviously larger near the column, with the result that greater depths are required in that area as compared to the outer parts of the footing. For very large footings, such as for bridge piers, stepped footings can give appreciable savings in concrete quantities.

Occasionally sloped footings [Figure 10.6(c)] are used instead of the stepped ones, but labor costs can be a problem. Whether stepped or sloped, it is considered necessary to place the concrete for the entire footing in a single pour to ensure the construction of a monolithic structure. If this procedure is not followed, it is desirable to use keys or shear friction reinforcing between the parts to insure monolithic action.

Before a column footing can be designed, it is necessary to make a few comments regarding shears and moments. This is done in the paragraphs to follow, while a related subject, load transfer from columns to footings, is discussed in the next section of this chapter.

Shears

There are two shear conditions that must be considered in column footings, regardless of their shapes. The first of these is one-way shear,

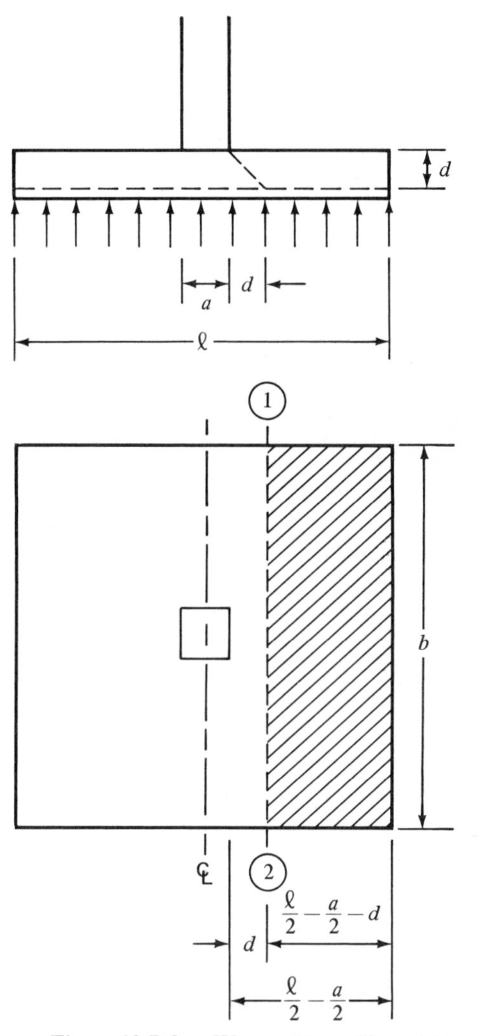

Figure 10.7 One-Way or Beam Shear

which is the same as that considered in wall footings in the preceding section. For this discussion reference is made to the footing of Figure 10.7. The total shear (V_u) to be taken along Section 1.2 equals the net soil pressure times the crosshatched area outside the section. In the expression to follow, b is the whole width of the footing from points 1 to 2. The maximum value of V_u if stirrups are not used equals ϕV_c equals $\phi 2\sqrt{f_c'}\, bd$ and the maximum depth required is as follows:

$$d = \frac{V_u}{\phi 2\sqrt{f_c'}\, b}$$

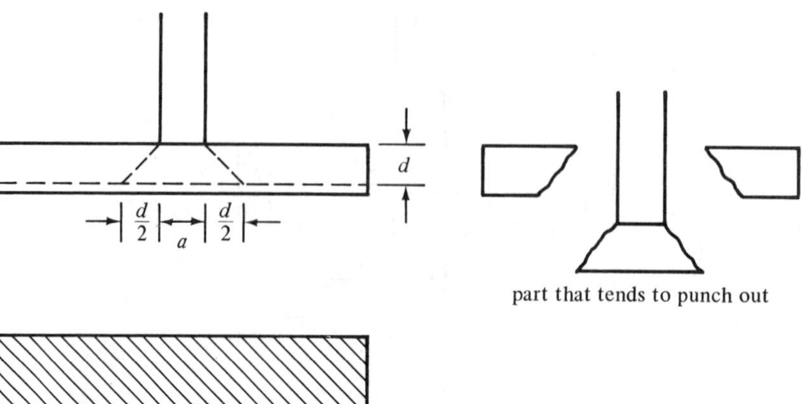

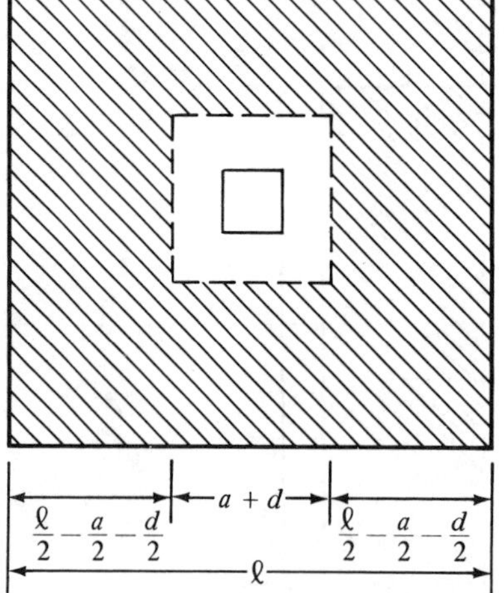

Figure 10.8 Two-Way or Punching Shear

The second shear condition is two-way or punching shear, with reference being made to Figure 10.8. The compression load from the column tends to spread out into the footing, opposing diagonal tension in that area, with the result that a square column tends to punch out a piece of the slab, which has the shape of a truncated pyramid. The ACI Code (11.11.1.2) states that the critical section for two-way shear is located at a distance $d/2$ from the face of the column and that the shear so computed must not exceed $\phi 4\sqrt{f_c'}b_o d$ if shear reinforcing is not provided.

The shear force V_u consists of all the net upward pressure on the cross-hatched area shown, that is, on the area outside the part tending to punch out. In the expression to follow, b_o is the perimenter around the punching area, equal to $4(a + d)$ in the figure, and the depth required for two-way shear is

$$d = \frac{V_u}{\phi 4\sqrt{f_c'}\, b_o}$$

Test have shown that when slabs are subjected to bending in two directions and when the long side of the supporting area is more than two times the short side the shear strength $V_c = 4\sqrt{f_c'}\, b_o d$ may be much too high. The Code 11.11.2 says that unless shear reinforcing is provided, V_c for punching or two-way shear may not exceed the following

$$V_c = \left(2 + \frac{4}{\beta_c}\right)\sqrt{f_c'}\, b_o d \nless 4\sqrt{f_c'}\, b_o d$$

In this expression β_c is the ratio of the long side of the loaded area to the short side of the loaded area. For nonrectangular loaded areas V_c can be obtained with the same expression if β_c is considered to equal the maximum length of the effective loaded area divided by the maximum width measured perpendicular to the length.

Moments

The bending moment in a square reinforced concrete footing is the same about both axes due to symmetry. It should be noted, however, that the effective depth of the footing cannot be the same in the two directions because the bars in one direction rest on top of the bars in the other direction. The effective depth used for calculations might be the average for the two directions or, more conservatively, the value for the bars on top. This lesser value is used for the examples in this text. Although the result is some excess of steel in one direction, it is felt that the steel in either direction must be sufficient to resist the moment in that direction. It should be clearly understood that having an excess of steel in one direction will not make up for a shortage in the other direction.

The critical section for bending is taken at the face of reinforced concrete columns or at a distance halfway from the edge of the base plate and the face of the column if structural steel columns are used (Code 15.4.2).

The determination of footing depths by the procedure described here will often require several cycles of a trial and error procedure. There are, however, many tables and handbooks available with which footing depths can be accurately estimated. One of these is the previously mentioned *CRSI Handbook*. In addition, there are many rules of thumb used by designers for making initial thickness estimates, such as 20% of the footing width or the column diameter plus 3 in. and so on.

The flexural steel percentage calculated for footings is very often less than $200/f_y$. The ACI Code (10.5.3), however, states that in slabs of uniform thickness the minimum area and maximum spacing of reinforcing bars in the direction of bending shall be as required for shrinkage and temperature reinforcement. Most designers feel that the combination of high shears and low ρ values that often occur in footings is not good practice and thus use $200/f_y$ as a minimum.

Example 10.2 illustrates the design of an isolated column footing.

EXAMPLE 10.2

Design a square column footing for a 16-in. square tied column that supports a dead load $D = 200^{k}$ and a live load $L = 160^{k}$. The column is reinforced with 8 #8 bars, the base of the footing is 5 ft below grade, and the soil weight is 100 #/ft³. $f_y = 40{,}000$ psi, $f_c' = 4{,}000$ psi, and $q_a = 5{,}000$ psf.

SOLUTION

Assume 22-in. footing ($d = 17.5$ in.)

$$q_e = 5{,}000 - \left(\tfrac{22}{12}\right)(150) - \left(\tfrac{38}{12}\right)(100) = 4408 \text{ psf}$$

$$A \text{ required} = \frac{360}{4.408} = 81.7 \text{ ft}^2$$

<u>Use 9′ 0″ × 9′ 0″ footing = 81.0 ft²</u>

Bearing pressure for strength design

$$q_u = \frac{(1.4)(200) + (1.7)(160)}{81.0} = 6.81 \text{ ksf}$$

Depth required for two-way or punching shear (Figure 10.9)

$$b_o = (4)(33.5) = 134''$$

$$V_u = (81.0 - 2.79^2)(6.81) = 498.6^{k}$$

$$d = \frac{V_u}{\phi 4\sqrt{f_c'}b_o} = \frac{498{,}600}{(0.85)(4\sqrt{4{,}000})(134)} = 17.30 + 4.5 = 21.8'' < 22'' \quad \underline{\text{ok}}$$

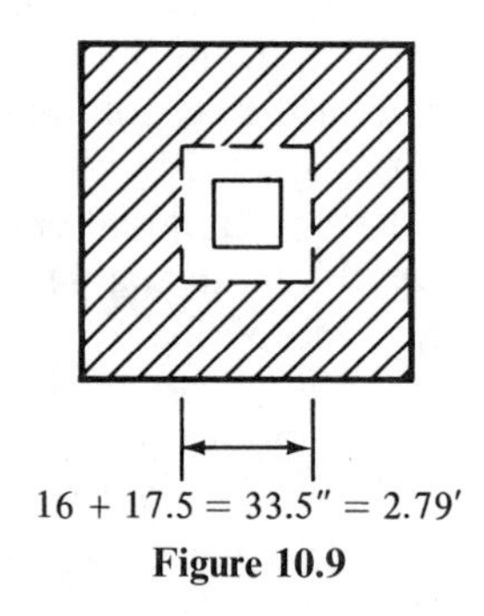

Figure 10.9

Depth required for one-way shear (Figure 10.10)

$$V_u = (9.00)(2.375)(6.81) = 145.6^{k}$$

$$d = \frac{V_u}{\phi 2\sqrt{f_c'}b} = \frac{145{,}600}{(0.85)(2\sqrt{4{,}000})(108)} = 12.54 + 4.5 = 17.04'' < 22'' \quad \underline{\text{ok}}$$

<u>use 22″ depth</u>

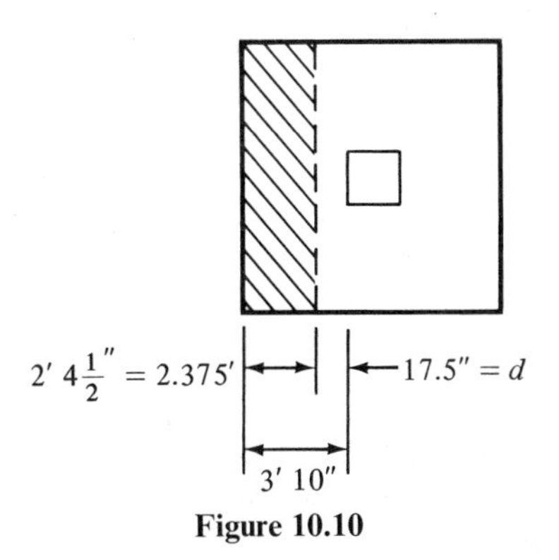

Figure 10.10

Steel area

$$M_u = (3.83)(9.00)(6.81)\left(\frac{3.83}{2}\right) = 449.5 \text{ ft-k}$$

$$\frac{M_u}{\phi bd^2} = \frac{(12)(449{,}500)}{(0.9)(108)(17.5)^2} = 181.2 \left(\rho < \frac{200}{f_y}\right)$$

$$\text{Use } \rho = \frac{200}{40{,}000} = 0.005$$

$$A_s = (0.005)(17.5)(108) = 9.45 \text{ in.}^2$$

use 10 #9 bars in both directions

Development length

required development $l_d = 25''$

actual development length furnished $= 46 - 3 = 43'' > 25''$ ok

10.7 LOAD TRANSFER FROM COLUMNS TO FOOTINGS

All forces acting at the base of a column must be satisfactorily transferred into the footing. Compressive forces can be transmitted directly by bearing, while uplift or tensile forces must be transferred by developed reinforcing.

A column transfers its load directly to the supporting footing over an area equal to the cross-sectional area of the column. The footing surrounding this contact area, however, supplies appreciable lateral support to the directly loaded part with the result that the loaded concrete in the footing can support more load. Thus for the same grade of concrete, the footing can carry more load than can the column.

At the base of the column, the permitted bearing strength is $\phi(0.85f_c'A_1)$ (where ϕ is 0.70), but it may be multiplied by $\sqrt{A_2/A_1} \ngtr 2$ for bearing

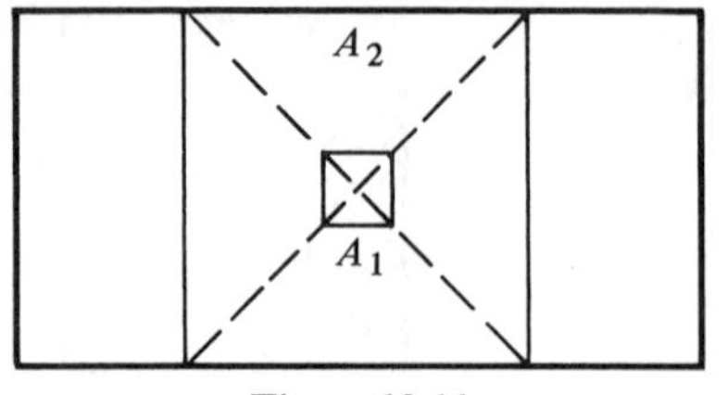

Figure 10.11

on the footing (ACI Code 10.16). In these expressions A_1 is the column area and A_2 is the area of the portion of the supporting footing that is geometrically similar and concentric with the columns. (See Figure 10.11.)

If the computed bearing force is higher than the smaller of the two allowable values, it will be necessary to carry the excess with dowels or with column bars extended into the footing. Should the computed bearing force be less than the allowable value, no dowels or extended reinforcing are theoretically needed, but the Code (15.8.4.1 and 15.8.4.2) states that there must be at least four bars or dowels and they must have an area no less than 0.005 times the cross-sectional area of the column or pedestal and the diameter of the dowels may not exceed the diameter of the column bars by more than 0.15 in. This diameter requirement insures sufficient tying together of the column and the footing over the whole contact area. The use of a few very large dowels spaced far apart might not do this very well. The development length of the bars must be sufficient to transfer the compression to the supporting member, as per the ACI Code (12.3). Should the bearing force be higher than the allowable values, developed reinforcing or dowels must be designed to carry the excess forces. In no case may the area of the designed reinforcement or dowels be less than the area specified for the case where the allowable bearing force was not exceeded. As a practical matter in placing dowels, it should be noted that regardless of how small a distance they theoretically need to be extended down into the footing, they are usually bent at their ends and set on the main footing reinforcing, as shown in Figure 10.12. There the dowels can be tied firmly in place and not be in danger of being pushed through the footing during construction, as might easily happen

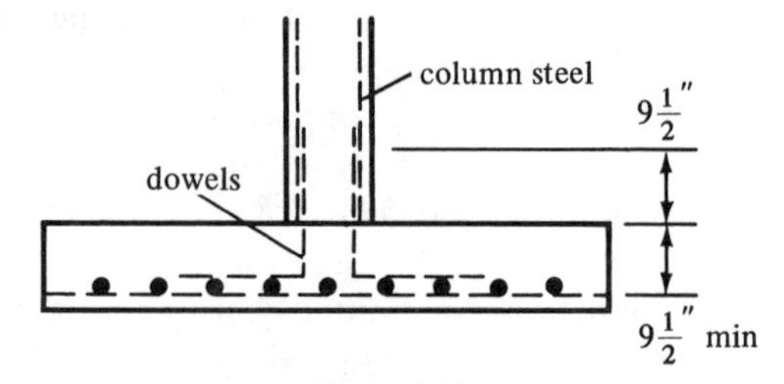

Figure 10.12

otherwise. The bent part of the bar does not count as part of the development length (ACI Code 12.5.3).

Should bending moments or uplift forces have to be transferred to a footing such that the dowels would be in tension, the development lengths must satisfy the requirements for tension bars.

Examples 10.3 and 10.4 provide brief examples of column-to-footing load transfer calculations.

EXAMPLE 10.3

Design for load transfer from the 16-in. × 16-in. column to the 9-ft-0-in. × 9-ft-0-in. footing of Example 10.2. $D = 200^k$, $L = 160^k$, $f_c' = 4{,}000$ psi for both footing and column, and $f_y = 40{,}000$ psi.

SOLUTION

bearing force at base of column

$$= (1.4)(200) + (1.7)(160) = 552^k$$

allowable bearing force in concrete at base of column

$$= \phi(0.85 f_c' A_1) = (0.70)(0.85)(4.0)(16 \times 16)$$
$$= 609.28^k > 552^k$$

allowable bearing force in footing concrete

$$= \phi(0.85 f_c' A_1)\sqrt{\frac{A_2}{A_1}} = (0.70)(0.85)(4.0)(16 \times 16)\sqrt{\frac{9 \times 9}{1.33 \times 1.33}}$$
$$= (0.70)(0.85)(4.0)(16 \times 16)(\text{Use } 2) = 1218.56^k > 552^k \quad \text{ok}$$

minimum A_s for dowels

$$= (0.005)(16 \times 16) = 1.28 \text{ in.}^2$$

use 4 #6 bars

Development lengths of dowels

$$l_d = \frac{0.02 f_y d_b}{\sqrt{f_c'}} = \frac{(0.02)(40{,}000)(0.75)}{\sqrt{4{,}000}} = 9.49''$$
$$l_d = 0.0003 f_y d_b = (0.0003)(40{,}000)(0.75) = 9.00''$$
$$l_d = 8.00''$$

use 4 #6 dowels extending $9\frac{1}{2}$ in. up into column and down into footing and set on top of reinforcing mat, as shown in Figure 10.12.

EXAMPLE 10.4

Design for load transfer from a 14-in. × 14-in. column to a 13-ft-0-in × 13-ft-0-in. footing with a P_u of 800^k. $f_c' = 3{,}000$ psi in the footing and 5,000 psi in the column and $f_y = 60{,}000$ psi. The column has 8 #8 bars.

SOLUTION

bearing force at base of column $= P_u = 800^k$

$= \phi$ allowable bearing force in concrete $+ \phi$ strength of dowels

allowable bearing force in concrete at base of column

$= (0.70)(0.85)(5.0)(14 \times 14) = 583.1^k < 800^k$ no good

allowable bearing force on footing concrete

$= (0.70)(0.85)(3.0)(14 \times 14)(\text{Use } 2) = 699.72^k < 800^k$ no good

Therefore, the dowels must be designed for excess load

Excess load $= 800 - 583.1 = 216.9^k$

$$A_s \text{ of dowels} = \frac{216.9}{(0.70)(60)} = 5.16 \text{ in.}^2 \text{ or } (0.005)(14)(14) = 0.98 \text{ in.}^2$$

use 6 #9 bars

Development length of dowels into column

$$l_d = \frac{(0.02)(60{,}000)(0.875)}{\sqrt{5{,}000}} = 14.85$$

$$l_d = (0.0003)(60{,}000)(0.875) = \underline{15.75}$$

$$l_d = 8''$$

Development length of dowels into footing (different from column values because f_c' values are different)

$$l_d = \frac{(0.02)(60{,}000)(0.875)}{\sqrt{3{,}000}} = \underline{19.17''}$$

$$l_d = (0.0003)(60{,}000)(0.875) = 15.75$$

use 6 #7 dowels extending 16 in. up into the column and 20 in. down into the footing.

10.8 RECTANGULAR ISOLATED FOOTINGS

As previously mentioned, isolated footings may be rectangular in plan if the column has a very pronounced rectangular shape or if the space available for the footing forces the designer into a rectangular shape.

The design procedure is almost identical with the one used for square footings. After the required area is calculated and the lateral dimensions

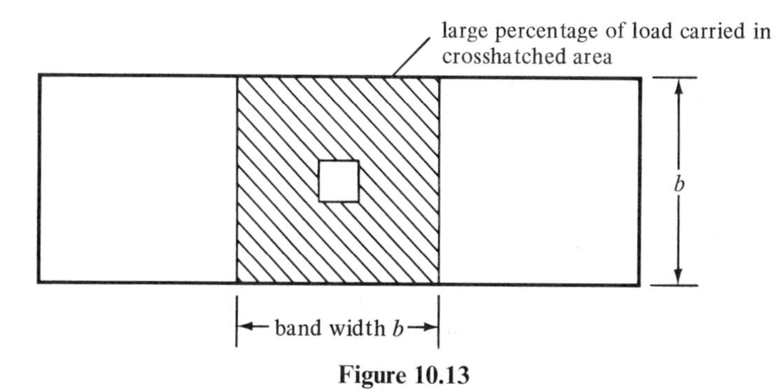

Figure 10.13

selected, the depths required for one-way and two-way shear are determined by the usual methods. One-way shear will very often control the depths for rectangular footings, whereas two-way shear normally controls the depths of square footings.

The next step is to select the reinforcing in the long direction. These bars are spaced uniformly across the footing, but such is not the case for the short-span reinforcing. With reference to Figure 10.13 it can be seen that the support provided by the footing to the column will be concentrated near the middle of the footing and thus the moment in the short direction will be concentrated somewhat in the same area near the column.

As a result of this concentration effect, it seems only logical to concentrate a large proportion of the short-span reinforcing in this area. The Code (15.4.4) states that a certain minimum percentage of the total short-span reinforcing should be placed in a band width equal to the length of the shorter direction of the footing. The amount of reinforcing in this band is to be determined with the following expression, in which β is the ratio of the length of the long side to the width of the short side:

$$\frac{\text{reinforcing in band width}}{\text{total reinforcing in short direction}} = \frac{2}{\beta + 1}$$

The remaining reinforcing in the short direction should be uniformly spaced over the ends of the footing but should at least meet the shrinkage and temperature requirements of the ACI Code (7.12).

Example 10.5 presents the partial design of a rectangular footing in which the depths for one- and two-way shears are determined and the reinforcement selected.

EXAMPLE 10.5

Design a rectangular footing for an 18 in. square column with a dead load of 185^k and a live load of 150^k. Make the length of the long side equal to twice the width of the short side. $f_y = 40{,}000$ psi, $f_c' = 4{,}000$ psi, and $q_a = 4{,}000$ psf. Assume the base of the footing is 5 ft 0 in. below grade.

SOLUTION

Assume 24-in. footing ($d = 19.5$ in.)

$$q_e = 4{,}000 - (\tfrac{24}{12})(150) - (\tfrac{36}{12})(100) = 3{,}400 \text{ psf}$$

$$A \text{ required} = \frac{335}{3.4} = 98.5 \text{ ft}^2 \qquad \underline{\text{use } 7'0'' \times 14'0'' = 98.0 \text{ ft}^2}$$

$$q_u = \frac{(1.4)(185) + (1.7)(150)}{98.0} = 5.24 \text{ ksf}$$

Checking depth for one-way shear (Figure 10.14)

$$b = 84''$$

$$V_u = (7.0)(4.62)(5.24) = 169.5^{\text{k}}$$

$$d = \frac{169{,}500}{(0.85)(2\sqrt{4{,}000})(84)} = 18.77 + 4.5 = 23.27'' \qquad \underline{\text{use } 24''}$$

Checking depth for two-way shear (Figure 10.15)

$$b_o = (4)(37.5) = 150''$$

$$V_u = [98.0 - (3.12)^2](5.24) = 462.5^{\text{k}}$$

$$d = \frac{462{,}500}{(0.85)(4\sqrt{4{,}000})(150)} = 14.34''$$

Design of longitudinal steel

$$M_u = (6.25)(7.0)(5.24)\left(\frac{6.25}{2}\right) = 716.4 \text{ ft-k}$$

$$\frac{M_u}{\phi bd^2} = \frac{(12)(716{,}400)}{(0.90)(84)(19.5)^2} = 299$$

$$p = 0.00784$$

$$A_s = (0.00784)(84)(19.5) = 12.84 \text{ in.}^2 \qquad \underline{\text{use } 13 \ \#9 \text{ bars}}$$

Design of steel in short direction (Figure 10.16)

$$M_u = (2.75)(14.0)(5.24)\left(\frac{2.75}{2}\right) = 277.4 \text{ ft-k}$$

$$\frac{M_u}{\phi bd^2} = \frac{(12)(277{,}400)}{(0.90)(168)(19.5)^2} = 57.9$$

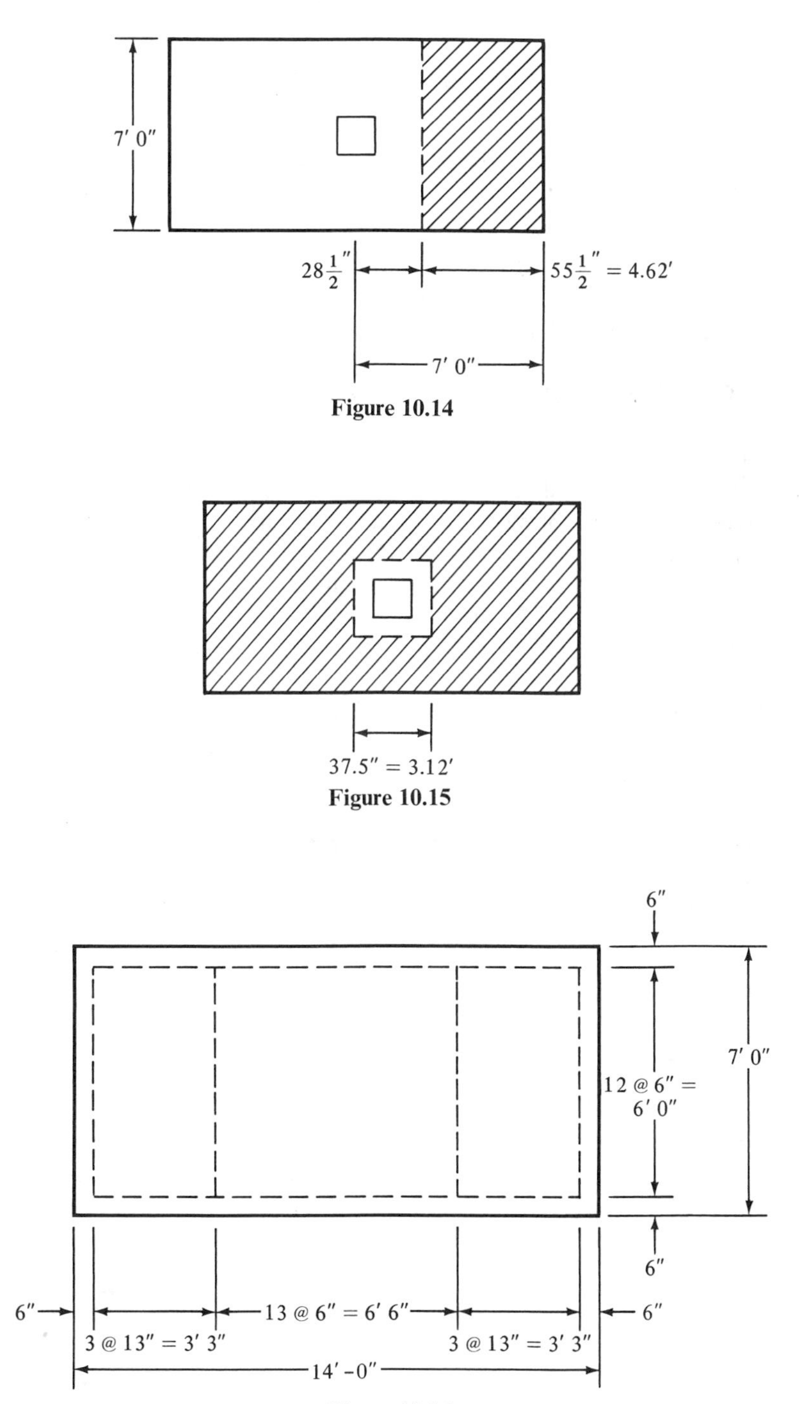

Figure 10.14

Figure 10.15

Figure 10.16

Use

$$p = \frac{200}{f_y} = \frac{200}{40{,}000} = 0.005$$

$$A_s = (0.005)(168)(19.5) = 16.38 \text{ in.}^2 \quad \underline{\text{use } 22 \ \#8 \text{ bars}}$$

$$\frac{\text{reinforcing in band width}}{\text{total reinforcing in short direction}} = \frac{2}{2+1} = \frac{2}{3}$$

Use $\frac{2}{3} \times 22$ or say 14 bars in band width

10.9 COMBINED FOOTINGS

Combined footings support more than one column. One situation where they may be used occurs when the columns are so close together that isolated individual footings would run into each other [Figure 10.17(a)]. Another frequent use of combined footings occurs where one column is very close to a property line, causing the usual isolated footing to extend across the line. For this situation the footing for the exterior column may be combined with the one for an interior column, as shown in Figure 10.17(b).

On some occasions where a column is close to a property line and where it is desired to combine its footing with that of an interior column, the interior column will be so far away as to make the idea impractical

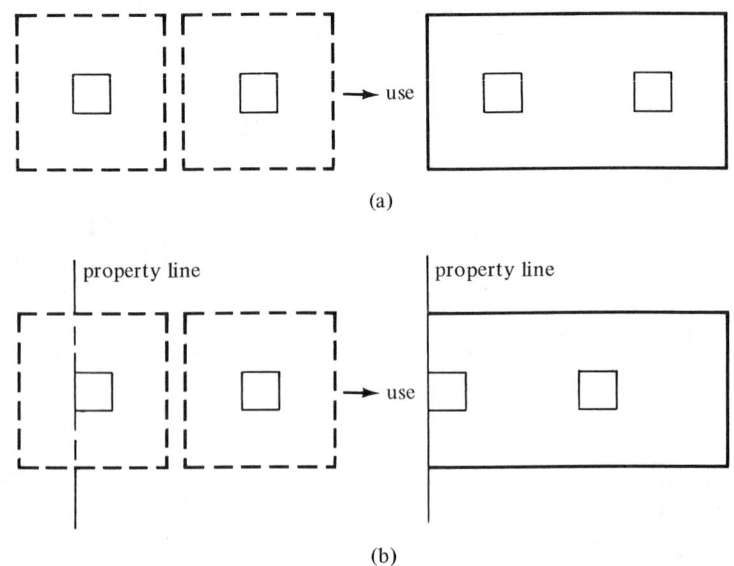

Figure 10.17

economically. For such a case counterweights or deadmen may be provided to take care of the eccentric loading.

As it is desirable to make bearing pressures uniform throughout the footing, the centroid of the footing should be made to coincide with the centroid of the column loads to attempt to prevent uneven settlements. This can be accomplished with combined footings that are rectangular in plan. Should the interior column load be greater than that of the exterior column, the footing may be so proportioned that its centroid will be in the correct position by extending the inward projection of the footing, as shown in the rectangular footing of Figure 10.17(b).

Other combined footing shapes that will enable the designer to make the centroids coincide are the trapezoid and strap or T footings shown in Figure 10.18(a) and (b).

You probably realize that a problem arises in establishing the centroids of loads and footings when deciding whether to use service or factored loads. The required centroid of the footing will be slightly different for the two cases. The author determines the footing areas and centroids with the service loads (ACI Code 15.2.4), but the factored loads could be used with reasonable results, too. The important item is to be consistent throughout the entire problem.

The design of combined footings has not been standardized as have the procedures used for the previous problems worked in this chapter.

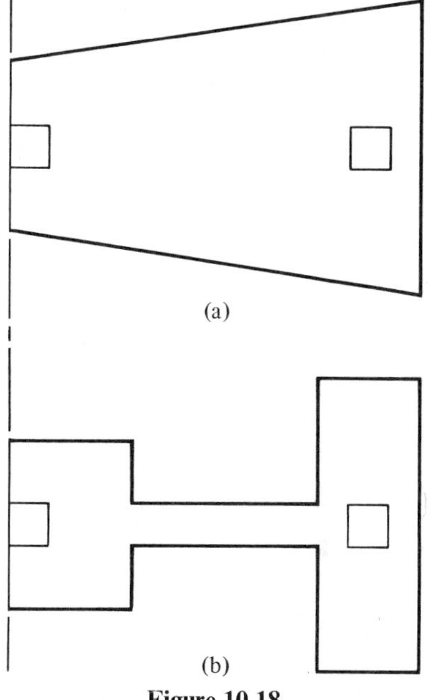

Figure 10.18

For this reason practicing concrete designers use slightly varying approaches. One of these methods is described in the paragraphs to follow.

First, the required area of the footing is determined for the service loads and the footing dimensions are selected such that the centroids coincide. Next, shear and moment diagrams are drawn along the long side of the footing. Then the required footing depth for one- and two-way shear is calculated, and the steel in the long direction is selected.

In the short direction it is assumed that each column load is spread over a width in the long direction equal to the column width plus $d/2$ on each side if that much footing is available. Then the steel is designed and a minimum amount of steel for temperature and shrinkage is provided in the remaining part of the footing.

Space is not taken here to design completely a combined footing, but Example 10.6 is presented to show those parts of the design that are different from the previous examples of this chapter. A comment should be made about the moment diagram. If the length of the footing is not selected so that its centroid is located exactly at the centroid of the column loads, the shear and moment diagrams will not close well at all since the numbers are very sensitive. Nevertheless, it is considered good practice to round off the footing lateral dimensions to the nearest 3 in.

EXAMPLE 10.6

Design a rectangular combined footing for the two columns shown in Figure 10.19. $q_a = 5$ ksf, $f_c' = 3{,}000$ psi, and $f_y = 60$ ksi. The bottom of the footing is to be 6 ft below grade.

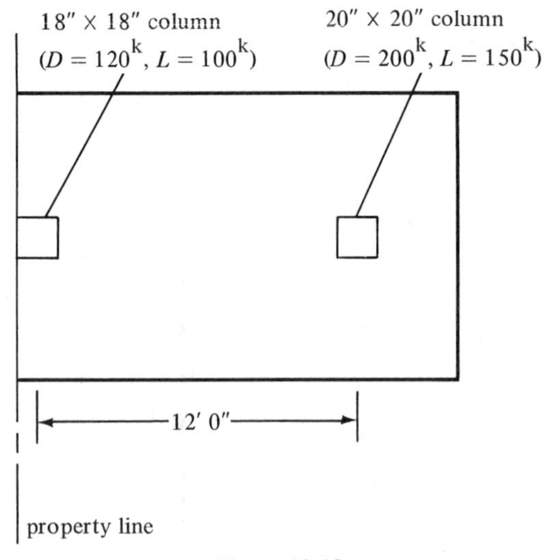

Figure 10.19

SOLUTION

Assume 30-in. footing (d = 25.5 in.)

$$q_e = 5{,}000 - (\tfrac{30}{12})(150) - (\tfrac{42}{12})(100) = 4{,}275 \text{ psf}$$

$$A \text{ required} = \frac{570}{4.275} = 133.3 \text{ ft}^2$$

Locate center of gravity of column service loads

$$X \text{ from c.g. of left column} = \frac{(350)(12)}{570} = 7.37'$$

$$\text{distance from property line to c.g.} = 0.75 + 7.37$$

$$= 8.12' \; (2 \times 8.12 = 16.24', \text{ say } 16'3'')$$

use 16′ 3″ × 8′ 3″ footing (A = 134 ft^2)

$$q_u = \frac{(1.4)(320) + (1.7)(250)}{134} = 6.515 \text{ ksf}$$

Drawing shear and moment diagrams (Figure 10.20)

Depth required for one-way shear

V_u a distance d from interior face of right column

$$= 302.5 - \left(\frac{25.5}{12}\right)(53.75) = 188.3^{\text{k}}$$

$$d = \frac{(188{,}300)}{(0.85)(2\sqrt{3{,}000})(99)} = 20.42''$$

Depth required for two-way shear

$$V_u \text{ at right column} = 535 - \left(\frac{45.5}{12}\right)^2(6.515) = 441.3^{\text{k}}$$

$$d = \frac{441{,}500}{(0.85)(4\sqrt{3{,}000})(4 \times 45.5)} = 13.02''$$

Design of longitudinal steel (without revising d from 25.5 in.)

$$-M_u = -809.6 \text{ ft-k}$$

$$\frac{M_u}{\phi b d^2} = \frac{(12)(809{,}600)}{(.90)(99)(25.5)^2} = 167.7$$

p = 0.0029 (by formula from Graph 1, Appendix)

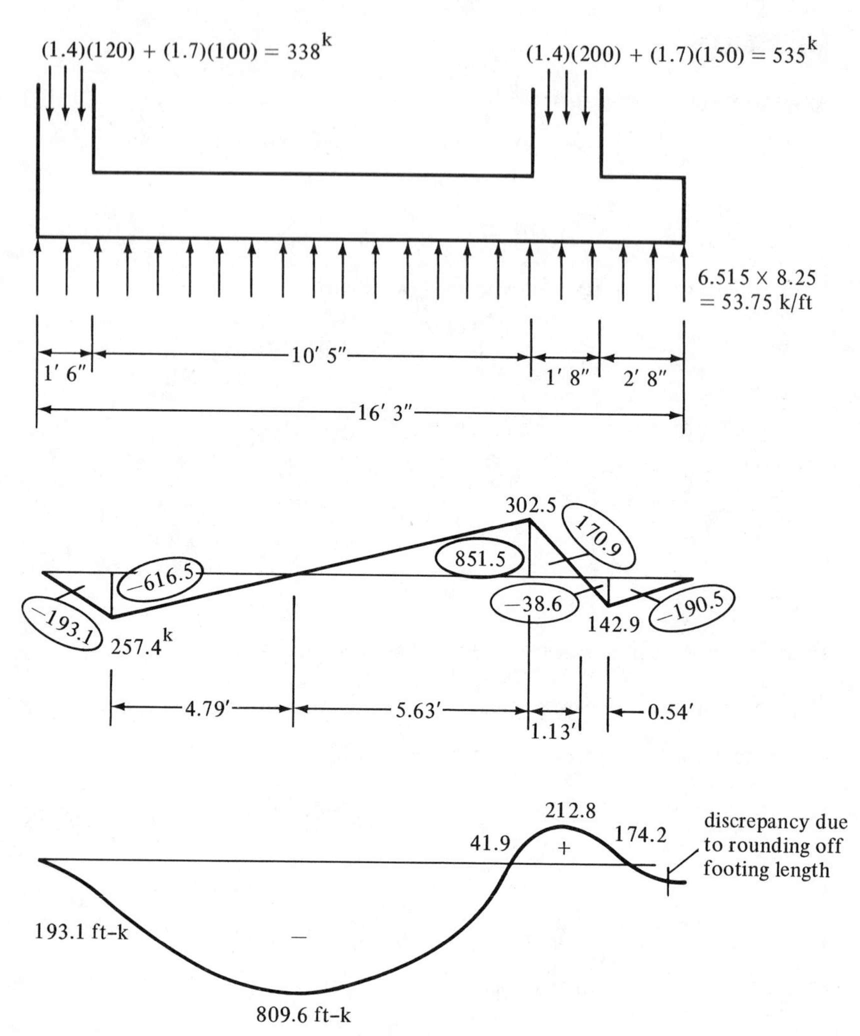

Figure 10.20

Use $200/f_y = 0.00333$.

$$-A_s = (0.00333)(99)(25.5) = 8.41 \text{ in.}^2 \qquad \underline{\text{say } 9 \ \#9}$$

$$+M_u = +190.5 \text{ ft-k (from shear diagram)}$$

$$\frac{M_u}{\phi b d^2} = \frac{(12)(190{,}500)}{(0.90)(99)(25.5)^2} = 39.46$$

$$p = < P_{\min}$$

Use $200/f_y = 0.0033$.

$$+A_s = (0.00333)(99)(25.5) = 8.41 \text{ in.}^2 \qquad \underline{\text{say } 9 \ \#9}$$

Design of short-span steel under interior column

$$\text{assuming steel spread over width} = \text{column width} + (2)\left(\frac{d}{2}\right)$$

$$= 20 + (2)\left(\frac{25.5}{2}\right) = 45.5''$$

Referring to Figure 10.21 and calculating M_u:

$$M_u = (3.29)(64.85)\left(\frac{3.29}{2}\right) = 350.97 \text{ ft-k}$$

$$\frac{M_u}{\phi bd^2} = \frac{(12)(350{,}970)}{(0.90)(45.5)(25.5)^2} = 158.2$$

$$p = 0.00282 \text{ (by formula Graph 1, Appendix)} < \frac{200}{f_y}$$

$$\underline{\text{Use } 0.00333.}$$

$$A_s = (0.00333)(45.5)(25.5) = 3.86 \text{ in.}^2$$

A similar procedure is used under the exterior column where the steel is spread over a width equal to 18 in. plus $d/2$ and not 18 in. plus $2(d/2)$, as sufficient room is not available on the property-line side of the column.

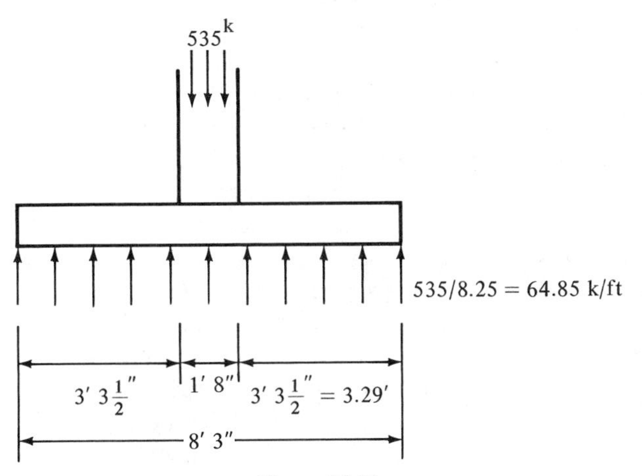

Figure 10.21

10.10 FOOTING DESIGN FOR EQUAL SETTLEMENTS

If three men are walking along a road carrying a log on their shoulders and one of them decides to lower his shoulder by 1 in., the result will be a drastic effect on the load supported by the other men. In the same way, if the footings of a building should settle by different amounts, the shears and moments throughout the structure will be greatly changed. In addition, there will be detrimental effects on the fitting of doors, windows, and partitions. Should all the footings settle by the same amount, however, these adverse effects will not occur. Thus equal settlement is the goal of the designer.

The footings considered in preceding sections have had their areas selected by taking the total dead plus live loads and dividing the sum by the allowable soil pressure. It would seem that if such a procedure were followed for all the footings of an entire structure, the result would be uniform settlements throughout—*but soils engineers have clearly shown that this assumption may be very much in error.*

A better way to handle the problem is to attempt to design the footings so that the *usual loads* on each footing will cause approximately the same pressures. The usual loads consist of the dead loads plus the average percentage of live loads normally present. The usual percentage of live loads present varies from building to building. For a church it might be almost zero, perhaps 25% to 30% for an office building, and maybe 75% or more for some warehouses. Furthermore, the percentage in one part of a building may be entirely different from that present for some other part (offices, storage, etc.).

One way to handle the problem is to design the footing that has the highest ratio of live to dead load, compute the usual soil pressure under that footing using dead load plus the average percentage of live load, and then to determine the areas required for the other footings so their usual soil pressures are all the same. It should be noted that the dead load plus 100% of the live load should not cause a pressure greater than the allowable soil pressure under any of the footings.

A student of soil mechanics will realize that the method of usual pressures, though not a bad design procedure, will not ensure equal settlements. This approach at best will only lessen the amounts of differential settlements. He will remember first that large footings tend to settle more than small footings, even though their soil pressures are the same, because the large footings exert compression on a larger and deeper mass of soil. There are other items that can cause differential settlements. Different types of soils may be present at different parts of the building; part of the area may be in fill and part in cut; there may be mutual influence of one footing on another; and so forth.

Example 10.7 illustrates the usual load procedure for a group of five-column footings.

EXAMPLE 10.7

Determine the footing areas required for the loads given in Table 10.2 so that the usual soil pressures are equal. Assume that the usual live load percentage is 30% for all the footings. $q_e = 4$ ksf.

Table 10.2 Footings

Footing	Dead Load	Live Load
A	150^k	200^k
B	120^k	100^k
C	140^k	150^k
D	100^k	150^k
E	160^k	200^k

SOLUTION

The largest percentage of live load to dead load occurs for footing *D*.

$$\text{area required for footing } D = \frac{100 + 150}{4} = 62.5 \text{ ft}^2$$

$$\text{usual soil pressure under footing } D = \frac{100 + (0.30)(150)}{62.5} = 2.32 \text{ ksf}$$

Computing the areas required for the other footings and their soil pressures under total service loads, the results are as shown in Table 10.3.

Table 10.3 Areas and Soil Pressures

Footing	Usual Load = $D + 0.30L$	Area Required = Usual Load ÷ 2.32	Total Soil Pressure
A	210^k	90.5 ft^2	3.87 ksf
B	150^k	64.7 ft^2	3.40 ksf
C	185^k	79.7 ft^2	3.64 ksf
D	145^k	62.5 ft^2	4.00 ksf
E	220^k	94.8 ft^2	3.80 ksf

10.11 FOOTINGS SUBJECTED TO LATERAL MOMENTS

Walls or columns often transfer moments as well as vertical loads to their footings. These moments may be due to wind, earthquake, lateral

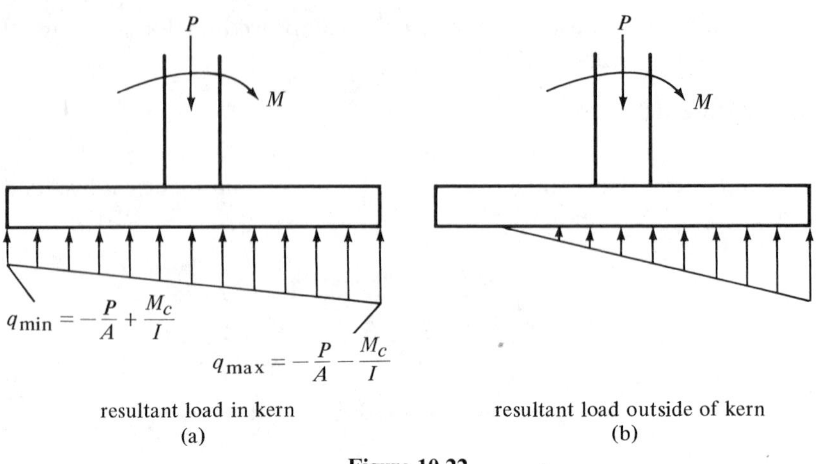

Figure 10.22

earth pressures, and so on. Such a situation is represented by the vertical load P and the bending moment M shown in Figure 10.22.

Because of this moment the resultant force will not coincide with the centroid of the footing. Of course, if the moment is constant in magnitude and direction, it will be possible to place the center of the footing under the resultant load and avoid the eccentricity, but lateral forces such as wind and earthquake can come from any direction and symmetrical footings will be needed.

The effect of the moment is to produce a uniformly varying soil pressure, which can be determined at any point with the expression

$$q = -\frac{P}{A} \pm \frac{Mc}{I}$$

In this discussion the term *kern* is used. If the resultant force strikes the footing base within the kern, the value of $-P/A$ is larger than $+Mc/I$ at every point and the entire footing base is in compression, as shown in Figure 10.22(a). If the resultant strikes the footing base outside the kern, the value of $+Mc/I$ will at some points be larger than $-P/A$ and there will be uplift or tension. The soil cannot resist tension, and the pressure variation will be as shown in Figure 10.22(b). The extent of the kern can be determined by replacing Mc/I with Pec/I, equating it to P/A, and solving for e.

Example 10.8 shows that the required area of a footing subjected to a vertical load and a lateral moment can be determined by trial and error. The procedure is to assume a size, calculate the maximum soil pressure, compare it with the allowable pressure, assume another size, and so on.

Once the area has been established, the remaining design will be handled as it was for other footings. Although the shears and moments are not uniform, the theory of design is unchanged. The factored loads are computed, the bearing pressures are determined, and the shears and moments are calculated. For strength design the footing must be proportioned for the worst of these situations (from ACI Code Section 9.2):

$$U = 1.4D + 1.7L$$
$$U = 0.75(1.4D + 1.7L + 1.7W)$$
$$U = 0.9D + 1.3W$$

EXAMPLE 10.8

Determine the width needed for a wall footing to support loads: $D = 18$ k/ft and $L = 12$ k/ft. In addition, a lateral load of 6 k/ft is assumed to be applied 5 ft above the top of the footing. Assume the footing is 18 in. thick and its base is 4 ft below the final grade. $q_a = 4$ ksf.

SOLUTION

First trial

neglecting moment $q_e = 4{,}000 - (\frac{18}{12})(150) - (\frac{30}{12})(100) = 3{,}525$ psf

$$\text{width required} = \frac{18 + 12}{3.525} = 8.51' \qquad \underline{\text{try } 9'0''}$$

$$M = 6 \times \text{distance to footing base} = 6 \times 6.5 = 39 \text{ ft-k}$$
$$A = (9)(1) = 9 \text{ ft}^2$$
$$I = (\tfrac{1}{12})(1)(9)^3 = 60.75 \text{ ft}^4$$

$$q_{max} = -\frac{P}{A} - \frac{Mc}{I} = -\frac{30}{9} - \frac{(39)(4.5)}{60.75}$$
$$= -6.22 \text{ ksf} > 3.525 \text{ ksf} \qquad \underline{\text{no good}}$$

$$q_{min} = -\frac{P}{A} + \frac{Mc}{I} = -\frac{30}{9} + \frac{(39)(4.5)}{60.75} = -0.44 \text{ ksf}$$

Second trial

Assume 14-ft-wide footing:

$$A = (14)(1) = 14 \text{ ft}^2$$
$$I = (\tfrac{1}{12})(1)(14)^3 = 228.7 \text{ ft}^4$$

$$q_{max} = -\frac{30}{14} - \frac{(39)(7)}{228.7} = -3.33 \text{ ksf} < 3.525 \text{ ksf}$$

$$q_{min} = -\frac{30}{14} + \frac{(39)(7)}{228.7} = -0.95 \text{ ksf}$$

$$\underline{\text{use } 14'0'' \text{ footing}}$$

10.12 PLAIN CONCRETE FOOTINGS

Occasionally plain concrete footings are used to support light loads if the supporting soil is of good quality. Very often the widths and thicknesses of such footings are determined by rules of thumb, such as: "the depth of a plain footing must be equal to no less than the projection beyond the edges of the wall." In this section, however, a plain concrete footing is designed in accordance with the requirements of the ACI Code.

The Code (15.11) states that when plain concrete footings are supported by soil, they cannot have an edge thickness less than 8 in. and they cannot be used on piles. The critical sections for shear and moment for plain concrete footings are the same as for reinforced concrete footings. Allowable bending and shear stresses are given for both strength design and allowable stress design. For strength design the maximum allowable flexural tension is $5.0\phi\sqrt{f_c'}$, with $\phi = 0.65$, and the average shear stress $(V_u/\phi bd)$ may not exceed $2.0\sqrt{f_c'}$ for one-way shear nor $4.0\sqrt{f_c'}$ for two-way shear, using the appropriate load and ϕ factors (Code 15.11.2).

The author feels very strongly that even though plain footings are designed in accordance with the Code requirements, they should at the very least be reinforced in the longitudinal direction with reinforcing to keep temperature and shrinkage cracks within reason and to enable the footing to bridge over soft spots in the underlying soil. Nevertheless, Example 10.9 presents the design of a plain concrete footing in accordance with the ACI Code provisions.

EXAMPLE 10.9

Design a plain concrete footing for a 12-in. reinforced concrete wall that supports a dead load of 12 k/ft, including the wall weight, and a 6 k/ft live load. The base of the footing is to be 5 ft below the final grade. $f_c' = 3{,}000$ psi and $q_a = 4{,}000$ psf.

SOLUTION

Assume 24-in. footing

$$q_e = 4{,}000 - (\tfrac{24}{12})(150) - (\tfrac{36}{12})(100) = 3{,}400 \text{ psf}$$

$$\text{width required} = \frac{18}{3.4} = 5.29' \qquad \underline{\text{say } 5'\,6''}$$

Bearing pressure for strength design

$$q_u = \frac{(1.4)(12) + (1.7)(6)}{5.5} = 4.91 \text{ ksf}$$

Checking bending stress neglecting bottom 3 in. of footing

$$M_u = (2.25)(4.91)(1.125) = 12.43 \text{ ft-k}$$
$$I = (\tfrac{1}{12})(12)(21)^3 = 9{,}261 \text{ in.}^4$$
$$f_t = \frac{(12)(12{,}430)(10.5)}{9{,}261} = 169 \text{ psi} < (5)(0.65)\sqrt{3{,}000} = 178 \text{ psi} \quad \underline{\text{ok}}$$

Checking shear stress at a distance *d* of 21 in. from face of wall

$$V_u = \left(\frac{5.5}{2} - \frac{6}{12} - \frac{21}{12}\right)(4.91) = 2.46^{\text{k}}$$
$$v_u = \frac{2{,}460}{(0.85)(12)(21)} = 11.5 \text{ psi} < 2.0\sqrt{3{,}000} = 110 \text{ psi} \quad \underline{\text{ok}}$$

PROBLEMS

For all problems assume concrete weighs 150 #/ft^3 and soil 100 #/ft^3.

10.1 to **10.4** Design wall footings for the values given. The walls are to consist of reinforced concrete.

Problem	Wall Thickness	D	L	f'_c	f_y	q_a	Distance from Bottom of Footing to Final Grade
10-1	12″	20^k/ft	30^k/ft	4 ksi	40 ksi	4 ksf	5′
10-2	12″	26^k/ft	20^k/ft	5 ksi	50 ksi	5 ksf	4′
10-3	14″	32^k/ft	26^k/ft	3 ksi	60 ksi	5 ksf	4′
10-4	15″	22^k/ft	24^k/ft	3 ksi	60 ksi	4 ksf	6′

10.5 Repeat Problem 10.3 if a masonry wall is used and $q_a = 4$ ksf.

10.6 to **10.10** Design square single-column footings for the values given.

Problem	Column Size	D	L	f'_c	f_y	q_a	Distance from Bottom of Footing to Final Grade
10-6	12″ × 12″	100^k	120^k	3 ksi	40 ksi	4 ksf	6′
10-7	14″ × 14″	155^k	250^k	4 ksi	60 ksi	4 ksf	5′
10-8	15″ × 15″	210^k	150^k	3 ksi	60 ksi	4 ksf	5′
10-9	16″ × 16″	160^k	150^k	4 ksi	50 ksi	5 ksf	4′
10-10	Round 18″ diameter	180^k	220^k	3 ksi	60 ksi	5 ksf	5′

10.11 Design for load transfer from an 18″ × 18″ column with 6 # 8 bars ($D = 250^k$, $L = 340^k$) to an 8-ft-0-in. × 8-ft-0-in. footing. $f_c' = 3$ ksi for footing and 5 ksi for column. $f_y = 60$ ksi.

10.12 Repeat Problem 10.6 if a rectangular footing with the length equal to twice the width is used.

10.13 Repeat Problem 10.7 if a rectangular footing with the length equal to 1.5 times the width is used.

10.14 Redesign the footing of Problem 10.8 if the width in one direction is limited to 7 ft 0 in.

10.15 Design a rectangular combined footing for the two columns shown in the accompanying illustration. The bottom of the footing is to be 5 ft below the final grade. $f_c' = 3.5$ ksi, $f_y = 50$ ksi, and $q_a = 5$ ksf.

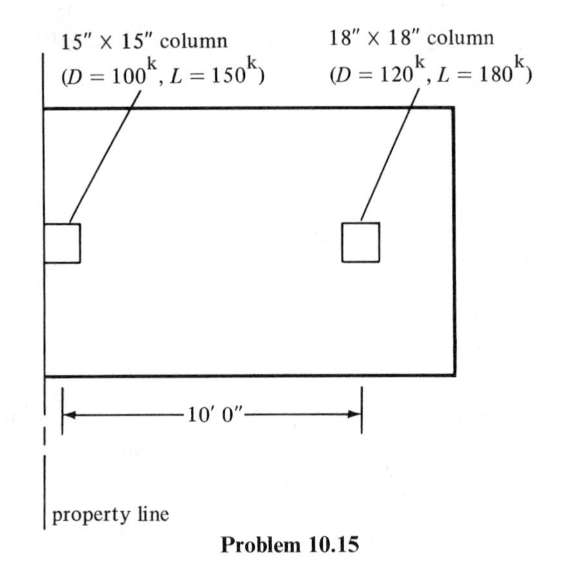

Problem 10.15

10.16 Determine the footing areas required for the loads given in the accompanying table so that the usual soil pressures are equal. Assume $q_e = 5$ ksf and a usual live load percentage of 25% for all of the footings.

Footing	Dead Load	Live Load
A	140^k	160^k
B	160^k	200^k
C	110^k	165^k
D	120^k	150^k
E	160^k	180^k
F	130^k	160^k

10.17 to **10.18** Determine the width required for the wall footings. Assume footings have 24 in. total thicknesses.

Problem	Reinforced Conc. Wall Thickness	D	L	f_c'	f_y	q_a	Lateral Load	Height of Lateral Load above Top of Footing	Distance from Bottom of Footing to Final Grade
10.17	12″	12 k/ft	20 k/ft	3 ksi	40 ksi	4 ksf	8 k/ft	4′	4′
10.18	14″	16 k/ft	22 k/ft	4 ksi	60 ksf	4 ksf	10 k/ft	5′	5′

10.19 to **10.20** Design plain concrete wall footings of uniform thickness.

Problem	Reinforced Conc. Wall Thickness	D	L	f_c'	q_a	Distance from Bottom of Footing to Final Grade
10.19	14″	8 k/ft	10 k/ft	3 ksi	4 ksf	4′
10.20	12″	12 k/ft	15 k/ft	4 ksi	5 ksf	5′

10.21 Repeat Problem 10.1 as a plain concrete footing of uniform thickness if the live load is reduced to 15 k/ft.

10.22 to **10.23** Design square plain concrete column footings of uniform thickness.

Problem	Reinforced Concrete Column Size	D	L	f_c'	q_a	Distance from Bottom of Footing to Final Grade
10.22	10″ × 10″	40^k	60^k	3 ksi	4 ksf	5′
10.23	12″ × 12″	80^k	100^k	3.5 ksi	4 ksf	5′

Problems with SI Units

10.24 to **10.25** Design wall footings for the values given. The walls are to consist of reinforced concrete. Concrete weight = 23.5 kN/m³. Soil weight = 16 kN/m³ $V_c = 0.166\sqrt{f_c'}bd$

Problem	Wall Thickness	D	L	f_c'	f_y	q_a	Distance from Bottom of Footing to Final Grade
10.24	300 mm	150 kN/m	200 kN/m	20.7 MPa	344.8 MPa	170 kN/m²	1.500 m
10.25	400 mm	180 kN/m	250 kN/m	27.6 MPa	413.7 MPa	210 kN/m²	1.200 m

10.26 to **10.28** Design square single column footings for the values given. $V_c = 0.166\sqrt{f_c'}bd$

Problem	Column Size	D	L	f_c'	f_y	q_a	Distance from Bottom of Footing to Final Grade
10.26	350 mm × 350 mm	400 kN	500 kN	20.7 MPa	275.8 MPa	170 kN/m^2	1.200 m
10.27	400 mm × 400 mm	650 kN	800 kN	27.6 MPa	413.7 MPa	170 kN/m^2	1.200 m
10.28	450 mm × 450 mm	750 kN	1000 kN	27.6 MPa	413.7 MPa	210 kN/m^2	1.600 m

10.29 Design a rectangular combined footing for the two columns shown in the accompanying illustration. The bottom of the footing is to be 1.800 m below the final grade. $f_c' = 27.6$ MPa and $f_y = 344.8$ MPa. $V_c = 0.166\sqrt{f_c'}bd$.

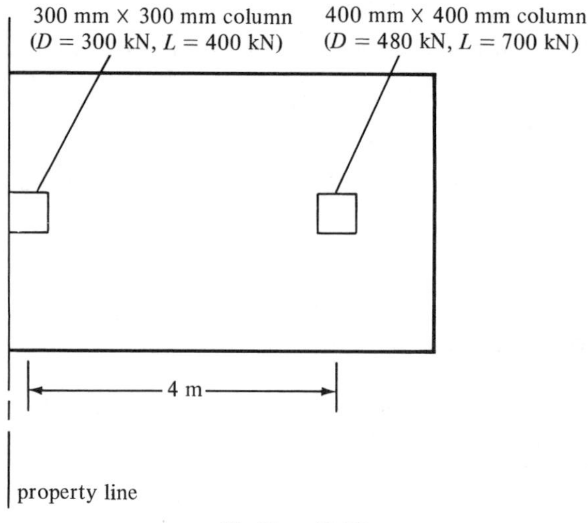

Problem 10.29

10.30 Design a plain concrete wall footing for a 300-mm-thick reinforced concrete wall that supports a 100-kN/m dead load (including its own weight) and a 120-kN/m live load. $f_c' = 20.7$ MPa, $f_y = 344.8$ MPa, and $q_a = 170$ kN/m^2. The base of the footing is to be 1.200 m below the final grade. Allowable $f_t = 0.415\sqrt{f_c'}$.

10.31 Design a square plain concrete column footing to support a 300-mm × 300-mm reinforced concrete column that in turn is supporting a 130-kN dead load and a 200-kN live load. $f_c' = 27.6$ MPa, $f_y = 344.8$ MPa, and $q_a = 210$ kN/m^2. The base of the footing is to be 1.500 m below the final grade. Allowable $f_t = 0.415\sqrt{f_c'}$.

CHAPTER 11

Retaining Walls

11.1 INTRODUCTION

A retaining wall is a structure built for the purpose of holding back or retaining or providing one-sided lateral confinement for soil or other loose material. The loose material being retained pushes against the wall, tending to overturn and slide it. Retaining walls are used in many design situations where there are abrupt changes in the ground slope. Perhaps the most obvious examples to the reader occur along highway or railroad cuts and fills. Often retaining walls are used in these locations to reduce the quantities of cut and fill as well as the right-of-way width required if the soils were allowed to assume their natural slopes. Retaining walls are used in many other locations as well, such as for bridge abutments, basement walls, and culverts.

Several different types of retaining walls are discussed in the next section, but whichever type is used, there will be three forces involved that must be brought into equilibrium. These include (1) the gravity loads of the concrete wall and any soil on top of the footing (the so-called *developed weight*), (2) the lateral pressure from the soil, and (3) the bearing resistance of the soil. In addition, the stresses within the structure have to be within permissible values and the loads must be supported in a manner such that undue settlements do not occur.

11.2 TYPES OF RETAINING WALLS

Retaining walls are generally classed as being gravity or cantilever types, with several variations possible. These are described in the paragraphs to follow, with reference being made to Figure 11.1.

The *gravity retaining wall*, Figure 11.1(a), is used for walls of up to about 10 to 12 ft in height. It is usually constructed with plain concrete and depends completely on its own weight for stability against sliding and overturning. Gravity walls may also be constructed with masonry, such as stone or block.

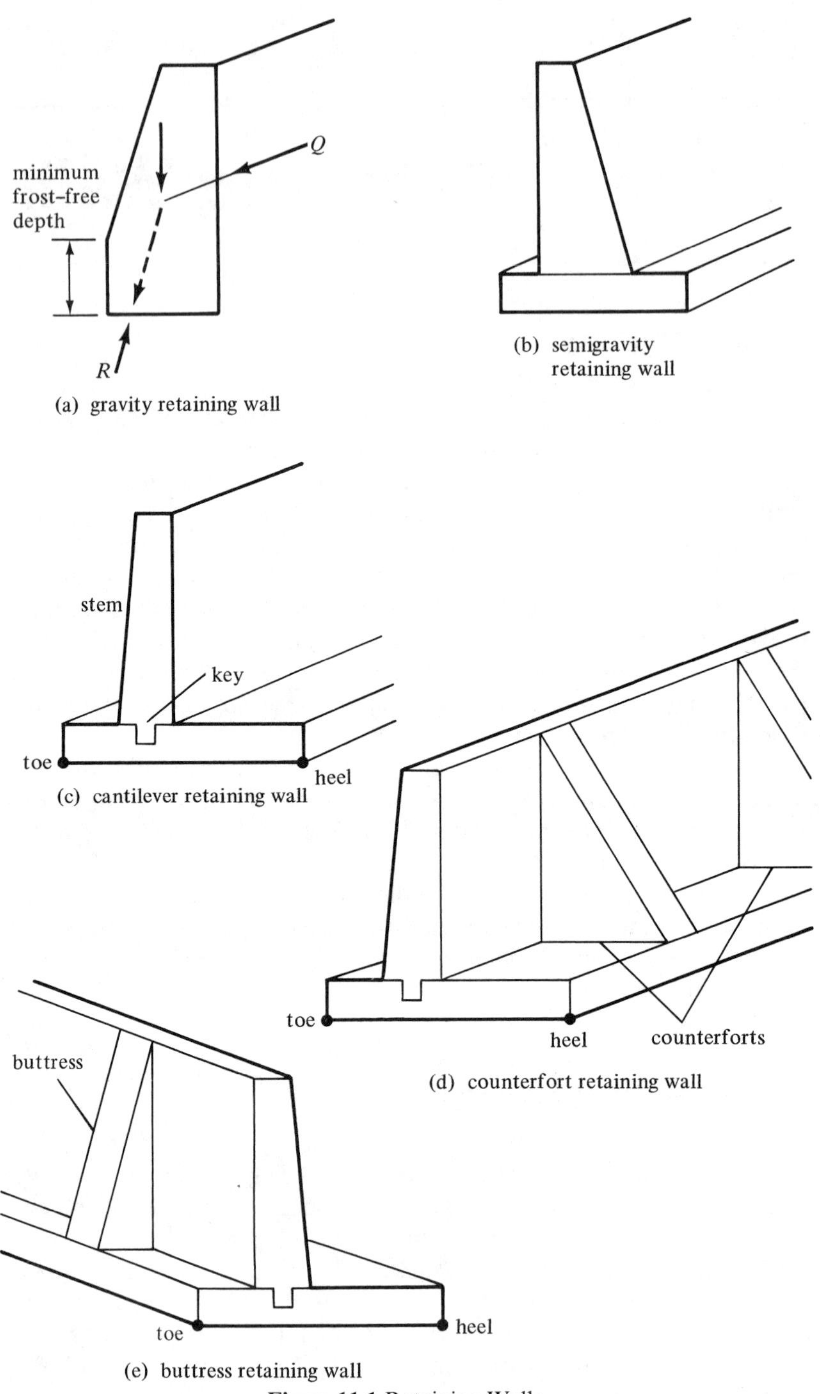

Figure 11.1 Retaining Walls

Semigravity retaining walls, Figure 11.1(b), fall in between the gravity and cantilever types (to be discussed in the next paragraph). They depend on their own weights plus the weight of some soil behind the wall to provide stability. Semigravity walls are used for approximately the same range of heights as the gravity walls and usually have some light reinforcement.

The *cantilever retaining wall* or one of its variations is the most common type of retaining wall. Such walls are generally used for heights from about 10 to 25 ft. In discussing retaining walls the vertical wall is referred to as the *stem*. The part of the footing that is pressed down into the soil is called the *toe*, while the part that tends to be lifted is called the *heel*. These parts are indicated for the cantilever retaining wall of Figure 11.1(c). The concrete and its reinforcing are so arranged that part of the material behind the wall is used along with the concrete weight to produce the necessary resisting moment against overturning. This resisting moment is generally referred to as the *righting moment*.

When it is necessary to construct retaining walls of greater heights than approximately 20 to 25 ft, the bending moments at the junction of the stem and footing become so large that the designer will, from economic necessity, have to consider other types of walls to handle the moments. This can be done by introducing cross walls on the front or back of the stem. If the cross walls are behind the stem (that is, inside the soil) and not visible, the retaining walls are called *counterfort walls*. Should the cross walls be visible (that is, on the toe side), the walls are called *buttress walls*. These walls are illustrated in parts (d) and (e) of Figure 11.1. The stems for these walls are continuous members supported at intervals by the buttresses or counterforts.

The counterfort type is more commonly used because it is normally thought to be more attractive as the cross walls or counterforts are not visible. Not only are the buttresses visible on the toe side but their protrusion on the outside or toe side of the wall will use up valuable space. Nevertheless, buttresses are somewhat more efficient than counterforts since they consist of concrete that is put in compression by the overturning moments, whereas the counterforts are concrete members used in a tension situation. Occasionally walls are designed with both buttresses and counterforts.

It seems highly probable that in the years to come the use of retaining walls built with precast concrete sections will become common. These may consist of sections erected much as are cast-in-place walls, or they may consist of walls or sheetings actually driven into the ground before excavation.[1]

[1] Kramish, F., 1964, "Structural Walls of Precast Concrete Sheet Piling," *Building Construction*, 5, no. 11, pp. 64–67.

11.3 DRAINAGE

One of the most important items in designing and constructing successful retaining walls is the prevention of water accumulation behind the walls. If water is allowed to build up there, the result can be great lateral water pressures against the wall and perhaps an even worse situation in cold weather climates due to frost action.

The best possible backfill for a retaining wall is a well-drained and cohesionless soil. Furthermore, this is the condition for which the designer normally plans and designs. In addition to a granular backfill material, weep holes of 4 in. or more in diameter (the large sizes are used for easy cleaning) are placed in the walls approximately 5 ft on center, horizontally and vertically, as shown in Figure 11.2(a). If the backfill consists of a coarse sand, it is desirable to put a few shovels of pea gravel around the weep holes to try to prevent the sand from stopping up the holes.

An even better method than weep holes (which discharge the water right at the areas of greatest footing bearing pressure on the soil beneath) involves the use of a 6- or 8-in. perforated pipe in a bed of gravel running along the base of the wall, as shown in Figure 11.2(b). Unfortunately, both weep holes and drainage pipes can become clogged, with the result that a condition of increased water pressure can occur.

The drainage methods described in the preceding paragraphs are also quite effective for reducing frost action in colder areas. Frost action can cause very large movements of the wall not just in terms of inches but perhaps even in terms of a foot or two, and over a period of time can lead to failures. Frost action, however, can be greatly reduced if coarse, properly drained materials are placed behind the walls. The thickness of the filled material perpendicular to the wall should equal at least the depth of frost penetration in the ground in that area.

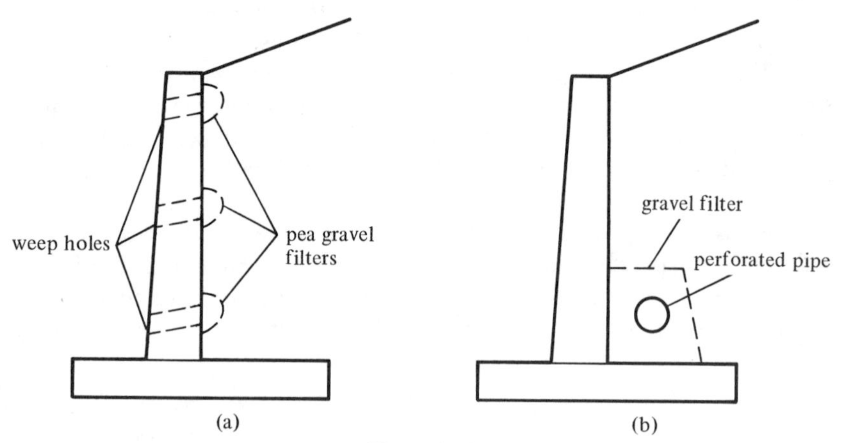

Figure 11.2

Gang-Formed Panels (Courtesy Burke Concrete Accessories, Inc.).

The best situation of all would be to keep the water out of the backfill altogether. Such a goal is normally impossible, but sometimes the surface of the backfill can be paved with asphalt or some other material, or perhaps a surface drain can be provided to remove the water, or sometimes it may be possible to interrupt the source of the water before it can get to the backfill.

11.4 FAILURES OF RETAINING WALLS

The number of failures or partial failures of retaining walls is rather alarming. The truth of the matter is that if large safety factors were not used, the situation would be even more severe. One reason for the large number of failures is the fact that designs are so often based on methods that are suitable for only certain special situations. For instance, if a wall that has a saturated clay behind it is designed by a method that is suitable for a dry granular material, future trouble can be the result.

11.5 LATERAL PRESSURES ON RETAINING WALLS

The actual pressures that occur behind retaining walls are quite difficult to estimate because of the large number of variables present. These include the kinds of backfill materials and their compactions and moisture contents, the types of materials beneath the footings, the presence or absence of surcharge, and other items. As a result, the detailed estimation of the lateral forces applied to various retaining walls is clearly a problem in theoretical soil mechanics. For this reason the discussion to follow is limited to a rather narrow range of cases.

If a retaining wall is constructed against a solid rock face, there will be no pressure applied to the wall by the rock. But if the wall is built to retain a body of water, hydrostatic pressure will be applied to the wall. At any point the pressure (p) will equal wh, where w is the unit weight of the water and h is the vertical distance from the surface of the water to the point in question.

If a wall is built to retain a soil, the soil's behavior will generally be somewhere in between that of rock and water. The pressure exerted against the wall will increase, as did the water pressure, with depth but not as rapidly. This pressure at any depth can be estimated with the following expression:

$$p = Cwh$$

In this equation w is the unit weight of the soil, h is the distance from the surface to the point in question, and C is a constant that is dependent on the characteristics of the backfill. Unfortunately, the value of C can vary quite a bit, being perhaps as low as 0.3 or 0.4 for loose granular soils and perhaps as high as 0.9 or even 1.0 for some clayey soils.

For this introductory discussion a retaining wall supporting a sloping earth fill is shown in Figure 11.3. Part of the earth behind the wall (shown by the crosshatched area) tends to slide along a curved surface (represented by the dotted line) and push against the retaining wall. The tendency of this soil to slide is resisted by friction along the soil underneath (called internal friction) and by friction along the vertical face of the retaining wall.

Internal friction is greater for a cohesive soil than for a noncohesive one, but the wetter such a soil becomes the smaller will be its cohesiveness and thus the flatter the plane of rupture. The flatter the plane of rupture, the greater is the volume of earth tending to slide and push against the

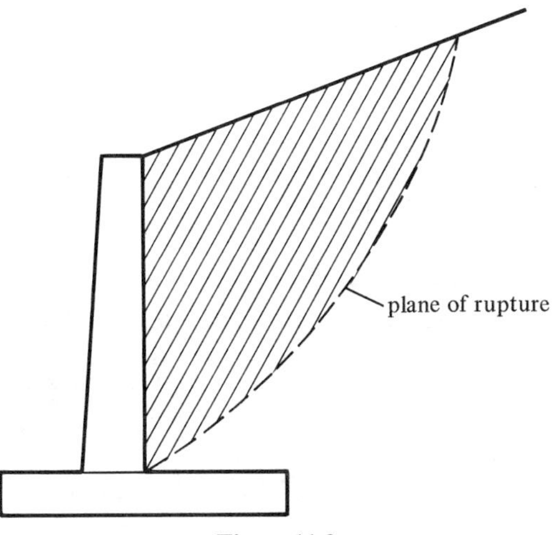

Figure 11.3

wall. Once again it can be seen that good drainage is of the utmost importance. Usually the designer assumes that he has a cohesionless granular backfill behind the walls.

Due to lateral pressure the usual retaining wall will give or deflect a little since it is constructed of elastic materials. Furthermore, unless the wall rests on a rock foundation, it will tilt or lean a small distance away from the soil due to the compressible nature of the supporting soils. For these reasons retaining walls are frequently constructed with a slight batter, or inclination, toward the backfill so that the deformations described are not obvious to the passersby.

Under the lateral pressures described, the usual retaining wall will move a little distance and *active soil pressure* will develop, as shown in Figure 11.4. For design purposes it is usually satisfactory to assume that this pressure varies linearly with the depth of the backfill. In other words, it is just as though, so far as lateral pressure is concerned, there is a liquid behind the wall of some weight less than that of water. For this reason the assumed lateral pressures are often referred to as *equivalent fluid pressures*. Values from 30 to 50 pcf are normally assumed.

If the wall moves away from the backfill and against the soil at the toe, a passive soil pressure will be the result. Passive pressure, which is also assumed to vary linearly with depth, is illustrated in Figure 11.4.

As long as the backfills are granular, noncohesive, and dry, the assumption of an equivalent liquid pressure is fairly satisfactory. Formulas based on an assumption of dry sand or gravel backfills are not satisfactory for soft clays or saturated sands. Actually, clays should not be used for backfills because their shear characteristics change easily and they may tend to creep against the wall, increasing pressures as time goes by.

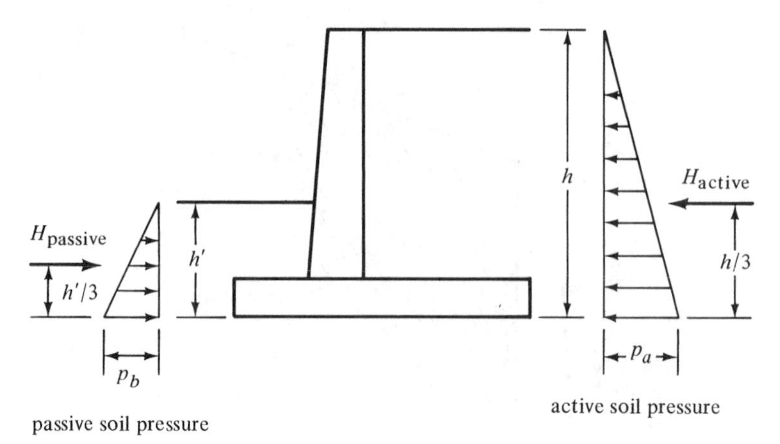

Figure 11.4

If a linear pressure variation is assumed, the active pressure at any depth can be determined from the expression to follow:

$$p_a = C_a wh$$

or, for passive pressure,

$$p_p = C_p wh'$$

In these expressions C_a and C_p are the approximate coefficients of active and passive pressures, respectively. These coefficients can be calculated by theoretical equations such as those of Rankine or Coulomb.[2] The Rankine equation (published in 1857) neglects the friction of the soil on the wall, while the Coulomb formula (published in 1776) takes it into consideration.

The Rankine equation is commonly used for ordinary retaining walls. It has been estimated that the cost of constructing retaining walls varies directly with the square of their heights. Thus as retaining walls become higher, the accuracy of the computed lateral pressures become more and more important in providing economical designs. As the Coulomb equation does take into account friction on the wall, it is thought to be the more accurate one and is often used for walls of over 20 ft. It is interesting to note that the two methods give identical results if the friction of the soil on the wall is neglected.

The Rankine expressions for the active and passive pressure coefficients are given at the end of this paragraph, with reference being made to Figure 11.5. In these expressions ϕ is the angle the backfill makes with the horizontal, while θ is the angle of internal friction of the soil. For well-drained sand or gravel backfills, the angle of internal friction is often

[2] Terzaghi, K., and Peck, R. B., 1948, *Soil Mechanics in Engineering Practice* (New York: Wiley), pp. 138–166.

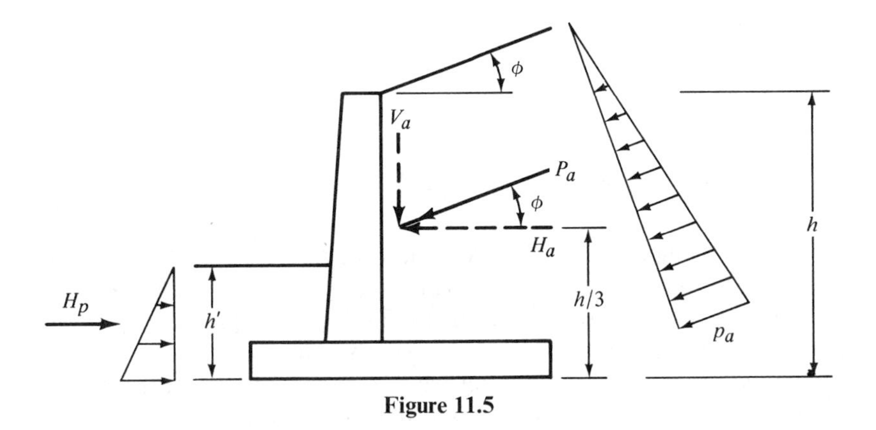

Figure 11.5

taken as the angle of repose of the slope. One common slope used is 1 vertically to $1\frac{1}{2}$ horizontally (33° 40′).

$$C_a = \cos\phi\left(\frac{\cos\phi - \sqrt{\cos^2\phi - \cos^2\theta}}{\cos\phi + \sqrt{\cos^2\phi - \cos^2\theta}}\right)$$

$$C_p = \cos\phi\left(\frac{\cos\phi + \sqrt{\cos^2\phi - \cos^2\theta}}{\cos\phi - \sqrt{\cos^2\phi - \cos^2\theta}}\right)$$

Should the backfill be horizontal—that is, should ϕ be equal to zero—the expressions become

$$C_a = \frac{1 - \sin\theta}{1 + \sin\theta}$$

$$C_p = \frac{1 + \sin\theta}{1 - \sin\theta}$$

One trouble with using these expressions is in the determinations of θ. It can be as small as 0° to 10° for soft clays and as high as 30° or 40° for some granular materials. As a result, the values of C_a can vary from perhaps 0.30 for some granular materials up to about 1.0 for some wet clays.

Once the values of C_a and C_p are determined, the total horizontal pressures, H_a and H_p, can be calculated as being equal to the areas of the respective triangular pressure diagrams. For instance, with reference made to Figure 11.5, the value of the active pressure is

$$H_a = (\tfrac{1}{2})(p_a)(h) = (\tfrac{1}{2})(C_a wh)(h)$$

$$H_a = \frac{C_a wh^2}{2}$$

and, similarly,

$$H_p = \frac{C_p w h'^2}{2}$$

11.6 FOOTING SOIL PRESSURES

Because of lateral forces the resultant force R intersects the soil underneath the footing as an eccentric load, causing a greater pressure at the toe. This toe pressure should be less than the permissible value q_a of the particular soil. It is desirable to keep the resultant force within the kern or the middle third of the footing base. This may not be economically possible in some cases, with the result that the pressure distribution will extend over only part of the base.

The soil pressure at any point can be calculated as can the stresses in an eccentrically loaded column:

$$f = -\frac{R_v}{A} \pm \frac{R_v e c}{I}$$

In this expression R_v is the vertical component of R or the total vertical load, e is the eccentricity of the load from the center of the footing, A is the area of a 1-ft-wide strip of soil of a length equal to the width of the footing base, and I is the moment of inertia of the same area about its centroid.

11.7 DESIGN OF SEMIGRAVITY RETAINING WALLS

As previously mentioned, semigravity retaining walls are designed to resist earth pressure by means of their own weight plus some developed soil weight. As they are normally constructed with plain concrete, stone, or perhaps some other type of masonry, their design is based on the assumption that tension cannot be permitted in the structure. If the resultant of the earth pressure and the wall weight (including any developed soil weight) falls within the middle third of the wall base, tensile stresses will probably be negligible in the wall.

The design procedure is a trial and error one. A wall size is assumed, safety factors against sliding and overturning are calculated, and the point where the resultant force strikes the base is determined and the soil pressures are calculated. It is normally felt that safety factors of 2.0 and 1.5 are the minimum values necessary against overturning and sliding, respectively. A suitable wall is probably obtained after two or three trials. Example 11.1 illustrates the calculations that would be made for each of the trial sizes.

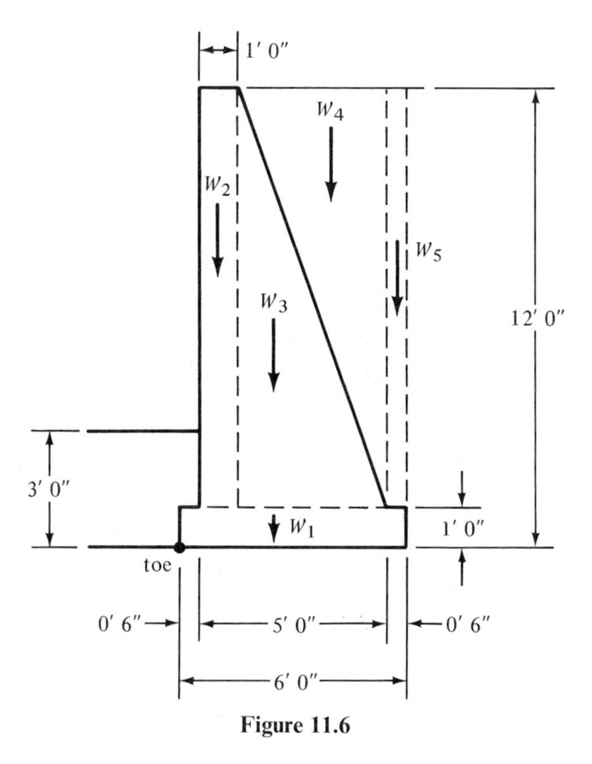

Figure 11.6

EXAMPLE 11.1

A semigravity retaining wall consisting of plain concrete (weight = 144 #/ft^3) is shown in Figure 11.6. The bank of supported earth is assumed to weigh 110 #/ft^3, to have a θ of 30°, and to have a coefficient of friction against sliding on soil of 0.5. Determine the safety factors against overturning and sliding and the bearing pressure underneath the toe of the footing. Use the Rankine expression for calculating the horizontal pressures.

SOLUTION

Computing the soil pressure coefficients

$$C_a = \frac{1 - \sin\theta}{1 + \sin\theta} = \frac{1 - 0.5}{1 + 0.5} = 0.333$$

$$C_p = \frac{1 + \sin\theta}{1 - \sin\theta} = \frac{1 + 0.5}{1 - 0.5} = 3.00$$

The value of H_a

$$H_a = \frac{C_a wh^2}{2} = \frac{(0.333)(110)(12)^2}{2} = 2{,}637\ \#$$

Overturning moment

$$\text{O.T.M.} = (2{,}637)(\tfrac{12}{3}) = 10{,}548 \text{ ft-lb}$$

Righting moments (taken about toe)

Force	Moment Arm	Moment
$W_1 = (6)(1)(144) = 864$ #	$\times$ 3.0′	= 2,592 ft-lb
$W_2 = (1)(11)(144) = 1584$	$\times$ 1.0	= 1,584 ft-lb
$W_3 = (\frac{1}{2})(4)(11)(144) = 3168$	$\times$ 2.83	= 8,965 ft-lb
$W_4 = (\frac{1}{2})(4)(11)(110) = 2420$	$\times$ 4.17	= 10,091 ft-lb
$W_5 = (0.5)(11)(110) = 605$	$\times$ 5.75	= 3,479 ft-lb
$R_v = 8641$ #		$M = 26{,}711$ ft-lb

Safety factor against overturning

$$\text{safety factor} = \frac{26{,}711}{10{,}548} = 2.53 > 2.00 \qquad \underline{\text{ok}}$$

Safety factor against sliding

Assuming soil above footing toe has eroded and thus the passive pressure is due only to soil of a depth equal to footing thickness,

$$H_p = \frac{C_p w h^2}{2} = \frac{(3.0)(110)(1)^2}{2} = 165 \text{ \#}$$

$$\text{safety factor against sliding} = \frac{(0.5)(8{,}641) + 165}{2{,}637} = 1.70 > 1.50 \qquad \text{ok}$$

Distance of resultant from toe

$$\text{distance} = \frac{26{,}711 - 10{,}548}{8{,}641} = 1.87' < 2.00' \qquad \underline{\text{just outside middle third}}$$

Soil pressure under toe

$$A = (1)(6.0) = 6.0 \text{ ft}^2$$

$$I = (\tfrac{1}{12})(1)(6)^3 = 18 \text{ ft}^4$$

$$f_{\text{toe}} = -\frac{R_v}{A} - \frac{R_v ec}{I} = -\frac{8{,}641}{6.0} - \frac{(8{,}641)(3.00 - 1.87)(3.00)}{18}$$

$$= -1{,}440 - 1{,}627 = -3{,}067 \text{ psf} < 4{,}000 \text{ psf} \qquad \underline{\text{ok}}$$

11.8 EFFECTS OF SURCHARGE

Should there be earth or other loads on the surface of the backfill, as shown in Figure 11.7, the horizontal pressure applied to the wall will be

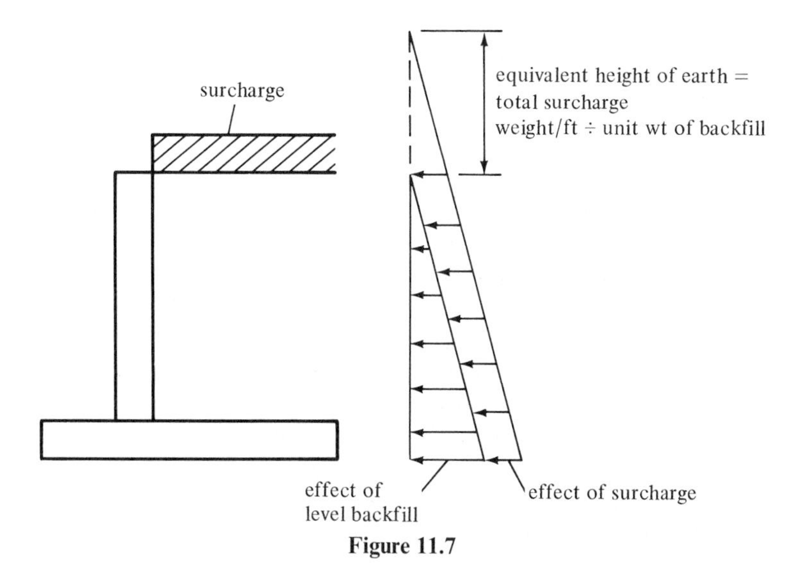

Figure 11.7

increased. If the surcharge is uniform over the sliding area behind the wall, the resulting pressure is assumed to equal the pressure that would be caused by an increased backfill height having the same total weight as the surcharge. It is usually easy to handle this situation by adding a uniform pressure to the triangular soil pressure for a wall without surcharge, as shown in the figure.

If the surcharge does not cover the area entirely behind the wall, some rather complex soil theories are available to consider the resulting horizontal pressures developed. As a consequence, the designer usually uses a rule of thumb to cover the case, a procedure that works reasonably well. He may assume, as shown in Figure 11.8, that surcharge cannot affect the pressure above the intersection of a 45° line from the edge of the surcharge

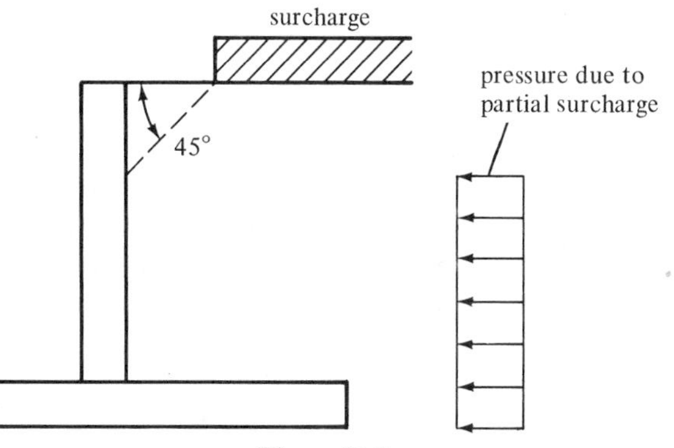

Figure 11.8

to the wall. The lateral pressure is increased, as by a full surcharge, below the intersection point. This is shown in the right side of the figure.

11.9 ESTIMATING SIZES OF CANTILEVER RETAINING WALLS

The statical design of retaining walls and consideration of their stability as to overturning and sliding is based on service-load conditions. In other words, the length of the footing and the position of the stem on the footing is based entirely on the actual soil backfill, estimated lateral pressure, coefficient of sliding friction of the soil, and so on.

On the other hand, the detailed designs of the stem and footing and their reinforcing are determined by the strength design method. To carry out these calculations, it is necessary to multiply the service loads and pressures by the appropriate load factors. From these factored loads the bearing pressures, moments, and shears are determined for use in the design.

Thus the initial part of the design consists of an approximate sizing of the retaining wall. Though this is actually a trial and error procedure, the values obtained are not too sensitive to slightly incorrect values and usually one or two trials are sufficient.

There are various rules of thumb available with which excellent initial size estimates can be made. In addition, various handbooks present the final sizes of retaining walls that have been designed for certain specific cases. This information will enable the designer to estimate very well the proportions of a wall he is preparing to design. The *CRSI Handbook* is one such useful reference.[3] In the next few paragraphs suggested methods are presented for estimating sizes without the use of a handbook. These approximate methods are very satisfactory as long as the conditions are not too much out of the ordinary.

Height of Wall

The necessary elevation at the top of the wall is normally obvious from the conditions of the problem. The elevation at the base of the footing should be selected so that it is below frost penetration in the particular area—about 3 or 4 ft below ground level in the northern states of the United States. From these elevations the overall height of the wall can be determined.

[3] *CRSI Handbook*, 1972 (Chicago: Concrete Reinforcing Steel Institute), pp. 14–3 through 14–23.

Stem Thickness

Stems are theoretically thickest at their bases because the shears and moments are greatest there. They will ordinarily have total thicknesses somewhere in the range of 8% to 12% of the overall heights of the retaining walls. The shears and moments in the stem decrease from the bottom to the top and, as a result, thicknesses and reinforcement can be reduced proportionately. Stems are normally tapered, as shown in Figure 11.9. The minimum thickness at the top of the stem is 8 in., with 12 in. preferable. As will be shown in Section 11.10, it is necessary to have a mat of reinforcing in the inside face of the stem and another mat in the outside face. To provide room for these two mats of reinforcing, for cover and spacing between the mats, a minimum total thickness of at least 8 in. is required.

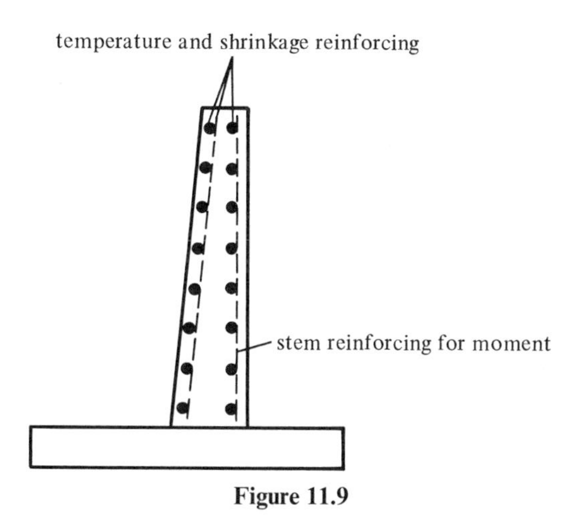

Figure 11.9

For heights up to about 12 ft, the stems of cantiliver retaining walls are normally made of constant thickness because the extra cost of setting the tapered formwork is usually not offset by the saving in concrete. Above 12-ft heights, concrete savings are usually sufficiently large to make tapering economical.

Actually the sloping face of the wall can be either the front or the back, but if the outside face is tapered it will tend to counteract somewhat the deflection and tilting of the wall due to lateral pressures.

Base Thickness

The final thickness of the base will be determined on the basis of shears and moments. For estimating, though, its total thickness will probably fall somewhere between 7% and 10% of the overall wall height. Minimum thicknesses of at least 10 to 12 in. are used.

Base Length

For preliminary estimates the base length can be taken to be about 40% to 60% of the overall wall height. A much better estimate, however, can be made by using the method described by Professor Ferguson in his reinforced concrete text.[4] For this discussion reference is made to Figure 11.10. In this figure, W is assumed to equal the weight of the material within area *abcd*. This area contains both concrete and soil but the author assumes here that it is all soil. This means a little larger safety factor will be developed against overturning. When surcharge is present, it will be added in as an additional depth of soil, as shown in the figure.

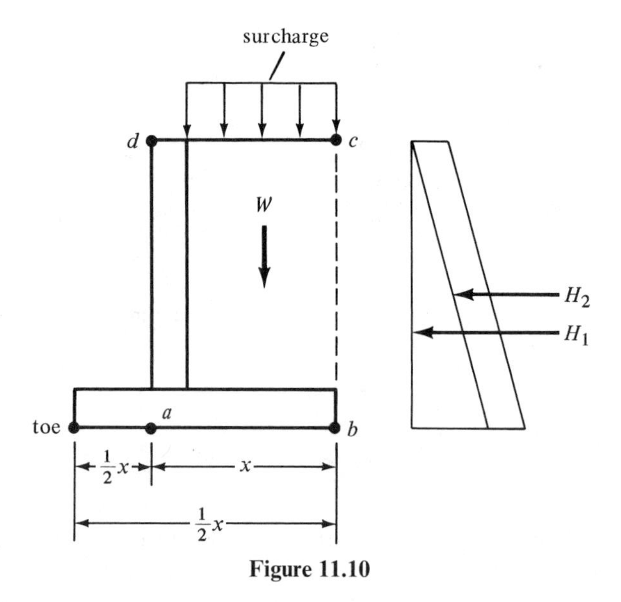

Figure 11.10

If the sum of moments about point a due to W and the lateral forces H_1 and H_2 equals zero, this will mean that the resultant force R will pass through point a. Such a moment equation can be written, equated to zero, and solved for x. Should the distance from the footing toe to point a be equal to one-half of the distance x in the figure and the resultant force R pass through point a, the footing pressure diagram will be triangular. In addition, if moments are taken about the toe of all the loads and forces

[4] Ferguson, P. M., 1973, *Reinforced Concrete Fundamentals*, 3d ed. (New York: Wiley), pp. 240–243.

for the conditions described, the safety factor against overturning will be approximately 2.

EXAMPLE 11.2

Using the approximate rules presented in this section, estimate the sizes of the parts of the retaining wall shown in Figure 11.11. The soil weighs 100 #/ft^3 and a surcharge of 300 psf is present. Assume $C_a = 0.32$.

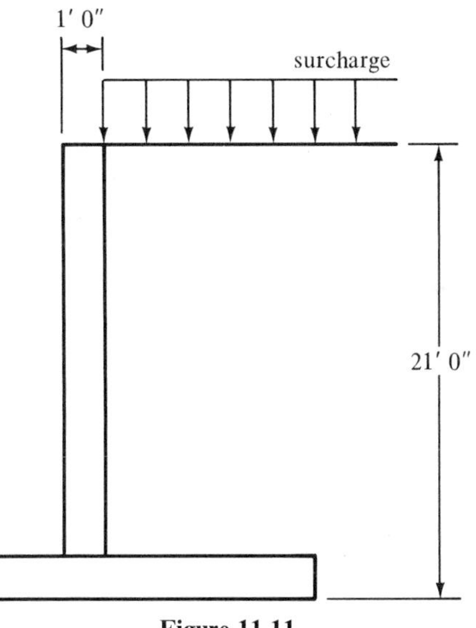

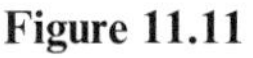
Figure 11.11

SOLUTION

Base thickness

Assume base t = 7% to 10% of overall wall height:

$$t = (0.07)(21) = 1.47' \qquad \underline{\text{say } 1'6''}$$

$$\text{height of stem} = 21'0'' \text{ minus } 1'6'' = \underline{19'6''}$$

Base length and position of stem

Calculating horizontal forces without load factors, as shown in Figure 11.12:

$$H_1 = (\tfrac{1}{2})(21)(672) = 7056 \ \#$$

$$H_2 = (21)(96) = 2016 \ \#$$

$$W = (x)(24)(100) = 2400x$$

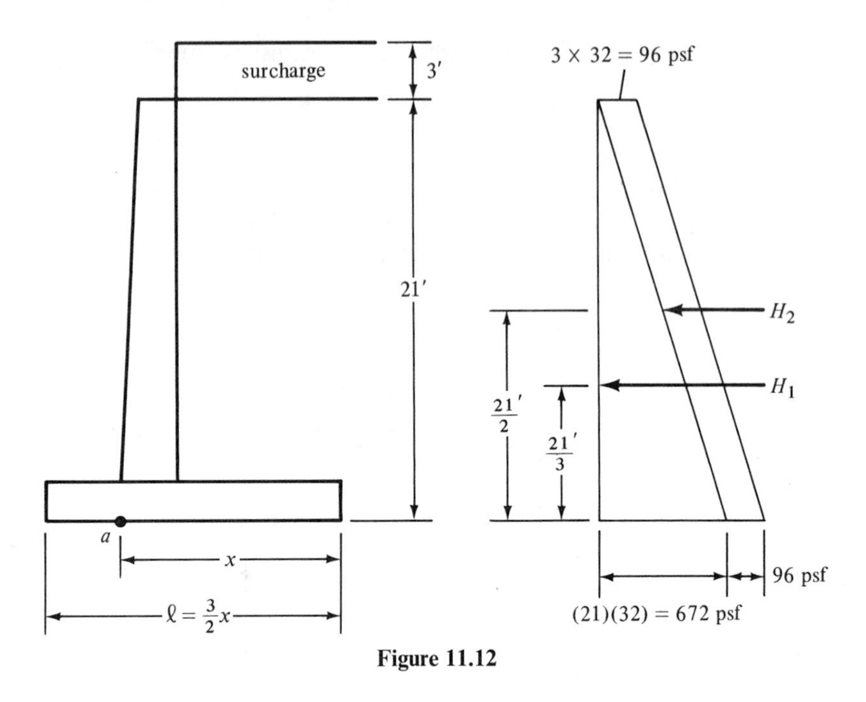

Figure 11.12

$\Sigma M_a = 0$

$$-(7056)(7.00) - (2016)(10.5) + (2400x)\left(\frac{x}{2}\right) = 0$$

$$x = 7.67'$$

$$l = (\tfrac{3}{2})(7.67) = 11.505' \text{ say } \underline{11'6''}$$

The final trial dimensions are shown in Figure 11.18.

11.10 DESIGN PROCEDURE FOR CANTILEVER RETAINING WALLS

This section is presented to describe in some detail the procedure used for designing a cantilever retaining wall. At the end of this section the complete design of such a wall is presented. Once the approximate size of the wall has been established, the stem, toe, and heel can be designed in detail. Each of these parts will be designed individually as a cantilever sticking out of a central mass, as shown in Figure 11.13.

Stem

The values of shear and moment at the base of the stem due to lateral earth pressures are computed and used to determine the stem thickness

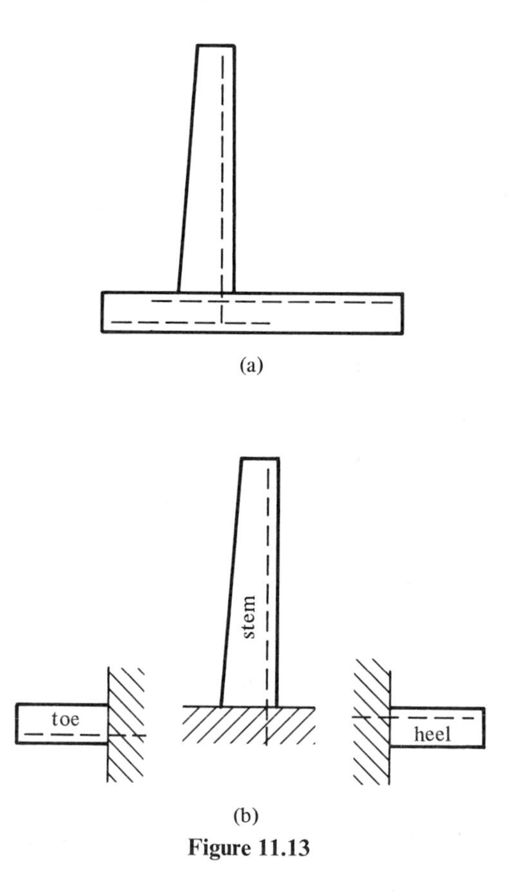

Figure 11.13

and necessary reinforcing. As the lateral pressures are considered to be live load forces a load factor of 1.7 is used.

It will be noted that the bending moment requires the use of vertical reinforcing bars on the soil side of the stem. In addition, temperature and shrinkage reinforcing must be provided. In Section 14.2 of the ACI Code, a minimum value of horizontal reinforcing equal to 0.0025 of the area of the wall *bt* is required as well as a minimum amount of vertical reinforcing (0.0015). These values may be reduced to 0.0020 and 0.0012 if the reinforcing is $\frac{5}{8}$ in. or less in diameter and if it consists of bars or welded wire fabric with f_y equal to or greater than 60,000 psi.

The major changes in temperature occur on the front or exposed face of the stem, and for this reason most of the horizontal reinforcing (perhaps two-thirds) should be placed on that face with just enough vertical steel used to support the horizontal bars. The concrete for a retaining wall should be placed in fairly short lengths not greater than 20 or 30 ft to reduce shrinkage stresses.

Factor of Safety Against Overturning

Moments are taken about the toe of the unfactored overturning and righting forces. Traditionally it has been felt that the safety factor against overturning should be at least equal to 2. In making these calculations backfill on the toe is usually neglected as it may very well be eroded. Of course, there are cases where there is a slab (for instance, a highway pavement on top of the toe backfill), which holds the backfill in place over the toe. For such situations it is reasonable to include the loads on the toe.

Factor of Safety Against Sliding

A consideration of sliding for retaining walls is a most important topic because a very large percentage of retaining wall failures occur due to sliding. To calculate the factor of safety against sliding, the estimated sliding resistance (equal to the coefficient of friction for concrete on soil times the resultant vertical force μR_v) is divided by the total horizontal force. The passive pressure against the wall is probably neglected and the unfactored loads are used.

It is usually felt that the factor of safety against sliding should be at least equal to 1.5. If it is less than this value, a lug or key is normally provided, as shown in Figure 11.14, with the front face cast directly against undisturbed soil. Keys are thought to be particularly necessary for moist clayey soils. The purpose of a key is to cause the development of passive pressure in front of and below the base of the footing, as shown by P_p in the figure. The actual theory involved, and thus the design of keys, is still a question among soil mechanics engineers. As a result, many designers select the sizes of keys by rules of thumb. One common practice

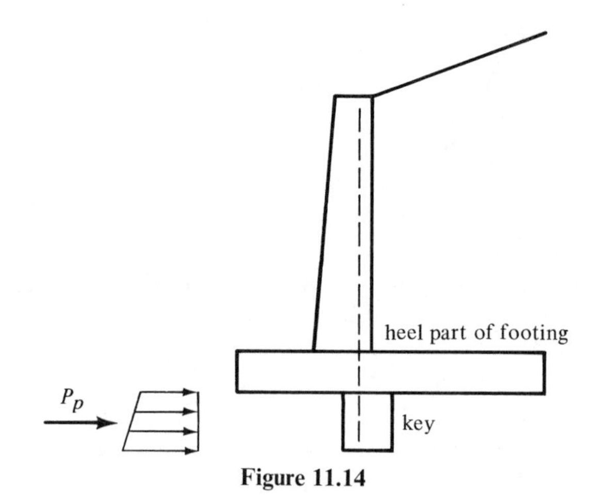

Figure 11.14

is to give them a depth equal to from two-thirds to the full depth of the footing. They are usually made approximately square in cross section and have no reinforcing provided other than perhaps the dowels mentioned in the next paragraph.

Keys are often located below the stem so that some dowels or extended vertical reinforcing may be extended into the key. If this procedure is used, the front face of the key needs to be at least 5 or 6 in. in front of the back face of the stem to allow room for the dowels. From a soil mechanics view, keys may be a little more effective if they are placed a little further toward the heel.

Heel Design

The major force applied to the heel of a retaining wall is the downward weight of the backfill behind the wall. Although it is true that there is some upward soil pressure, many designers choose to neglect it since it is relatively small. The downward loads tend to push the heel of the footing down, and the necessary upward reaction to hold it attached to the stem is provided by the vertical tensile steel in the stem, which is extended down into the footing.

As the reaction in the direction of the shear does not introduce compression into the heel part of the footing in the region of the stem, it is not permissible to determine V_u at a distance d from the face of the stem, as provided in Section 11.1.3.1 of the ACI Code. The value of V_u is determined instead at the face of the stem due to the downward loads. This shear is often of such magnitude as to control the thickness, but the moment at the face of the stem should be checked also. Since the load here consists of soil and concrete, a load factor of 1.4 is used for making the calculations.

It will be noted that the bars in the heel will be in the top of the footing. As a result, the required development length of these "top bars" is equal to 1.4 times the basic development value and may be rather large.

Toe Design

The toe is assumed to be a beam cantilevered from the front face of the stem. The loads it must support include the weight of the cantilever slab and the upward soil pressure beneath. Usually any earth fill on top of the toe is neglected (as though it had been eroded). Obviously, such a fill would increase the upward soil pressure beneath the footing, but as it acts downward and cancels out the upward pressure, it produces no appreciable changes in the shears and moments in the toe.

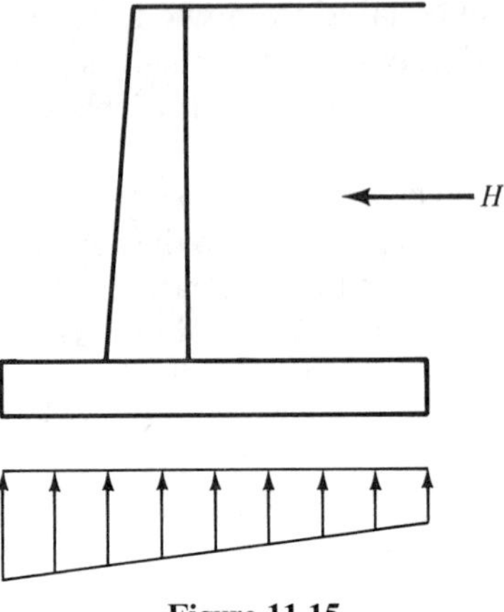

Figure 11.15

A study of Figure 11.15 shows that the upward soil pressure is the major force applied to the toe. As this pressure is primarily caused by the lateral force H, a load factor of 1.7 is used for the calculations. The maximum moment for design is taken at the face of the stem, while the maximum shear for design is assumed to occur at a distance d from the face of the stem since the reaction in the direction of the shear does introduce compression into the toe of the footing. The average designer makes the thickness of the toe the same as the thickness of the heel, although such a practice is not essential.

It is the usual practice in retaining wall construction to provide a shear keyway between the base of the stem and the footing. This practice, though definitely not detrimental, is of questionable value. The keyway is normally formed by pushing a bevelled 2×4 or 2×6 into the top of the footing, as shown in Figure 11.16. After the concrete hardens, the wood member is removed, and when the stem is cast in place above, a keyway is formed. It is becoming more and more common just to use a

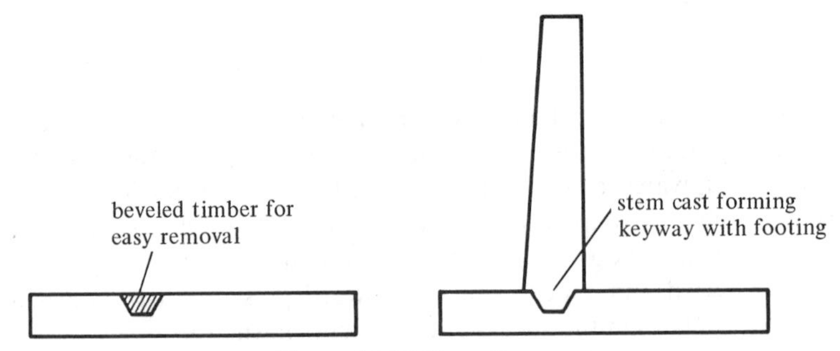

Figure 11.16 Shear Keyway

roughened surface on the top of the footing where the stem will be placed. This practice seems to be just as satisfactory as the use of a keyway.

In Example 11.3, #8 bars 6 in. on center are selected for the vertical steel at the base of the stem. Either these bars need to be embedded into the footing for development purposes or dowels equal to the stem steel need to be used for the transfer. This latter practice is quite common as it is rather difficult to hold the stem steel in position while the base concrete is placed.

The required development length of the #8 bars down into the footing or for #8 dowels is 35 in. by Table A.15 (see Appendix) when $f_y =$ 60,000 psi and $f_c' = 3{,}000$ psi). This length cannot be obtained vertically in the 2-ft-0-in. footing used unless the bars or dowels are either bent as shown in Figure 11.17(a) or extended through the footing and into the base key as shown in Figure 11.17(b). Actually, the required development length can be reduced if more but smaller dowels are used. For #7 dowels l_d is 26 in. and for #6 dowels it is 19 in.

If instead of dowels the vertical stem bars are embedded into the footing, they should not extend up into the wall more than 8 or 10 ft before they are spliced, because they are difficult to handle in construction and may easily be bent out of place or even broken. Actually, you can see after examining Figure 11.17(a) that such an arrangement of stem steel can on some occasions be very advantageous economically.

The bending moment in the stem decreases rapidly above the base and, as a result, the amount of reinforcing can be similarly reduced. It is to be remembered that these bars can only be cut off in accordance with the ACI Code development length requirements.

Example 11.3 illustrates the detailed design of a cantilever retaining wall. Several important descriptive remarks are presented in the solution and these should be carefully read.

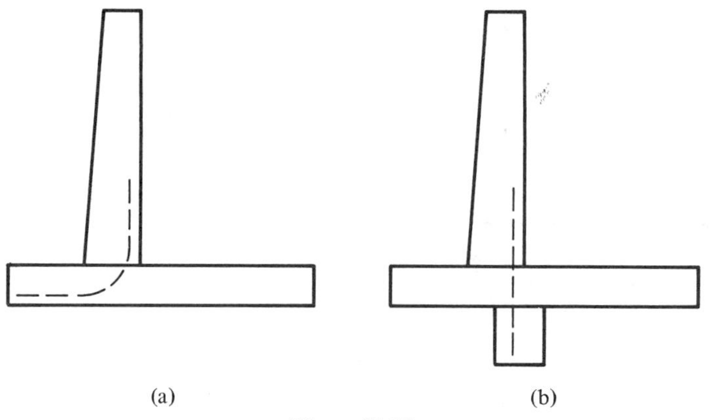

Figure 11.17

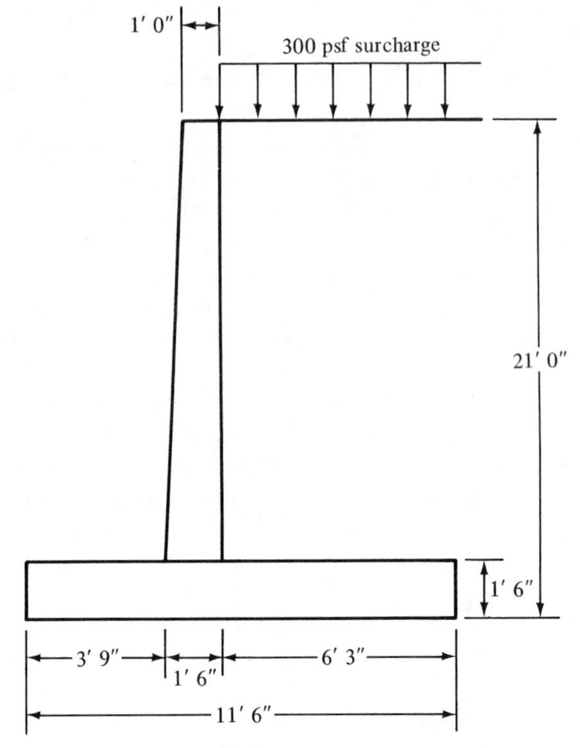

Figure 11.18

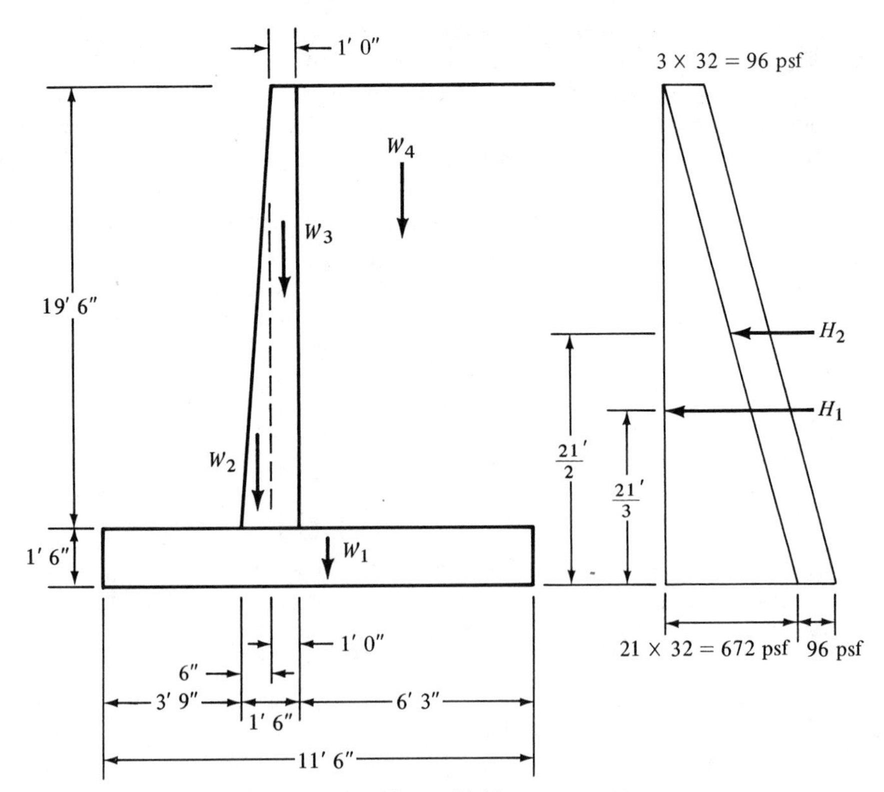

Figure 11.19

EXAMPLE 11.3

Complete the design of the cantilever retaining wall whose dimensions were estimated in Example 11.2 and are shown in Figure 11.18. $f_c' =$ 3,000 psi, $f_y = 60{,}000$ psi, $q_a = 4{,}000$ psf, and the coefficient of sliding friction equals 0.50 for concrete on soil. Use ρ approximately equal to $0.18 f_c'/f_y$ to maintain reasonable deflection control.

SOLUTION

The safety factors against overturning and sliding and the soil pressures under the heel and toe are computed using the actual unfactored loads.

Safety factor against overturning (with reference to Figure 11.19)

Overturning Moment

Force	Moment Arm	Moment
$H_1 = (\frac{1}{2})(21)(672) = 7056\ \#$	$\times\ 7.00 =$	49,392 ft-lb
$H_2 = (21)(96) = 2016\ \#$	$\times\ 10.50 =$	21,168 ft-lb
Total		70,560 ft-lb

Righting Moment

Force	Moment Arm	Moment
$w_1 = (1.5)(11.5)(150) = 2588\ \#$	$\times\ 5.75 =$	14,881 ft-lb
$w_2 = (\frac{1}{2})(19.5)(\frac{6}{12})(150) = 731\ \#$	$\times\ 4.08 =$	2,982 *ft-lb*
$w_3 = (19.5)(\frac{12}{12})(150) = 2925\ \#$	$\times\ 4.75 =$	13,894 *ft-lb*
$w_4 = (22.5)(6.25)(100) = 14{,}062\ \#$	$\times\ 8.37 =$	117,699 ft-lb
$R_v = 20{,}306\ \#$		149,456 ft-lb

$$\text{safety factor against overturning} = \frac{149{,}456}{70{,}560} = 2.12 > 2.00 \qquad \underline{\text{ok}}$$

Factor of safety against sliding

Here the passive pressure against the wall is neglected. Normally it is felt that the factor of safety should be at least 1.5. If it is not satisfactory, a shear key against sliding is normally used. The key should develop a passive pressure that is sufficient to resist the excess lateral force.

$$\text{force causing sliding} = H_1 + H_2 = 9{,}072\ \#$$

$$\text{resisting force} = \mu R_v = (0.50)(20{,}306) = 10{,}153\ \#$$

$$\text{safety factor} = \frac{10{,}153}{9{,}072} = 1.12 < 1.50$$

Use 1-ft-6-in. × 1-ft-6-in. key (size selected to provide sufficient development length for dowels selected later in solution).

Footing soil pressures

R_v = 20,306 # and is located a distance $\bar{x}$ from the toe of the footing.

$$\bar{x} = \frac{149{,}456 - 70{,}560}{20{,}306} = \frac{78{,}896}{20{,}306}$$

3.89′ <u>just inside middle third</u>

$$\text{soil pressure} = -\frac{R_v}{A} \pm \frac{Mc}{I}$$

$$A = (1)(11.5) = 11.5 \text{ ft}^2$$

$$I = (\tfrac{1}{12})(1)(11.5)^3 = 126.74 \text{ ft}^4$$

$$f_{\text{toe}} = -\frac{20{,}306}{11.5} - \frac{(20{,}306)(5.75 - 3.89)(5.75)}{126.74}$$

$$= -1{,}766 - 1{,}714 = -3{,}480 \text{ psf}$$

$$f_{\text{heel}} = -1{,}766 + 1{,}714 = -52 \text{ psf}$$

Design of stem

The lateral forces applied to the stem are calculated using a load factor of 1.7, as shown in Figure 11.20.

Design of stem for moment

$$M_u = (H_1)(6.50) + (H_2)(9.75) = (10{,}345)(6.50) + (3{,}182)(9.75)$$

$$M_u = 98{,}267 \text{ ft-lb}$$

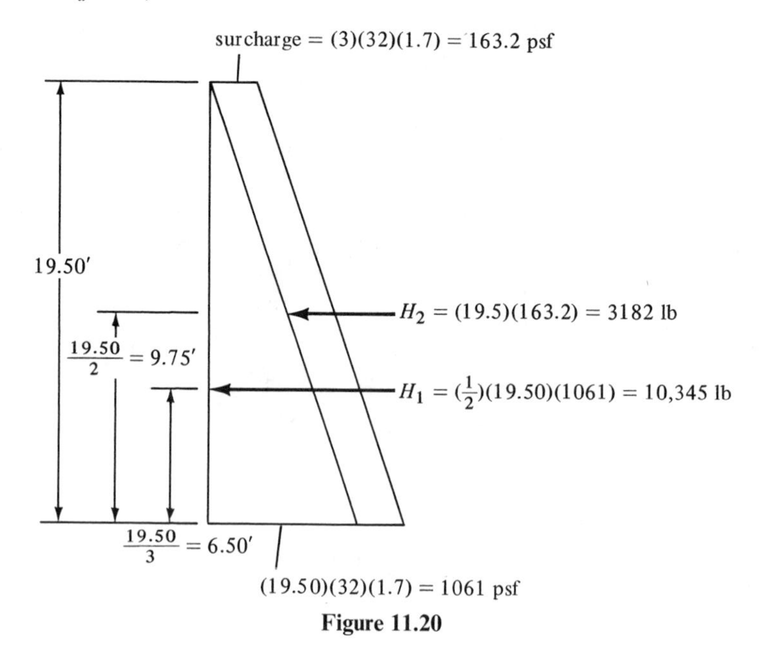

Figure 11.20

Use

$$\rho = \text{approximately} \frac{0.18 f_c'}{f_y} = \frac{(0.18)(3{,}000)}{60{,}000} = 0.009$$

$$\frac{M_u}{\phi b d^2} (\text{from chart}) = 482.6$$

$$bd^2 = \frac{(12)(98{,}267)}{(0.9)(482.6)} = 2{,}715$$

$$d = \sqrt{\frac{2{,}715}{12}} = 15.04'' + 2'' + \frac{1''}{2}$$

$$= 17.54'' \qquad \underline{\text{say } 18'' \ (d = 15.50'')}$$

$$\frac{M_u}{\phi b d^2} = \frac{(12)(98{,}267)}{(0.90)(12)(15.5)^2} = 454.5$$

$$\rho = 0.00841 \text{ (from chart)}$$

$$A_s = (0.00841)(12)(15.5) = 1.56 \text{ in.}^2 \qquad \text{use \#8 at } 6''$$

minimum vertical A_s by ACI Section 14.2.11 = 0.0015 ok

temperature and shrinkage $A_s = (0.0025)(12)(\text{average stem } t)$

$$= (0.0025)(12)\left(\frac{12 + 18}{2}\right) = 0.450 \text{ in.}^2$$

(Say one-third inside face and two-thirds outside face.)

use #4 at $7\frac{1}{2}''$ outside face and #4 at 15″ inside face

Checking shear stress in stem

Actually V_u at a distance d from the top of the footing can be used, but for simplicity,

$$V_u = H_1 + H_2 = 10{,}345 + 3{,}182 = 13{,}527 \ \#$$

$$\phi V_c = \phi 2\sqrt{f_c'} bd = (0.85)(2\sqrt{3000})(12)(15.5)$$

$$= 17{,}319 > 13{,}527 \qquad \underline{\text{ok}}$$

Design of heel

The upward soil pressure is conservatively neglected and a load factor of 1.4 is used for calculating the shear and moment since soil and concrete make up the load.

$$V_u = (22.5)(6.25)(100)(1.4) + (1.5)(6.25)(150)(1.4) = 21{,}655 \ \#$$

$$\phi V_c = (0.85)(2\sqrt{3000})(12)(14.5) = 16{,}202 < 21{,}655 \qquad \underline{\text{no good}}$$

Try 24-in. depth (d = 20.5 in.)

Neglecting slight change in V_u,

$$V_u = (0.85)(2\sqrt{3000})(12)(20.5)$$
$$= 22{,}906 > 21{,}655 \qquad \underline{\text{ok}}$$

$$M_u \text{ at face of stem} = (21{,}655)\left(\frac{6.25}{2}\right) = 67{,}672 \text{ ft-lb}$$

$$\frac{M_u}{\phi b d^2} = \frac{(12)(67{,}672)}{(0.9)(12)(20.5)^2} = 178.9$$

$$\rho = 0.00310 < \frac{200}{f_y} = 0.00333$$

Use 0.00333.

$$A_s = (0.00333)(12)(20.5) = 0.82 \text{ in.}^2/\text{ft} \qquad \underline{\text{use \#8 @ 11''}}$$
$$l_d \text{ required} = 4'\,0'' < 6'\,0'' \text{ available} \qquad \underline{\text{ok}}$$

Heel reinforcing is shown in Figure 11.21.

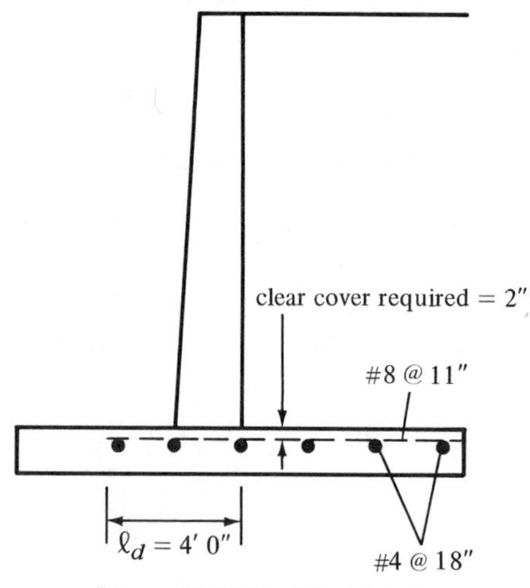

Figure 11.21 Heel Reinforcing

Note: Temperature and shrinkage steel is normally considered to be unnecessary in the heel and toe. However, the author has placed #4 bars at 18 in. on center in the long direction, as shown in Figures 11.21 and 11.23, to serve as spacers for the flexural steel and to form mats out of the reinforcing.

Design of toe

The soil pressures previously determined, for service loads are multiplied by a load factor of 1.7 since they are primarily caused by the lateral forces and shown in Figure 11.22.

$$V_u = 11{,}092 + 7{,}529 = 18{,}621$$

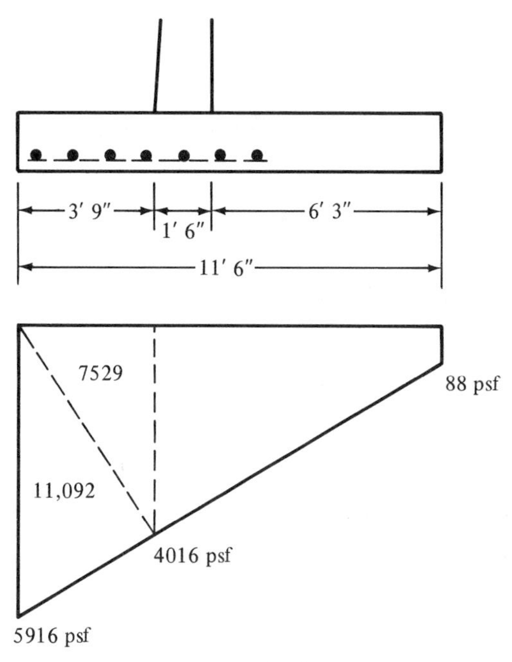

Figure 11.22

(The shear can be calculated a distance d from the face of the stem since the reaction in the direction of the shear does introduce compression into the toe of the slab, but this advantage is neglected because 18,621 # is already less than the 21,655 # shear in the heel, which was satisfactory.

$$M_u \text{ at face of stem} = (7{,}529)\left(\frac{3.75}{3}\right) + (11{,}092)\left(\frac{2}{3} \times 3.75\right) = 37{,}141 \text{ ft-lb}$$

$$\frac{M_u}{\phi bd^2} = \frac{(12)(37{,}141)}{(0.9)(12)(20.5)^2} = 98.2$$

$$\rho = \text{less than } \rho_{\min}$$

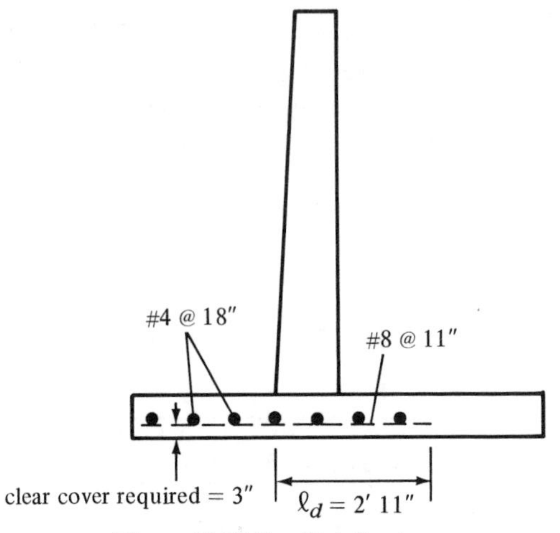

Figure 11.23 Toe Reinforcing

Therefore, use

$$\frac{200}{60{,}000} = 0.00333$$

$$A_s = (0.00333)(12)(20.5) = 0.82 \text{ in.}^2/\text{ft} \quad \underline{\text{use } \#8 \text{ at } 11''}$$

$$l_d \text{ required} = 35'' < 3'6'' \text{ available} \quad \underline{\text{ok}}$$

Toe reinforcing is shown in Figure 11.23.

Selection of dowels and lengths of vertical stem reinforcing

The detailed selection of vertical bar lengths in the stem is not given here to save space and only a few general comments are presented. First, Table 11.1 shows the reduced bending moments up in the stem and the corresponding reductions in reinforcing required.

Table 11.1

Distance from Top of Stem	M_u (ft-lb)	Effective Stem d (in.)	ρ	A_s Required (in.^{2}/ft)	Bars Needed
5′	3,176	11.04	Use ρ_{min} = 0.00333	0.44	#8 at 21″ or #7 at 16″
10′	17,218	12.58	Use ρ_{min} = 0.00333	0.50	#8 at 18″ or #7 at 14″
15′	48,960	14.12	0.00482	0.82	#8 at 11″ or #7 at $8\frac{1}{2}''$
19.5′	98,267	15.50	0.00841	1.56	#8 at 6″

After considering the possible arrangements of the steel in Figure 11.17 and the required areas of steel at different elevations in Table 11.1, the author decided to use dowels for load transfer at the stem base.

Use #8 dowels at 6 in. extending 35 in. down into footing and key.

If these dowels are spliced to the vertical stem reinforcing with no more than one-half the bars being spliced within the required lap length, the splices will fall into the class B category (ACI Code 12.16) and their lap length should at least equal $1.3l_d = (1.3)(35) = 45.5$ in. Therefore, two dowel lengths are used—half 3 ft 10 in. up into the stem and the other half 7 ft 8 in.—and the #7 bars are lapped over them, half running to the top of the wall and the other half to middepth. Actually, a much more refined design can be made that involves more cutting of bars. For such a design a diagram comparing the theoretical steel area required at various elevations in the stem and the actual steel furnished is very useful. It is to be remembered that the bars cut off must run at least a distance d or 12 diameters beyond their theoretical cutoff points.

PROBLEMS

11.1 to **11.4** Using the Rankine equation calculate the total horizontal active force and the overturning moment for the walls given. Assume that $\sin \theta = 0.5$, and the soil weighs 100 #/ft^3. Neglect the fill on the toe for each wall.

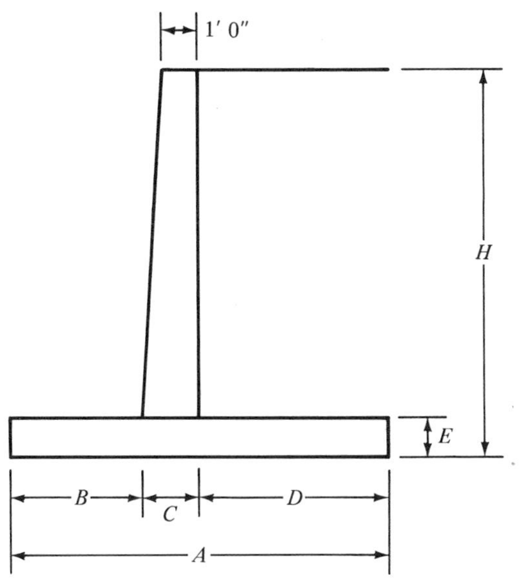

Problem 11.1 to 11.4

Problem	A	B	C	D	E	H
11.1	8′0″	2′0″	1′6″	4′6″	1′6″	15′0″
11.2	10′6″	2′6″	1′9″	6′3″	1′8″	20′0″
11.3	11′0″	3′6″	1′6″	6′0″	1′6″	22′0″
11.4	12′6″	4′0″	1′6″	7′0″	2′0″	24′0″

11.5 Repeat Problem 11.1 if ϕ is 20°.

11.6 Repeat Problem 11.3 if ϕ is 33° 40′.

11.7 to **11.9** Determine the safety factors against overturning and sliding for the gravity and semigravity walls shown if $\theta = 30°$ and the coefficient of friction(concrete on soil) is 0.5. Compute also the soil pressure under the toe and heel of each footing. The soil weighs 100 #/ft^3 while the plain concrete used in the footing weighs 144 #/ft^3. Determine horizontal pressures using the Rankine equation.

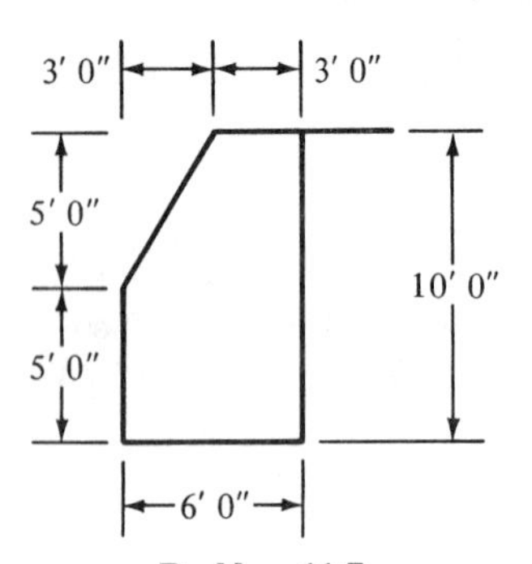

Problem 11.7

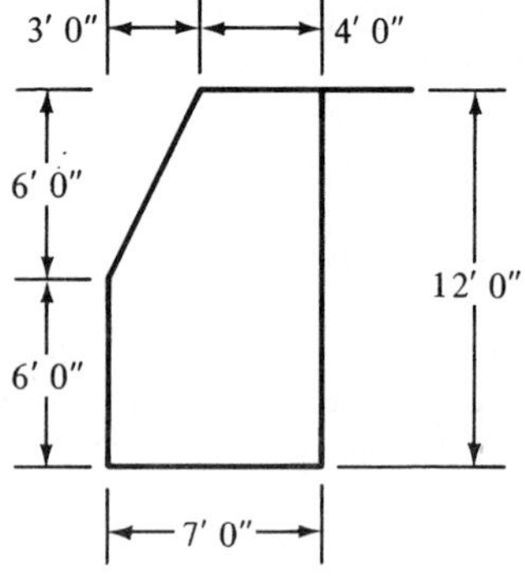

Problem 11.8

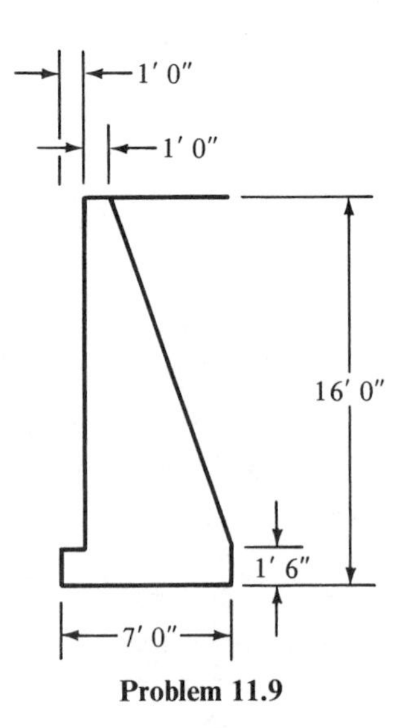

Problem 11.9

11.10 to **11.13** If Rankine's coefficient C_a is 0.30, the soil weight 110 #/ft^3, the concrete weight 150 #/ft^3, and the coefficient of friction (concrete on soil) is 0.45, determine the safety factors against overturning and sliding for the walls shown.

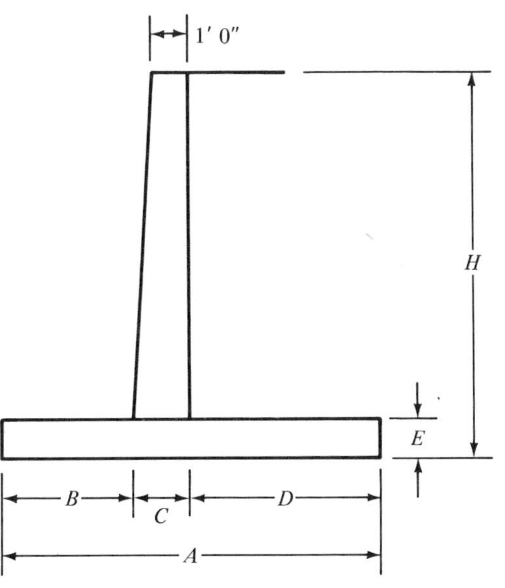

Problem 11.10 to 11.13

Problem	A	B	C	D	E	H
11.10	8′ 0″	2′ 0″	1′ 0″	5′ 0″	1′ 3″	12′ 0″
11.11	9′ 0″	2′ 6″	1′ 6″	5′ 0″	1′ 6″	16′ 0″
11.12	10′ 6″	3′ 0″	1′ 6″	6′ 0″	1′ 6″	20′ 0″
11.13	11′ 0″	3′ 6″	1′ 6″	6′ 0″	1′ 9″	22′ 0″

11.14 Repeat Problem 11.4 assuming a surcharge of 200 psf.
11.15 Repeat Problem 11.9 assuming a surcharge of 200 psf.
11.16 Repeat Problem 11.12 assuming a surcharge of 330 psf.

11.17 to **11.20** Determine approximate dimensions of retaining walls, check safety factors against overturning and sliding, and calculate soil pressures. Also determine required stem thickness at their bases and select vertical reinforcing there, using $f_y = 60{,}000$ psi, $f_c' = 3{,}000$ psi, $q_a = 5{,}000$ psf, $\rho =$ approximately $0.18\ f_c'/f_y$, angle of internal friction = 33°40′, and coefficient of sliding friction (concrete on soil) = 0.45.

Problem	H	Surcharge
11.17	12′ 0″	None
11.18	16′ 0″	None
11.19	18′ 0″	None
11.20	15′ 0″	200 psf

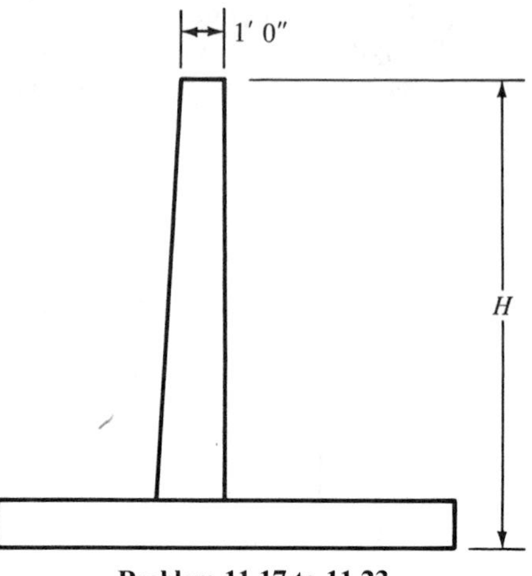

Problem 11.17 to 11.23

11.21 to **11.23** Determine same information required for Problems 11.17 to 11.20 with same data. Design heels instead of stems.

Problem	*H*	Surcharge
11.21	14′ 0″	None
11.22	18′ 0″	300 psf
11.23	20′ 0″	300 psf starting 4′ 0″ from inside face of wall

PROBLEMS WITH SI UNITS

11.24 to **11.26** Using the Rankine equation calculate the total horizontal force and the overturning moment for the walls given. Assume $\sin\theta = 0.5$, the soil weighs 16 kN/m^3, and the concrete weighs 23.5 kN/m^3. Neglect the fill on the toe for each wall.

Problem	*A*	*B*	*C*	*D*	*E*	*H*
11.24	2.400 m	600 mm	500 mm	1 300 mm	450 mm	4 m
11.25	2.700 m	700 mm	500 mm	1 500 mm	500 mm	6 m
11.26	3.150 m	800 mm	550 mm	1 800 mm	500 mm	8 m

11.27 to **11.28** Determine the safety factors against overturning and sliding for the gravity and semigravity walls shown if $\theta = 30°$ and the coefficient of sliding (concrete on soil) is 0.45. Compute also the soil pressure under the toe and heel of each footing. The soil weighs 16 kN/m^3 while the plain concrete used in the footing weighs 22.5 kN/m^3.

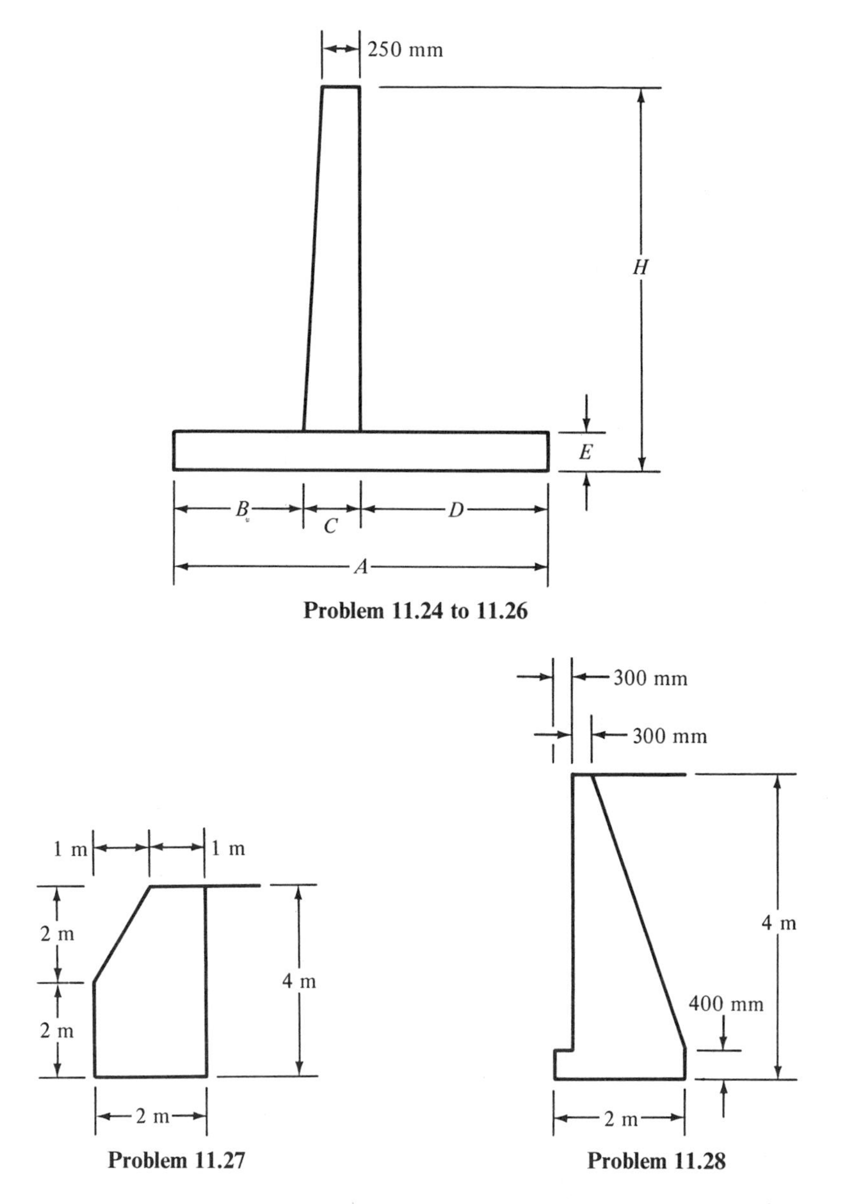

Problem 11.24 to 11.26

Problem 11.27

Problem 11.28

11.29 to **11.30** If Rankine's coefficient is 0.35, the soil weight 16 kN/m^3, the concrete weight 23.5 kN/m^3, and the coefficient of friction (concrete on soil) is 0.50, determine the safety factors against overturning and sliding for the walls shown.

Problem	A	B	C	D	E	H
11.29	4 m	1,500 mm	300 mm	2 200 mm	700 mm	5 m
11.30	5 m	1,500 mm	500 mm	3 000 mm	800 mm	7 m

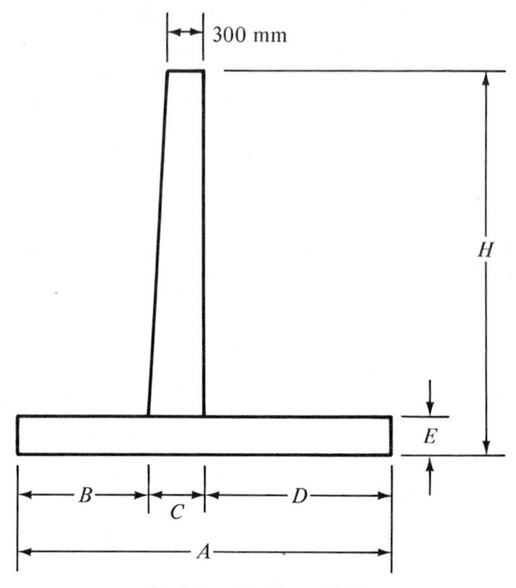

Problem 11.29 to 11.30

11.31 to **11.33** Select approximate dimensions for cantilever retaining walls and determine reinforcing required at base of stem using the following data: $f_c' = 20.7$ MPa, $f_y = 413.7$ MPa, $\rho =$ approximately $\frac{1}{2}\rho_{\max}$, angle of internal friction $= 33°40'$, coefficient of sliding friction (concrete on soil) $= 0.50$.

Problem	H	Surcharge
11.31	4 m	None
11.32	6 m	None
11.33	7 m	4 kN/m

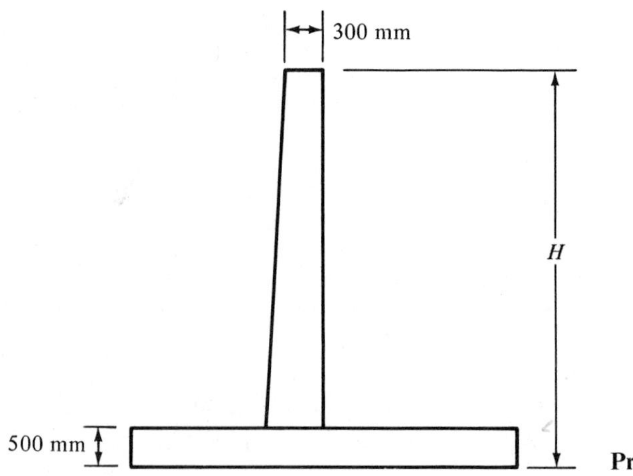

Problem 11.31 to 11.33

CHAPTER 12

Continuous Reinforced Concrete Structures

12.1 INTRODUCTION

During the construction of reinforced concrete structures, as much concrete as possible is placed in each pour. For instance, the concrete for a whole or a large part of a concrete floor, including the supporting beams and girders and parts of the columns, may be placed at the same time. The reinforcing bars extend from member to member, as from one span of a beam into the next. When there are construction joints, the reinforcing bars are left protruding from the older concrete so they may be lapped or spliced to the bars in the newer concrete. In addition, the old concrete is cleaned so that the newer concrete will bond to it as well as possible. The result of all these facts is that reinforced concrete structures are generally monolithic or continuous and thus statically indeterminate.

A load placed in one span of a continuous structure will cause shears, moments and deflections in the other spans of that structure. Not only are the beams of a reinforced concrete structure continuous but the entire structure is continuous. In other words, loads applied to a column affect the beams, slabs, and other columns and vice versa.

12.2 QUALITATIVE INFLUENCE LINES

There are many possible methods that might be used to analyze continuous structures. The most common hand calculation method is moment distribution, but other methods are frequently used, such as matrix methods, computer solutions, and others. Whichever method is used, you should understand that to determine maximum shears and moments at different sections in the structure, it is necessary to consider different positions of the live loads. As a background for this material, a brief review of *qualitative influence lines* is presented.

One Shell Plaza, Second Highest Reinforced Concrete Building in World, Houston, Texas (Courtesy of Master Builders).

Qualitative influence lines are based on a principle introduced by the German professor Heinrich Müller-Breslau. This principle follows: *The deflected shape of a structure represents to some scale the influence line for a function such as reaction, shear, or moment if the function in question is allowed to act through a small distance.* In other words, the structure draws its own influence line when the proper displacement is made.

The shape of the usual influence line needed for continuous structures is so simple to obtain with the Müller-Breslau principle that in many situations it is unnecessary to compute the numerical values of the ordinates. It is possible to roughly sketch the diagram with sufficient accuracy to locate the critical positions for live loads for various functions of the structure. These diagrams are referred to as qualitative influence lines, while those with numerical values are referred to as quantitative influence lines.[1]

If the influence line is desired for the left reaction of the continuous beam of Figure 12.1(a), its general shape can be determined by letting the

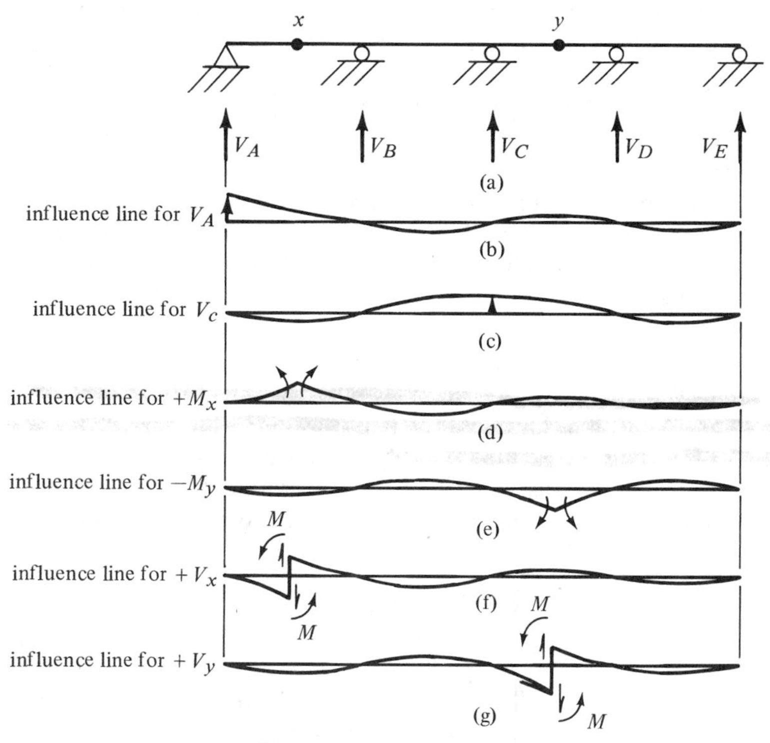

Figure 12.1 Influence Lines

[1] McCormac, J. C., 1975, *Structural Analysis*, 3d ed. (New York: Intext), pp. 185–188, 400–402.

reaction act upward through a unit distance, as shown in part (b) of the figure. If the left end of the beam is pushed up, the beam will take the shape shown. This distorted shape can be easily sketched by remembering that the other supports are unyielding. The influence line for V_c, drawn in a similar manner, is shown in Figure 12.1(c).

Figure 12.1(d) shows the influence line for positive moment at point x near the center of the left-hand span. The beam is assumed to have a pin or hinge inserted at x and a couple is applied adjacent to each side of the pin, which will cause compression in the top fibers. Bending the beam on each side of the pin causes the left span to take the shape indicated, and the deflected shape of the remainder of the beam may be roughly sketched. A similar procedure is used to draw the influence line for negative moment at point y in the third span, except that a moment couple is applied at the assumed pins, which will tend to cause compression in the bottom beam fibers, corresponding with negative moment.

Finally, qualitative influence lines are drawn for positive shear at points x and y. At point x the beam is assumed to be cut, and the two vertical forces of the nature required to give positive shear are applied to the beam on the sides of the cut section. The beam will take the shape shown in Figure 12.1(f). The same procedure is used in Figure 12.1(g) to draw the diagram for positive shear at point y. [Theoretically, for qualitative shear influence lines, it is necessary to have a moment on each side of the cut section sufficient to maintain equal slopes. Such moments are indicated in parts (f) and (g) of the figure by the letter M.]

From these diagrams considerable information is available concerning critical live-loading conditions. If a maximum positive value of V_A were desired for a uniform live load, the load would be placed in spans 1 and 3, where the diagram has positive ordinates; if maximum negative moment were required at point y, spans 2 and 4 would be loaded; and so on.

Qualitative influence lines are particularly valuable for determining critical load positions for buildings, as illustrated by the moment influence line for the building of Figure 12.2. In drawing diagrams for an entire frame, the joints are assumed to be free to rotate, but the members at each joint are assumed to be rigidly connected to each other so that the

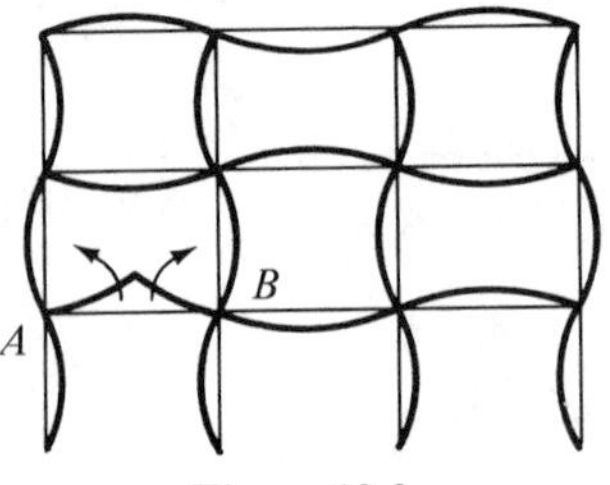

Figure 12.2

angles between them do not change during rotation. The influence line shown in the figure is for positive moment at the center of beam *AB*.

The spans that should be loaded to cause maximum positive moment are obvious from the diagram. It should be realized that loads on a member more than approximately three spans away have little effect on the function under consideration.

12.3 LIMIT DESIGN

Throughout the preceding eight chapters of this book, ultimate design methods have been used to proportion members that were analyzed by elastic methods. This procedure certainly must seem inconsistent to you because members that are subjected to loads greater than about half their ultimate capacities do not behave elastically. Although the method is inconsistent, it does nevertheless produce safe and conservative designs. Furthermore, the evaluation of the true collapse load or limit strength of an entire reinforced concrete frame requires a rather difficult analysis.

It can be clearly shown that a statically indeterminate beam or frame will normally not collapse when its ultimate moment capacity is reached at just one section. Instead, there is a redistribution of the moments in the structure. Its behavior is rather similar to the case where three men are walking along with a log on their shoulders and one of the men gets tired and lowers his shoulder just a little. The result is a redistribution of loads to the other men and thus changes in the shears and moments throughout the log.

It might be well at this point to attempt to distinguish between the terms *plastic design* as used in structural steel and *limit design* as used in reinforced concrete. In structural steel, plastic design involves the increased resisting moment of a member after the extreme fiber of the member is stressed to its yield point as well as the redistribution or change in the moment pattern in the member. In reinforced concrete the increase in resisting moment of a section after part of the section has been stressed to its yield point has already been accounted for in the strength design procedure. Therefore, limit design for reinforced concrete structures is concerned only with the change in the moment pattern after the steel reinforcing at some cross section is stressed to its yield point.

The basic assumption used for limit design of reinforced concrete structures and for plastic design of steel structures is the ability of these materials to resist a so-called yield moment while an appreciable increase in local curvature occurs. In effect, if one section of a statically indeterminate member reaches this moment, it begins to yield but does not fail. It rather acts like a hinge (called a *plastic hinge*) and throws the excess load off to those sections of the members that have lesser stresses. The

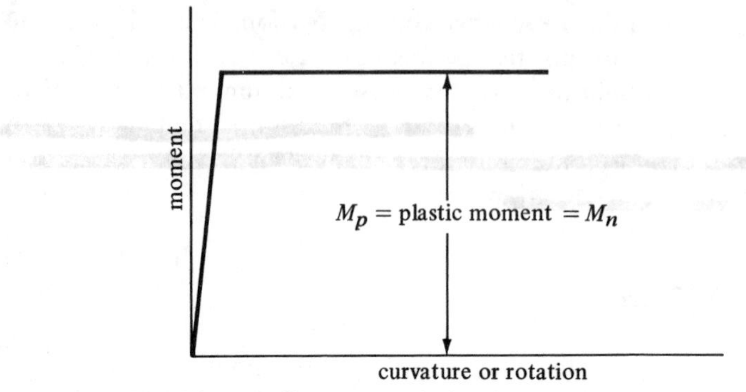

Figure 12.3 Moment Curvature Relationship for an Ideal Plastic Material

resulting behavior is much like that of the log supported by three men when one man lowered his shoulder.

To apply the limit design or plastic theory to a particular structure, it is necessary for that structure to behave plastically. For this initial discussion it is assumed that an ideal plastic material, such as a ductile structural steel, is involved. Figure 12.3 shows the relationship of moment to the resulting curvature of a short length of a ductile steel member.

Though the moment-to-curvature relationship for reinforced concrete is quite different from the ideal one pictured in Figure 12.3, the actual curve can be approximated reasonably well by the ideal one, as shown in Figure 12.4. The dotted line in the figure represents the ideal curve, while the solid line is a typical one for reinforced concrete. Tests have shown

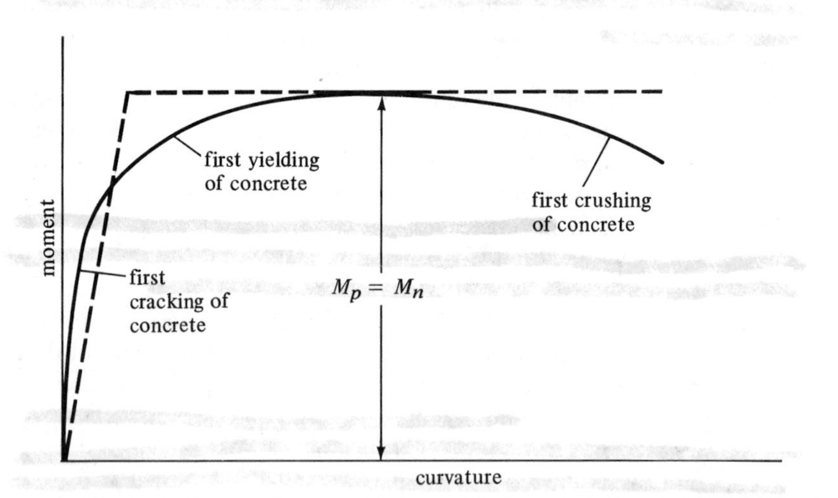

Figure 12.4 Typical Moment-Curvature Relationship for a Reinforced Concrete Member

that the lower the reinforcing percentage in the concrete ρ or $\rho - \rho'$ (where ρ' is the percentage of compressive reinforcing), the closer will the concrete curve approach the ideal curve. This is particularly true when very ductile reinforcing steels such as grade 40 are used. Should a large percentage of steel be present in a reinforced concrete member, the yielding that actually occurs before failure will be so limited that the ultimate or limit behavior of the member will not be greatly affected by yielding.

The Collapse Mechanism

To consider moment redistribution in steel or reinforced concrete structures, it is felt necessary to first consider the location and number of plastic hinges required to cause a structure to collapse. A statically determinate beam will fail if one plastic hinge develops. To illustrate this fact the simple beam of constant cross section loaded with a concentrated load at midspan shown in Figure 12.5(a) is considered. Should the load be increased until a plastic hinge is developed at the point of maximum moment (underneath the load in this case), an unstable structure will have been created, as shown in Figure 12.5(b). Any further increase in load will cause collapse.

The plastic theory is of little advantage for statically determinate beams and frames, but it may be of decided advantage for statically indeterminate beams and frames. For a statically indeterminate structure to fail, it is necessary for more than one plastic hinge to form. The number of plastic hinges required for failure of statically indeterminate structures will be shown to vary from structure to structure but may never be less

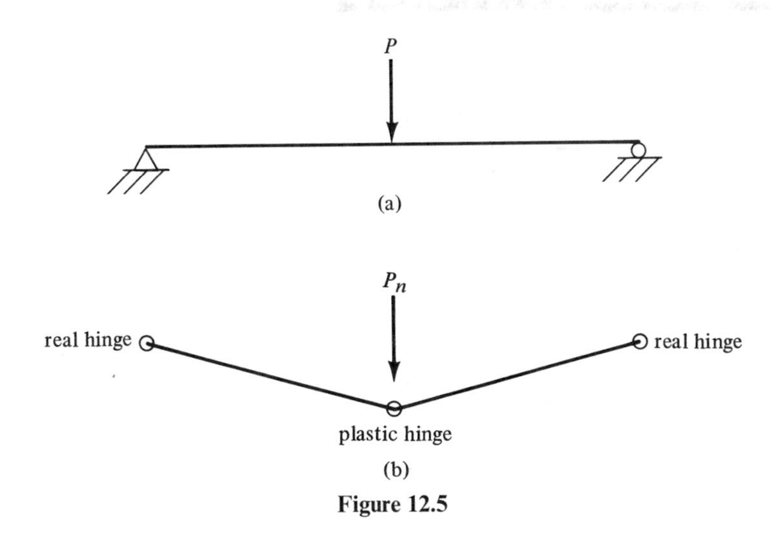

Figure 12.5

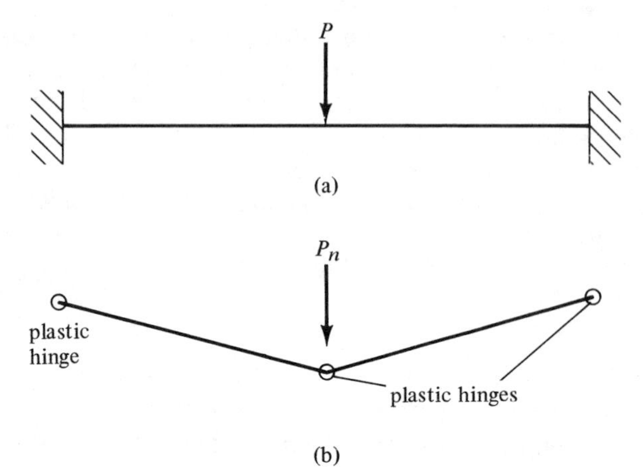

Figure 12.6

than two. The fixed-end beam of Figure 12.6 cannot fail unless the three plastic hinges shown in the figure are developed.

Although a plastic hinge may have formed in a statically indeterminate structure, the load can still be increased without causing failure if the geometry of the structure permits. The plastic hinge will act like a real hinge insofar as increased loading is concerned. As the load is increased, there is a redistribution of moment because the plastic hinge can resist no more moment. As more plastic hinges are formed in the structure, there will eventually be a sufficient number of them to cause collapse.

The propped beam of Figure 12.7 is an example of a structure that will fail after two plastic hinges develop. Three hinges are required for collapse, but there is a real hinge on the right end. In this beam the largest elastic

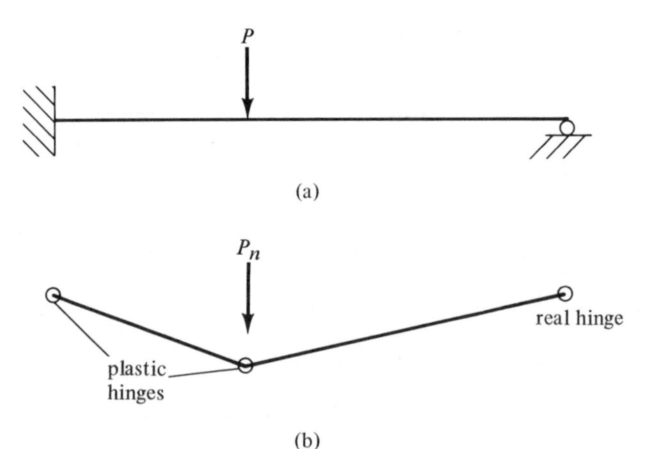

Figure 12.7

moment caused by the design-concentrated load is at the fixed end. As the magnitude of the load is increased, a plastic hinge will form at that point.

The load may be further increased until the moment at some other point (here it will be at the concentrated load) reaches the plastic moment. Additional load will cause the beam to collapse. The arrangement of plastic hinges, and perhaps real hinges that permit collapse in a structure, is called the *mechanism*. Parts (b) of Figures 12.5, 12.6, and 12.7 show mechanisms for various beams.

Plastic Analysis by the Equilibrium Method

To plastically analyze a structure, it is necessary to compute the plastic or ultimate moments of the sections, to consider the moment redistribution after the ultimate moments develop, and finally to determine the ultimate loads that exist when the collapse mechanism is created. The method of plastic analysis known as the *equilibrium method* will be illustrated in this section.

As the first illustration the fixed-end beam of Figure 12.8 is considered. It is desired to determine the value of w_n, the theoretical ultimate load the beam can support. The maximum moments in a uniformly loaded fixed-end beam in the elastic range occur at the fixed ends, as shown in the figure.

If the magnitude of the uniform load is increased, the moments in the beam will be increased proportionately until a plastic moment is eventually developed at some point. Due to symmetry, plastic moments will be developed at the beam ends, as shown in Figure 12.9(b). Should the loads be further increased, the beam will be unable to resist moments larger than M_n at its ends. Those points will rotate through large angles

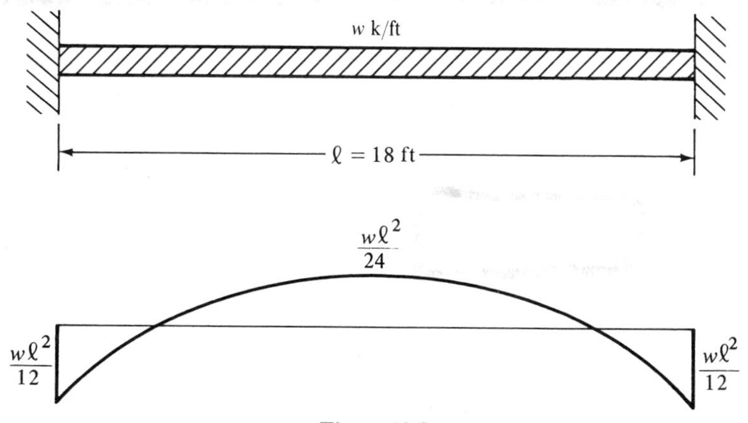

Figure 12.8

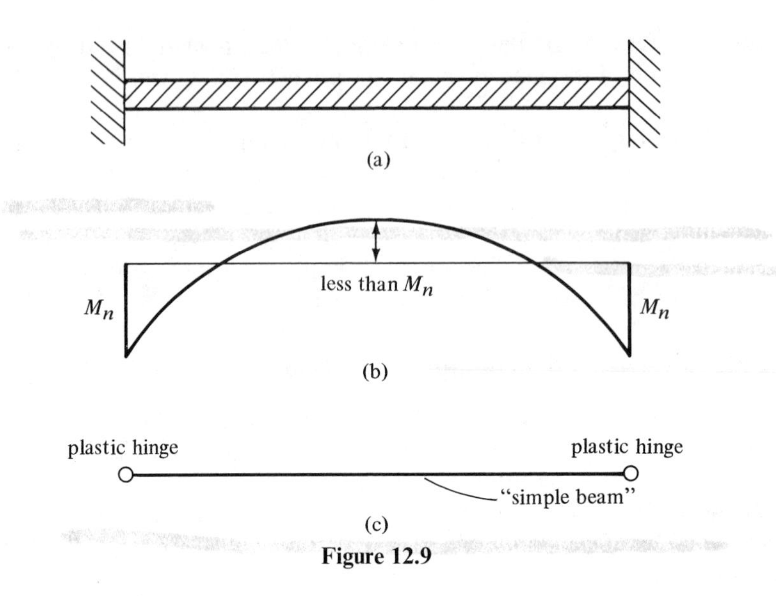

Figure 12.9

and thus the beam will be permitted to deflect more and permit the moments to increase out in the span. Although the plastic moment has been reached at the ends and plastic hinges formed, the beam cannot fail as it has, in effect, become a simple end-supported beam, for further load increases as shown in Figure 12.9(c).

The load can now be increased on this "simple" beam and the moments at the ends will remain constant; however, the moment out in the span will increase as it would in a uniformly loaded simple beam. This increase is shown by the dotted line in Figure 12.10(b). The load may be increased until the moment at some other point (here the beam center line) reaches the plastic moment. When this happens, a third plastic hinge will have developed and a mechanism will have been created permitting collapse.

One method of determining the value of w_n is to take moments at the center line of the beam (knowing the moment there is M_n at collapse). Reference is made here to Figure 12.10(a) for the beam reactions.

$$M_n = -M_n + \left(w_n \frac{l}{2}\right)\left(\frac{l}{2} - \frac{l}{4}\right) = \frac{w_n l^2}{16}$$

$$w_n = \frac{16M_n}{l^2}$$

The same value could be obtained by considering the diagrams shown in Figure 12.11. You will remember that a fixed-end beam can be replaced with a simply supported beam plus a beam with end moments. Thus the final moment diagram for the fixed-end beam equals the moment diagram if the beam had been simply supported plus the end moment diagram.

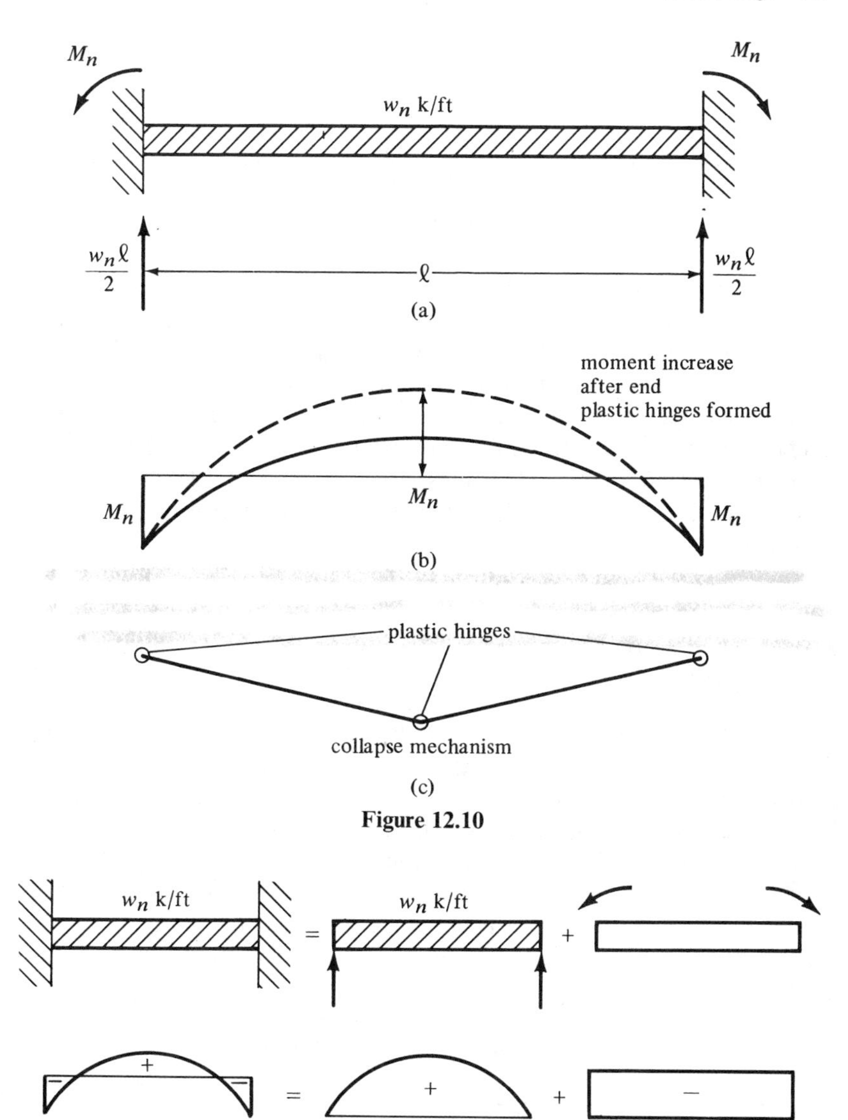

Figure 12.10

Figure 12.11

For the beam under consideration, the value of M_n can be calculated as follows (see Figure 12.12).

$$2M_n = \frac{w_n l^2}{8}$$

$$M_n = \frac{w_n l^2}{16}$$

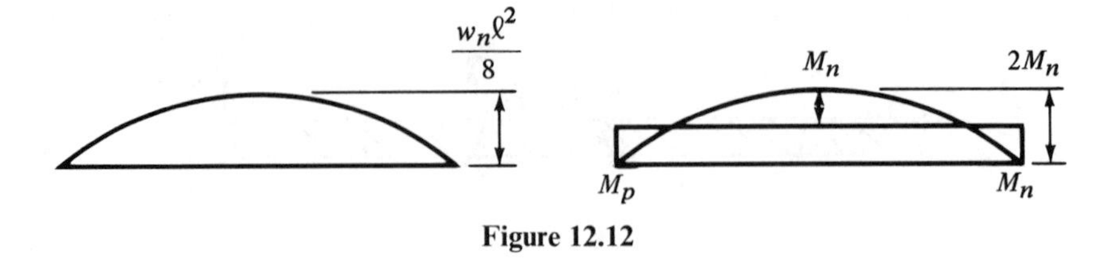

Figure 12.12

The propped beam of Figure 12.13, which supports a concentrated load, is presented as a second illustration of plastic analysis. It is desired to determine the value of P_n, the theoretical ultimate load the beam can support before collapse. The maximum moment in this beam in the elastic range occurs at the fixed end, as shown in the figure. If the magnitude of the concentrated load is increased, the moments in the beam will increase proportionately until a plastic moment is eventually developed at some point. This point will be at the fixed end where the elastic moment diagram has its largest ordinate.

After this plastic hinge is formed, the beam will act as though it is simply supported insofar as load increases are concerned, because it will have a plastic hinge at the left end and a real hinge at the right end. An increase in the magnitude of the load P will not increase the moment at the left end but will increase the moment out in the beam, as it would in a simple beam. The increasing simple beam moment is indicated by the dotted line in Figure 12.13(c). Eventually the moment at the concentrated load will reach M_n and a mechanism will form consisting of two plastic hinges and one real hinge, as shown in Figure 12.13(d).

The value of the theoretical maximum concentrated load P_n that the beam can support can be determined by taking moments to the right or left of the load. Figure 12.13(e) shows the beam reactions for the conditions existing just before collapse. Moments are taken to the right of the load as follows:

$$M_n = \left(\frac{P_n}{2} - \frac{M_n}{20}\right)10$$

$$P_n = 0.3M_n$$

The subject of plastic analysis can be continued for different types of structures and loadings, as described in several textbooks on structural analysis or steel design.[2] The method has been proved to be satisfactory for ductile structural steels by many tests. Concrete, however, is a relatively brittle material, and the limit design theory has not been fully accepted by the ACI Code. The Code does recognize that there is some redistribution of moments and permits partial redistribution based on a rule of thumb that is presented in the next section of this chapter.

[2] McCormac, J. C., 1971, *Structural Steel Design*, 2d ed. (New York: Intext), pp. 554–592.

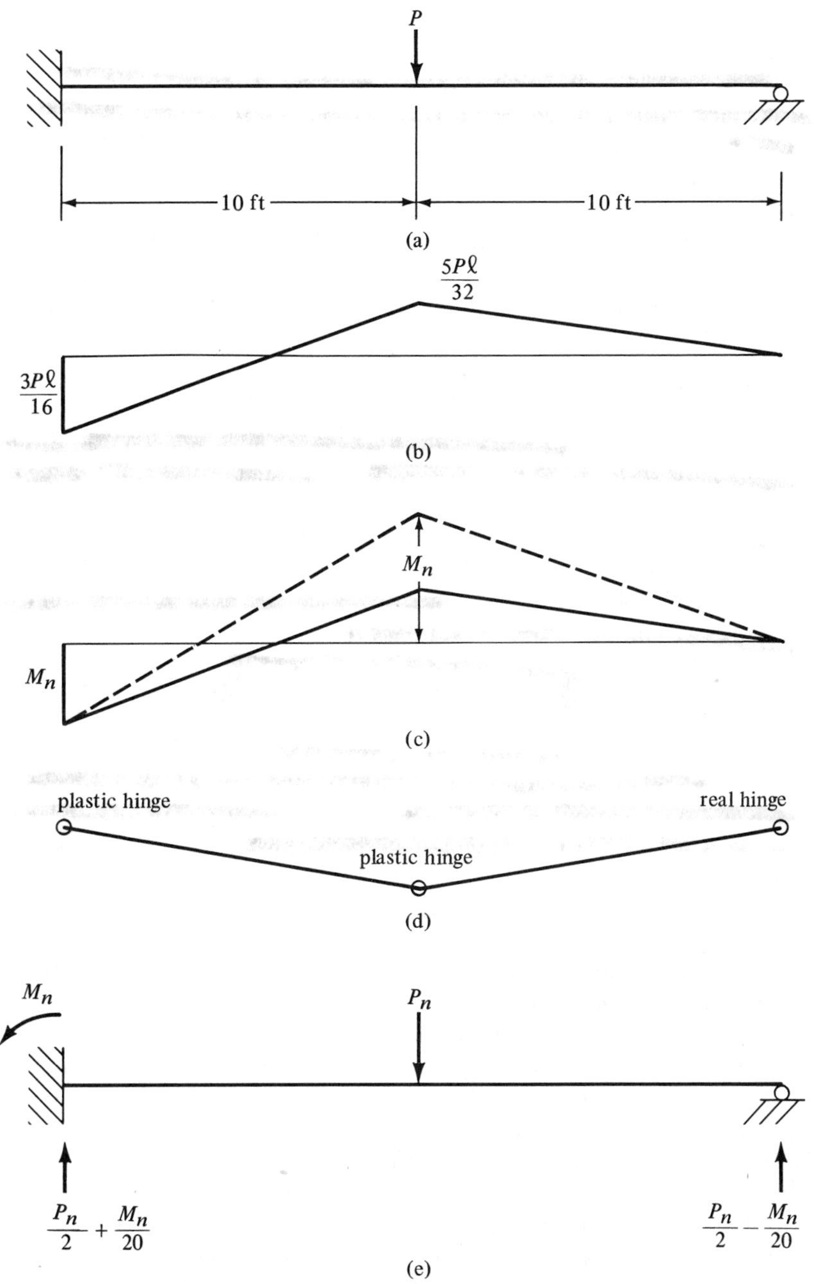
P
10 ft
10 ft
(a)
5Pℓ/32
3Pℓ/16
(b)
M_n
M_n
(c)
plastic hinge
real hinge
plastic hinge
(d)
M_n
P_n
$\frac{P_n}{2} + \frac{M_n}{20}$
$\frac{P_n}{2} - \frac{M_n}{20}$
(e)

Figure 12.13

12.4 LIMIT DESIGN UNDER THE ACI CODE

Tests of reinforced concrete frames have shown that under certain conditions there is definitely a redistribution of moments before collapse occurs. Recognizing this fact, the ACI Code (8.4.1) permits the maximum negative moments in a continuous flexural member as determined by a theoretical elastic analysis (*not by an approximate analysis*) to be increased or decreased by not more than the percentage given at the end of this paragraph. The altered negative moments are then used to determine the distribution of moments throughout the rest of the member.

$$20\left(1 - \frac{\rho - \rho'}{\rho_b}\right)$$

In this expression ρ is the percentage of tensile steel A_s/bd, ρ' is the percentage of compression steel A_s'/bd, and ρ_b is the balanced steel ratio. The Code states that the net reinforcement ratio ρ, or $\rho - \rho'$, at the section where the moment is reduced may not exceed $0.50\rho_b$. The value of ρ_b, which is determined by the following expression, was previously developed in Section 4.8 of this book. Table A.8 (see Appendix) show calculated ρ_b values for different concretes and steels.

$$\rho_b = \frac{0.85\beta_1 f_c'}{f_y}\left(\frac{87{,}000}{87{,}000 + f_y}\right)$$

The permitted increase or decrease of negative moments is not applicable to prestressed concrete sections nor is it applicable to members designed by the alternate design method nor to slabs designed by the Direct Design Method described in Chapter 14 of this text. A different value for such members is permitted in Section 18.10.4 of the ACI Code. The ACI Code's percentage of moment redistribution has purposely been limited to a very conservative value to be sure that excessive-size concrete cracks do not occur at high steel stresses and to ensure the provision of adequate ductility for moment redistribution at the plastic hinges.

It is probable that the ACI Code will expand their presently conservative redistribution method after the behavior of plastic hinges is better understood, particularly as regards shears, deflections, and development of reinforcing. It is assumed here that the sections are satisfactorily reinforced for shear so that the ultimate moments can be reached without shear failure occurring. The adjustments are applied to the moments resulting from each of the different loading conditions. The member in question will then be proportioned on the basis of the resulting moment envelope. Figures 12.14 through 12.17 illustrate the application of the moment redistribution permitted by the Code to a three-span continuous beam with a service dead load of 2 k/ft and a service live load of 3 k/ft.

It will be noted in these figures that factored loads are used for all the calculations.

Three different live-load conditions are considered in these figures. To determine the maximum positive moment in span 1, a live load is placed in spans 1 and 3 (Figure 12.14). Similarly, to produce maximum positive moment in span 2, the live load is placed in that span only (Figure 12.15). Finally, maximum negative moment at the first interior support from the left end is caused by placing the live load in spans 1 and 2 (Figure 12.16).

For this particular beam it is assumed that the Code permits a 10% up or down adjustment in the negative support moments. The result

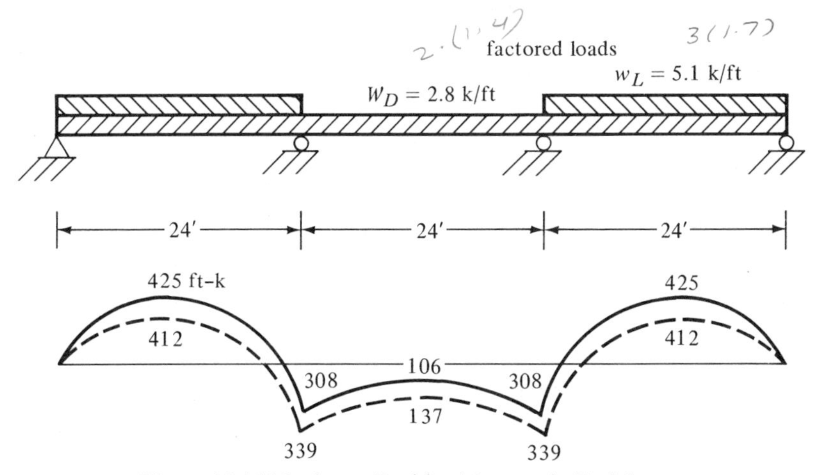

Figure 12.14 Maximum Positive Moment in End Spans

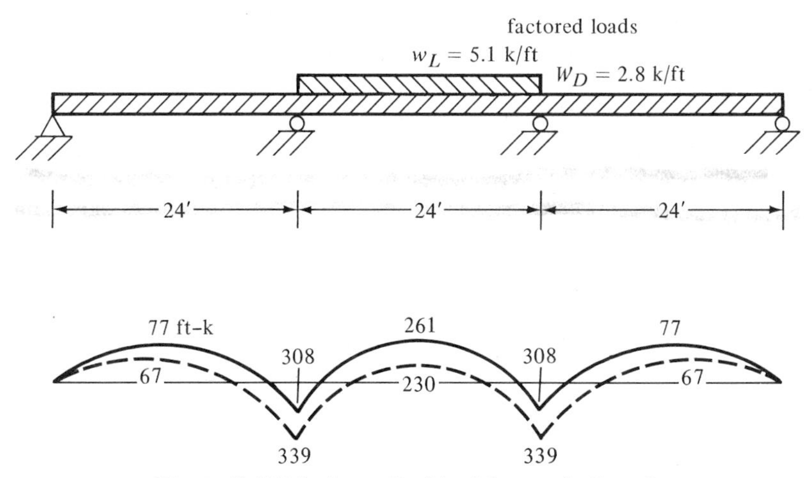

Figure 12.15 Maximum Positive Moment in Span 2

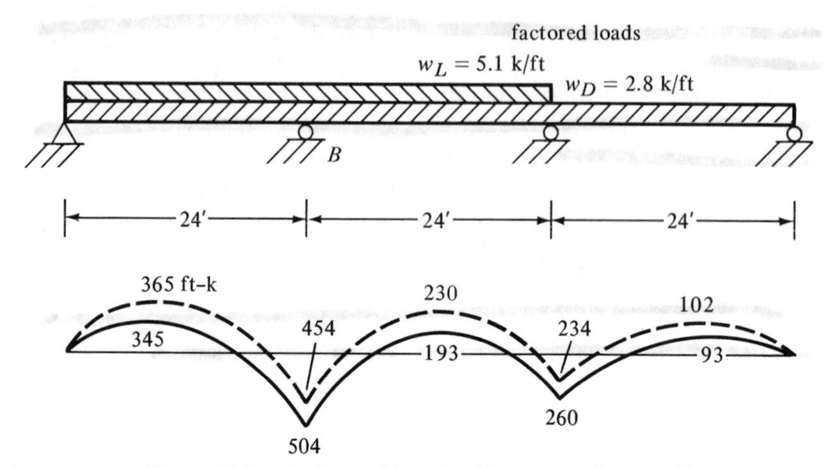

Figure 12.16 Maximum Negative Moment at Support B

will be smaller design moments at the critical sections. Initially the loading for maximum positive moment in span 1 is considered as shown in Figure 12.14. If the maximum calculated negative moments of 308 ft-k at the interior supports are each increased by 10% to 339 ft-k, the maximum positive moments in spans 1 and 3 will be reduced proportionately to 412 ft-k.

In the same fashion in Figure 12.15, where the beam is loaded to produce maximum positive moment in span 2, an increase in the negative moments at the supports from 308 ft-k to 339 ft-k will reduce the maximum positive moment in span 2 from 261 ft-k to 230 ft-k.

Finally, in Figure 12.16 the live-load placement causes a maximum negative moment at the first interior support of 504 ft-k. If this value is reduced by 10%, the maximum moment there will be −454 ft-k. In this figure the author has reduced the negative moment at the other interior support by 10% also. Should it be of advantage, however, it can be assumed that one negative support moment is decreased and the other one increased.

It will be noticed that the net result of all of the various increases or decreases in the negative moments is a net reduction in both the maximum

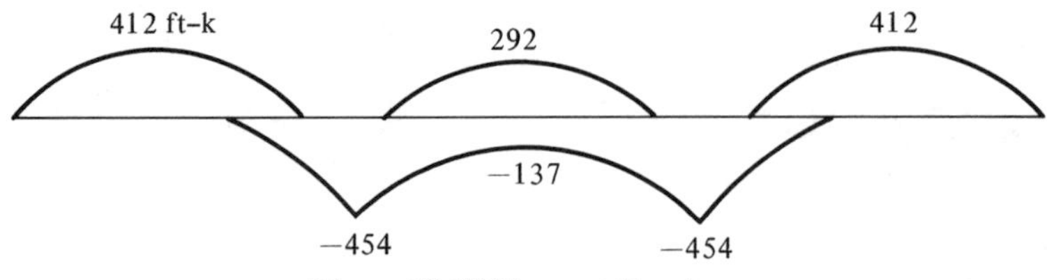

Figure 12.17 Moment Envelope

positive and the maximum negative values. The result of these various redistributions is actually an envelope of the extreme values of the moments at the critical sections. The envelope for the three-span beam considered in this section is presented in Figure 12.17. You can see at a glance the parts of the beams that need positive reinforcement, negative reinforcement, or both.

The reductions in bending moments due to moment redistribution as described here do not mean that the safety factors for continuous members will be less than those for simple spans. Rather, the excess strength that such members have due to this continuity is reduced so that the overall factors of safety are nearer but not less than those of simple spans.

Various studies have shown that cracking and deflection of members selected by the limit design process are no more severe than for the same members designed without taking advantage of the permissible redistributions.[3,4]

12.5 APPROXIMATE ANALYSIS OF CONTINUOUS FRAMES FOR VERTICAL LOADS

Statically indeterminate structures may be analyzed "exactly" or "approximately." Some approximate methods involving the use of simplifying assumptions are presented in this section. Despite the increased use of computers for making "exact analyses," approximate methods are used about as much or more than ever for several reasons. These include the following:

1. The structure may be so complicated that no one who has the knowledge to make an exact analysis is available.
2. For some structures either method may be subject to so many errors and imperfections that approximate methods may yield values as accurate as those obtained with an exact analysis. A specific example is the analysis of a building frame for wind loads where the walls, partitions, and floors contribute an indeterminate amount to wind resistance. Wind forces calculated in the frame by either method are not accurate.
3. To design the members of a statically indeterminate structure, it is necessary to make an estimate of their sizes before structural analysis

[3] Cohn, M. Z., 1964, "Rotational Compatibility in the Limit Design of Reinforced Concrete Continuous Beams," *Proceedings of the International Symposium on the Flexural Mechanics of Reinforced Concrete*, ASCE-ACI (Miami), pp. 359–382.

[4] Mattock, A. H., 1959, "Redistribution of Design Bending Moments in Reinforced Concrete Continuous Beams," *Proceedings of the Institution of Civil Engineers*, 13, pp. 35–46.

can begin by an exact method. Approximate analysis of the structure will yield forces from which reasonably good initial estimates can be made as to member sizes.

4. Approximate analyses are quite useful in rough checking exact solutions.

From the discussion of influence lines in Section 12.2, you can see that unless a computer is used (a rather practical alternative today) an exact analysis involving several different placements of the live loads would be a long and tedious affair. For this reason it is common when a computer is not readily available to use some approximate method of analysis, such as the ACI moment and shear coefficients, the equivalent rigid frame method, the assumed point-of-inflection-location method, and others discussed in the pages to follow.

ACI Coefficients for Continuous Beams and Slabs

A very common method used for the design of continuous reinforced concrete structures involves the use of the ACI coefficients given in Section 8.3.3 of the Code. These coefficients, which are reproduced in Table 12.1, provide estimated maximum shears and moments for buildings of normal proportions. The values calculated in this manner will usually be somewhat larger than those that would be obtained with an exact analysis. As a result, appreciable economy can normally be obtained by taking the

Table 12.1

ACI Coefficients

Positive moment	
end spans	
If discontinuous end is unrestrained	$\frac{1}{11}wl_n^2$
If discontinuous end is integral with the support	$\frac{1}{14}wl_n^2$
Interior spans	$\frac{1}{16}wl_n^2$
Negative moment at exterior face of first interior support	
Two spans	$\frac{1}{9}wl_n^2$
More than two spans	$\frac{1}{10}wl_n^2$
Negative moment at other faces of interior supports	$\frac{1}{11}wl_n^2$
Negative moment at face of all supports for (a) slabs with spans not exceeding 10 ft and (b) beams and girders where ratio of sum of column stiffnesses to beam stiffness exceeds eight at each end of the span	$\frac{1}{12}wl_n^2$
Negative moment at interior faces of exterior supports for members built integrally with their supports	
Where the support is a spandrel beam or girder	$\frac{1}{24}wl_n^2$
Where the support is a column	$\frac{1}{16}wl_n^2$
Shear in end members at face of first interior support	$1.15(wl_n/2)$
Shear at face of all other supports	$wl_n/2$

time or effort to make such an analysis. In this regard it should be realized that these coefficients are considered to have their best application for continuous frames having more than three or four continuous spans.

In developing the coefficients the negative moment values were reduced to take into account the usual support widths and also some moment redistribution, as described in Section 12.4 of this text. In addition, the positive moment values have been increased somewhat to account for the moment redistribution. It will also be noted that the coefficients account for the fact that in monolithic construction the supports are not simple and moments are present at end supports, such as where those supports are beams or columns.

In applying the coefficients, w is the design load while l_n is the clear span for calculating positive moments and the average of the adjacent clear spans for calculating negative moments. These values were developed for members with approximately equal spans (the larger of two adjacent spans not exceeding the smaller by more than 20%) and for cases where the ratio of the uniform service live load to the uniform service dead load is not greater than three. In addition, the values are not applicable to to prestressed concrete members. Should these limitations not be met a more precise method of analysis must be used.

For the design of a continuous beam or slab, the moment coefficients provide in effect two sets of moment diagrams for each span of the structure. One diagram is the result of placing the live loads so that they will cause maximum positive moment out in the span, while the other is the result of placing the live loads so as to cause maximum negative moments at the supports. To be truthful, however, it is not possible to produce maximum negative moments at both ends of a span simultaneously. It takes one placement of the live loads to produce maximum negative moment at one end of the span and another placement to produce maximum negative moment at the other end. The assumption of both maximums occurring at the same time is on the safe side, however, because the resulting diagram will have greater critical values than are produced by either one of the two separate loading conditions.

The ACI coefficients give maximum points for a moment envelope for each span of a continuous frame. Typical envelopes are shown in Figure 12.18 for a continuous slab, which is assumed to be constructed integrally with its exterior supports, which are spandrel girders.

Example 12.1 presents the design of the slab of Figure 12.19 using the moment coefficients of the ACI Code. The calculations for this problem can be conveniently set up in some type of table such as the one shown in Figure 12.20. For this particular slab the author used an arrangement of reinforcement that included bent bars. It is quite common, however, in slabs, particularly those of 5 in. or less in thickness, to use straight bars only in the top and bottom of the slab.

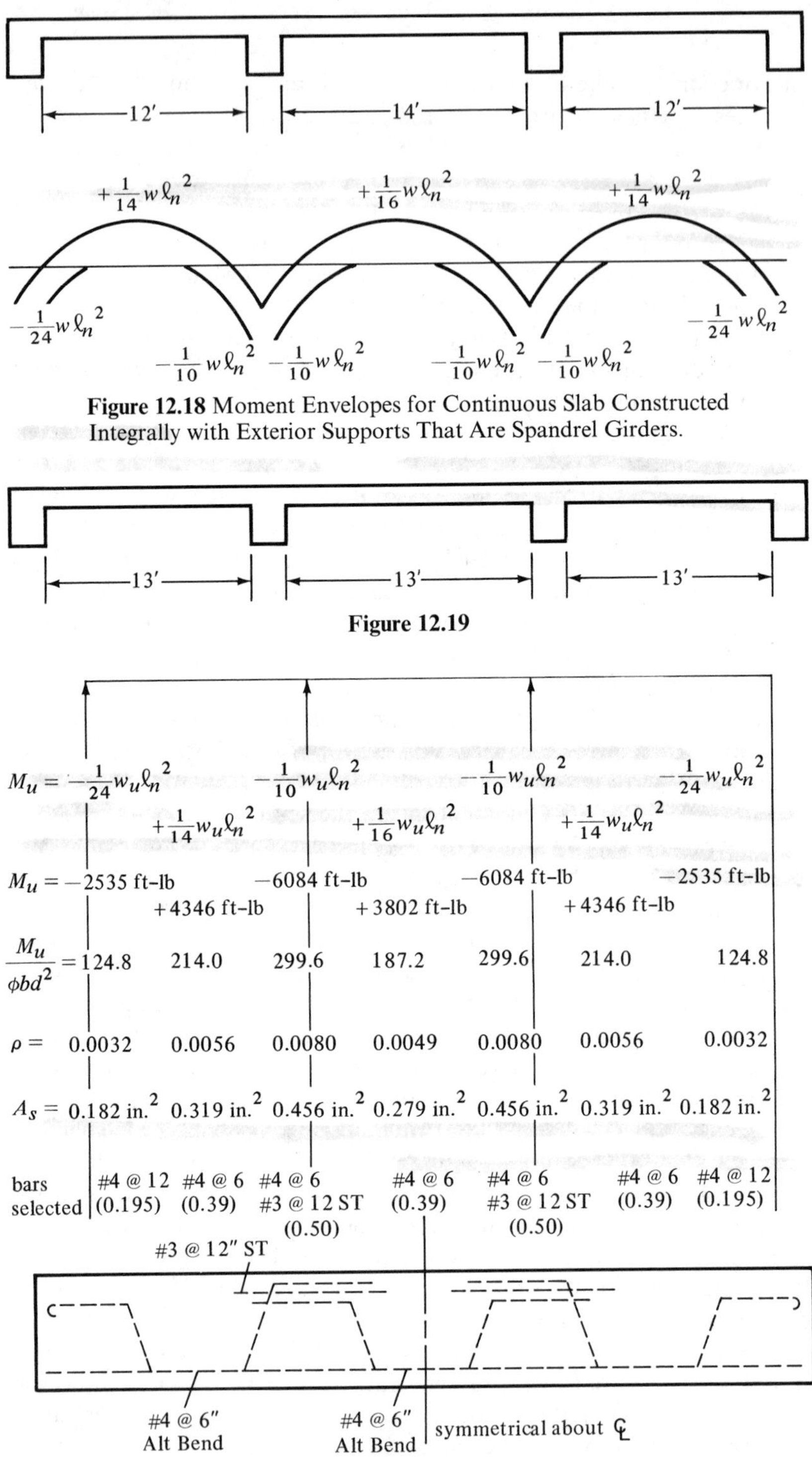

Figure 12.18 Moment Envelopes for Continuous Slab Constructed Integrally with Exterior Supports That Are Spandrel Girders.

Figure 12.19

Figure 12.20

EXAMPLE 12.1

Design the continuous slab of Figure 12.19 for moments calculated with the ACI coefficients. The slab is to support a service live load of 150 psf in addition to its own dead weight. $f_c' = 3{,}000$ psi and $f_y = 40{,}000$ psi.

SOLUTION

Minimum t for deflection by ACI Code (9.5.2)

$$\text{deflection multiplier for 40,000-psi steel} = 0.4 + \frac{40{,}000}{100{,}000} = 0.80$$

$$\text{minimum } t \text{ for end span} = 0.8\frac{l}{24} = \frac{(0.8)(12 \times 13)}{24} = 5.2''$$

$$\text{minimum } t \text{ for interior span} = 0.8\frac{l}{28} = \frac{(0.8)(12 \times 13)}{28} = 4.46''$$

Loads and maximum moment assuming 6-in. slab ($d = 4\frac{3}{4}$ in.)

$$w_D = \text{slab weight} = (\tfrac{6}{12})(150) = 75 \text{ psf}$$

$$w_L = 150 \text{ psf}$$

$$w_u = (1.4)(75) + (1.7)(150) = 360 \text{ psf}$$

$$M_{ax}M_u = \tfrac{1}{10}w_u l_n^2 = (\tfrac{1}{10})(360)(13)^2 = 6{,}084 \text{ ft-lb}$$

Computing moments, ρ values, A_s requirements, and selecting bars at each section, as shown in Figure 12.20.

The design of web reinforcement is theoretically based on the maximum shears occurring at various sections along the span. For instance, to determine the maximum shear occurring at Section 1.1 in the beam of Figure 12.21, the dead load would extend all across the span while the live load would be placed from the section to the most distant support.

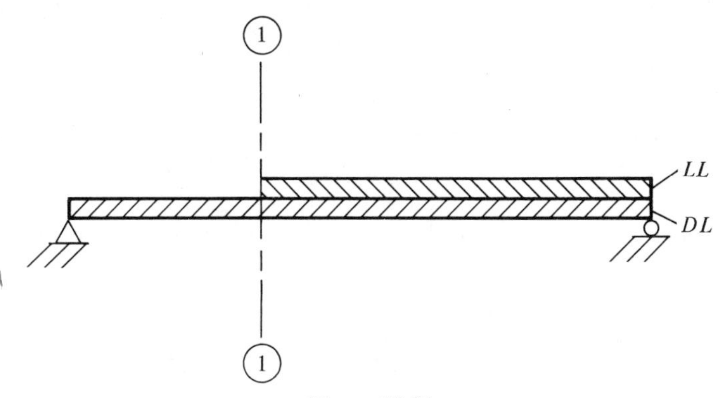

Figure 12.21

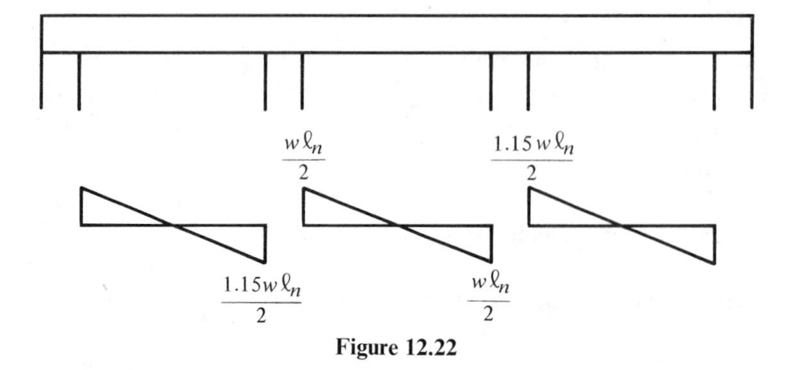

Figure 12.22

If the live load is placed so as to cause maximum shears at various points along the span and the shear calculated for each point, a maximum shear curve can be drawn. Practically speaking, however, it is unnecessary to go through such a lengthy process for buildings of normal proportions, as the values that would be obtained do not vary significantly from the values given by the ACI Code, which are shown in Figure 12.22.

Equivalent Rigid-Frame Method

When continuous beams frame into and are supported by girders, the normal assumption made is that the girders provide only vertical support. Thus they are analyzed purely as continuous beams, as shown in Figure 12.23. The girders do provide some torsional stiffness, and if the calculated torsional moments exceed $\phi(0.5\sqrt{f_c'}\sum x^2 y)$ they must be considered, as will be described in Chapter 13.

Where continuous beams frame into columns, the bending stiffnesses of the columns together with the torsional stiffnesses of the girders are of such magnitude that they need to be considered. An approximate method frequently used for analyzing such reinforced concrete members is the *equivalent rigid-frame method.* In this method which is applicable only to gravity loads the loads are assumed to be applied only to the floor or roof under consideration and the far ends of the columns are assumed to be fixed, as shown in Figure 12.24. The sizes of the members are estimated, and an analysis is made on the basis of moment distribution.

For this type of analysis it is necessary to estimate the sizes of the members and compute their relative stiffnesses or I/l values. From these values distribution factors can be computed and the method of moment

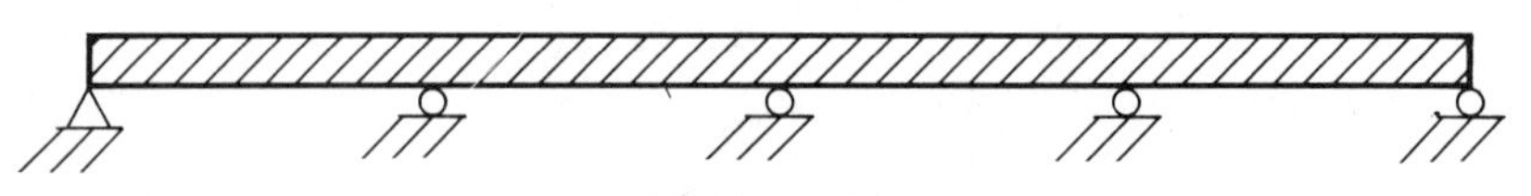

Figure 12.23

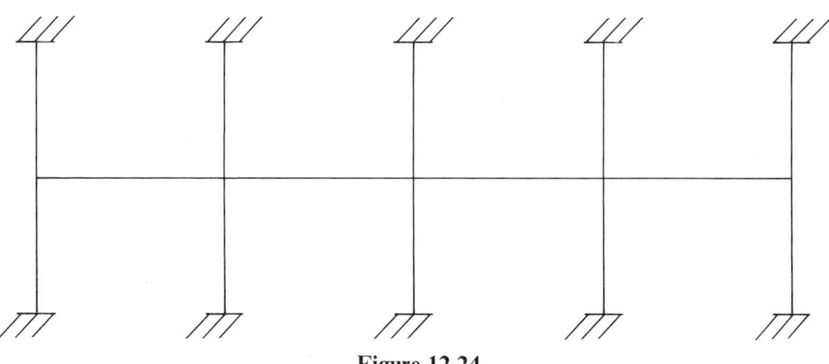

Figure 12.24

distribution applied. The moments of inertia of both columns and beams are normally calculated on the basis of gross concrete sections, with no allowance made for reinforcing.

There is a problem involved in determining the moment of inertia to be used for continuous T beams. The moment of inertia of a T beam is much greater where there is positive moment with the flanges in compression than where there is negative moment with the flanges cracked due to tension. Since the moment of inertia varies along the span, it is necessary to use an equivalent value. A practice often used is to assume the equivalent moment of inertia equals twice the moment of inertia of the web, assuming the web depth equals the full effective depth of the beam.[5] Some designers use other equivalent values, such as assuming an equi valent T section with flanges of effective widths equal to so many (say 2 to 6) times the web width. These equivalent sections can be varied over a rather wide range without appreciably affecting the final moments.

The ACI Code (8.9.2) states that for such an approximate analysis, only two live-load combinations need to be considered. These are (1) live load placed on two adjacent spans and (2) live load placed on alternate spans. Example 12.2 illustrates the application of the equivalent rigid frame method to a continuous T beam.

Computer results appear to indicate that the model shown in Figure 12.24 (as permitted by the ACI Code) may not be trustworthy for unsymmetrical loading. Differential column shortening can completely redistribute the moments obtained from the model (i.e., positive moments can become negative moments). As a result, the designer should take into account possible axial deformations in his designs.

EXAMPLE 12.2

Using the equivalent rigid-frame method, draw the shear and moment diagrams for the continuous T beam of Figure 12.25. The beam is assumed

[5] *Continuity in Concrete Building Frames*, 4th ed., 1959 (Chicago: Portland Cement Association), pp. 17–20.

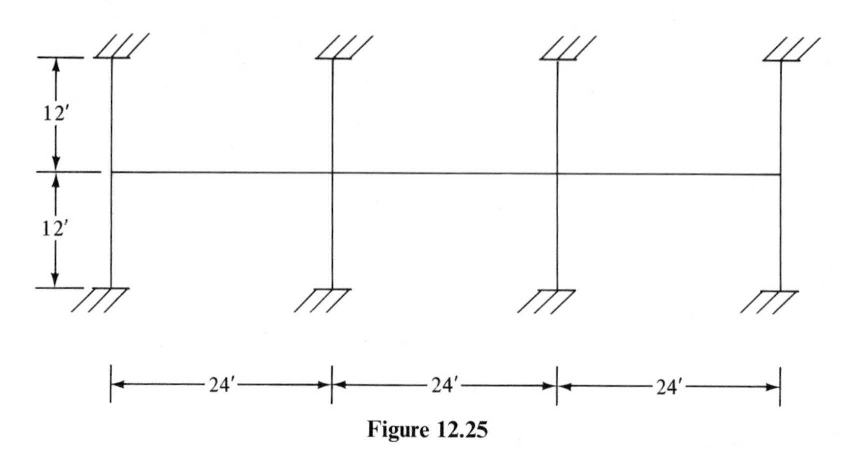

Figure 12.25

to be framed into 16-in. × 16-in. columns and is to support a service dead load of 2 k/ft (including beam weight) and a service live load of 3 k/ft. Assume the live load is applied in the center span only. The girders are assumed to have a depth of 24 in. and a web width of 12 in. Assume the I of the T beam equals two times the I of its web.

SOLUTION

Computing fixed-end moments

$$w_u \text{ in first and third spans} = (1.4)(2) = 2.8 \text{ k/ft}$$

$$M = \frac{(2.8)(24)^2}{12} = 134.4 \text{ ft-k}$$

$$w_u \text{ in center span} = (1.4)(2) + (1.7)(3) = 7.9 \text{ k/ft}$$

$$M_u = \frac{(7.9)(24)^2}{12} = 379.2 \text{ ft-k}$$

Computing stiffness factors

$$I \text{ of columns} = (\tfrac{1}{12})(16)(16)^3 = 5{,}461 \text{ in.}^4$$

$$k \text{ of columns} = \frac{I}{l} = \frac{5{,}461}{12} = 455$$

$$\text{equivalent } I \text{ of T beam} = (2)(\tfrac{1}{12}b_w h^3) = (2)(\tfrac{1}{12})(12)(24)^3 = 27{,}648 \text{ in.}^4$$

$$k \text{ of T beam} = \frac{27{,}648}{24} = 1{,}152$$

Record stiffness factors on frame and compute distribution factors, as shown in Figure 12.26.

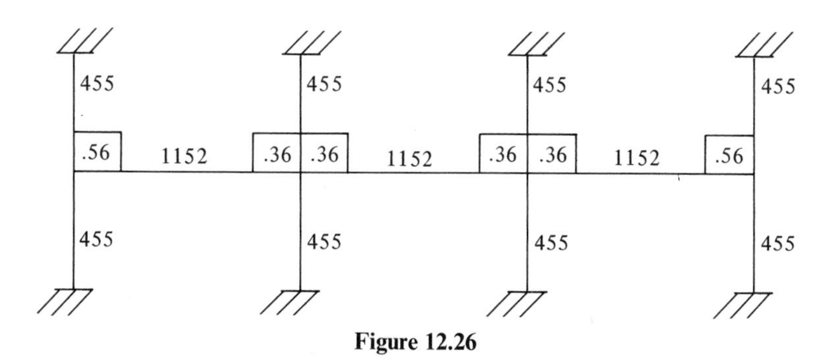

Figure 12.26

Balance fixed-end moments and draw shear and moment diagrams, as shown here and in Figure 12.27.

.56	.36	.36	.36	.36	.56
−134.4	+134.4	−379.2	+379.2	−134.4	+134.4
{+ 75.3	+ 37.6			− 37.6	− 75.3
+ 37.3	+ 74.6	+ 74.6	+ 37.3		
− 20.9	− 10.4	− 44.0	− 88.0	− 88.0	− 44.0
+ 9.8	+ 19.6	+ 19.6	+ 9.8	+ 12.3	+ 24.6
− 5.5	− 2.7	− 4.0	− 8.0	− 8.0	− 4.0
+ 1.2	+ 2.4	+ 2.4	+ 1.2	+ 1.1	+ 2.2
− 0.7	− 0.4	− 0.4	− 0.8	− 0.8	− 0.4
− 37.9	+ 0.3	+ 0.3	+330.7	−255.4	+ 0.2
	+255.4	−330.7			+ 37.7
↑33.60	33.60↑	↑94.80	94.80↑	↑33.60	33.60↑
↓ 9.06	9.06↑			↑ 9.06	9.06↓
↑24.54	42.66↑	↑94.80	94.80↑	↑42.66	24.54↑

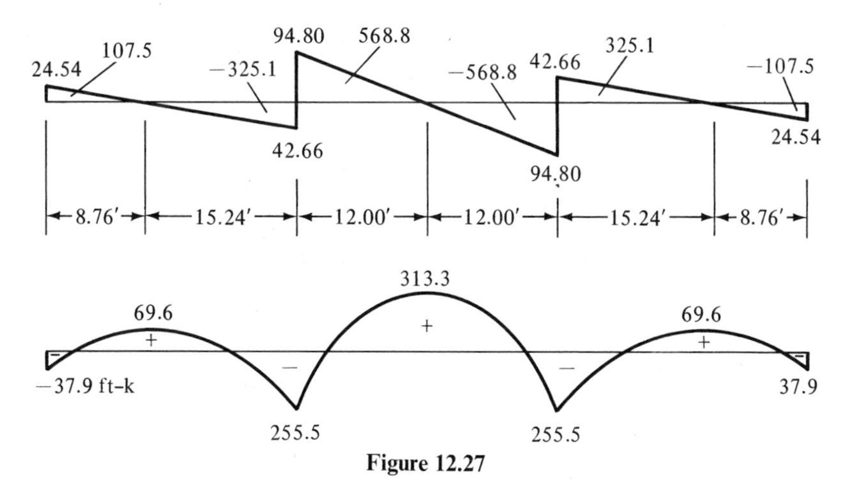

Figure 12.27

Assumed Points of Inflection

Another approximate method of analyzing statically indeterminate building frames is to assume the locations of the points of inflection in the members. Such assumptions have the effect of creating simple beams between the points of inflection in each span, and the positive moments in each span can be determined by statics. Negative moments occur in the girders between their ends and the points of inflection. They may be computed by considering the portion of the beam out to the point of inflection to be a cantilever. The shear at the end of each of the girders contributes to the axial forces in the columns. Similarly, the negative moments at the ends of the girders are transferred to the columns.

In Figure 12.28 beam *AB* of the building frame shown is analyzed by assuming points of inflection at the one-fifth points and fixed supports at the beam ends.

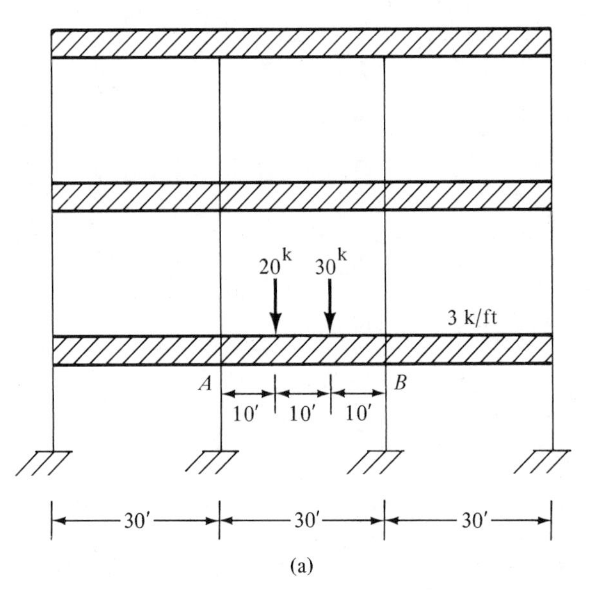

(a)

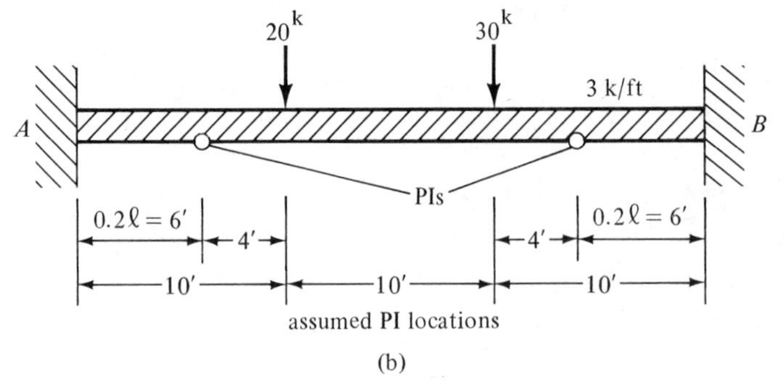

(b)

Figure 12.28

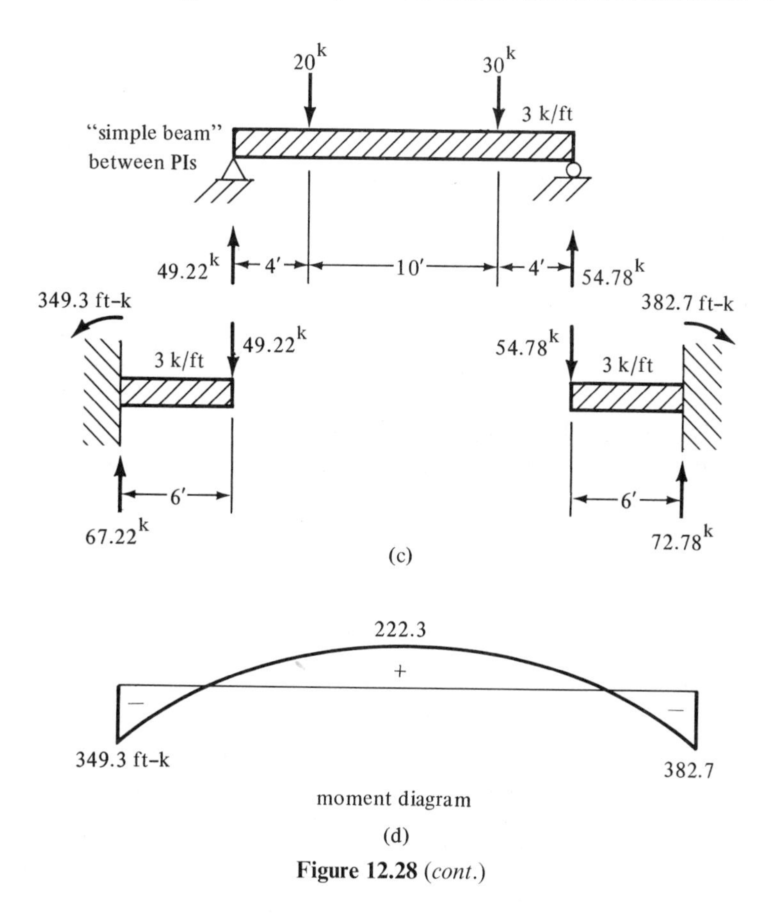

Figure 12.28 *(cont.)*

12.6 APPROXIMATE ANALYSIS OF CONTINUOUS FRAMES FOR LATERAL LOADS

Building frames are subjected to lateral loads as well as to vertical loads. The necessity for careful attention to these forces increases as buildings become taller. Not only must buildings have sufficient lateral resistance to prevent failure but they also must have a sufficient resistance to deflections to prevent injuries to their various parts. Rigid-frame buildings are highly statically indeterminate, and their analysis by exact methods (unless computers are used) is so lengthy as to make the approximate methods very popular.

The approximate method presented here is called the portal method. Because of its simplicity it has probably been used more than any other approximate method for determining wind forces in building frames. This method, which was presented by Albert Smith in the *Journal of the Western Society of Engineers* in April 1915, is said to be satisfactory for most buildings up to 25 stories in height. Another method very similar to the portal method is the cantilever method. It is thought to give slightly better results for high narrow buildings and can be used for buildings not in excess of 25 to 35 stories.[6]

The portal method is merely a variation of the method described in Section 12.5 for analyzing beams in which the location of the points of inflection were assumed. For the portal method the loads are assumed to be applied at the joints only. If this loading condition is correct, the moments will vary linearly in the members and points of inflection will be located fairly close to member midpoints.

No consideration is given in the portal method to the elastic properties of the members. These omissions can be very serious in unsymmetrical frames and in very tall buildings. To illustrate the seriousness of the matter, the changes in member sizes are considered in a very tall building. In such a building there will probably not be a great deal of change in beam sizes from the top floor to the bottom floor. For the same loadings and spans, the changed sizes would be due to the large wind moments in the lower floors. The change, however, in column sizes from top to bottom would be tremendous. The result is that the relative sizes of columns and beams on the top floors are entirely different from the relative sizes on the lower floors. When this fact is not considered, it causes large errors in the analysis.

In the portal method the entire wind loads are assumed to be resisted by the building frames, with no stiffening assistance from the floors, walls, and partitions. Changes in the lengths of girders and columns are assumed to be negligible. They are not negligible, however, in tall slender buildings, the height of which is five or more times the least horizontal dimension.

If the height of a building is roughly five or more times its least lateral dimension, it is generally felt that a more precise method of analysis should be used. There are several approximate methods that make use of the elastic properties of the structures and that give values closely approaching the results of the "exact" methods. These include the factor method,[7] the Witmer method of K percentages,[8] and the Spurr method.[9]

The building frame shown in Figure 12.29 is analyzed by the portal method, as described in the following paragraphs.

At least three assumptions must be made for each individual portal or for each girder. In the portal method, the frame is theoretically divided

[6] American Society of Civil Engineers, 1940, "Wind Bracing in Steel Buildings," *Transactions of the American Society of Civil Engineers*, 105, p. 1723.

[7] Norris, C. H., and Wilbur, J. B., 1976, *Elementary Structural Analysis*, 3rd ed. (New York: McGraw-Hill), pp. 207–212.

[8] American Society of Civil Engineers, 1940, "Wind Bracing in Steel Buildings," *Transactions of the American Society of Civil Engineers*, 105, pp. 1723–1727.

[9] American Society of Civil Engineers, 1940, "Wind Bracing in Steel Buildings," *Transactions of the American Society of Civil Engineers*, 105, pp. 1723–1727.

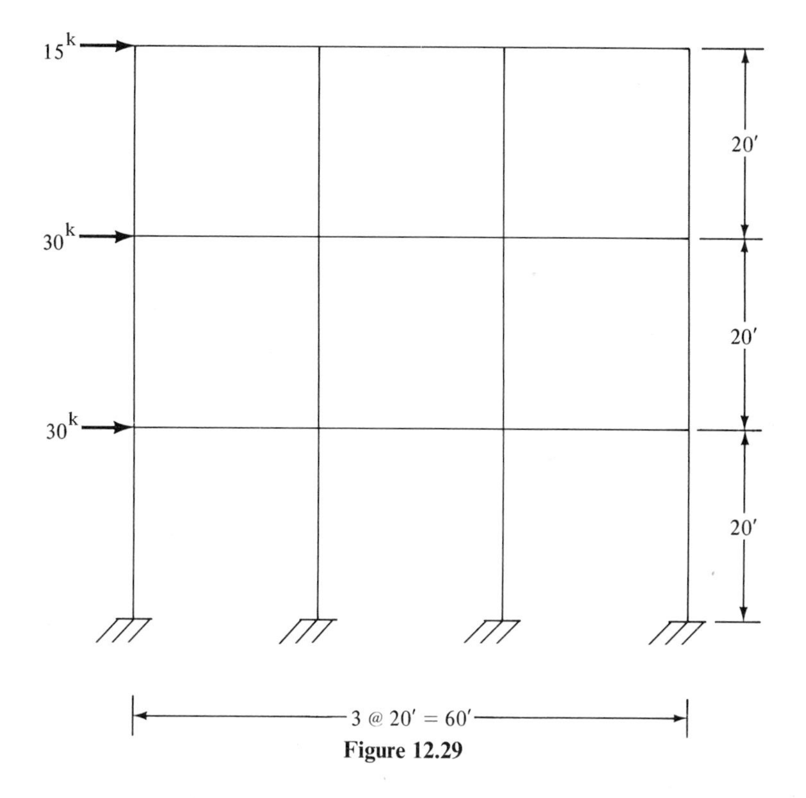

Figure 12.29

into independent portals (Figure 12.30) and the following three assumptions are made:

1. The columns bend in such a manner that there is a point of inflection at middepth.
2. The girders bend in such a manner that there is a point of inflection at their center lines.
3. The horizontal shears on each level are arbitrarily distributed between the columns. One commonly used distribution (and the one illustrated here) is to assume that the shear divides among the columns in the ratio of one part to exterior columns and two parts to interior columns.

The reason for the ratio in assumption 3 can be seen in Figure 12.30. Each of the interior columns is serving two bents, whereas the exterior

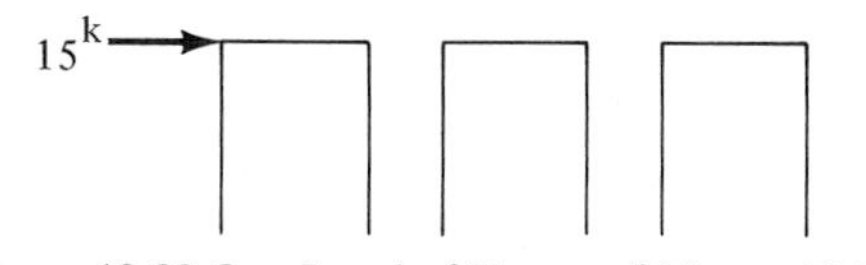

Figure 12.30 One Level of Frame of Figure 12.29

columns are serving only one. Another common distribution is to assume that the shear V taken by each column is in proportion to the floor area it supports. The shear distribution by the two procedures would be the same for a building with equal bays, but for one with unequal bays the results would differ with the floor area method probably giving more realistic results.

Frame Analysis

The frame is analyzed in Figure 12.31 on the basis of these assumptions. The arrows shown on the figure give the direction of the girder shears and the column axial forces. You can visualize the stress condition of the frame if you assume the wind is tending to push it over from left to right, stretching the left exterior columns and compressing the right exterior columns. Briefly the calculations were made as follows:

1. Column shears. The shears in each column on the various levels were first obtained. The total shear on the top level is 15^k. Because there are two exterior and two interior columns, the following expression may

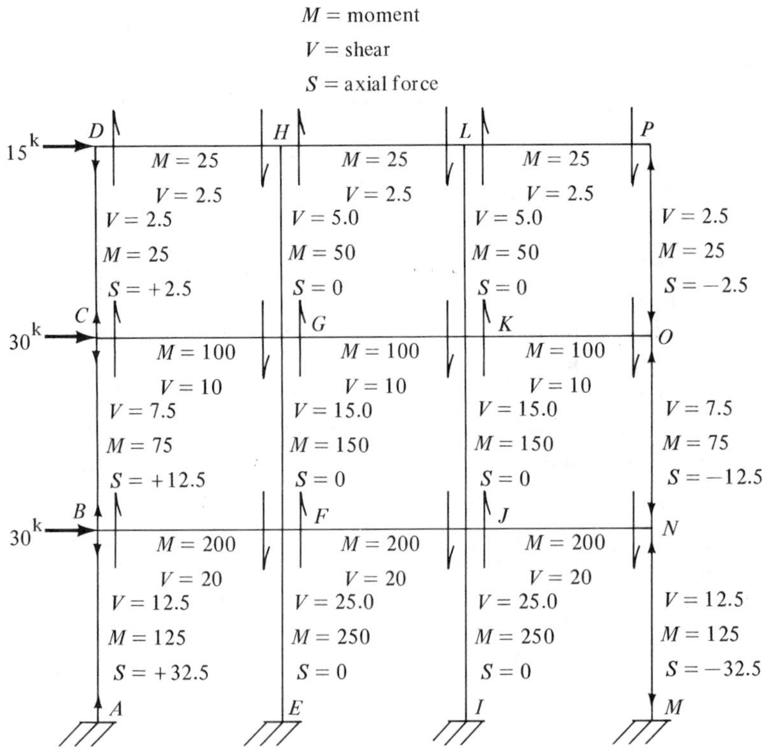

M = moment

V = shear

S = axial force

Figure 12.31

be written:

$$x + 2x + 2x + x = 15^k$$
$$x = 2.5^k$$
$$2x = 5.0^k$$

The shear in column *CD* is 2.5^k; in GH it is 5.0^k and so on. Similarly, the shears were determined for the columns on the first and second levels, where the total shears are 75^k and 45^k, respectively.

2. Column moments. The columns are assumed to have points of inflection at their middepths; therefore, their moments, top and bottom, equal the column shears times half the column heights.

3. Girder moments and shears. At any joint in the frame, the sum of the moments in the girders equals the sum of the moments in the columns. The column moments have been previously determined. By beginning at the upper left-hand corner of the frame and working across from left to right, adding or subtracting the moments as the case may be, the girder moments were found in this order: *DH*, *HL*, *LP*, *CG*, *GK*, and so on. It follows that with points of inflection at girder centerlines, the girder shears equal the girder moments divided by half-girder lengths.

4. Column axial forces. The axial forces in the columns may be directly obtained from the girder shears. Starting at the upper left-hand corner, the column axial force in CD is equal to the shear in girder DH. The axial force in column *GH* is equal to the difference between the two girder shears *DH* and *HL*, which equals zero in this case. (If the width of each of the portals is the same, the shears in the girder on one level will be equal, and the interior columns will have no axial force, since only lateral loads are considered.)

12.7 COMPUTER ANALYSIS OF BUILDING FRAMES

Truthfully speaking, all structures are three-dimensional, but theoretical analyses of such structures by hand calculation methods are so lengthy as to be impractical. As a result, such systems are normally assumed to consist of two-dimensional or planar systems and they are analyzed independently of each other. The methods of analysis presented in this chapter were handled in this manner.

Today, modern electronic computers have greatly changed the picture, as it is possible to analyze complete three-dimensional structures. As a result, more realistic analyses are available and the necessity for high safety factors is reduced. The application of computers is not restricted merely to analysis; they are used in almost every phase of concrete work from analysis to design, to detailing, to specification writing, to material takeoffs, to cost estimating, and so on.

PROBLEMS

12.1 to **12.3** Draw qualitative influence lines for the functions indicated in the structures shown.

12.1 Reactions at A and C, positive moment and positive shear at X.

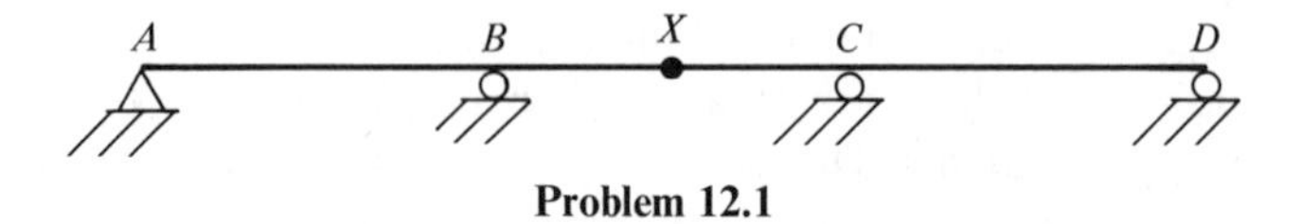

Problem 12.1

12.2 Reaction at D, negative moment at X, negative moment at D.

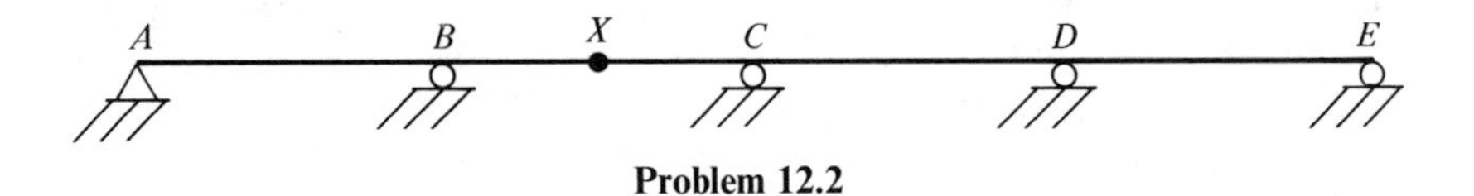

Problem 12.2

12.3 Positive moment at X, positive shear at X, negative moment just to the left of Y.

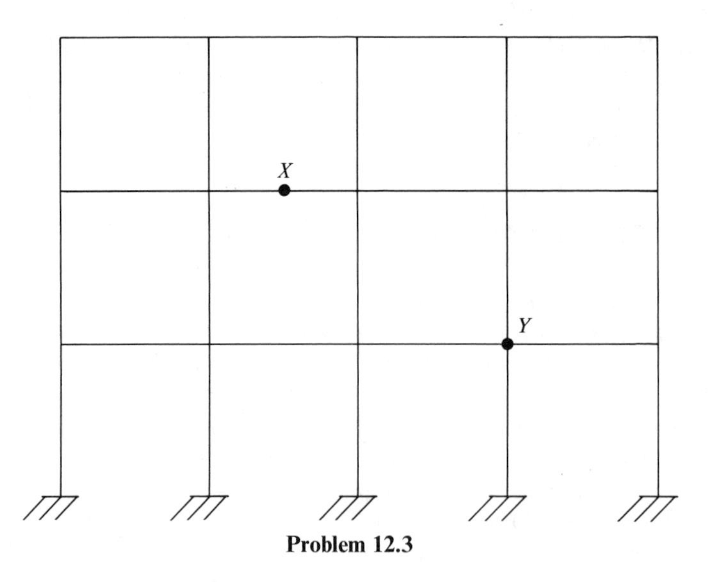

Problem 12.3

12.4 For the continuous beam shown and for a service dead load of 1.5 k/ft and a service live load of 2.4 k/ft, draw the moment envelope assuming a permissible 10% up or down redistribution of the maximum negative moment.

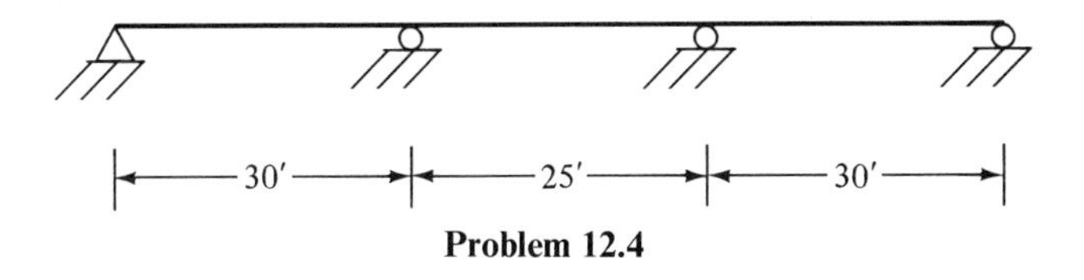

Problem 12.4

12.5 to **12.6** Design the continuous slabs shown using the ACI moment coefficients, assuming a service live load of 200 psf is to be supported in addition to the weight of the slabs. The slabs are to be built integrally with the end supports, which are spandrel beams. $f_y = 60{,}000$ psi. $f_c' = 3{,}000$ psi. Clear spans are shown in figure.

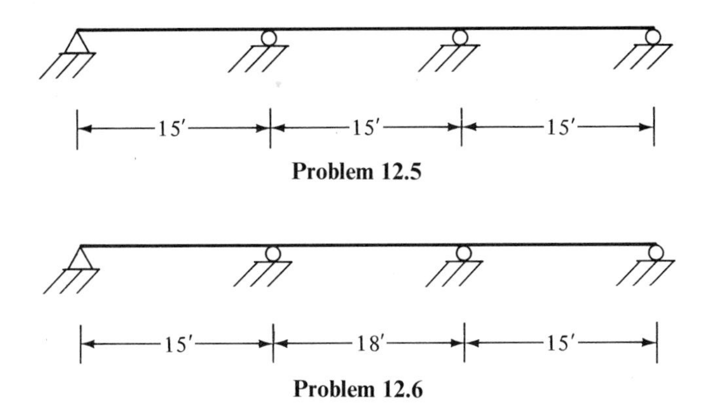

Problem 12.6

12.7 With the equivalent rigid-frame method, draw the shear and moment diagrams for the continuous beam shown using factored loads. Service dead load including beam weight is 1.5 k/ft and service live load is 2 k/ft. Place the live load in the center span only. Assume the I_s of the T beams equal two times the I_s of their webs.

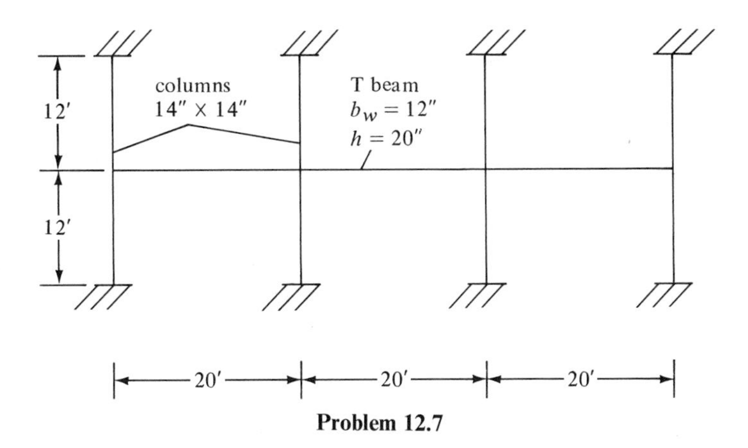

Problem 12.7

12.8 With the equivalent rigid-frame method, draw the shear and moment diagrams for the continuous beam shown using factored loads. Service dead load including beam weight is 2.4 k/ft and service live load is 3.2 k/ft. Place the live load in spans 1 and 2 only. Assume the I_s of the T beams equal two times the I_s of their webs.

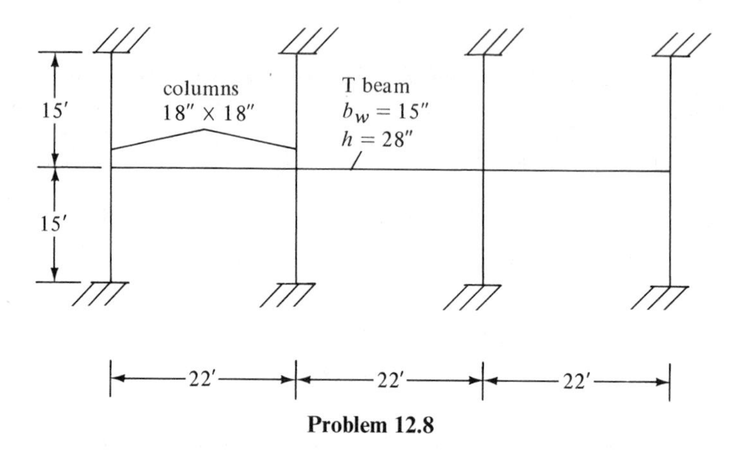

Problem 12.8

12.9 Draw the shear and moment diagrams for member AB of the frame shown in the accompanying illustration if points of inflection are assumed to be located at $0.15L$ from each end of the span.

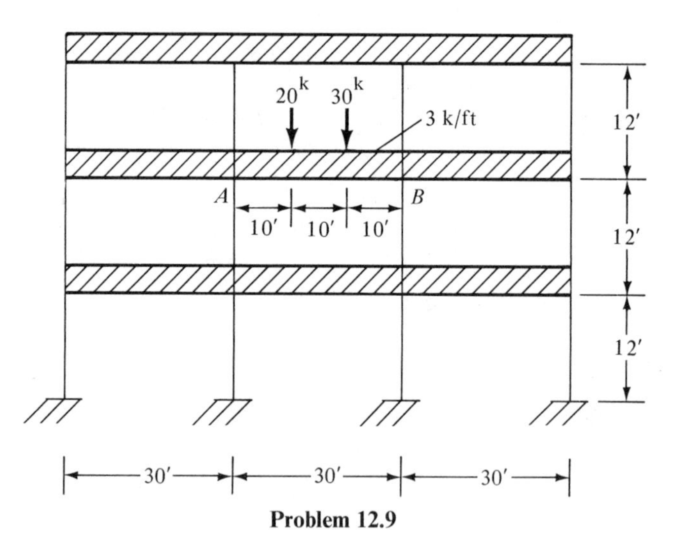

Problem 12.9

12.10 to **12.12** Compute moments, shears, and axial forces for all the members of the frames shown using the portal method.

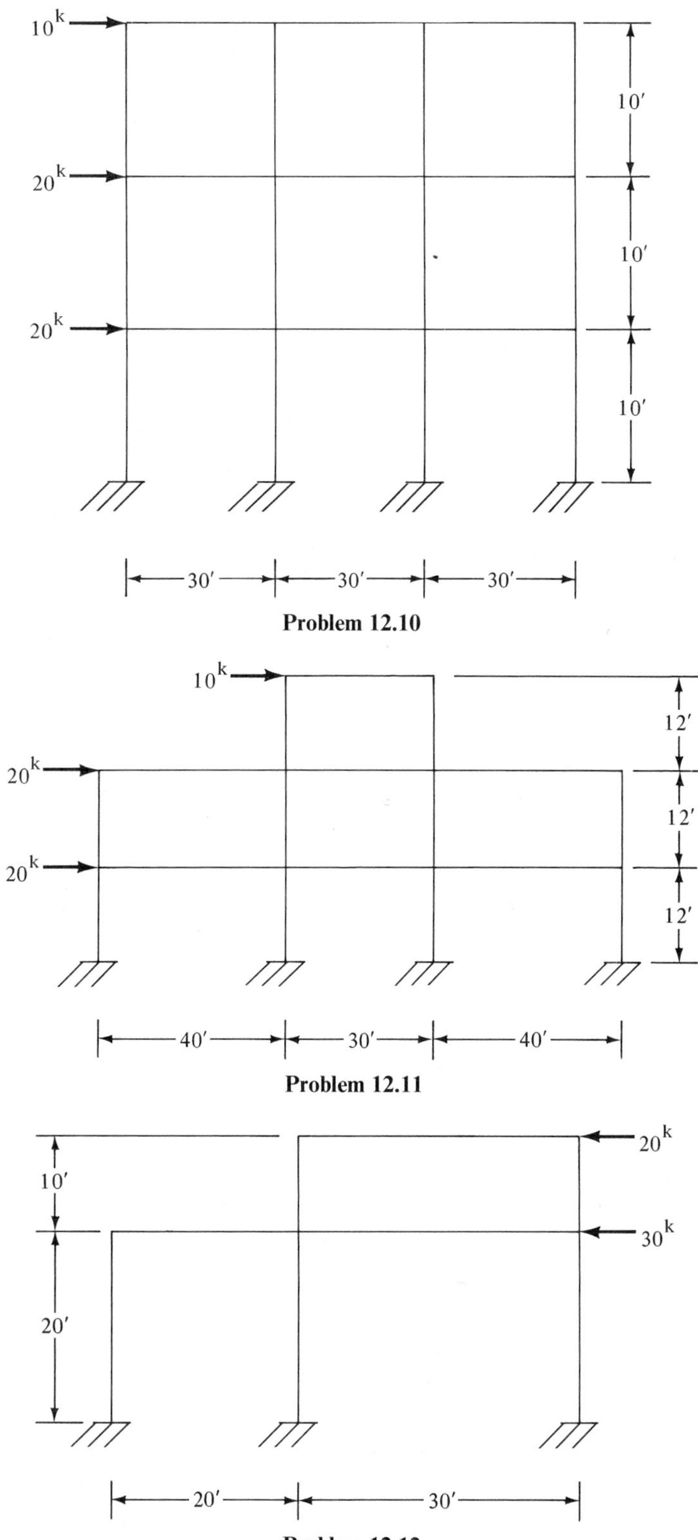

Problem 12.12

PROBLEMS WITH SI UNITS

12.13 For the continuous beam shown and for a service dead load of 26 kN/m and a service live load of 32 kN/m, draw the moment envelope assuming a permissible 10% up or down redistribution of the maximum negative moment.

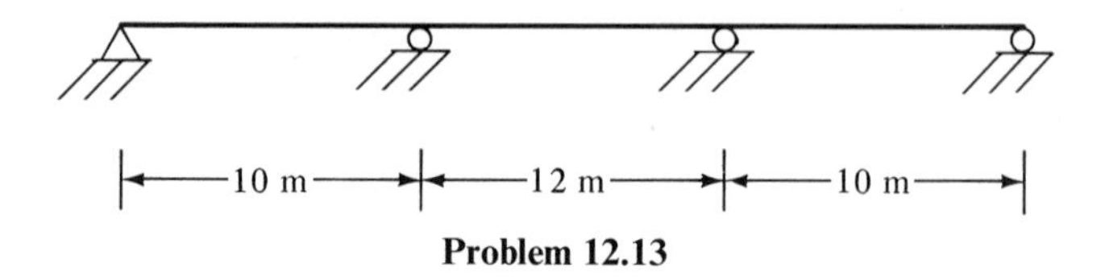

Problem 12.13

12.14 Design the continuous slab shown using the ACI moment coefficients, assuming the concrete weighs 23.5 kN/m^3 and assuming a service live load of 9 kN/m is to be supported. The slabs are built integrally with the end supports, which are spandrel beams. $f_y = 275.8$ MPa, $f_c' = 20.7$ MPa. Clear spans are given.

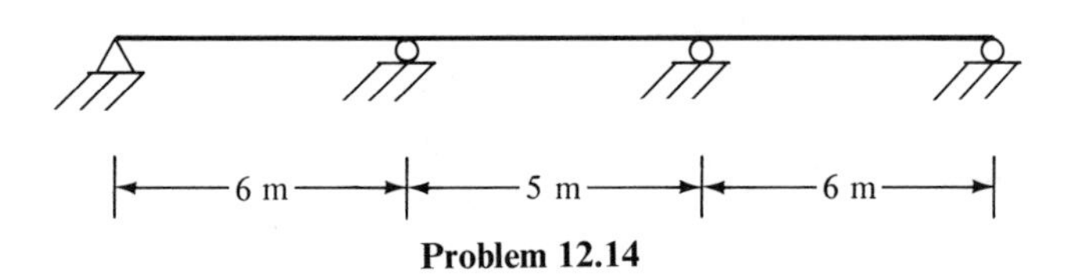

Problem 12.14

12.15 Using the equivalent rigid-frame method, draw the shear and moment diagrams for the continuous beams shown using factored loads. Service dead load including beam weight is 20 kN/m and service live load is 28 kN/m. Place the live load in spans 1 and 3 only. Assume the I_s of T beams equal 1.5 times the I_s of their webs.

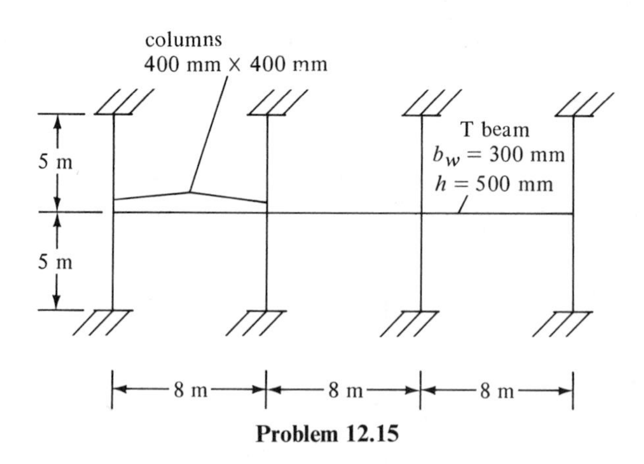

Problem 12.15

12.16 Compute moments, shears, and axial forces for all the members of the frame shown using the portal method.

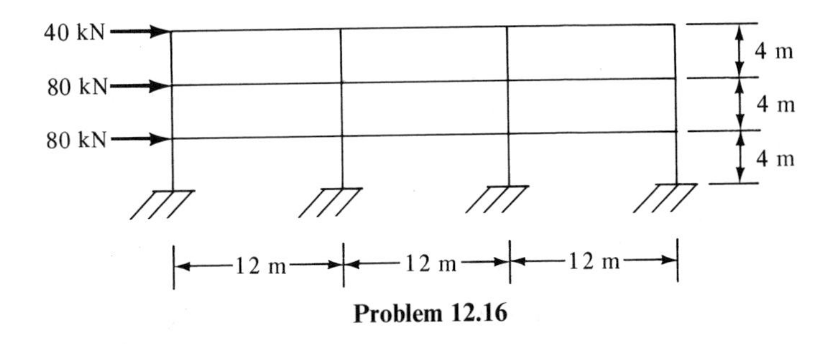

Problem 12.16

CHAPTER 13

Torsion

13.1 INTRODUCTION

The average designer probably does not worry about torsion very much. He thinks almost exclusively of axial forces, shears, and bending moments, and yet most reinforced concrete structures are subject to some degree of torsion. Until recent years the safety factors required by codes for the design of reinforced concrete members for shear, moment, and so forth were so large that the effects of torsion could be safely neglected in all but the most extreme cases. Today, however, overall safety factors are less than they used to be and members are smaller, with the result that torsion is a more common problem.

Appreciable torsion does occur in many structures, such as in the main girders of bridges which are twisted by transverse beams or slabs. It occurs in buildings where the edge of a floor slab and its beams are supported by a spandrel beam running between the exterior columns. This situation is illustrated in Figure 13.1, where the floor beams tend to twist the spandrel beam laterally. Other cases where torsion may be particularly significant

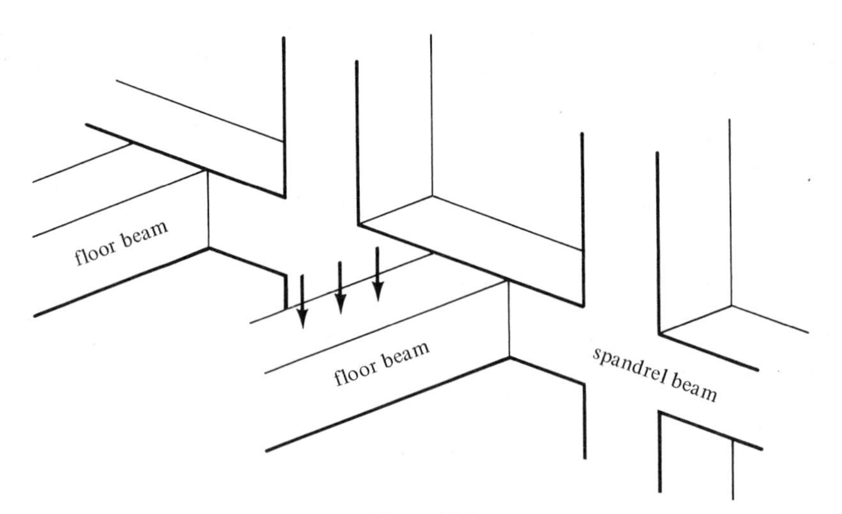

Figure 13.1

Christmas Common Bridge Over the M40 in England
(Courtesy Cement and Concrete Association).

occur in curved bridge girders, spiral stairways, and whenever large loads are applied to any beams "off center." It should be realized that if the supporting member is able to rotate, the resulting torsion stresses will be small. If, however, the member is restrained, the torsion stresses can be quite large.

In recent years there have been more reports of structural failures attributed to torsion. As a result a rather large amount of research has been devoted to the subject, and thus there is a much improved understanding of the behavior of structural members subjected to torsion. On the basis of this rather extensive experimental work, the 1977 ACI Code includes very specific requirements for the design of reinforced concrete members subjected to torsion or to torsion combined with shear and bending. It should be realized that maximum shears and torsional forces may occur in those areas where bending moments are small. For such cases the interaction of shear and torsion can be particularly important as it relates to design.

13.2 CALCULATION OF TORSION STRESSES

Should a plain concrete member be subjected to pure torsion, it will crack and fail along 45° spiral lines due to the diagonal tension corresponding to the torsional stresses. For a very effective demonstration of this type of failure, you can take a piece of chalk in your hands and twist it until it breaks. Although the diagonal tension stresses produced by twisting are very similar to those caused by shear, they will occur on all faces of a member. As a result, they add to the stresses caused by shear on one side and subtract from them on the other.[1]

[1] White, R. N.; Gergely, P.; and Sexsmith, R. G., 1974, *Structural Engineering*, vol. 3 (New York: Wiley), pp. 423–424.

For members with solid cross sections composed of rectangles, the *elastic theory of torsion* says that the maximum torsion stresses can be calculated approximately with the following expression:

$$v_{tu} = \frac{3T_u}{\sum x^2 y}$$

In this equation T_u is the ultimate torque produced by the factored loads and $\sum x^2 y$ is a value calculated from the component rectangles of the beam cross section. The short dimension of each rectangle is x and the long one is y. This elastic theory expression is only accurate for computing torsional stresses where y is considerably larger than x for each rectangle. As a matter of fact, the torsion stress can be shown to equal about 5 $T_u/(\sum x^2 y)$ for a rectangular section with $x = y$.

Despite the fact that results can be decidedly in error for sections of certain proportions, the ACI Code does not specify any limits on the use of this equation. The equation is written as shown at the end of this paragraph and is the same as the preceding equation except that ϕ, the capacity reduction factor with a value of 0.85, has been added to the denominator.

$$v_{tu} = \frac{3T_u}{\phi \sum x^2 y}$$

The calculation of torsional stresses for compound sections such as T- and L-shapes is more complex. For such cases, however, a satisfactory approximation can be made by subdividing the sections into their component rectangles and assuming that each rectangle resists a portion of the total torque in proportion to its torsional rigidity.[2] Tests on such members have shown that such an assumption is conservative as long as the overhanging flange widths do not exceed three times the flange thicknesses.

Figure 13.2 shows the determination of $\sum x^2 y$ for several different beam cross sections. Part (a) of the figure shows the calculations for a rectangular section while part (b) shows the calculations for a T beam that has torsional reinforcing in the web and in the flange directly above the web. For this case $\sum x^2 y$ is calculated for the web rectangle extending to the top of the flange plus the flange rectangles on either side whose widths may not exceed $3h_f$.

Figure 13.2(c) shows a T beam that has torsional reinforcement provided for both the web and the overhanging flanges. For such a case $\sum x^2 y$ is calculated as being the larger of (1) the value obtained for the web rectangle extending up through the flange plus the value obtained for the

[2] Bach, C., 1911, *Elastizität und Festigkeit* (Berlin: Springer).

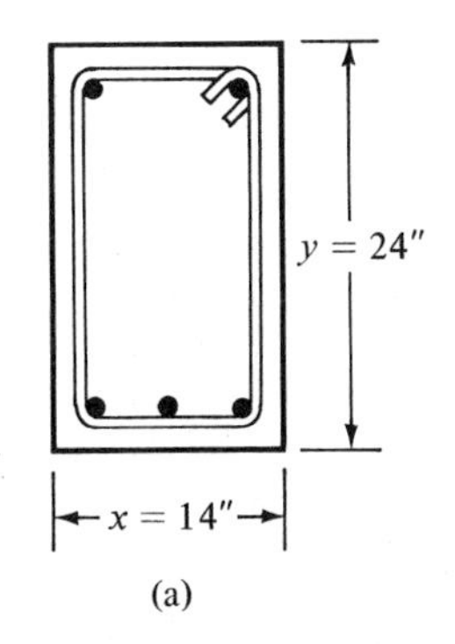

$$\Sigma x^2 y = (14)^2(24) = 4704 \text{ in.}^3$$

(a)

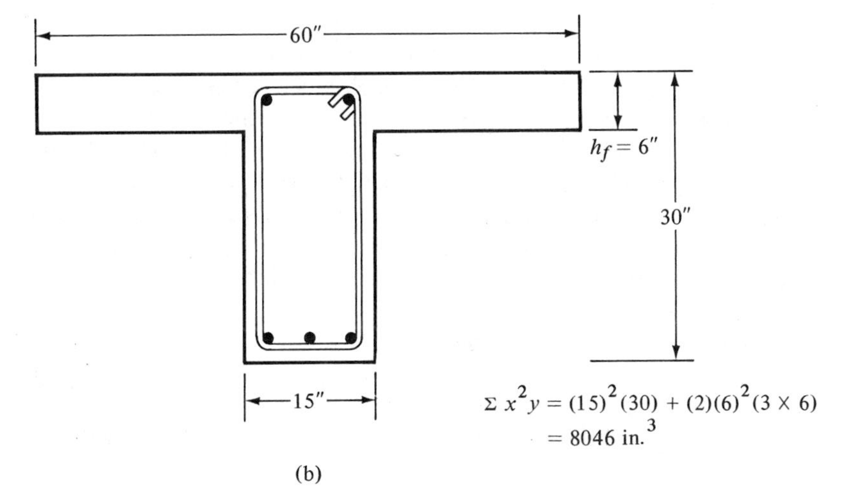

$$\Sigma x^2 y = (15)^2(30) + (2)(6)^2(3 \times 6)$$
$$= 8046 \text{ in.}^3$$

(b)

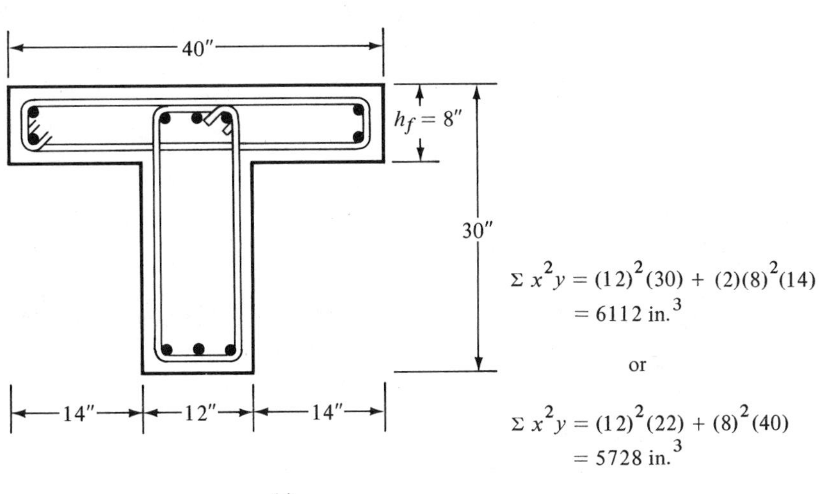

$$\Sigma x^2 y = (12)^2(30) + (2)(8)^2(14)$$
$$= 6112 \text{ in.}^3$$

or

$$\Sigma x^2 y = (12)^2(22) + (8)^2(40)$$
$$= 5728 \text{ in.}^3$$

(c)

Figure 13.2

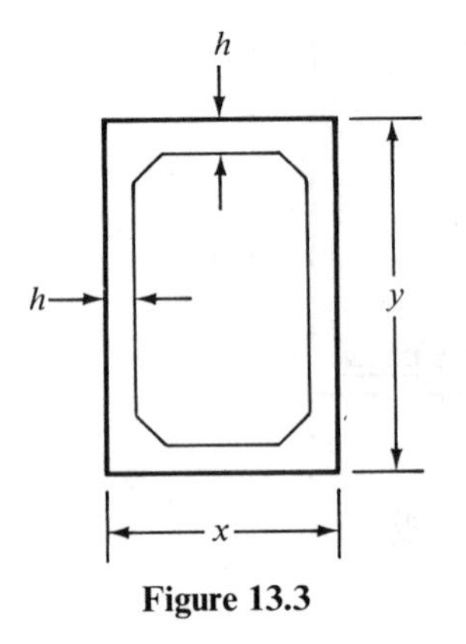

Figure 13.3

overhanging rectangles or (2) the value obtained for the web rectangle below the flange plus the value obtained for the whole flange rectangle.[3]

The ACI Code (11.6.1.2) states that the same equation used for solid sections may be used for hollow sections provided the wall thickness h is at least equal to $x/4$ (refer to Figure 13.3). If the wall thickness is less than $x/4$, the torsional strength of the hollow section is appreciably less than it is for a comparable solid section. If the wall thickness is less than $x/4$ but greater than $x/10$, the section can still be taken as a solid one but the expression for v_{tu} must be modified, as shown. Torsional test results of hollow sections show this reduction to be rather conservative, but such conservatism seems wise as hollow beams with thin walls subjected to torsion fail in a brittle manner whereas solid sections fail in a ductile manner.[4]

Should h be less than $x/10$, the stiffness of the wall must be considered. Actually such walls are usually avoided because of their greater flexibility and susceptibility to buckling.

$$v_{tu} = \frac{3T_u}{\sum x^2 y} \quad \text{if} \quad h \geq \frac{x}{4}$$

$$v_{tu} = \left(\frac{x}{4h}\right)\left(\frac{3T_u}{\sum x^2 y}\right) \quad \text{if} \quad h < \frac{x}{4} > \frac{x}{10}$$

Note: A fillet is required for the corners of all box sections. See the ACI Commentary (11.6.1.2) for limiting dimensions.

[3] ACI Committee 318, 1977, *Commentary on Building Code Requirements for Reinforced Concrete* (ACI 318-77) (Detroit: American Concrete Institute), p. 60.

[4] ACI Committee 318, 1977, *Commentary on Building Code Requirements for Reinforced Concrete* (ACI 318-77) (Detroit: American Concrete Institute), p. 60.

If torsional stresses are less than about $1.5\sqrt{f_c'}$, they will not appreciably reduce either the shear or the flexural strengths of reinforced concrete members. As a result, torsion design is rarely necessary for interior beams supporting slabs on each side that have approximately equal spans. Should the torsional stress be higher than $1.5\sqrt{f_c'}$, it will be necessary to provide additional reinforcing. This reinforcing consists of transverse or spiral reinforcing used to restrict the development of cracks and to prevent torsional failure.

The $1.5\sqrt{f_c'}$ value is roughly equal to about one fourth of the torsion strength of a member without concrete reinforcement. If this value is substituted into the equation for v_{tu} the limiting torsional moment T_u (above which torsional reinforcing is required) is obtained as given in Section 11.6.1 of the Code.

$$1.5\sqrt{f_c'} = \frac{3T_u}{\phi \sum x^2 y}$$

$$T_u = \phi(0.5\sqrt{f_c'} \sum x^2 y)$$

Torsion causes stresses on all faces of a beam and thus can cause cracks on all those faces even on the compression side. As a result when torsional reinforcing is needed it must be provided for all faces of a beam. The normal ⊔ shaped stirrups are not satisfactory and stirrups must be closed either by welding their ends together to form a continuous loop as illustrated in Figure 13.4 or by bending their ends around a longitudinal bar as shown in Figure 13.2. Actually the Code (12.14.5) does permit the use of closed stirrups made with two ⊔ shaped stirrups arranged with their legs lapping over each other ⧈ but it will be noticed that the lap lengths required for such cases may be no less than 1.7 times the required development lengths for the stirrup bar sizes being used.

In addition to closed stirrups it is necessary to use additional longitudinal bars placed in the corners and perhaps in between to produce a cage. The strength of the closed stirrups cannot be developed unless this additional longitudinal reinforcing is provided.

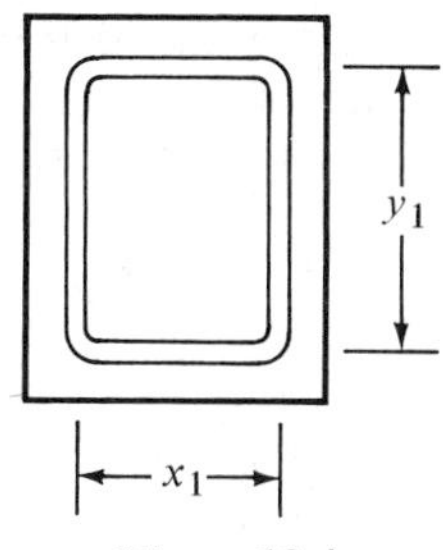

Figure 13.4

13.3 TORSIONAL STRENGTH OF CONCRETE MEMBERS

The torsional strengths of sections made with homogeneous materials can be estimated quite accurately, but the torsional strength of nonhomogeneous reinforced concrete sections is another matter entirely. Although many investigations have been made into the subject, there is as yet no clear-cut theory available to handle torsion in reinforced concrete members. To further complicate the matter, pure torsion is seldom found in such members because it is usually combined with bending and shear and perhaps axial forces.

Several excellent textbooks contain a great deal of theoretical information concerning the present theories used for estimating the torsional strengths of reinforced concrete members.[5,6] In these books torsional strength is considered for several different types of situations. These include torsion in plain concrete, torsion and shear in reinforced concrete beams without web reinforcement, torsion and shear in members with web reinforcement, and so on. The next few paragraphs summarize a little of the information obtained from these studies.

Torsion in Plain Concrete

Initially plain concrete members subject to torsion are considered. The reason for considering such members is that they behave *almost* identically with longitudinally *or* transversely reinforced beams before cracking begins because the contribution of their reinforcing to torsional behavior is negligible. Actually if *both* transverse and longitudinal steel are present, their effect on torsional behavior is appreciable.

A failure in torsion occurs by bending in a plane inclined at about 45° with the longitudinal axis of the beam. Quite a few theories have been presented for estimating the strength of plain concrete members subject to torsion, including the plastic theory, the membrane analogy, and others.[7]

The magnitude of the torsional resistance of a plain concrete member cannot be precisely related to a particular strength of the concrete such as its f_c' value. A large number of tests have been made on such members, however, and they seem to indicate that the torsional cracking stress will generally be between $4.0\sqrt{f_c'}$ and $7.0\sqrt{f_c'}$. The size of the specimens used in the tests affected the results somewhat. It is thought that torsional

[5] Bresler, B., ed., 1974, *Reinforced Concrete Engineering*, vol. 1 (New York: Wiley), pp. 248–272.

[6] Park, R., and Paulay, T., 1975, *Reinforced Concrete Structures* (New York: Wiley), chap. 8.

[7] Bresler, B., ed., 1974, *Reinforced Concrete Engineering*, vol. 1 (New York: Wiley), pp. 249–252.

stresses can go about this high in reinforced concrete members before torsion cracks begin to develop. As a result, ACI Committee 438 uses $6.0\sqrt{f_c'}$ as the estimated cracking stress.

Reinforced Concrete Beams without Web Reinforcement Subjected to Flexure and Torsion

As a result of flexure and torsion, some diagonal cracks are produced much as in beams subjected to flexure and shear. If the members are in flexure there will be a compression zone on one side of the beam that restricts the cracks. As a result, though diagonal torsion and flexure cracks have developed, the other part of the beam is still capable of resisting some torsion. Again the manner in which the torsion is resisted is not clearly understood. The compression side of the beam can resist some torsion, and additional torsional resistance is provided due to the dowel action of the reinforcing bars across the cracks on the tensile side of the member.

When the factored torsion moment $T_u = \phi T_n$ exceeds $\phi(0.5\sqrt{f_c'}\sum x^2 y)$ it is necessary to include torsion effects with those of moment and shear when the beam is designed. The nominal torsional strength of a member is furnished by the concrete and by the torsional reinforcing.

$$T_n = T_c + T_s$$

When torsional reinforcing is needed the Code permits the torsion stress resisted by the concrete to go to $2.4\sqrt{f_c'}$ which is approximately 40% of its torsional cracking stress. This reduced value is used to estimate the torsional moment which can be resisted by the concrete when bending moment is simultaneously applied. The Code (11.6.6.1) says the torsional moment strength provided by the concrete is to be calculated by

$$T_c = \frac{0.8\sqrt{f_c'}\sum x^2 y}{\sqrt{1 + \left(\dfrac{0.4V_u}{C_t T_u}\right)^2}}$$

In the preceding expression V_u is the factored shear force at the section in question and T_u is the factored torsional moment at the same section. In developing the equation the value $b_w d/\sum x^2 y$ is replaced with C_t. Actually the effect of bending is not directly considered but as the permissible torsion stress is less than half of the torsional cracking stress the resulting T_c value is rather conservative for combinations of torsion, shear, and bending.

If T_u is $> \phi(0.5\sqrt{f_c'}\sum x^2 y)$ but less than the value of T_c calculated by this expression it would seem that no torsion reinforcing is required. However, this is not correct because for such members (other than certain slabs, footings, joists, and beams described in Section 11.5.5.1 of the Code) a certain minimum amount of shear and torsion reinforcing is specified by the Code (11.5.5.5) as will be described.

Combined Shear and Torsion in Members with Web Reinforcement

To simplify the work of the reinforced concrete designer the ACI has attempted to make the design procedure for torsion reinforcing as close as possible to the procedure used for shear reinforcing. If the factored torsional moment T_u exceeds ϕT_c it will be necessary to provide torsion reinforcing. The nominal torsional moment strength T_s which must be supplied can be determined as follows

$$T_u = \phi T_c + \phi T_s$$

$$T_s = \frac{T_u - \phi T_c}{\phi}$$

The nominal strength of torsion reinforcing can be determined with the following expression given in Section 11.6.9.1 of the Code.

$$T_s = \frac{A_t \alpha_t x_1 y_1 f_y}{s}$$

In this expression α_t is a coefficient equal to $[0.66 + 0.33\,(y_1/x_1)]$ but not more than 1.50. The values of x_1 and y_1 are illustrated in Figure 13.4. The design yield strength of the stirrups is limited to a maximum of 60,000 psi in order to control crack sizes. Higher strength steels can be brittle near sharp bends as at stirrup corners, thus providing another reason for holding down yield strengths.

Where torsion reinforcing is required it is to be provided in addition to the reinforcing supplied for shear, flexure, and axial forces. The purpose ot web reinforcement in members subject to torsion is quite similar to the purpose of web reinforcement in members subject to shear forces. Once the diagonal cracks begin to occur, the torsion cannot be resisted unless a different mechanism other than just the concrete is available. Such a mechanism is furnished by forming a type of space truss usually consisting of closed stirrups together with additional longitudinal steel in the corners of the stirrups and perhaps in between.

It is necessary to have both longitudinal and closed transverse reinforcement to resist the diagonal tension stresses caused by torsion. If one of these two types of torsional reinforcing is omitted the other type is relatively ineffective. The stirrups must be closed because of the fact that torsional cracks can occur on all faces of a member. It is thus necessary for both the torsion and shear reinforcing to be adequately anchored at both ends so they will be fully effective on either side of a crack.

The area of longitudinal bars required for flexural reinforcing must be increased by the larger of the following two values (Code 11.6.9.3).

$$A_l = 2A_t\left(\frac{x_1 + y_1}{s}\right)$$

$$A_l = \left[\frac{400xs}{f_y}\left(\frac{T_u}{T_u + (V_u/3C_t)}\right) - 2A_t\right]\left(\frac{x_1 + y_1}{s}\right)$$

In the second expression the value of A_l need not exceed the value obtained if $50b_w s/f_y$ is substituted in place of $2A_t$.

The additional longitudinal bars must be at least #3 in size, they must be spread around the perimeter of the section with at least one bar placed in each corner of the closed stirrups, and their center-to-center spacing must not exceed 12 in. These bars are placed in the stirrup corners to provide anchorage for the stirrup legs and also to aid in fabricating the reinforcing cage. They have proved to be very effective in controlling cracks and developing torsional strength.

13.4 TORSION DESIGN PROCEDURE

Torsion stresses occur at the same time as flexural and shear stresses with the result that it is necessary to consider the interaction of all three. Though several methods have been proposed to do just this, there is presently no rigorous method available that has been universally accepted by the design community.

In this section the approximate and conservative method required by the Code (11.6) for combined shear and torsion design is presented. The procedure involves the determination of the web reinforcing required by shear, the determination of the web reinforcing required by torsion, the addition of the two, and finally the determination of the additional longitudinal steel needed. To simplify the calculations it is convenient to calculate each of the two sets of web reinforcing in terms of required steel areas per inch of beam length. Although the design process is simple and straightforward the following detailed list of steps may be helpful to you.

1. The first step is to see if shear and torsion reinforcing are required. The factored shear V_u and the factored torsion T_u are computed. As for shear reinforcing, those sections located less than a distance d from the face of the support may be designed for T_u at a distance d (Code 11.6.4). Shear reinforcing is needed if V_u is greater than $\phi V_c = \phi 2\sqrt{f_c'}b_w d$ while torsion reinforcing is required if T_u is greater than $\phi T_c = \phi(0.5\sqrt{f_c'}\sum x^2 y)$.

2. The nominal shearing strength provided by the concrete V_c is reduced to the following value (Code 11.3.1.4) when the factored torsion T_u exceeds $\phi(0.5\sqrt{f_c'}\sum x^2 y)$.

$$V_c = \frac{2\sqrt{f_c'}b_w d}{\sqrt{[1 + 2.5C_t(T_u/V_u)^2}}$$

The value of $V_s = (V_u - \phi V_c)/\phi$ is determined. Its maximum permissible value is $8\sqrt{f_c'}b_w d$ as stated by the Code (11.5.6.8). The area of the shear reinforcing required can be calculated with the following expression

$$V_s = \frac{A_v f_y d}{s}$$

It is however convenient to compute the shear stirrup area required per inch of beam as follows

$$\frac{A_v}{s} = \frac{V_s}{f_y d} = \frac{V_u - \phi V_c}{\phi f_y d}$$

You will remember that the area so obtained is for both of the legs of the stirrup. In Section 11.5.5.3 of the Code a minimum value for $A_v/s = 50b_w/f_y$ is required if $T_u \lessdot \phi(0.5\sqrt{f_c'}\sum x^2 y)$.

3. The torsional moment strength contributed by the concrete is computed with the following expression.

$$T_c = \frac{0.8\sqrt{f_c'}\sum x^2 y}{\sqrt{1 + (0.4V_u/C_t T_u)^2}}$$

The value of the nominal torsional moment strength T_s which must be provided by the torsion reinforcement and which Section 11.6.9.4 of the Code says cannot exceed $4T_c$ is as follows

$$T_s = \frac{T_u - \phi T_c}{\phi}$$

As for shear reinforcing it is convenient to obtain the stirrup area required per inch of beam. You will note that the area obtained is for one leg of the stirrup.

$$T_s = \frac{A_t \alpha_t x_1 y_1 f_y}{s}$$

$$\frac{A_t}{s} = \frac{T_u - \phi T_c}{\phi f_y \alpha_t x_1 y_1}$$

4. The total area of stirrups per inch of beam for both shear and torsion can be determined by adding A_t/s and A_v/s taking into account the fact that A_t is for one leg and A_v is for two. The area obtained may be no less than the value computed from the following formula given in Section 11.5.5.5 of the Code, $A_v + 2A_t = 50b_w s/f_y$.

The spacing of the closed stirrups selected may not be greater than $(x_1 + y_1)/4$ nor 12 in., nor the maximum spacings previously required for shear stirrups.

Section 11.6.7.5 of the Code requires that stirrups and other bars and wires used as torsion reinforcing must be continued a distance d beyond the point where they theoretically are no longer required. This distance which is larger than normally used for shear and flexural reinforcing is considered necessary because the diagonal tension cracks caused by torsion occur in a rather long helical form.

It should be noted that if an interior beam is loaded to produce maximum shear and moment (that is, with full load on the slab on both sides of the beam), the torsion will be reduced, or if it is loaded to produce maximum torsion (that is, with live load placed on the slab on one side of the beam only), the maximum shears and moments will be reduced. This is not the case for exterior beams, and as a result, it is necessary to add more longitudinal reinforcing. For interior beams extra bars are not usually added except for corner bars added on the compression sides to form the steel cages with the closed stirrups.

Example 13.1 illustrates the design of torsion reinforcing for a rectangular beam.

EXAMPLE 13.1

In addition to the service live and dead uniform loads shown, the beam of Figure 13.5 is subjected to a 10 ft-k service dead-load torsion and a 12 ft-k service live-load torsion at the face of the support. It is assumed that the magnitude of these torsional moments decrease uniformly from the beam end to the ℄. Design the shear and torsion reinforcement for this beam if $f_c' = 4{,}000$ psi and $f_y = 60{,}000$ psi.

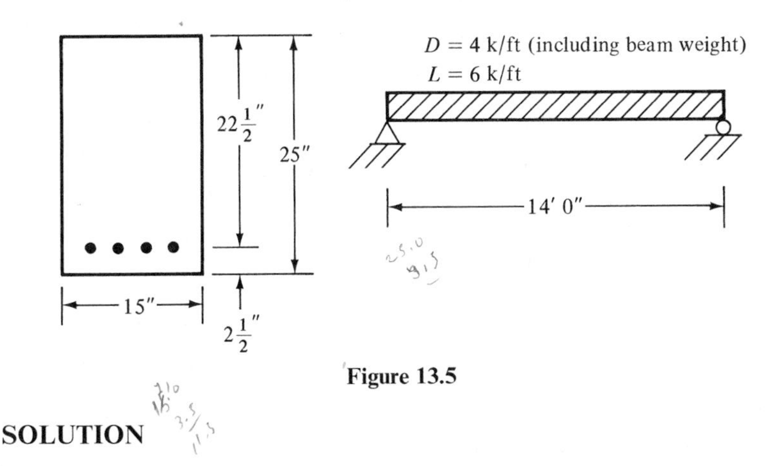

Figure 13.5

SOLUTION

Calculating shear and torsion values

$$w_u = (1.4)(4) + (1.7)(6) = 15.8 \text{ k/ft}$$
$$V_u = (7)(15.8) = 110.6^{\text{k}}$$

V_u @ a distance d from face of support

$$= \left(\frac{61.5}{84}\right)(110.6) = 80.975^k = 80{,}975\#$$

$$T_u = (1.4)(10) + (1.7)(2) = 34.4 \text{ ft-k}$$

T_u @ a distance d from face of support

$$= \left(\frac{61.5}{84}\right)(34.4) = 25.186 \text{ ft-k} = 302{,}232 \text{ in.-lb}$$

This is a conservative value which does not account for any rotation of the beam.

Checking to see if torsion reinforcement is required

$$\phi(0.5\sqrt{f_c'}\textstyle\sum x^2 y) = (0.85)(0.5\sqrt{4{,}000})(15)^2(25)$$
$$= 151{,}196 \text{ in.-lb} < T_u \text{ of } 302{,}232 \text{ in.-lb}$$

$\therefore$ torsion effects must be considered

Design of shear reinforcement

$$C_t = \frac{b_w d}{\sum x^2 y} = \frac{(15)(22.5)}{(15)^2(25)} = 0.06$$

$$V_c = \frac{2\sqrt{f_c'}b_w d}{\sqrt{1 + [2.5C_t(T_u/V_u)]^2}} \quad \text{when} \quad T_u > \phi(0.5\sqrt{f_c'}\textstyle\sum x^2 y)$$

$$= \frac{(2\sqrt{4000})(15)(22.5)}{\sqrt{1 + [2.5 \times 0.06 \times (302{,}232/80{,}975)]^2}} = 37{,}250\#$$

$$V_u = \phi V_s + \phi V_c$$

$$V_s = \frac{V_u - \phi V_c}{\phi} = \frac{80{,}975 - (0.85)(37{,}250)}{0.85} = 58{,}015\#$$

Maximum V_s permitted by Code (11.5.6.8)

$$= 8\sqrt{f_c'}b_w d = (8\sqrt{4000})(15)(22.5) = 170{,}763\# > 58{,}015\# \quad \text{ok}$$

$$\frac{A_v}{s} = \frac{V_u - \phi V_c}{\phi f_y d} = \frac{80{,}975 - (0.85)(37{,}250)}{(0.85)(60{,}000)(22.5)} = 0.0430 \text{ in.}^2/\text{in.}$$

$\frac{A_v}{s}$ minimum by Code (11.5.5.3)

$$= \frac{50b_w}{f_y} = \frac{(50)(15)}{60{,}000} = 0.0125 \text{ in.}^2/\text{in.} < 0.0430 \text{ in.}^2/\text{in.} \quad \text{ok}$$

Torsional moment strength contributed by concrete

$$T_c = \frac{0.8\sqrt{f'_c}\sum x^2y}{\sqrt{1 + \left(\frac{0.4V_u}{C_tT_u}\right)^2}} = \frac{(0.8\sqrt{4000})(15)^2(25)}{\sqrt{1 + \left(\frac{0.4 \times 80{,}975}{0.06 \times 302{,}232}\right)^2}} = 139{,}033 \text{ in.-lb}$$

$$T_s = \frac{T_u - \phi T_c}{\phi} = \frac{302{,}232 - (0.85)(139{,}033)}{0.85} = 216{,}532 \text{ in.-lb} < 4T_c \ \underline{\text{ok}}$$

Determining torsion reinforcing required per inch

$$\alpha_t = \left[0.66 + (0.33)\left(\frac{y_1}{x_1}\right)\right] = \left[0.66 + (0.33)\left(\frac{21.5}{11.5}\right)\right]$$

$$= 1.28'' < 1.50''$$

$$\frac{A_t}{s} = \frac{T_s}{\alpha_t x_1 y_1 f_y} = \frac{216{,}532}{(1.28)(11.5)(21.5)(60{,}000)} = 0.0146 \text{ in.}^2/\text{in.}$$

Selection of web reinforcing

It will be noted that A_v in the expression to follow is the cross sectional area of two legs for a closed stirrup while A_t is the area of one leg of the stirrup. Thus the equation gives the total area of one leg of the stirrup.

$$\frac{A_t}{s} + \frac{A_v}{2s} = 0.0146 + \frac{0.0430}{2} = 0.0361 \text{ in.}^2/\text{in.}$$

For #3 stirrups

$$s = \frac{0.11}{0.0361} = 3.05''$$

For #4 stirrups

$$s = \frac{0.20}{0.0361} = 5.54''$$

For #5 stirrups

$$s = \frac{0.31}{0.0361} = 8.59''$$

Maximum permissible spacing

$$= \frac{x_1 + y_1}{4} = \frac{11.5 + 21.5}{4} = 8.25'' \qquad \text{nor} \qquad 12''$$

Use #5 closed stirrup at 8″ O.C.

By proportions the required spacing of the stirrups can be computed at other distances from the support. It will also be noted that further out

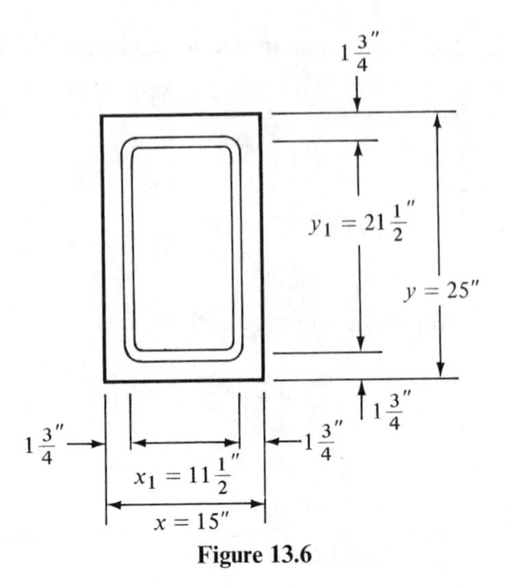

Figure 13.6

in the beam where T_u becomes less than $\phi(0.5\sqrt{f_c'}\sum x^2 y) = 151{,}196$ in.-lb the maximum spacing goes up to the maximum value for shear alone.

Check beam for minimum permissible area of stirrups by Section 11.5.5.5 of the Code

$$A_v + 2A_t = 2A_{\text{bar}} = 50\,\frac{b_w s}{f_y}$$

$$2A_{\text{bar}} = (50)\left(\frac{15 \times 8}{60{,}000}\right) = 0.10 \text{ in.}^2$$

$$\min A_{\text{bar}} = 0.05 \text{ in.}^2 < 2 \times 0.31 \text{ in.}^2 \text{ furnished} \qquad \text{ok}$$

Increase in longitudinal steel area

$$A_l = 2A_t\left(\frac{x_1 + y_1}{s}\right) = (2)(0.31)\left(\frac{11.5 + 21.5}{8.00}\right) = 2.56 \text{ in.}^2$$

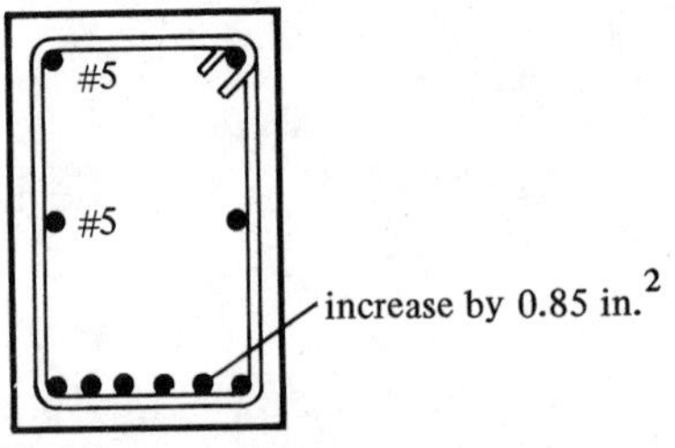

#5

#5

increase by 0.85 in.^{2}

Figure 13.7 Additional Longitudinal Reinforcing Required for Tension

The second equation for A_l should also be checked but in this case it did not control.

Add $2.56/3 = 0.853$ in.^{2} (say 2 #6 = 0.88 in.^{2}) at the top and mid-depth of the member and increase the required longitudinal reinforcing in the bottom by 0.853 in.^{2} as shown in Figure 13.7

PROBLEMS

13.1 The beam shown in the accompanying illustration is subjected to a 12 ft-k service dead-load torsion and a 16 ft-k service live-load torsion at the face of the support. Assuming that torsion decreases uniformly from the beam end to the beam ℄, design shear and torsion reinforcement. $f_c' = 3{,}000$ psi and $f_y = 50{,}000$ psi.

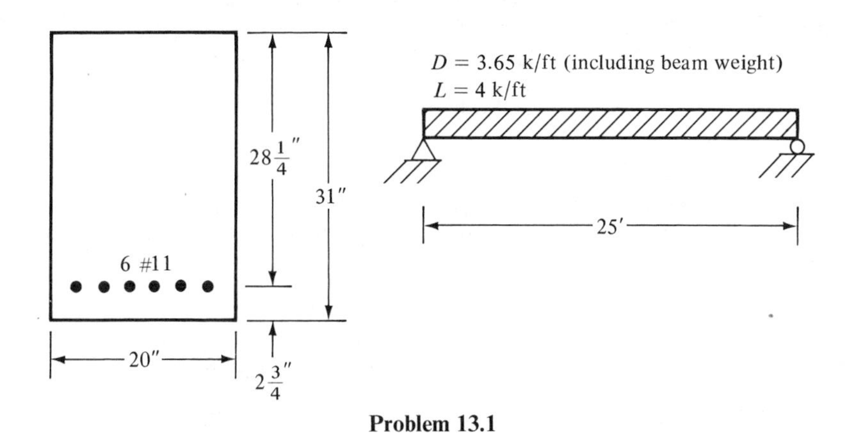

Problem 13.1

13.2 The reinforced concrete spandrel beam shown in the accompanying illustration is integrally connected with the 5-in. slab. If $f_c' = 4{,}000$ psi and $f_y = 60{,}000$ psi, select the torsional steel reinforcing required at a distance d from the face of the support where $V_u = 32^k$ and $T_u = 10$ ft-k.

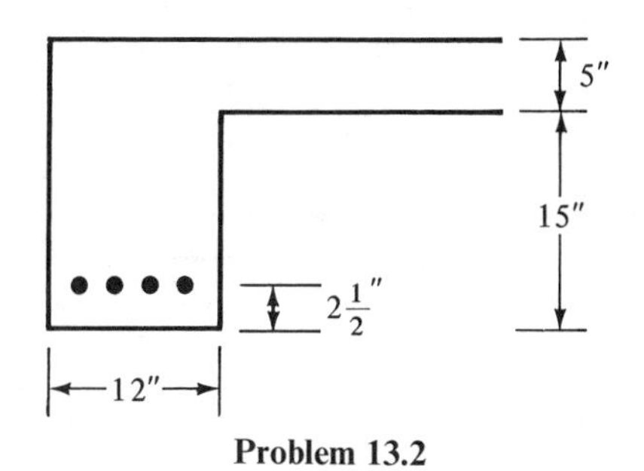

Problem 13.2

13.3 Design the torsional reinforcement for the beam shown in the accompanying figure at a section a distance d from the face of the support for a torsional moment of 24 ft-k and $V_u = 90^k$. $f_c' = 3{,}000$ psi and $f_y = 60{,}000$ psi.

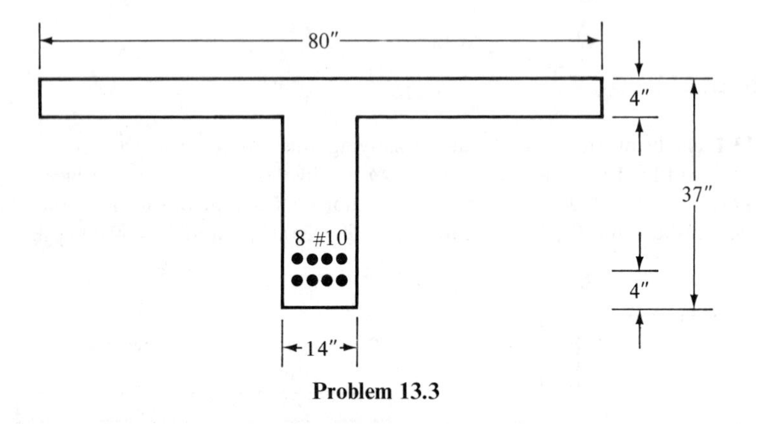

Problem 13.3

13.4 Select the torsional reinforcing required for the edge beam shown in the accompanying illustration if $f_c' = 4{,}000$ psi and $f_y = 50{,}000$ psi. T_u equals 30 ft-k at the face of the support and is assumed to vary along the beam in proportion to the shear.

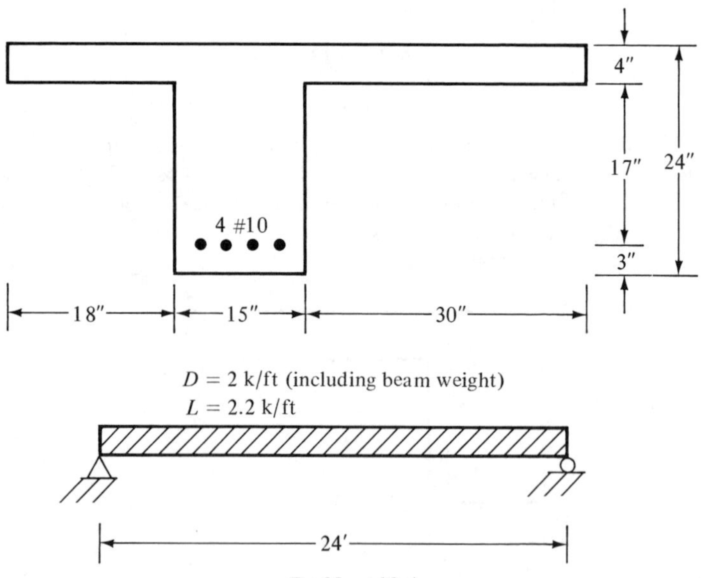

Problem 13.4

13.5 A 12-in. × 22-in. spandrel beam ($d = 19$ in.) with a 20-ft simple span has a 4-in. slab on one side acting as a flange. It must carry a maximum V_u of 60^k and a maximum T_u of 20 ft-k at the face of the support. Assuming these values are zero at the beam ℄, design torsional reinforcing. $f_c' = 3{,}000$ psi and $f_y = 60{,}000$ psi.

13.6 Design the web reinforcing for the beam shown if the load is acting 3 in. off center of the beam. Assume the torsion equals the uniform load times the 3 in. $f_c' = 3{,}000$ psi, and $f_y = 60{,}000$ psi. Use #4 stirrups. Assume the torsion value varies from a maximum at the support to 0 at the beam ℄, as does the shear.

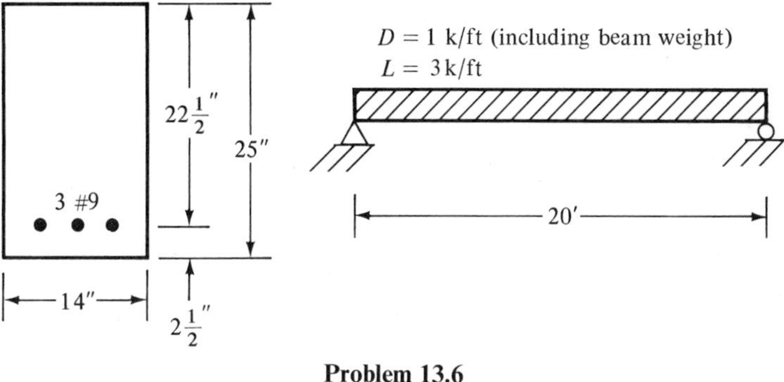

Problem 13.6

13.7 Is the beam shown satisfactory to resist a T_u of 20 ft-k and a V_u of 50^k if $f_c' = 3{,}000$ psi and $f_y = 60{,}000$ psi? The bars shown are used in addition to those provided for bending moment.

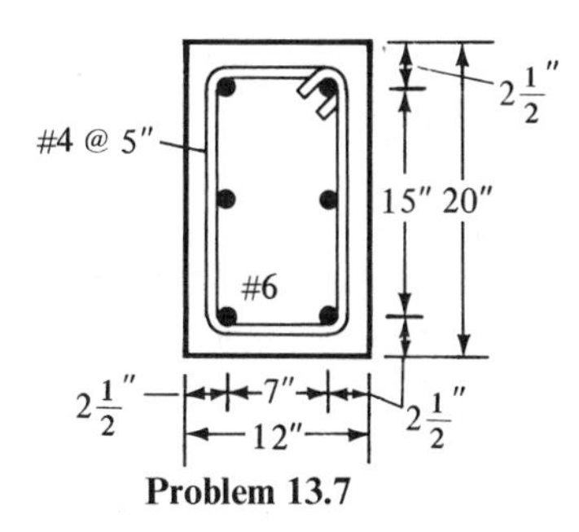

Problem 13.7

PROBLEMS WITH SI UNITS

13.8 Design shear and torsion reinforcement for the beam shown in the accompanying figure assuming that the torsion decreases uniformly from the beam end to the beam ℄. The member is subjected to a 34-kN·m service dead-load torsion and a 40-kN·m service live-load torsion at the face of the support. $f_y = 344.8$ MPa and $f_c' = 20.7$ MPa.

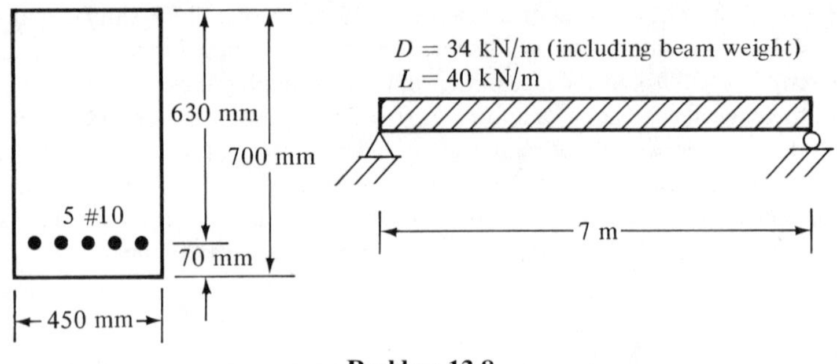

Problem 13.8

13.9 Design the torsional reinforcement for the beam shown in the accompanying figure at a section located at a distance d from the face of the support for a torsional moment of 40 kN·m and $V_u = 500$ kN. $f_c' = 20.7$ MPa and $f_y = 275.8$ MPa.

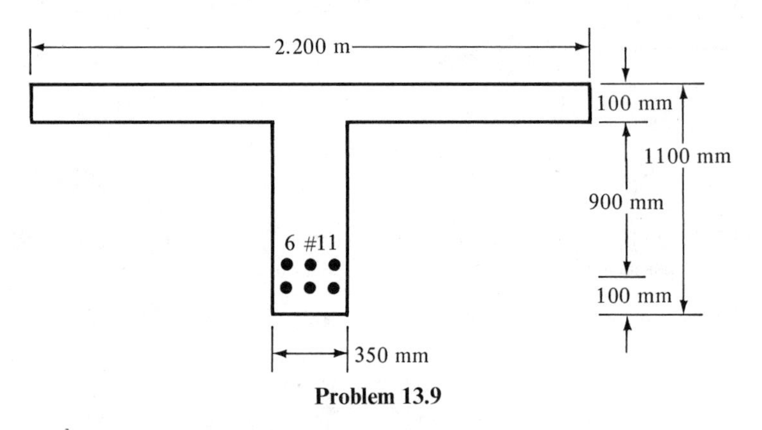

Problem 13.9

13.10 Design the web reinforcing for the beam shown if the load is acting 100 mm off center of the beam. Assume the torsion equals the uniform load times 100 mm. $f_c' = 17.2$ MPa and $f_y = 275.8$ MPa. Use #4 stirrups and assume that the torsion and shear vary from a maximum at the support to 0 at the beam ℄.

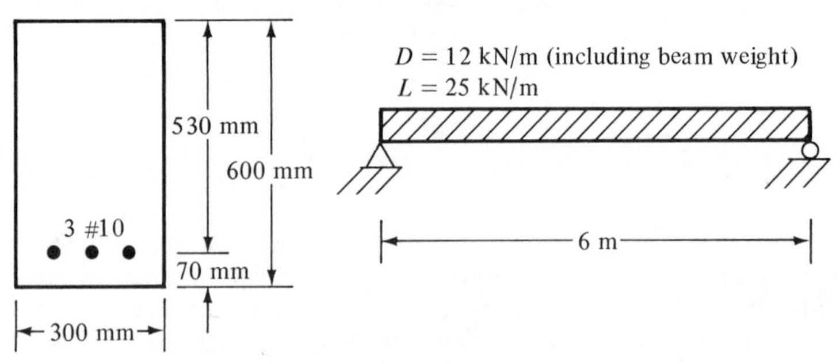

Problem 13.10

CHAPTER 14

Design of Two-Way Slabs

14.1 INTRODUCTION

In general, slabs are classified as being one-way or two-way. Slabs that primarily deflect in one direction are referred to as *one-way slabs* [see Figure 14.1(a)]. Simple-span, one-way slabs have previously been discussed in Sections 3.4 and 4.10 of this text, while the design of continuous one-way slabs was considered in Section 12.5. When slabs are supported by columns arranged generally in rows, so that the slabs can deflect in two directions, they are usually referred to as *two-way slabs*.

Two-way slabs may be strengthened by the addition of beams between the columns, by thickening the slabs around the columns (*drop panels*), and by flaring the columns under the slabs (*column capitals*). These situations are shown in Figure 14.1 and discussed in the next several paragraphs.

Flat plates [Figure 14.1(b)] are solid concrete slabs of uniform depths that transfer loads directly to the supporting columns without the aid of beams or capitals or drop panels. Flat plates can be quickly constructed due to their simple formwork and reinforcing bar arrangements. They need the smallest overall story heights to provide specified headroom requirements, and they also give the most flexibility in the arrangement of columns and partitions.

For flat plates there may be a problem in transferring the shear at the perimenter of the columns. In other words, there is a danger that the columns may punch through the slabs. As a result, it is frequently necessary to increase column sizes or slab thicknesses or to use *shearheads*. Shearheads consist of steel *I* or channel shapes placed in the slab over the columns, as discussed in Section 14.5 of this chapter. Though such procedures may seem to be expensive, it is noted that the simple formwork required for flat plates will usually result in such economical construction that the extra costs required for shearheads are more than canceled. For heavy industrial loads or long spans, however, some other type of floor system may be required.

Flat slabs [Figure 14.1(c)] include two-way reinforced concrete slabs with capitals, drop panels, or both. These slabs are very satisfactory for heavy loads and long spans. Although the formwork is more expensive than for flat plates, flat slabs will require lesser amounts of concrete and

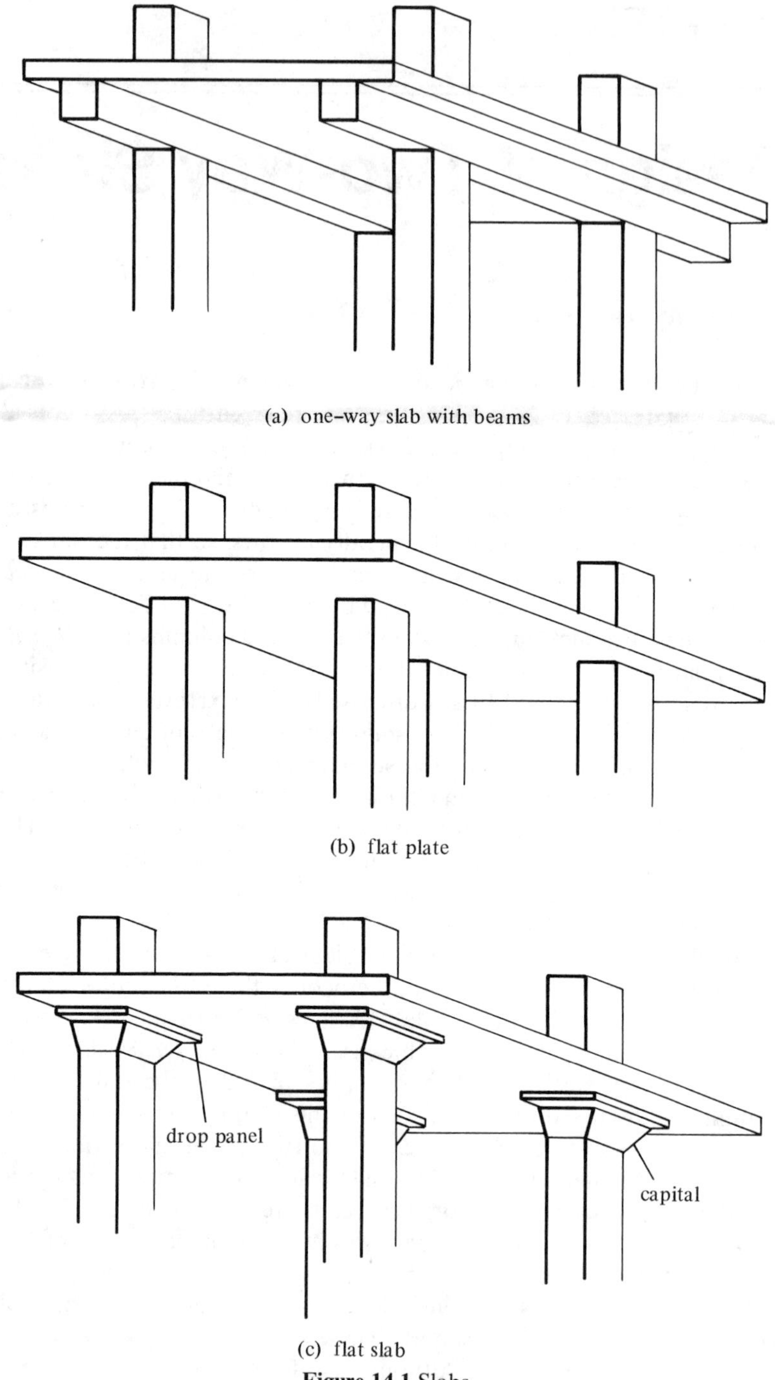

Figure 14.1 Slabs

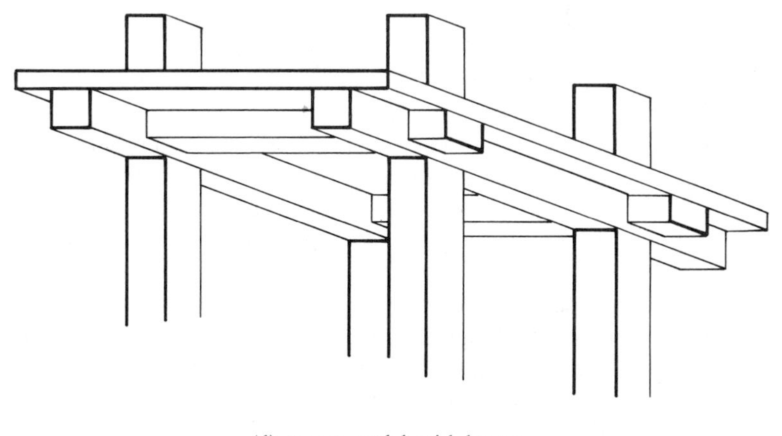

(d) two-way slab with beams

Figure 14.1 (*cont.*)

reinforcing for the same loads and spans. They are particularly economical for warehouses, parking and industrial buildings, and similar structures where exposed drop panels or capitals are acceptable.

Finally, in Figure 14.1(d) a *two-way slab with beams* is shown. This type of floor system is obviously used where its cost is less than the costs of flat plates or flat slabs. In other words, when the loads or spans or both become quite large, the slab thickness and column sizes required for flat plates or flat slabs are of such magnitude that it is more economical to use two-way slabs with beams, despite the higher formwork costs.

14.2 ANALYSIS OF TWO-WAY SLABS

Two-way slabs bend under load into dished shape surfaces so there is bending in both principal directions. As a result, they must be reinforced in both directions by layers of bars that are perpendicular to each other. A theoretical elastic analysis for such slabs is a very complex problem due to their highly indeterminate nature. Numerical techniques such as finite difference and finite elements are required, but such methods are not really practical for routine design.

Actually, the fact that a great deal of stress redistribution can occur in such slabs at high loads makes it unnecessary to make designs based on theoretical analyses. As a result, the design of two-way slabs is generally based on empirical moment coefficients, which, though they might not accurately predict stress variations, result in slabs with satisfactory overall safety factors. In other words, if too much reinforcing is placed in one part of a slab and too little somewhere else, the resulting slab behavior will probably still be satisfactory. *The total amount of reinforcement in a slab seems more important than its exact placement.*

You should clearly understand that though this chapter is devoted to two-way slab design based on approximate methods of analysis, there is no intent to prevent the designer from using more exact methods. He may design slabs on the basis of numerical solutions, or yield-line analysis, or other theoretical methods, providing he can clearly demonstrate that he has met all the necessary safety and serviceability criteria required by the ACI Code.

Although it has been the practice of designers for many years to use approximate analyses for design and to use average moments rather than maximum ones, two-way slabs so designed have proved to be very satisfactory under service loads. Furthermore, they have proved to have appreciable overload capacity.

14.3 DESIGN OF TWO-WAY SLABS BY THE ACI CODE

The 1977 ACI Code (13.3.2) specifies in detail two methods for designing two-way slabs. These are the direct design method and the equivalent frame method.

1. *Direct design method.* The Code (13.6) provides a procedure with which a set of moment coefficients can be determined. The method, in effect, involves a single-cycle moment distribution analysis of the structure based on the estimated flexural stiffnesses of the slabs, beams (if any), and columns and also the torsional stiffnesses of the slabs and beams (if any) transverse to the direction in which flexural moments are being determined. Some type of moment coefficients have been used satisfactorily for many years for slab design. They do not, however, give very satisfactory results for slabs with unsymmetrical dimensions and loading patterns.

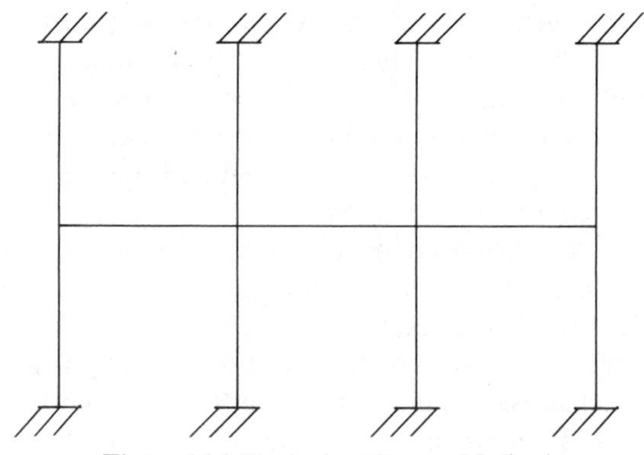

Figure 14.2 Equivalent Frame Method

2. *Equivalent frame method.* In this method a portion of the structure is taken out by itself, as shown in Figure 14.2, and analyzed much as a portion of a building frame was handled in Example 12.2. The same detailed stiffness values used for the direct design method are used for the equivalent frame method. The method that is very satisfactory for symmetrical frames as well as for those with unusual dimensions or loadings is presented in Section 13.7 of the Code.

14.4 COLUMN AND MIDDLE STRIPS

After the design moments have been determined by either the direct design method or the equivalent frame method, they are distributed across each panel. The panels are divided into column and middle strips, as shown in Figure 14.3, and positive and negative moments are estimated in each strip. The *column strip* is a slab with a width on each side of the column center line equal to one-fourth the smaller of the panel dimensions l_1 and l_2. It includes beams if they are present. The middle strip is the part of the slab in between the two column strips.

The part of the moments assigned to the column and middle strips may be assumed to be uniformly spread over the strips. As will be later described in this chapter, the percentage of the moment assigned to a

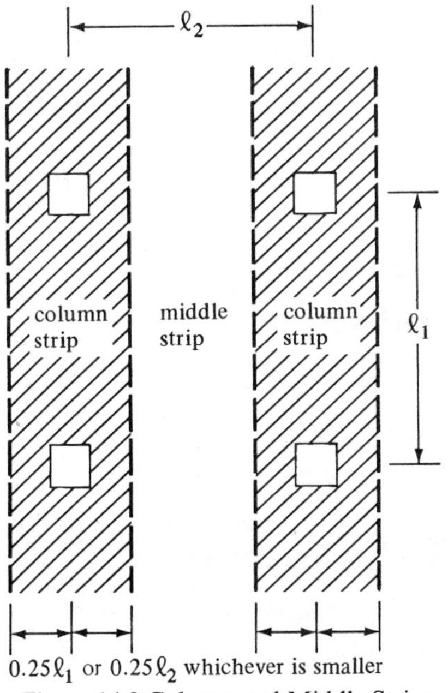

Figure 14.3 Column and Middle Strips

column strip depends on the effective stiffness of that strip and on its *aspect ratio* l_2/l_1 (where l_1 is the length of span, center-to-center, of supports in the direction in which moments are being determined and l_2 is the span length, center-to-center, of supports in the direction transverse to l_1).

14.5 SHEAR RESISTANCE OF SLABS

For two-way slabs supported by beams or walls, shears are calculated at a distance d from the face of the walls or beams. The value of ϕV_c is, as for beams, $\phi 2\sqrt{f_c'}b_w d$. Shear is not usually a problem for these types of slabs.

For flat slabs and flat plates supported directly by columns, shear may be the critical factor in design. In almost all tests of such structures, failures have been due to shear or perhaps shear and torsion. These conditions are particularly serious around exterior columns.

There are two kinds of shear that must be considered in the design of flat slabs and flat plates. These are the same two that were considered in column footings—one-way and two-way shears (that is, beam shear and punching shear). For beam shear analysis the slab is considered to act as a wide beam running between the supports. The critical sections

Police Station, Philadelphia, Pennsylvania (Courtesy Portland Cement Association).

are taken at a distance d from the face of the column or capital. For punching shear the critical section is taken at a distance $d/2$ from the face of the column, capital, or drop panel and the shear strength as in footings, is $\phi 4\sqrt{f_c'}b_w d$.

If shear stresses are too large around interior columns, it is possible to increase the shearing strength of the slabs by as much as 75% by using shearheads. A shearhead, as defined in Section 11.11.4 of the Code, consists of four steel I or channel shapes fabricated into cross arms and placed in the slabs as shown in Figure 14.4(a). The Code states that

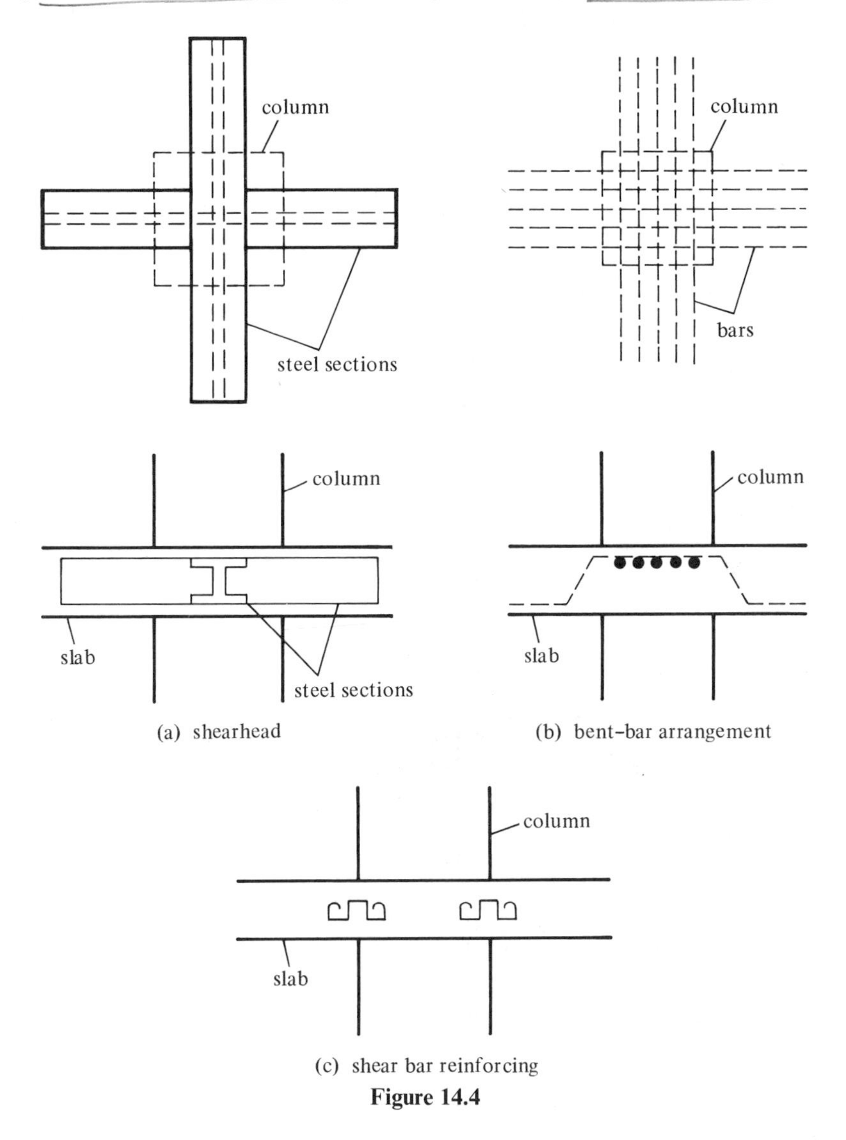

(a) shearhead

(b) bent-bar arrangement

(c) shear bar reinforcing

Figure 14.4

shearhead designs of this type do not apply at exterior columns. Thus special designs are required and the Code does not provide specific requirements. Shearheads increase the effective b_o for two-way shear and they also increase the negative moment resistance of the slab, as described in the Code (11.11.4.9). The negative moment reinforcing bars in the slab are usually run over the top of the steel shapes, while the positive reinforcing is normally stopped short of the shapes.

Another type of shear reinforcement permitted in slabs by the Code (11.11.3) involves the use of groups of bent bars or wires. One possible arrangement of such bars is shown in Figure 14.4(b). The bars are bent

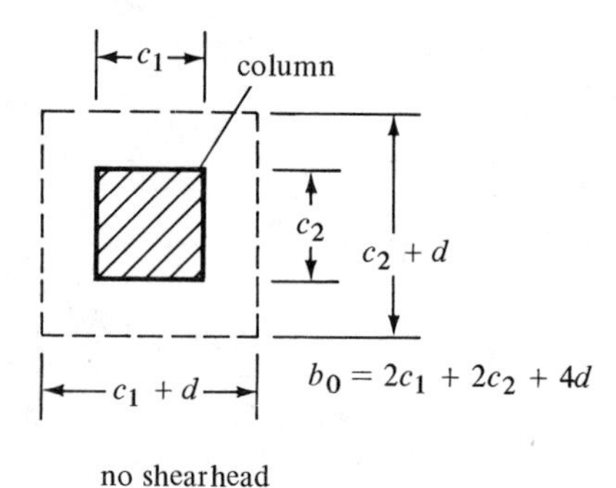

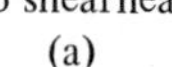
no shearhead
(a)

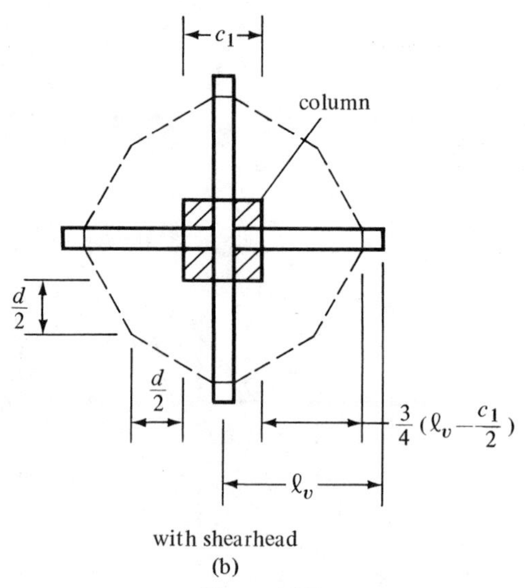

with shearhead
(b)

Figure 14.5

across the potential diagonal tension cracks at 45° angles and they are run along the bottoms of the slabs for the distances needed to fully develop the bar strengths. Another type of bar arrangement that might be used is shown in Figure 14.4(c). When bars (or wires) are used as shear reinforcement, the Code (11.11.3.2) states that the maximum two-way shear strength permitted on the critical section at a distance $d/2$ from the face of the column may be increased from $4\sqrt{f_c'}b_od$ to $6\sqrt{f_c'}b_od$.

The main advantage of shearheads is that they push the critical sections for shear further out from the columns, thus giving a larger perimeter to resist the shear, as illustrated in Figure 14.5. In this figure l_v is the length of the shearhead arm from the centroid of the concentrated load or reaction and c_1 is the dimension of the rectangular or equivalent rectangular column or capital or bracket measured in the direction in which moments are being calculated. The Code (11.11.4.7) states that the critical section for shear shall cross the shearhead arm at a distance equal to $\frac{3}{4}[l_v - (c_1/2)]$ from the column face, as shown in Figure 14.5(b). Although this critical section is to be located so that its perimeter is a minimum, it does not have to be located closer to the column face than $d/2$ at any point. When steel *I* or channel shapes are used as shearheads, the maximum shear strength can be increased to $7\sqrt{f_c'}b_od$ at a distance $d/2$ from the column. This is only permissible if the maximum computed shear does not exceed $4\sqrt{f_c'}b_od$ along the dotted critical section for shear shown in Figure 14.5(b).

14.6 DEPTH LIMITATIONS AND STIFFNESS REQUIREMENTS

For most slabs the theoretical calculation of deflections is too complicated for practical use. As a result, the usual practice is to follow certain minimum thickness rules that were developed on the basis of measurements of deflections in actual two-way structures. If these rules are followed, it is considered unnecessary to compute deflections. Unfortunately, attempts to make these limiting values as accurate as possible have caused the expressions to become rather complex. In these somewhat frightening equations, the following terms are used:

l_n = the clear span in the long direction measured face to face of columns for slabs without beams and face to face of beams for slabs with beams

β = the ratio of the long to the short clear span

β_s = the ratio of the sum of the lengths of the sides of the panel that are continuous with other panels to the total perimeter of the panel

α_m = the average value of the ratios of beam-to-slab stiffness on all sides of a panel

Throughout this chapter the letter α is used to represent the ratio of the flexural stiffness ($E_{cb}I_b$) of a beam section to the flexural stiffness of the slab ($E_{cs}I_s$) whose width equals the distance between the center lines of the panels on each side of the beam.

$$\alpha = \frac{E_{cb}I_b}{E_{cs}I_s}$$

in which

E_{cb} = the modulus of elasticity of the beam concrete

E_{cs} = the modulus of elasticity of the column concrete

I_b = the gross moment of inertia about the centroidal axis of a section made up of the beam and the slab on each side of the beam extending a distance equal to the projection of the beam above or below the slab (whichever is greater) but not exceeding four times the slab thickness

I_s = the moment of inertia of the gross section of the slab taken about the centroidal axis and equal to $h^3/12$ times the slab width, where the width is the same as for α

The thickness of slabs or other two-way construction may not be less than the larger value obtained by substituting into two equations which are given in Section 9.5.3.1 of the ACI Code. In the first of these two equations (numbered 14.1 here), the complicated denominator in effect requires greater slab thicknesses by taking into account the smaller stiffnesses present when high-strength steels are used. (High-strength steels are associated with larger amounts of cracking in the concrete.) The quantities β and β_s are used in an attempt to take into account the effects of the shape and location of the panel on its deflection, while the effect of beams on deflection is represented by α_m.

$$h = \frac{l_n(800 + 0.005f_y)}{36{,}000 + 5{,}000\beta\{\alpha_m - 0.5[1 - \beta_s][1 + (1/\beta)]\}} \tag{14.1}$$

When very stiff beams are used, the thickness determined with equation 14.1 may be smaller than is desirable and, as a result, another minimum thickness must be calculated from equation 14.2, which follows:

$$h = \frac{l_n(800 + 0.005f_y)}{36{,}000 + 5{,}000\beta(1 + \beta_s)} \tag{14.2}$$

When very flexible beams are used, or no beams at all, the term in braces in the denominator of Equation 14.1 becomes negative, and undesirably large slab thicknesses may result. To limit this value the Code specifies a maximum slab thickness to be determined from the following expression:

$$h = \frac{l_n(800 + 0.005f_y)}{36{,}000} \tag{14.3}$$

For panels with discontinuous edges, the Code (9.5.3.3) requires that edge beams be used, which have a minimum α equal to 0.8, or else that the minimum slab thicknesses, as determined by Equations 14.1 through 14.3, must be increased by 10%. A 10% decrease in minimum thickness is provided by the Code (9.5.3.2) for slabs without beams but with drop panels extending in each direction from center line of support a distance no less than one-sixth of the span length in that direction measured center-to-center of supports and a projection below the slab at least equal to $h/4$.

Regardless of the values obtained on the basis of the preceding rules, the thicknesses shall not be less than the following:

For slabs without beams or drop panels	5 in.
For slabs without beams but with drop panels	4 in.
For slabs with beams on all four edges with a value of α at least equal to 2.0	$3\frac{1}{2}$ in.

Should the various rules for minimum thickness be followed but the resulting slab be insufficient to provide the shear capacity required for the particular column size, column capitals will probably be required. Beams running between the columns may be used for some slabs where partitions or heavy equipment loads are placed near column lines. A very common case of this type occurs where exterior beams are used when the exterior walls are supported directly by the slab. Another situation where beams may be used occurs where there is concern about the magnitude of slab vibrations.[1] Example 14.1 illustrates the application of the minimum slab thickness rules for a flat plate.

EXAMPLE 14.1

Using the ACI Code, determine the minimum thickness required for each of the numbered panels of the flat plate floor shown in Figure 14.6. Edge beams are not used along the exterior floor edges. $f_c' = 3{,}000$ psi and $f_y = 60{,}000$ psi.

[1] Fintel, M., ed., 1974, *Handbook of Concrete Engineering* (New York: Van Nostrand Reinhold), p. 58.

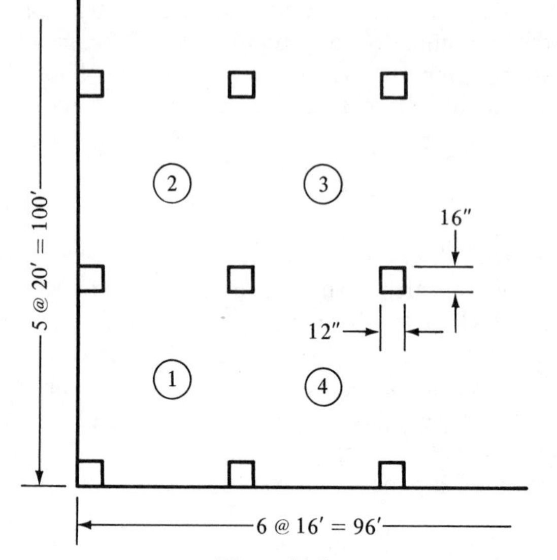

Figure 14.6

SOLUTION

$$\beta = \frac{20 - \frac{16}{12}}{16 - \frac{12}{12}} = 1.245$$

$$\alpha_m = 0$$

Panel 1

Substituting into Equations 14.1 and 14.2,

$$\beta_s = \tfrac{36}{72} = 0.500$$

$$h = \frac{(18.67)(800 + 0.005 \times 60{,}000)}{36{,}000 + (5{,}000)(1.245)\{0 - [0.5][1 - 0.5][1 + (1/1.245)]\}}$$

$$= 0.619' = 7.42''$$

$$h = \frac{(18.67)(800 + 0.005 \times 60{,}000)}{36{,}000 + (5{,}000)(1.245)(1 + 0.5)} = 0.453' = 5.44''$$

Panel 2

$$\beta_s = \tfrac{52}{72} = 0.722$$

$$h = \frac{(18.67)(800 + 0.005 \times 60{,}000)}{36{,}000 + (5{,}000)(1.245)\{0 - [0.5][1 - 0.722][1 + (1/1.245)]\}}$$

$$= 0.597' = 7.17''$$

$$h = \frac{(18.67)(800 + 0.005 \times 60{,}000)}{36{,}000 + (5{,}000)(1.245)(1 + 0.722)} = 0.440' = 5.27''$$

Panel 3

$$\beta_s = \tfrac{72}{72} = 1.00$$

$$h = \frac{(18.67)(800 + 0.005 \times 60{,}000)}{36{,}000 + (5{,}000)(1.245)\{0 - [0.5][1 - 1][1 + (1/1.245)]\}}$$

$$= 0.570' = 6.85''$$

$$h = \frac{(18.67)(800 + 0.005 \times 60{,}000)}{36{,}000 + (5{,}000)(1.245)(1 + 1)} = 0.424' = 5.09''$$

Panel 4

$$\beta_s = \tfrac{56}{72} = 0.778$$

$$h = \frac{(18.67)(800 + 0.005 \times 60{,}000)}{36{,}000 + (5{,}000)(1.245)\{[0 - 0.5][1 - 0.778][1 + (1/1.245)]\}}$$

$$= 0.592' = 7.10''$$

$$h = \frac{(18.67)(800 + 0.005 \times 60{,}000)}{36{,}000 + (5{,}000)(1.245)(1 + 0.778)} = 0.436' = 5.24''$$

Conclusion

1. Substituting into Equation 14.3, as it is the same for all four panels,

$$\text{maximum } h = \frac{(18.67)(800 + 0.005 \times 60{,}000)}{36{,}000} = 0.570' = 6.85''$$

2. The minimum h for slabs without drop panels is 5″ (Code 9.5.3.1).
3. Since no edge beams are used, α is less than 0.8 and the Code (9.5.3.3) states that the controlling value determined with equations 14.1 through 14.3 must be increased by 10%.

$$\text{Use } h = (1.10)(6.85) = 7.54'' \qquad \underline{\text{say } 7\tfrac{1}{2}''}$$

14.7 LIMITATIONS OF DIRECT DESIGN METHOD

For the moment coefficients determined by the direct design method to be applicable, the Code (13.6.1) says that the following limitations must be met, unless a theoretical analysis shows that the strength furnished after the appropriate load reduction or ϕ factors are applied is sufficient to support the anticipated loads and provided that all serviceability conditions such as deflection limitations are met:

1. There must be at least three continuous spans in each direction.
2. The panels must be rectangular, with the length of the longer side of any panel not being more than 2.0 times the length of its shorter side.

3. Span lengths of successive spans in each direction may not differ in length by more than one-third of the longer span.
4. Columns may not be offset by more than 10% of the span length in the direction of the offset from either axis between center lines of successive columns.
5. The live load shall not be more than three times the dead load.
6. If a panel is supported on all sides by beams, the relative stiffness of those beams in the two perpendicular directions, as measured by the following expression, shall not be less than 0.2 nor greater than 5.0:

$$\frac{\alpha_1 l_2^{\,2}}{\alpha_2 l_1^{\,2}}$$

14.8 DISTRIBUTION OF MOMENTS IN SLABS—DIRECT DESIGN METHOD

The total moment M_o that is resisted by a slab equals the sum of the maximum positive and negative moments in the span. It is the same as the total moment that occurs in a simply supported beam. For a uniform load it is as follows:

$$M_o = \frac{(wl_2)(l_1)^2}{8} \tag{14.4}$$

In this expression l_1 is the span length, center to center, of supports in the direction in which moments are being taken and l_2 is the length of the span transverse to l_1, measured center to center of the supports.

The moment that actually occurs in such a slab has been shown by experience and tests to be somewhat less than the value determined by equation 14.4. For this reason l_1 is replaced with l_n, the clear span measured face to face of the supports in the direction in which moments are taken. The Code (13.6.2.5) states that l may not be taken less than 65% of the span center-to-center of supports. Using l_n the total moment becomes:

$$M_o = \frac{(wl_2)(l_n)^2}{8} \tag{14.5}$$

When the moment is being calculated in the long direction, the total static moment is written here as M_{ol} and in the short direction as M_{os}.

It is next necessary to know what proportion of these total moments are positive and what proportion are negative. If a slab was completely fixed at the end of each panel, the division would be as it is in a fixed-end beam, two-thirds negative and one-third positive, as shown in Figure 14.7.

This division is reasonably accurate for interior panels where the slab is continuous for several spans in each direction with equal span lengths

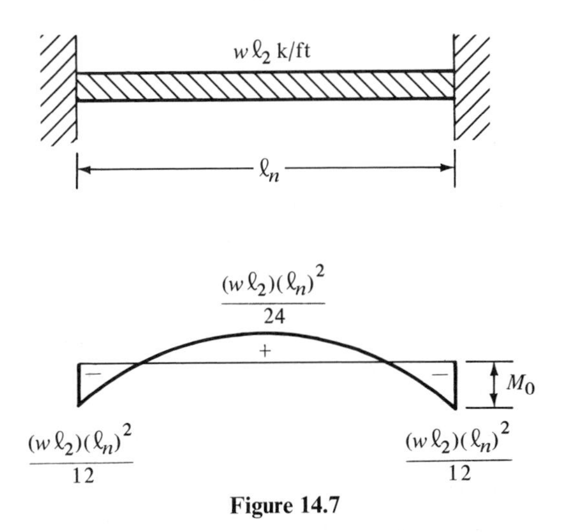

Figure 14.7

and loads. In effect, the rotation of the interior columns is assumed to be small and moment values of $0.65M_o$ for negative moment and $0.35M_o$ for positive moment are specified by the Code (13.6.3.2). For cases where the span lengths and loadings are different, the proportion of positive and negative moments may vary appreciably and the use of a more detailed method of analysis is desirable. The equivalent frame method will provide rather good approximations for such situations.

The relative stiffnesses of the columns and slabs of exterior panels are of far greater significance in their effect on the moments than is the case for interior panels. The magnitudes of the moments are very sensitive to the amount of torsional restraint supplied at the discontinuous edges. This restraint is provided both by the flexural stiffness of the slab and by the flexural stiffness of the exterior column.

Should the stiffness of an exterior column be quite small, the end negative moment will be very close to zero. If the stiffness of the exterior column is very large, the positive and negative moments will still not be the same as those in an interior panel unless an edge beam with a very large torsional stiffness is provided that will substantially prevent rotation of the discontinuous edge of the slab.

If a 2-ft-wide beam were to be framed into a 2-ft-wide column of infinite flexural stiffness in the plane of the beam, the joint would behave as would a perfectly fixed end, and the negative beam moment would equal the fixed-end moment.

If a two-way slab 24 ft wide were to be framed into this same 2-ft-wide column of infinite stiffness, the situation of no rotation would occur only along the part of the slab at the column. For the remaining 11-ft widths of slab on each side of the column, there would be rotation varying from zero at the side face of the column to maximums 11 ft on each side of the column. As a result of this rotation, the negative moment at the face of the column would be less than the fixed-end moment. Thus the stiffness of the exterior column is reduced by the rotation of the attached transverse slab.

To take into account the fact that the rotation of the edge of the slab is different at different distances from the column, the exterior columns and slab edge beam are replaced with an equivalent column that has the same flexibility as the column plus the edge beam. From the preceding discussion it can be seen that the equivalent column stiffness (called K_{ec} here) is always less at a joint with a two-way slab and a column than is the stiffness of the column itself (called K_c here).[2]

If no edge beam is used, the torsional stiffness of a piece of slab as wide as the column depth is considered. The equivalent column is assumed to consist of the actual column plus an attached torsional member running transverse to the direction in which the moments are being calculated, as shown in Figure 14.8. This same equivalent column is used if the equivalent frame method is being applied.

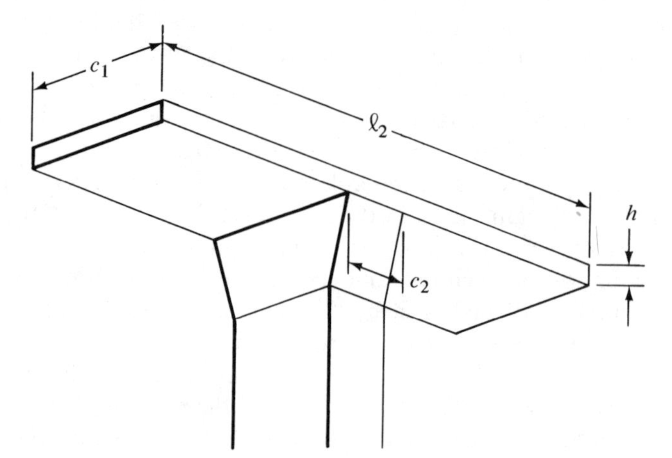

Figure 14.8 Equivalent Column

The flexibility (that is, the reciprocal of the stiffness) of the equivalent column is written as follows:

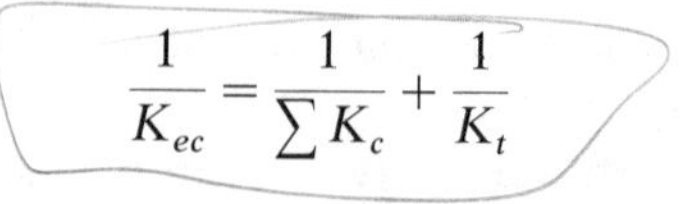

$$\frac{1}{K_{ec}} = \frac{1}{\sum K_c} + \frac{1}{K_t}$$

[2] Rice, P. F., and Hoffman, E. S., 1972., *Structural Design Guide to the ACI Building Code* (New York: Van Nostrand Reinhold), pp. 91–92.

where

K_{ec} = flexural stiffness of the equivalent column
K_c = flexural stiffness of actual column
K_t = torsional stiffness of the edge beam

Section 13.7.5.2 of the Code states that K_t is to be calculated with the following expression:

$$K_t = \sum \frac{9E_{cs}C}{l_2[1 - (c_2/l_2)]^3}$$

In this expression c_2 is the size of the rectangular or equivalent rectangular column, capital, or bracket measured transverse to the direction in which moments are being calculated and C is a cross-sectional constant determined from the following expression:

$$C = \sum\left(1 - 0.63\frac{x}{y}\right)\frac{x^3y}{3}$$

In this expression x is the shorter dimension of a rectangular part of the cross section and y is the longer dimension. In calculating C the expression is applied to each part of the cross section and the results are summed.

For an end span the Code (13.6.3.3) states that the static design moment is to be distributed in accordance with the expressions to follow, in which α_{ec} is the ratio of the flexural stiffness of the equivalent column to the sum of the flexural stiffnesses of the slab plus the beams at a joint taken in the direction in which moments are being determined and equals $K_{ec}/\sum(K_s + K_b)$. The percentages are as follows and are summarized in Figure 14.9:

$$\text{interior negative design moment} = 0.75 - \frac{0.10}{1 + (1/\alpha_{ec})}$$

$$\text{positive design moment} = 0.63 - \frac{0.28}{1 + (1/\alpha_{ec})}$$

$$\text{exterior negative design moment} = \frac{0.65}{1 + (1/\alpha_{ec})}$$

The next problem is to estimate what proportion of these moments are taken by the column strips and what proportions are taken by the middle strips. For this discussion a flat plate structure is assumed and the moment resisted by the column strip is estimated by considering the tributary areas shown in Figure 14.10.

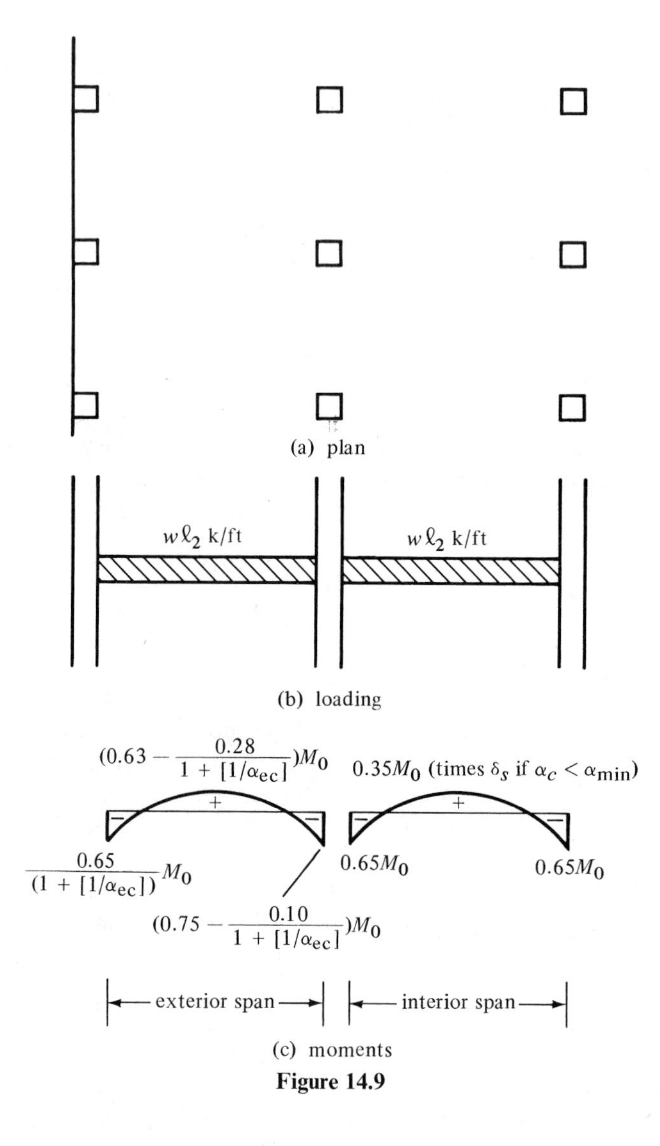

Figure 14.9

To simplify the mathematics the load to be supported is assumed to fall within the dotted lines shown in either parts (a) or (b) of Figure 14.11. The corresponding load is placed on the simple span and its center-line moment determined as an estimate of the portion of the static moment taken by the column strip.[3]

[3] White, R. N.; Gergely, P.; and Sexsmith, R. G., 1974 *Structural Engineering*, vol. 3, *Behavior of Members and Systems* (New York: Wiley), pp. 456–461.

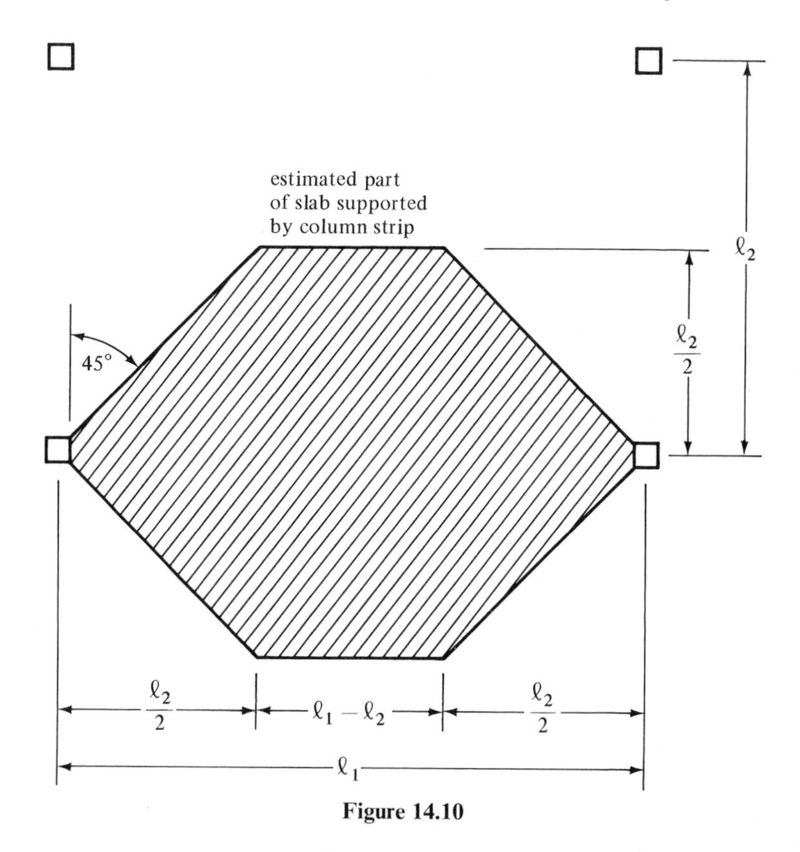

Figure 14.10

In the first case (part a) the load is concentrated near the ℄ of the beam, thus causing the moment to be overestimated a little, while in the second case (part b) the load is spread uniformly from end to end, causing the moment to be underestimated. Based on these approximations the estimated moments in the column strips for square panels will vary from $0.5M_o$ to $0.75M_o$. As l_1 becomes larger than l_2, the column strip takes a larger proportion of the moment. For such cases about 60% to 70% of M_o will be resisted by the column strip.

If you sketch in the approximate deflected shape of a panel, you will see that a larger portion of the positive moment is carried by the middle strip than by the column strip, and vice versa for the negative moments. As a result, about 60% of the positive M_o is expected to be resisted by the column strip and about 70% of the negative M_o.[4]

[4] White, R. N.; Gergely, P.; and Sexsmith, R. G., 1974 *Structural Engineering*, vol. 3, *Behavior of Members and Systems* (New York: Wiley), pp. 459–461.

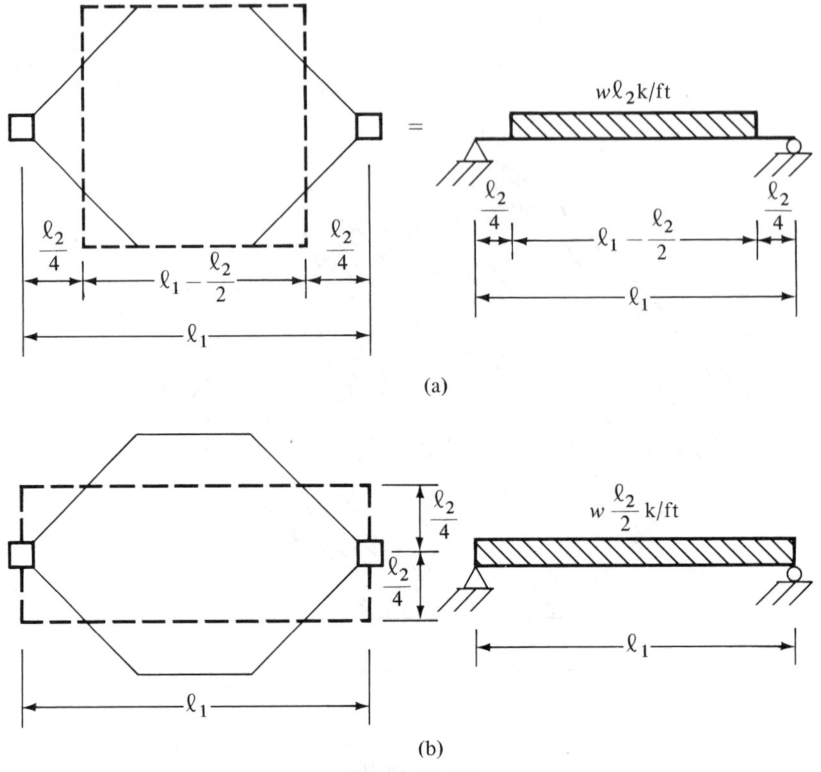

Figure 14.11

Section 13.6.4.1 of the Code states that the column strip shall be proportioned to resist the percentages of the total interior negative design moment given in Table 14.1.

In the table α_1 is again the ratio of the stiffness of a beam section to the stiffness of a width of slab bounded laterally by the ℄ of the adjacent panel, if any, on each side of the beam and equals $= E_{cb}I_b/E_{cs}I_s$.

Table 14.1 Percentages of Interior Negative Design Moments to be Resisted by Column Strip

$\frac{l_2}{l_1}$	0.5	1.0	2.0
$\frac{\alpha_1 l_2}{l_1} = 0$	75	75	75
$\frac{\alpha_1 l_2}{l_1} \geq 1.0$	90	75	45

Table 14.2 Percentages of Exterior Negative Design Moment to be Resisted by Column Strip

$\frac{l_2}{l_1}$		0.5	1.0	2.0
$\frac{\alpha_1 l_2}{l_1} = 0$	$\beta_t = 0$	100	100	100
	$\beta_t \geq 2.5$	75	75	75
$\frac{\alpha_1 l_2}{l_1} \geq 1.0$	$\beta_t = 0$	100	100	100
	$\beta_t \geq 2.5$	90	75	45

Section 13.6.4.2 of the Code states that the column strip is to be assumed to resist percentages of the exterior negative design moment, as given in Table 14.2.

The column strip (Section 13.6.4.4 of the Code) is to be proportioned to resist the portion of the positive moments given in Table 14.3.

Table 14.3 Percentages of Positive Design Moment to be Resisted by Column Strip

$\frac{l_2}{l_1}$	0.5	1.0	2.0
$\frac{\alpha_1 l_2}{l_1} = 0$	60	60	60
$\frac{\alpha_1 l_2}{l_2} \geq 1.0$	90	75	45

In Section 13.6.5 the Code requires that the beam be allotted 85% of the column strip moment if $\alpha_1(l_2/l_1) \geq 1.0$. Should $\alpha_1(l_2/l_1)$ be between 1.0 and zero, the moment allotted to the beam is determined by linear interpolation from 85% to 0%. The part of the moment not given to the beam is allotted to the slab in the column strip.

Finally, the Code (13.6.6) requires that the portion of the design moments not resisted by the column strips as previously described is to be allotted to the corresponding half middle strip. The middle strip will be designed to resist the total of the moments assigned to its two half middle strips.

14.9 DESIGN OF AN INTERIOR FLAT PLATE

In this section an interior flat plate is designed by the direct design method. The procedure specified in Chapter 13 of the ACI Code is applicable not only to flat plates but also to flat slabs, waffle flat slabs, and two-way slabs with beams. The steps necessary to perform the designs are briefly summarized at the end of this paragraph. The order of the steps may have to be varied somewhat for different types of slab designs. Either the direct design method or the equivalent frame method may be used to determine the design moments. The design steps are as follows:

1. Estimate the slab thickness to meet the Code requirements.
2. Determine the depth required for shear.
3. Calculate the total static moments to be resisted in the two directions.
4. Estimate the percentages of the static moments that are positive and negative, and proportion the resulting values between the column and middle strips.
5. Select the reinforcing.

Example 14.2 illustrates this method of design applied to a flat plate. In the solution it will be noted that the $M_u/\phi bd^2$ values are quite small and thus most of the ρ values do not fall within Table A.13 (see Appendix). For such cases the author assumes a value of $z = 0.95d$ and solves the following expression for A_s:

$$A_s = \frac{M_u}{\phi f_y z}$$

After this steel area is determined, it is theoretically necessary to compute the value of z and recalculate the steel area and so forth. The author did not make these last calculations as he felt the values would change by negligible amounts. Such a trial and error procedure is illustrated, however, at the end of this paragraph for the positive moment in the long-span column strip of Example 14.2. In this example the width of the column strip is 8 ft, M_u is 42.5 ft-k, f_y is 60,000 psi, and $f_c' = 3{,}000$ psi.

Assume $z = (0.95)(6.5) = 6.18$ in.

$$A_s = \frac{M_u}{\phi f_y z} = \frac{(12)(42.5)}{(0.9)(60)(6.18)} = 1.53 \text{ in.}^2$$

$$a = \frac{(1.53)(60)}{(0.85)(3)(12 \times 8)} = 0.375''$$

$$z = 6.50 - \frac{0.375}{2} = 6.31''$$

$$A_s = \frac{(12)(42.5)}{(0.9)(60)(6.31)} = 1.50 \text{ in.}^2$$

Therefore, negligible change from 1.53 in.^{2}

For several locations in Example 14.2, the steel areas calculated as described here for moment are less than the temperature and shrinkage minimum $0.0018bh$ for grade 60 bars. The larger value of the two must be used. Actually, the temperature percentage includes bars in the top and bottom of the slab. In the negative moment region, some of the positive steel bars have been extended into the support region and are also available for temperature and shrinkage steel. If desirable, these positive bars can be lapped instead of being stopped in the support.[5]

EXAMPLE 14.2

Design an interior flat plate for the structure considered in Example 14.1. This plate is shown in Figure 14.12. Assume a service live load equal to 80 psf and a service dead load equal to 110 psf (including slab weight). $f_y = 60{,}000$ psi and $f_c' = 3{,}000$ psi. Assume column heights are 12 ft.

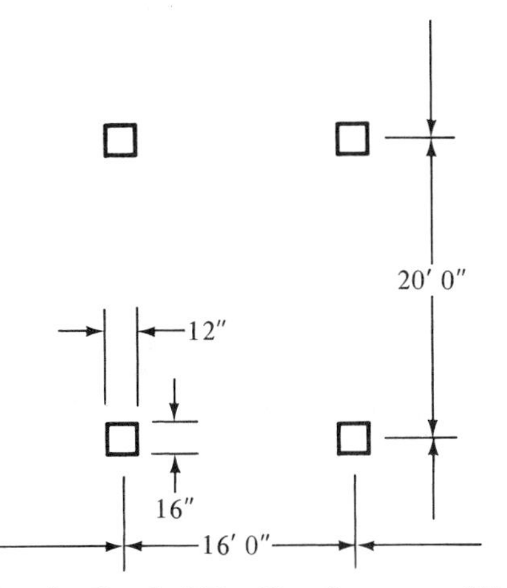

Figure 14.12 Interior Panel of Flat Plate Structure of Example 14.1

SOLUTION

1. Estimate slab thickness. Try a $7\frac{1}{2}$-in. slab as calculated in Example 14.1.

2. Determine depth required for shear. Using d for shear equal to the estimated average of the d values in the two directions,

$$d = 7.50'' - \tfrac{3}{4}'' \text{ cover} - 0.50 = 6.25''$$

$$w_u = (1.4)(110) + (1.7)(80) = 290 \text{ psf}$$

[5] Ferguson, P. M., 1973, *Reinforced Concrete Fundamentals*, 3d ed. (New York: Wiley), p. 348.

Checking one-way or beam shear
(rarely checked as it seldom controls)

Using the dimensions shown in Figure 14.13,

$$V_u = (8.81)(290) = 2555 \ \# \text{ for a } 12'' \text{ width}$$

$$\phi V_c = \phi 2\sqrt{f_c'}b_o d$$
$$= (0.85)(2\sqrt{3000})(12)(6.25)$$
$$= 6983 \ \# > 2555 \ \# \qquad \underline{\text{ok}}$$

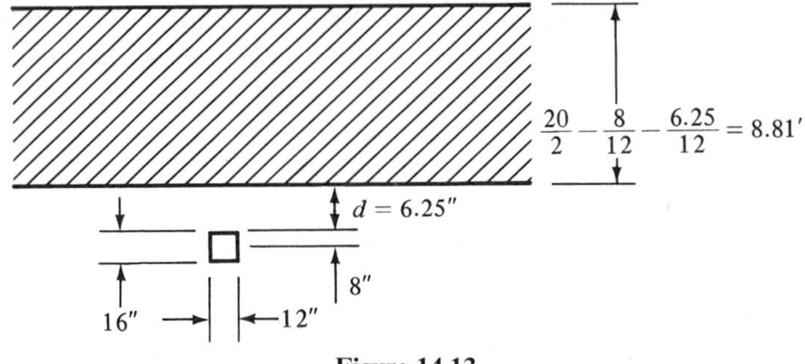

Figure 14.13

Checking two-way or punching shear around the column

$$b_o = (2)(16 + 6.25) + (2)(12 + 6.25) = 81''$$

$$V_u = \left[(20)(16) - \left(\frac{16 + 6.25}{12}\right)\left(\frac{12 + 6.25}{12}\right)\right](0.290) = 91.98^{\text{k}} = 91{,}980\#$$

$$\phi V_c = (0.85)(4\sqrt{3000})(81)(6.25)$$
$$= 94{,}277 \ \# > 91{,}980 \ \# \qquad \underline{\text{ok}}$$

<u>use $h = 7\frac{1}{2}''$</u>

3. Calculate static moments:

$$M_{ol} = \frac{w_u l_2 l_n^2}{8} = \frac{(0.290)(16)(18.67)^2}{8} = 202.2 \text{ ft-k}$$

$$M_{os} = \frac{w_u l_1 l_n^2}{8} = \frac{(0.290)(20)(15)^2}{8} = 163.1 \text{ ft-k}$$

4. and 5. Proportion the static moments to the column and middle strips and select the reinforcing. These calculations can be conveniently arranged, as in Table 14.4 which follows. This table is very similar to the one used for the design of the continuous one-way slab in Chapter 12.

The selection of the reinforcing bars is the final step taken in the design of this flat plate. The Code in their Figure 13.4.8 (repeated in Figure 14.14

Table 14.4

	Long Span (estimate $d = 6.50''$)				Short Span (estimate $d = 6.00''$)			
	Column strip (8′)		Middle strip (8′)		Column strip (8′)		Middle strip (12′)	
	−	+	−	+	−	+	−	+
M_u	(.65)(.75)(202.2) = −98.6 ft-k	(.35)(.60)(202.2) = +42.5 ft-k	(.65)(202.2) − 98.6 = −32.8 ft-k	(.35)(202.2) − 42.5 = +28.3 ft-k	(.65)(.75)(163.1) = −79.5 ft-k	(.35)(.60)(163.1) = +34.3 ft-k	(.65)(163.1) − 79.5 = −26.5 ft-k	(.35)(163.1) − 34.3 = +22.8 ft-k
$\frac{M_u}{\phi bd^2}$	324.1	139.7	107.8	93.0	306.7	132.3	68.2	58.6
ρ	0.00580	not in table; assume $z = 6.18''$	not in table; assume $z = 6.18''$	not in table; assume $z = 6.18''$	0.00546	not in table; assume $z = 5.70''$	not in table; assume $z = 5.70''$	not in table; assume $z = 5.70''$
A_s	3.62 in.^{2}	1.53 in.^{2}	temp. controls 1.30 in.^{2}	temp. controls 1.30 in.^{2}	3.14 in.^{2}	1.34 in.^{2}	temp. controls 1.94 in.^{2}	temp. controls 1.94 in.^{2}
Bars selected	19 #4	8 #4	7 #4	7 #4	16 #4	7 #4	10 #4	10 #4

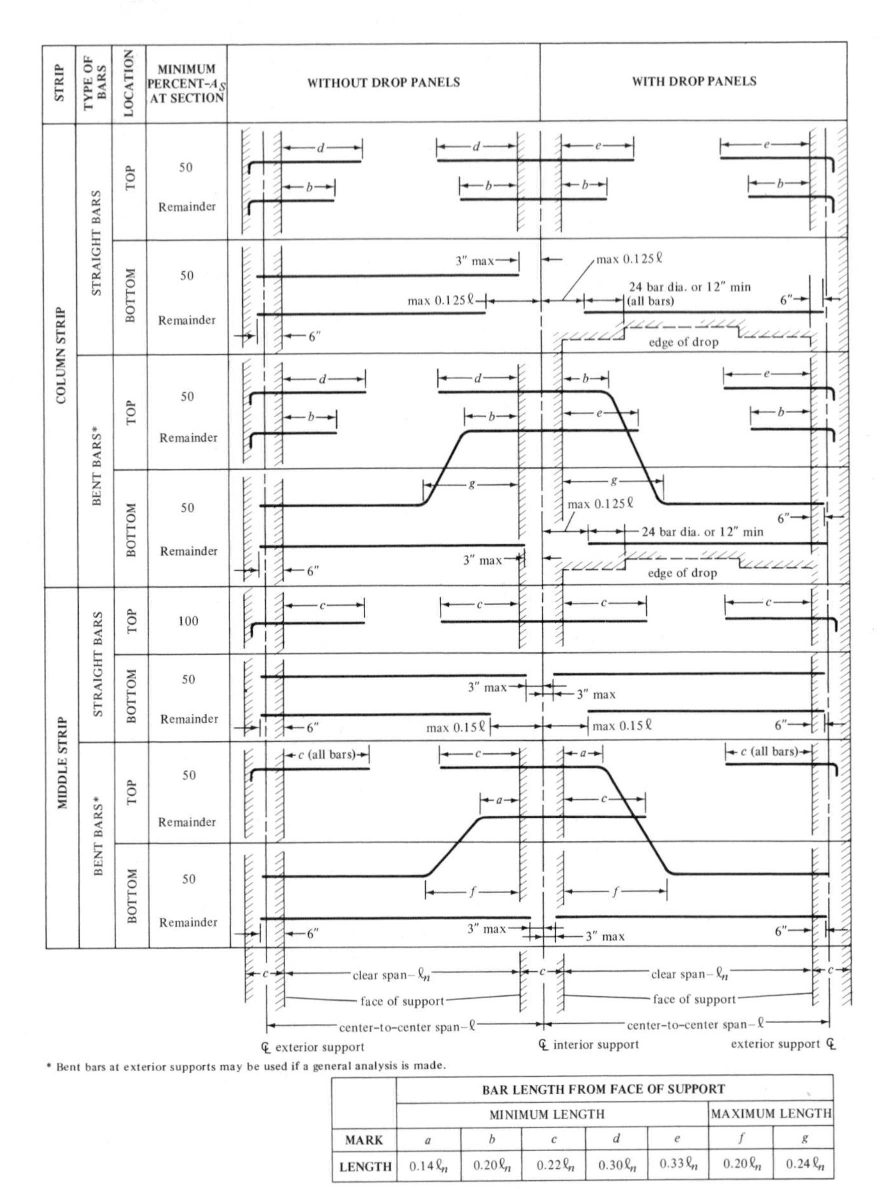

* Bent bars at exterior supports may be used if a general analysis is made.

	BAR LENGTH FROM FACE OF SUPPORT						
	MINIMUM LENGTH					MAXIMUM LENGTH	
MARK	a	b	c	d	e	f	g
LENGTH	$0.14\ell_n$	$0.20\ell_n$	$0.22\ell_n$	$0.30\ell_n$	$0.33\ell_n$	$0.20\ell_n$	$0.24\ell_n$

Figure 14.14 Minimum Length of Slab Reinforcement, Slabs Without Beams

here) shows the minimum lengths of slab reinforcing bars for flat plates and for flat slabs with drop panels. This figure shows that some of the positive reinforcing must be run into the support area within 3 in. of the support ℄.

The bars selected for this flat plate are shown in Figure 14.15. Bent bars are used in this example but straight bars could have been used just

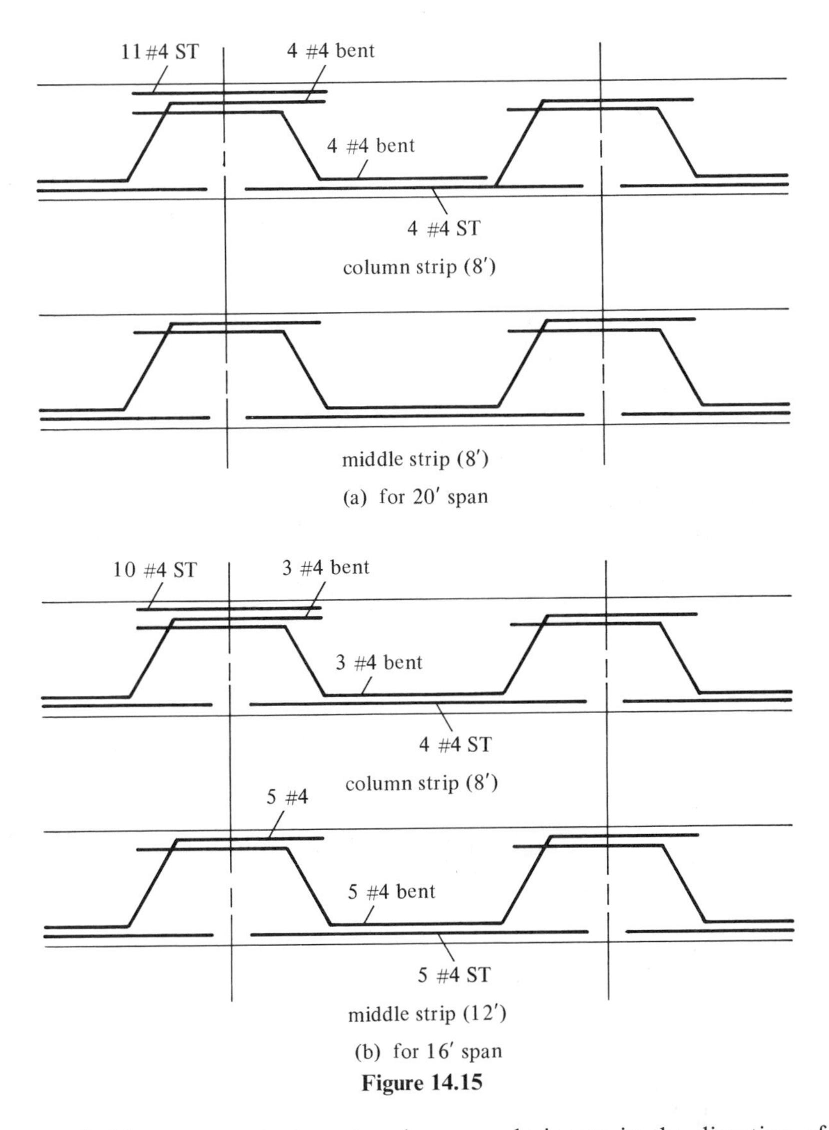

Figure 14.15

as well. There seems to be a trend among designers in the direction of using more straight bars in slabs and fewer bent bars.

14.10 MINIMUM COLUMN STIFFNESSES

The moments in a continuous floor slab are appreciably affected by different positions or patterns of the live loads. The usual procedure, however, is to calculate the total static moments assuming all panels are subjected to full live load. When different loading patterns are used, the moments can be changed so much that overstressing may occur in the slab. For such situations the Code permits overstresses as high as 33%.

To stay within the 33% limit, the Code requires that when β_a (the ratio of the unfactored dead load to the unfactored live load) is less than 2.0, column stiffnesses (and thus column sizes) may not be less than certain values. The reasoning for such limits should be obvious after you read the following statements:

1. The larger the columns at the ends of a span, the closer the ends of that span approach fixed conditions and thus the smaller become the effects on the moments in the span when live loads are changed in in other spans.
2. The larger the ratio of dead load to live load, the smaller the effect of live load variations.

Section 13.6.10 of the Code states that α_c [the ratio of the stiffness of the columns above and below a slab to the combined stiffness of the slabs and beams at a joint taken in the direction in which moments are being taken and equal to $K_{ec}/\sum(K_s + K_b)$] may not be less than $\alpha_{\min}$, as given in Table 14.1 (Code Table 13.6.10). Should this requirement not be met, the positive design moment in the panels supported by the columns in question must be increased. This is done by multiplying them by the coefficient δ_s determined from the following expression:

$$\delta_s = 1 + \left(\frac{2 - \beta_a}{4 + \beta_a}\right)\left(1 - \frac{\alpha_c}{\alpha_{\min}}\right) \tag{14.6}$$

It will be noted in Table 14.5 that if β_a is 2,0 or more, $\alpha_{\min}$ is 0. As β_a falls below 2.0, the value of $\alpha_{\min}$ increases. For the flat plate designed in Example 14.2, with $L = 80$ psf and $D = 110$ psf, $\beta_a = \frac{110}{80} = 1.375$, the aspect ratio $(l_2/l_1) = \frac{20}{16} = 1.25$, and $\alpha = 0$ (with no beams present). The values to follow can be determined from the table:

with $\beta_a = 2.0$, $\dfrac{l_2}{l_1}$ = from 0.5 to 2.0, $\alpha = 0$, $\alpha_{\min} = 0$

with $\beta_a = 1.0$, $\dfrac{l_2}{l_1} = 1.25$, $\alpha = 0$, $\alpha_{\min} = 0.8$

with $\beta_a = 1.375$, $\dfrac{l_2}{l_1} = 1.25$, $\alpha = 0$, $\alpha_{\min} = 0.5$ (by interpolation between the two preceding $\alpha_{\min}$ values)

From structural analysis texts the stiffness of a prismatic member such as a flat plate may be taken as $K_s = 4EI_s/l$, where I_s is the moment of inertia of the gross section of the slab about the centroidal axis. Should beams or drop panels or column capitals be present, the slab will not be prismatic and thus I will vary and it will be necessary to calculate a more

Table 14.5 Minimum α_{min}

β_a	Aspect Ratio l_2/l_1	Relative Beam Stiffness α				
		0	0.5	1.0	2.0	4.0
2.0	0.5–2.0	0	0	0	0	0
1.0	0.5	0.6	0	0	0	0
	0.8	0.7	0	0	0	0
	1.0	0.7	0.1	0	0	0
	1.25	0.8	0.4	0	0	0
	2.0	1.2	0.5	0.2	0	0
0.5	0.5	1.3	0.3	0	0	0
	0.8	1.5	0.5	0.2	0	0
	1.0	1.6	0.6	0.2	0	0
	1.25	1.9	1.0	0.5	0	0
	2.0	4.9	1.6	0.8	0.3	0
0.33	0.5	1.8	0.5	0.1	0	0
	0.8	2.0	0.9	0.3	0	0
	1.0	2.3	0.9	0.4	0	0
	1.25	2.8	1.5	0.8	0.2	0
	2.0	13.0	2.6	1.2	0.5	0.3

correct stiffness value or take it from tables for that purpose. For the slab of Example 14.2,

$$I_s = (\tfrac{1}{12})(12 \times 16)(7.5)^3 = 6{,}750 \text{ in.}^4$$

$$K_s = \frac{4EI_s}{l} = \frac{(4E)(6{,}750)}{(12)(20)} = 112.5E$$

$K_b = 0$ as there is no beam.

Referring back to α_{min}

$$\alpha_{min} = \frac{K_c}{K_s}$$

$$0.50 = \frac{K_c}{K_s} = \frac{K_c}{112.5E}$$

From which

$$K_c = 56.25E$$

Remembering that

$$K_c = \frac{4EI_c}{l} = \frac{4EI_c}{(12)(12)} = \frac{EI_c}{36}$$

$$56.25E = \frac{EI_c}{36}$$

$$I_c = 2{,}025 \text{ in.}^4 < I \text{ of column used here}$$

$$= (\tfrac{1}{12})(16)(12)^3 = 2{,}304 \text{ in.}^4 \qquad \underline{\text{ok}}$$

14.11 ANALYSIS OF TWO-WAY SLABS WITH BEAMS—DIRECT DESIGN METHOD

In this section the moments are determined by the direct design method for an exterior panel of a two-way slab with beams. It is felt that if you can handle this problem, you can handle any other case that may arise in flat plates, flat slabs, or two-way slabs with beams using the direct design method.

The requirements of the Code are so lengthy and complex that in the example to follow (14.3), the steps and appropriate code sections are spelled out in detail. The practicing designer should obtain a copy of the *CRSI Handbook*, as the tables therein will be of tremendous help in slab design.

EXAMPLE 14.3

Determine the negative and positive moments required for the design of the exterior panel of the two-way slab with beam structure as shown in Figure 14.16. The slab is to support a live load of 120 psf and a dead load of 100 psf including the slab weight. The columns are 15 in. × 15 in. and

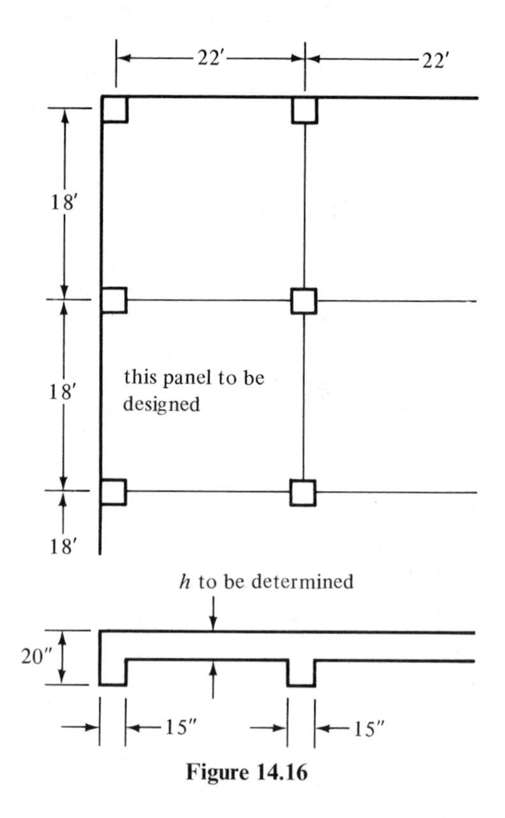

Figure 14.16

12 ft long. The slab is supported by beams along the column line with the cross section shown. Determine the slab thickness and check the shear stress. $f_c' = 3{,}000$ psi and $f_y = 60{,}000$ psi.

SOLUTION

1. Checking ACI Code limitations (13.6.1). These conditions are met.
2. Minimum thickness as required by Code (9.5.3.1):
   (a) Assume $h = 6$ in.
   (b) Effective flange projection of column line beam as specified by Code (13.2.4)

$$= 4h_f = (4)(6) = 24'' \quad \text{or} \quad h - h_f = 20 - 6 = \underline{14''}$$

   (c) Estimated moments of inertia of T beams:

$$I \text{ for edge beams} = \text{approximately } (\tfrac{1}{12})(15)(20)^3(1.5) = 15{,}000 \text{ in.}^4$$
$$I \text{ for interior beams} = \text{approximately } (\tfrac{1}{12})(15)(20)^3(2.0) = 20{,}000 \text{ in.}^4$$

   (d) Estimated moments of inertia for the slab strips. For the strip parallel to the edge beam (with a width $= \frac{22}{2} + \frac{7.5}{12} = 11.62$ ft),

$$I = (\tfrac{1}{12})(12 \times 11.62)(6)^3 = 2{,}510 \text{ in.}^4$$
$$\text{For the 18-ft-wide strip } I = (\tfrac{1}{12})(12 \times 18)(6)^3 = 3{,}888 \text{ in.}^4$$
$$\text{For the 22-ft-wide strip } I = (\tfrac{1}{12})(12 \times 22)(6)^3 = 4{,}752 \text{ in.}^4$$

   (e) Calculating α values (where α is the ratio of the stiffness of the beam section to the stiffness of a width of slab bounded laterally by the center line of the adjacent panel, if any, on each side of the beam):

$$\alpha \text{ for edge beam} = \frac{15{,}000}{2{,}510} = 5.98$$

$$\alpha \text{ for 18' interior beam (with 22' slab width)} = \frac{20{,}000}{4{,}752} = 4.21$$

$$\alpha \text{ for 22' interior beam (with 18' slab width)} = \frac{20{,}000}{3{,}888} = 5.14$$

$$\text{average } \alpha = \alpha_m = \frac{5.98 + 4.21 + 5.14}{3} = 5.11$$

$$\beta_s = \text{ratio of length of continuous edges to total perimeter} = \tfrac{62}{80} = 0.78$$

$$\beta = \text{ratio of long to short clear span} = \frac{22 - \frac{15}{12}}{18 - \frac{15}{12}} = 1.24$$

(f) Thickness limits by Code (9.5.3.1):

$$h = \frac{(20.75)(800 + 0.005 \times 60{,}000)}{36{,}000 + (5{,}000)(1.24)\{5.11 - 0.5[1 - 0.78][1 + (1/1.24)]\}}$$

$$= 0.482' = 5.78''$$

$$h = \frac{(20.75)(800 + 0.005 \times 60{,}000)}{36{,}000 + (5{,}000)(1.24)(1 + 0.78)} = 0.485' = 5.82''$$

But h need not be more than

$$h = \frac{(20.75)(800 + 0.005 \times 60{,}000)}{36{,}000} = 0.634' = 7.61''$$

Also, h may not be less than $3\frac{1}{2}''$.

use $h = 6''$ (will check shear later)

3. Moments for the short-span direction centered on interior column line

$$w_u = (1.4)(100) + (1.7)(120) = 344 \text{ psf}$$

$$M_o = \frac{(wl_2)(l_n)^2}{8} = \frac{(0.344)(22)(16.75)^2}{8} = 265 \text{ ft-k}$$

(a) Dividing this static design moment into negative and positive portions, as per Section 13.6.3.2 of Code:

$$\text{negative design moment} = (0.65)(265) = -172 \text{ ft-k}$$

$$\text{positive design moment} = (0.35)(265) = +93 \text{ ft-k}$$

(b) Allotting these interior moments to beam and column strips, as per Section 13.6.4 of Code:

$$\frac{l_2}{l_1} = \frac{22}{18} = 1.22$$

$$\alpha_1 = \alpha \text{ in direction of short span} = 4.21$$

$$\alpha_1 \frac{l_2}{l_1} = (4.21)(1.22) = 5.14$$

The portion of the interior negative moment to be resisted by the column strip, as per Table 14.1 of this chapter, by interpolation is $(0.68)(-172) = 117$ ft-k. This -117 ft-k is allotted 85% to the beam (Code 13.6.5), or -99 ft-k, and 15% to the slab or -18 ft-k. The remaining negative moment, $172 - 117 = 55$ ft-k, is allotted to the middle strip.

The portion of the interior positive moment to be resisted by the column strip, as per Table 14.3 of this chapter, by interpolation is

(0.68)(+93) = +63 ft-k. This 63 ft-k is allotted 85% to the beam (Code 13.6.5), or +54 ft-k, and 15% to the slab, or +9 ft-k. The remaining positive moment, 93 − 63 = 30 ft-k, goes to the middle strip.

4. Moments for the short-span direction centered on the edge beam:

$$M_o = \frac{(wl_2)(l_n)^2}{8} = \frac{(0.344)(11.62)(16.75)^2}{8} = 140 \text{ ft-k}$$

(a) Dividing this static design moment into negative and positive portions, as per Section 13.6.3.2 of the Code:

$$\text{negative design moment} = (0.65)(140) = -91 \text{ ft-k}$$
$$\text{positive design moment} = (0.35)(140) = +49 \text{ ft-k}$$

(b) Allotting these exterior moments to beam and column strips, as per Section 13.6.4 of the Code:

$$\frac{l_2}{l_1} = \frac{22}{18} = 1.22$$

$$\alpha_1 = \alpha \text{ for edge beam} = 5.98$$

$$\alpha_1 \frac{l_2}{l_1} = (5.98)(1.22) = 7.30$$

The portion of the exterior negative moment going to the column strip, from Table 14.2 of this chapter, by interpolation is (0.68)(−91) = −62 ft-k. This −62 ft-k is allotted 85% to the beam (Code 13.6.5), or −53 ft-k, and 15% to the slab, or −9 ft-k. The remaining negative moment, 91 − 62 = −29 ft-k, is allotted to the middle strip.

The portion of the exterior positive moment to be resisted by the column strip, as per Table 14.3, is (0.68)(+49) = +33 ft-k. This 33 ft-k is allotted 85% to the beam, or +28 ft-k, and 15% to the slab, or +5 ft-k. The remaining positive moment, 49 − 33 = +16 ft-k, goes to the middle strip.

A summary of the short-span moments is presented in Table 14.6.

Table 14.6 Short-span Moments (ft kip)

	Column Strip Moments		Middle Strip Slab Moments
	Beam	Slab	
Interior slab-beam strip			
Negative	−99	−18	−55
Positive	+54	+9	+30
Exterior slab-beam strip			
Negative	−53	−9	−29
Positive	+28	+5	+16

5. Moments for the long-span direction:

$$M_o = \frac{(wl_2)(l_n)^2}{8} = \frac{(0.344)(18)(22.75)^2}{8} = 401 \text{ ft-k}$$

(a) Calculate α_{ec} so that the static design moment may be proportioned into negative and positive portions as per Code (13.6.3.3). The value of E_c is dropped out of the values to follow since $f_c' = 3{,}000$ psi for all concrete.

$$K_s = \text{stiffness of slab} = \frac{4E_cI_s}{l} = \frac{(4)(3{,}888)}{12 \times 22} = 58.9$$

$$K_b = \text{stiffness of beam} = \frac{4E_cI_b}{l} = \frac{(4)(20{,}000)}{12 \times 22} = 303.0$$

$$I \text{ of column} = (\tfrac{1}{12})(15)(15)^3 = 4{,}219 \text{ in.}^4$$

$$K_c = \text{stiffness of column} = \frac{(4E_c)(4{,}219)}{12 \times 12} = 117.2$$

C is the torsional constant for the edge beam and for the rectangular flange projection (as given by ACI equation 13.8 in Section 13.7.5.3 of Code). Dimensions for this case are given in Figure 14.17.

$$C = \sum\left(1 - 0.63\frac{x}{y}\right)\frac{x^3y}{3}$$

$$C = \left(1 - 0.63\frac{15}{20}\right)\left(\frac{15^3 \times 20}{3}\right) + \left(1 - 0.63\frac{6}{14}\right)\left(\frac{6^3 \times 14}{3}\right) = 12{,}605$$

K_t is the torsional stiffness of the edge beam (as given by ACI equation 13.7 from Section 13.7.5.2 of Code).

$$K_t = \frac{9E_{cs}C}{l_2[1 - (c_2/l_2)]^3} = \frac{(2)(9)(12{,}605)}{(12 \times 18)[1 - (15/12 \times 18)]^3} = 1{,}304$$

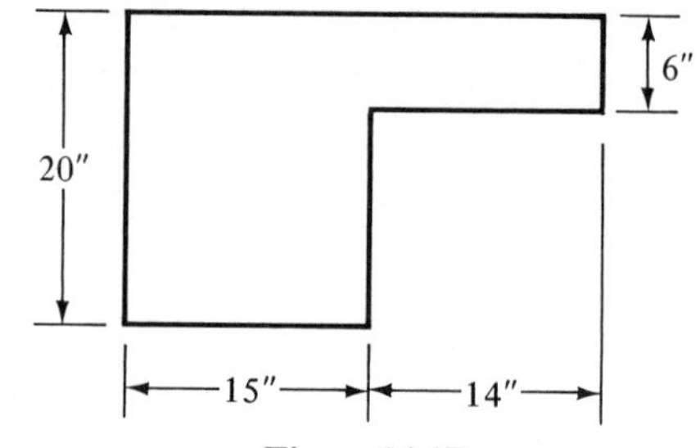

Figure 14.17

Section 13.7.5.4 of the Code states that where beams frame into a column in the direction in which moments are being determined, K_t is to be multiplied by the ratio of the I of the slab with the column line beam to the I of the slab alone.

Use

$$K_t = \text{approximately}\left(\frac{20{,}000}{3{,}888}\right)(1{,}304) = 5.14 \times 1{,}304 = 6{,}702$$

with equation 13.6 of the Code (13.7.4.2). The flexibility of the equivalent column is

$$\frac{1}{K_{ec}} = \frac{1}{\sum K_c} + \frac{1}{K_t}$$

$$\frac{1}{K_{ec}} = \frac{1}{2 \times 117.2} + \frac{1}{6{,}702}$$

$$K_{ec} = 226.5$$

The ratio of stiffness of the equivalent column to the combined stiffness of the slabs and beams at a joint is α_{ec}.

$$\alpha_{ec} = \frac{K_{ec}}{\sum(K_s + K_b)} = \frac{226.5}{58.9 + 303.0} = 0.626$$

(b) With the equations of Section 13.6.3.3 of the Code for an end span, the static design moment is distributed as follows:

$$\text{interior negative design moment} = \left[0.75 - \frac{0.10}{1 + (1/\alpha_{ec})}\right](M_o)$$

$$= \left[0.75 - \frac{0.10}{1 + (1/0.626)}\right](401)$$

$$= -285 \text{ ft-k}$$

$$\text{positive design moment} = \left[0.63 - \frac{0.28}{1 + (1/\alpha_{ec})}\right](M_o)$$

$$= \left[0.63 - \frac{0.28}{1 + (1/0.626)}\right](401)$$

$$= +209 \text{ ft-k}$$

$$\text{exterior negative design moment} = \left[\frac{0.65}{1 + (1/\alpha_{ec})}\right](M_o)$$

$$= \left[\frac{0.65}{1 + (1/0.626)}\right](401)$$

$$= -100 \text{ ft-k}$$

(c) Allotting these moments to beam and column strips:

$$\frac{l_2}{l_1} = \frac{18}{22} = 0.818$$

$$\alpha_1 = \alpha_1 \text{ for the } 22' \text{ beam} = 5.14$$

$$\alpha_1 \frac{l_2}{l_1} = (5.14)(0.818) = 4.21$$

$$\beta_t = \frac{E_{cb}C}{2E_{cs}I_s} = \frac{(E_c)(12{,}605)}{(2)(E_c)(3{,}888)} = 1.62$$

The portion of the interior negative design moment allotted to the column strip, from Table 14.1, by interpolation is (0.80)(−285) = −228 ft-k. This −228 ft-k is allotted 85% to the beam (Code 13.6.5), or −194 ft-k, and 15% to the slab, or −34 ft-k. The remaining negative moment, −57 ft-k, is allotted to the middle strip.

The portion of the positive design moment to be resisted by the column strip, as per Table 14.3, is (0.80)(209) = +167 ft-k. This 167 ft-k is allotted 85% to the beam, or (0.85)(167) = 142 ft-k, and 15% to the slab, or +25 ft-k. The remaining positive moment, 209 − 167 = 42 ft-k, goes to the middle strip.

The portion of the exterior negative moment allotted to the column strip, from Table 14.2, is (0.86)(−100) = −86 ft-k. This −86 ft-k is allotted 85% to the beam, or −73 ft-k, and 15% to the slab, or −13 ft-k. The remaining negative moment, −14 ft-k, is allotted to the middle strip.

A summary of the long-span moments is presented in Table 14.7.

Table 14.7 Long Span Moments (ft kip)

	Column Strip Moments		Middle Strip Slab Moments
	Beam	Slab	
Interior negative	−194	−34	−57
Positive	+142	+25	+42
Exterior negative	−73	−13	−14

6. Check shear strength in the slab at a distance d from the face of the beam. Shear is assumed to be produced by the load on the tributary area shown in Figure 14.18, working with a 12-in.-wide strip as shown.

$$V_u = (0.344)\left(9 - \frac{7.5}{12} - \frac{5}{12}\right) = 2.738^{\text{k}} = 2738\#$$

$$\phi V_c = (0.85)(2\sqrt{3000})(12)(5)$$

$$= 5587\# > 2738\# \qquad \underline{\text{ok}}$$

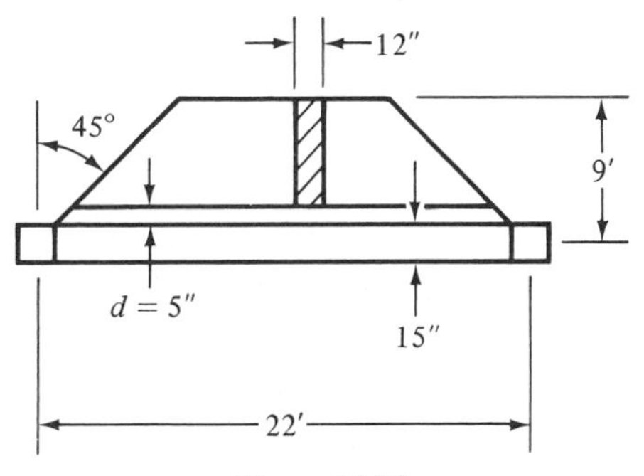

Figure 14.18

14.12 EQUIVALENT FRAME METHOD

The only difference between the direct design method and the equivalent frame method is in the determination of the longitudinal moments in the spans of the equivalent rigid frame. Whereas the direct design method involves a one-cycle moment distribution, the equivalent frame method involves a normal moment distribution of several cycles. The design moments obtained by either method are distributed to the column and middle strips in the same fashion.

It will be remembered that the range in which the direct design method can be applied is limited by a maximum 3-to-1 ratio of live-to-dead load and a maximum ratio of the longitudinal span length to the transverse span length of 2 to 1. In addition, the columns may not be offset by more than 10% of the span length in the direction of the offset from either axis between center lines of successive columns. There are no such limitations on the equivalent frame method. This is a very important matter because so many floor systems do not meet the limitations specified for the direct design method.

Analysis by either method will yield almost the same moments for those slabs which meet the limitations required for application of the direct design method. For such cases it is then simpler to use the direct design method.

The equivalent frame method merely involves the elastic analysis of a structural frame consisting of a row of equivalent columns and horizontal slab members that are each one panel long and have a transverse width equal to the distance between center lines of the panels on each side of the columns in question. For vertical loads each floor, together with the columns above and below, is analyzed separately. For such an analysis the far ends of the columns are considered to be fixed. Figure 14.19 shows a typical equivalent slab beam as described in Section 14.8 of this chapter. For lateral loads it is necessary to consider an equivalent frame which extends for the entire height of the building.

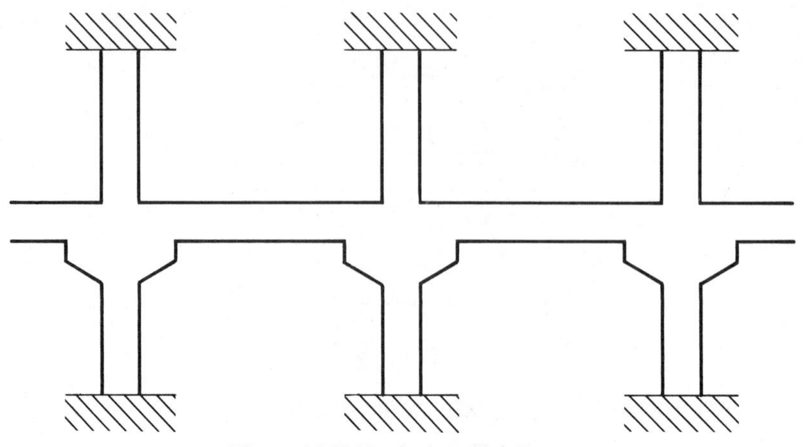

Figure 14.19 Equivalent Slab Beam

It can be seen that to perform the moment distribution involved, it is necessary to calculate the fixed-end moments, stiffness factors, and carry-over factors for the equivalent frame with its varying moments of inertia. The evaluation of these factors is a very tedious task, though various tables are available that will appreciably simplify the work.[6] One set of such tables is given in Section 13.7.4 of the ACI Commentary. When tables are not available covering the specific case, a method that is often used is the column analogy method.[7]

The designer selects trial sizes of the concrete members and then checks their adequacy by a review. For the slab thickness the same ACI minimum h values must be met as in the direct design method.

The analysis of the frame is made for the full design live load applied to all spans unless the design live load exceeds 0.75 times the actual dead load (ACI Code 13.7.6). When the live load is greater than 0.75 times the dead load, a pattern loading with three-fourths times the live load is used for calculating moments and shears.

The maximum positive moment in the middle of a span is assumed to occur when three-fourths of the full design load is applied in that panel and in alternate spans. The maximum negative moment in the slab at a support is assumed to occur when three-fourths of the full design live load is applied to the adjacent panels. The values so obtained may not be less than those calculated assuming full live loads in all spans. Numerical illustrations of the equivalent frame method are presented in several texts.[8]

[6] Simmonds, S. H., and Misic, J., 1971, "Design Factors for the Equivalent Frame Method," *Journal ACI*, 68, no. 11, pp. 825–831.

[7] McCormac, J. C., 1975, *Structural Analysis*, 3d ed. (New York: Intext), pp. 534–548.

[8] Fintel, M., ed., 1974, *Handbook of Concrete Engineering* (New York: Van Nostrand Reinhold), pp. 65–80.

14.13 REFERENCES

Even though this chapter has been rather lengthy, several important items have not been included. Such items as column capitals, drop panels, shear stresses at exterior columns, design of shearheads, and yield line analysis have not been included. For a detailed consideration of these and other items the reader is referred to the excellent concrete text by Wang and Salmon.[9]

[9] Wang, C. K., and Salmon, C. G. 1978. *Reinforced Concrete Design* 3d ed. New York: Harper & Row (to be published).

CHAPTER 15

Prestressed Concrete

15.1 INTRODUCTION

Prestressing can be defined as the imposition of internal stresses into a structure that are of opposite character to those that will be caused by the service or working loads. A rather common method used to describe prestressing is shown in Figure 15.1, where a row of books have been squeezed together by a person's hands. The resulting "beam" can carry a downward load as long as the compressive stress due to squeezing at the bottom of the "beam" is greater than the tensile stress there due to the moment produced by the weight of the books and the superimposed loads. Such a beam has no tensile strength and thus no moment resistance until it is squeezed together or prestressed. You might very logically now expand your thoughts to a beam consisting of a row of concrete blocks squeezed together and then to a plain concrete beam with its negligible tensile strength similarly prestressed.

The theory of prestressing is quite simple and has been used for many years in various kinds of structures. For instance, wooden barrels have long been made by putting tightened metal bands around them, thus compressing the staves together and making a tight container with resistance to the outward pressures of the enclosed liquids. Prestressing is primarily used for concrete beams to counteract tension stresses caused by the weight of the members and the superimposed loads. Should these loads cause a positive moment in a beam, it is possible by prestressing to introduce a negative moment that can counteract part or all of the

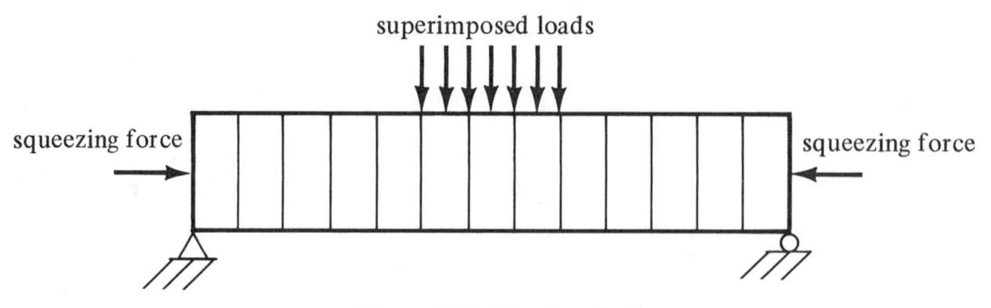

Figure 15.1 Prestressing

Prestressed Concrete Channels, John A. Denies Sons Company Warehouse #4, Memphis, Tennessee (Courtesy Master Builders).

positive moment. An ordinary beam has to have sufficient strength to support itself as well as the other loads, but it is possible with prestressing to produce a negative loading that will eliminate the effect of the beam's weight, thus producing a "weightless beam."

From the preceding discussion it is easy to see why prestressing has captured the imagination of so many persons and why it has all sorts of possibilities now and in the future.

In the earlier chapters of this book, only a portion of the cross sections of members in bending could be considered effective in resisting loads because a large part of those cross sections were in tension and thus cracked. If, however, flexural members can be prestressed so that their entire cross sections are kept in compression, then the properties of the entire sections are available to resist the applied forces.

For a little more detailed illustration of prestressing, reference is made to Figure 15.2. It is assumed that the following steps have been taken with regard to this beam:

1. Steel strands (represented by the dotted lines) were placed in the lower part of the beam form.
2. The strands were tensioned to a very high stress.

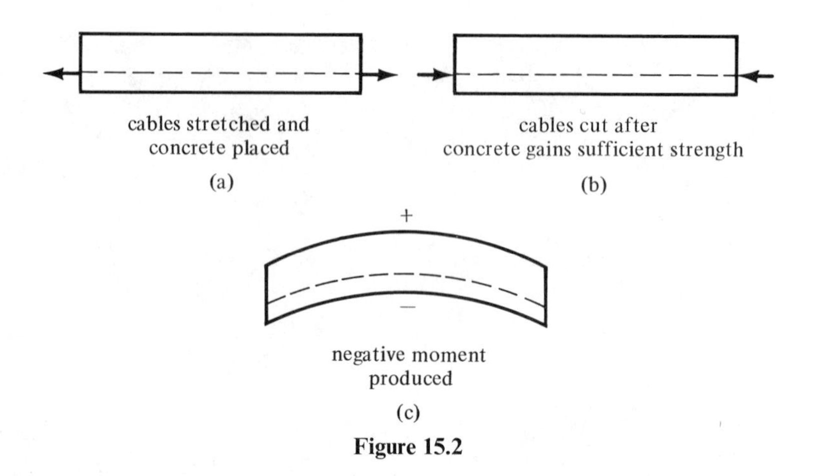

Figure 15.2

3. The concrete was placed in the form and allowed to gain sufficient strength for the prestressed strands to be cut.
4. The strands were cut.

The cut strands tend to resume their original length, thus compressing the lower part of the beam and causing a negative bending moment. The positive moment caused by the beam weight and any superimposed gravity loads is directly opposed by the negative moment. Another way of explaining this is to say that a compression stress has been produced in the bottom of the beam opposite in character to the tensile stress that is caused there by the working loads.

15.2 ADVANTAGES AND DISADVANTAGES OF PRESTRESSED CONCRETE

Advantages

As described in the last section, it is possible with prestressing to utilize the entire cross sections of members to resist loads. Thus smaller members can be used to support the same loads or the same size members can be used for longer spans. This is a particularly important advantage because member weights make up a substantial part of the total design loads of concrete structures.

Prestress members are crack free under working loads and, as a result, look better and are more watertight, thus providing better corrosion protection for the steel. Furthermore, crack-free prestressed members require less maintenance and last longer than cracked reinforced concrete members. Therefore, for a large number of structures, prestressed concrete provides the lowest first-cost solution, and when its reduced maintenance is considered, prestressed concrete provides the lowest overall cost for many additional cases.

The negative moments caused by prestressing produce camber in the members, with the result that total deflections are reduced. Other advantages of prestressed concrete include the following: reduction in diagonal tension stresses, sections with greater stiffnesses under working loads, and increased fatigue and impact resistance as compared to ordinary reinforced concrete.

Disadvantages

Prestressed concrete requires the use of higher-strength concretes and steels and the use of more complicated formwork with resulting higher labor costs. Other disadvantages include the following: closer control required in manufacture, losses in the initial prestressing forces, additional stress conditions that must be checked in design, and end anchorages and end bearing plates that may be required.

15.3 PRETENSIONING AND POSTTENSIONING

The two general methods of prestressing are pretensioning and posttensioning. *Pretensioning* was illustrated in Section 15.1, where the prestress tendons were tensioned before the concrete was placed. After the concrete had hardened sufficiently, the tendons were cut and the prestress force transmitted to the concrete by bond. This method is particularly well suited for mass production because the casting beds can be constructed several hundred feet long. The tendons can be run for the entire bed lengths and used for casting several beams in a line at the same time, as shown in Figure 15.3.

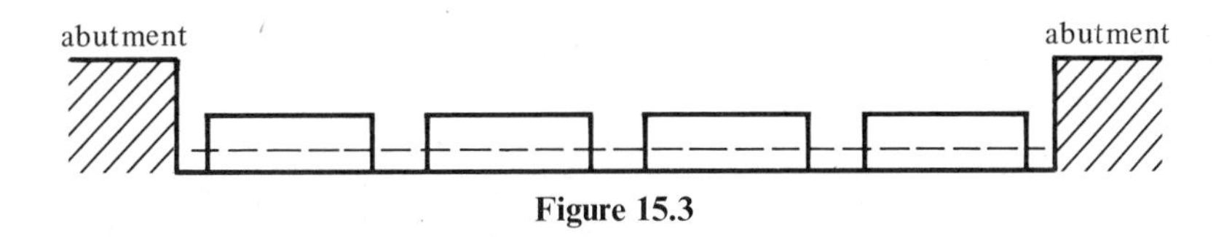

Figure 15.3

In *posttensioned* construction (see Figure 15.4) the tendons are stressed after the concrete is placed and has gained the desired strength. Plastic or metal tubes or conduits or sleeves or similar devices with unstressed tendons inside (or later inserted) are located in the form and the concrete placed. After the concrete has sufficiently hardened, the tendons are stretched and mechanically attached to end anchorage devices to keep

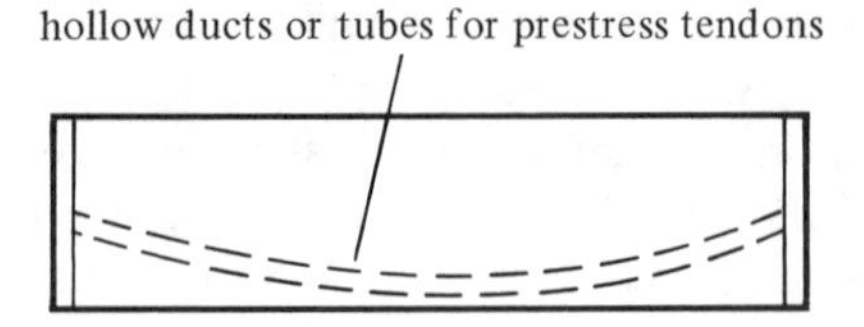

Figure 15.4 Posttensioned Beam

the tendons in their stretched positions. Thus in posttensioning the prestress forces are transferred to the concrete not by bond but by end bearing.

It is actually possible in posttensioning to have either bonded or unbonded tendons. If bonded, the conduits are often made of aluminum, steel, or other metal sheathing. In addition, it is possible to use steel tubing or rods or rubber cores that are cast in the concrete and removed a few hours after the concrete is placed. After the steel is tensioned, cement grout is injected into the duct for bonding. The grout is also useful in protecting the steel from corrosion. If the tendons are to be unbonded, they should be greased to facilitate tensioning and to protect them from corrosion.[1]

15.4 MATERIALS USED FOR PRESTRESSED CONCRETE

The materials ordinarily used for prestressed concrete are concrete and high-strength steels. The concrete used is probably of a higher strength than that used for reinforced concrete members for several reasons. These include the following:

1. The modulus of elasticity of such concretes is higher, with the result that the elastic strains in the concretes are smaller when the tendons are cut. Thus the relaxations or losses in the tendon stresses are smaller.
2. In prestressed concrete the entire members are kept in compression and thus all the concrete is effective in resisting forces. Thus it is reasonable to pay for a more expensive but stronger concrete if all of it is going to be used. (In ordinary reinforced concrete members, more than half of the cross sections are in tension and thus assumed to be cracked. As a result, more than half of a higher-strength concrete used there would be wasted.)
3. Most prestressed work in the United States is of the precast, pretensioned type done at the prestress yard where the work can be

[1] Lin, T. Y., 1963, *Prestressed Concrete Structures*, 2d ed. (New York: Wiley), pp. 54–56.

carefully controlled; consequently, dependable higher-strength concrete can readily be obtained.

4. For pretensioned work the higher-strength concretes permit the use of higher bond stresses between the cables and the concrete.

High-strength steels are necessary to produce and keep satisfactory prestress forces in members. The strains that occur in these steels during stressing are much greater than those that can be obtained with ordinary reinforcing steels. As a result, when the concrete elastically shortens in compression and also shortens due to creep and shrinkage, the losses in strain in the steel (and thus stress) is a smaller percentage of the total stress. Another reason for using high-strength steels is that a large prestress force can be developed in a small area.

Early work with prestressed concrete using ordinary strength bars to induce the prestressing forces in the concrete resulted in failure because the low stresses that could be put into the bars were completely lost due to the concrete's shrinkage and creep. Should a prestress of 20,000 psi be put into such rods, the resulting strains would be equal to $20{,}000/29 \times 10^6) = 0.00069$. This value is less than the long-term creep and shrinkage strain normally occurring in concrete, roughly 0.0008, which would completely relieve the stress in the steel. Should a high-strength steel be stressed to about 150,000 psi and have the same creep and shrinkage, the stress reduction will be of the order of $(0.0008)(29 \times 10^6) = 23{,}200$ psi, leaving $150{,}000 - 23{,}200 = 126{,}800$ psi in the steel (a loss of only 15.47% of the steel stress).[2]

Three forms of prestressing steel are used: single wires, wire strands, and bars. The greater the diameter of the wires, the smaller become their strengths and bond to the concrete. As a result, wires are manufactured with diameters from 0.192 in. up to a maximum of 0.276 in. (about $\frac{9}{32}$ in.). In posttensioning work large numbers of wires are grouped in parallel into tendons. Strands that are made by twisting wires together are used for most pretensioned work. They are of the seven-wire type, where a center wire is tightly surrounded by twisting the other six wires helically around it. Strands are manufactured with diameters from $\frac{1}{4}$ to $\frac{1}{2}$ in. Sometimes large-size, high-strength, heat-treated alloy steel bars are used for posttensioned sections. They are available with diameters running from $\frac{3}{4}$ to $1\frac{3}{8}$ in.

High-strength prestressing steels do not have distinct yield points (see Figure 15.10) as do the structural carbon reinforcing steels. The practice of considering yield points, however, is so firmly embedded in

[2] Winter, G., and Nilson, A. H., 1972, *Design of Concrete Structures*, 8th ed. (New York: McGraw-Hill), pp. 495–496.

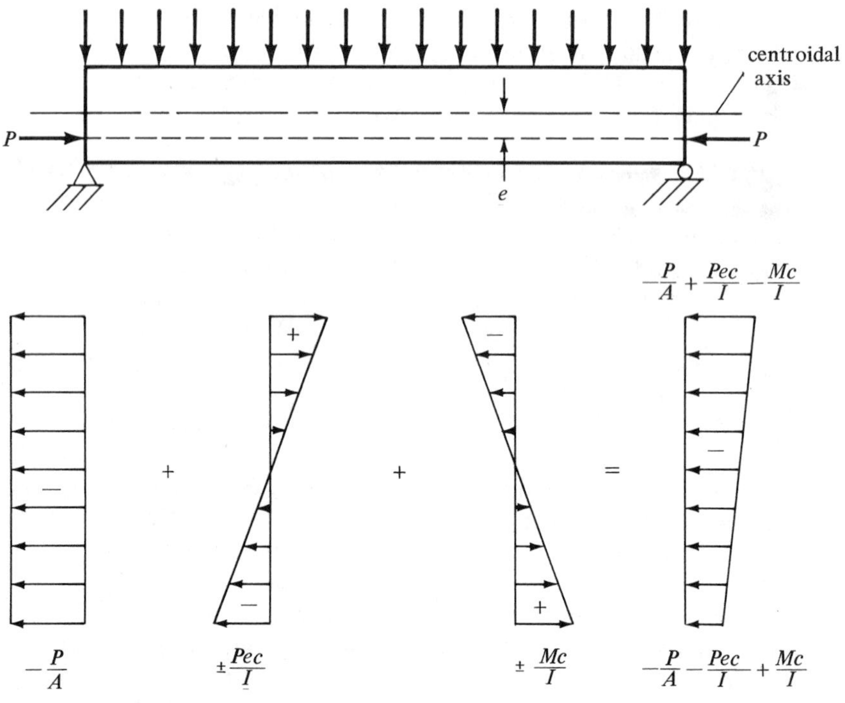

Figure 15.5

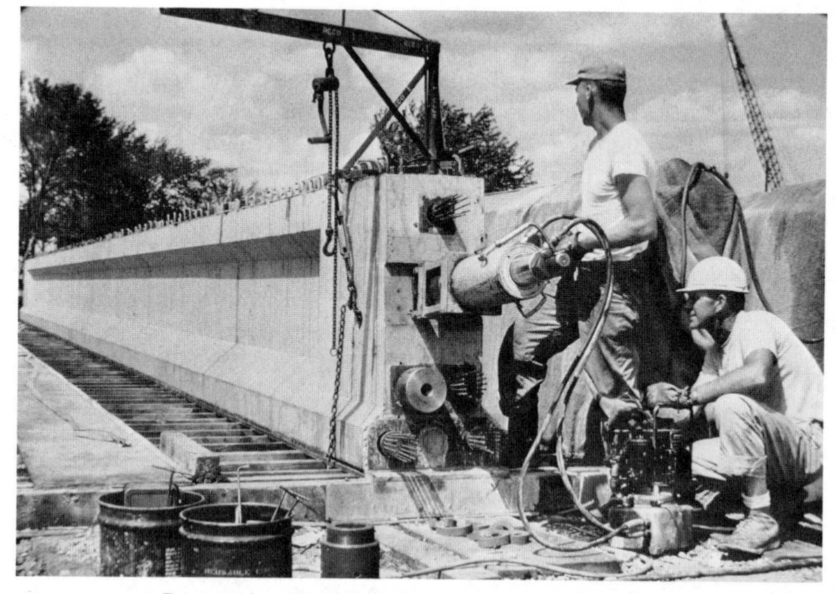

Prestressing Girders for Butler, Pennsylvania Bridge
(Courtesy Portland Cement Association).

the average designer's mind that high-strength steels are probably given an arbitrary yield point anyway. The yield stress for wires and strands is usually assumed to be the stress when a total elongation of 1% has occurred in the steel. For high-strength bars the yield stress is assumed to occur when a 0.2% permanent strain occurs.

15.5 STRESS CALCULATIONS

For a consideration of stresses in a prestressed rectangular beam, reference is made to Figure 15.5. For this example the prestress tendons are assumed to be straight, although it will later be shown that a curved shape is more practical for most beams. The tendons are assumed to be located an eccentric distance e below the centroidal axis of the beam. As a result, the beam is subjected to a combination of direct compression and a moment due to the eccentricity of the prestress. In addition, there will be a moment due to the external load including the beam's own weight. The resulting stress at any point in the beam caused by these three factors can be written as

$$f = -\frac{P}{A} \pm \frac{Pec}{I} \pm \frac{Mc}{I}$$

In Figure 15.5 a stress diagram is drawn for each of these three items and all three are combined to give the final stress diagram.

The usual practice is to base the stress calculations in the elastic range on the properties of the gross concrete section. The gross section consists of the concrete external dimensions with no additions made for the transformed area of the steel tendons nor subtractions made for the duct areas in posttensioning. This method is considered to give satisfactory results because the changes in stresses obtained if net or transformed properties are used are usually not significant.

Example 15.1 illustrates the calculations needed to determine the stresses at various points in a simple-span prestressed rectangular beam. It will be noted that, as there are no moments at the ends of a simple beam due to the external loads nor to the beam's own weight, the Mc/I part of the stress equation is zero there and the equation reduces to

$$f = -\frac{P}{A} \pm \frac{Pec}{I}$$

EXAMPLE 15.1

Calculate the stresses in the top and bottom fibers at the center line and ends of the beam shown in Figure 15.6.

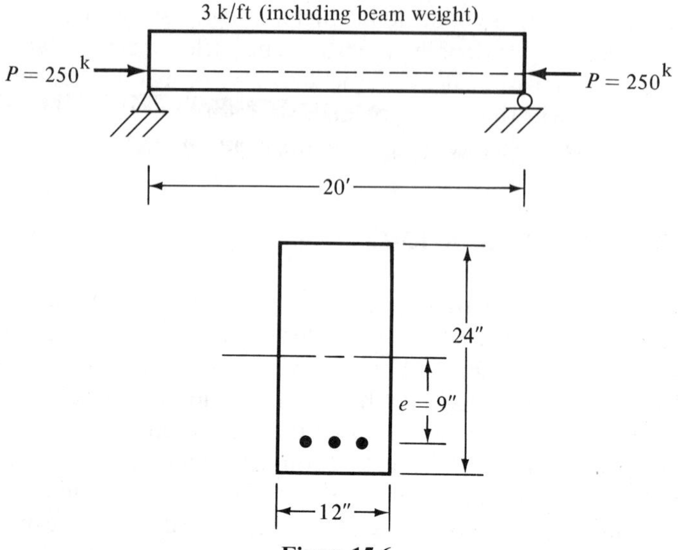

Figure 15.6

SOLUTION

Section properties

$$I = (\tfrac{1}{12})(12)(24)^3 = 13{,}824 \text{ in.}^4$$

$$A = (12)(24) = 288 \text{ in.}^2$$

$$M = \frac{(3)(20)^2}{8} = 150 \text{ ft-k}$$

Stresses at beam center line

$$f_{\text{top}} = -\frac{P}{A} + \frac{Pec}{I} - \frac{Mc}{I} = -\frac{250}{288} + \frac{(250)(9)(12)}{13{,}824} - \frac{(12)(150)(12)}{13{,}824}$$

$$= -0.868 + 1.953 - 1.562 = -0.477 \text{ ksi}$$

$$f_{\text{bottom}} = -\frac{P}{A} - \frac{Pec}{I} + \frac{Mc}{I} = -0.868 - 1.953 + 1.562 = -1.259 \text{ ksi}$$

Stresses at beam ends

$$f_{\text{top}} = -\frac{P}{A} + \frac{Pec}{I} = -0.868 + 1.953 = +1.085 \text{ ksi}$$

$$f_{\text{bottom}} = -\frac{P}{A} - \frac{Pec}{I} = -0.868 - 1.953 = -2.821 \text{ ksi}$$

In Example 15.1 it was shown that when the prestress tendons are straight, the tensile stress at the top of the beam at the ends will be quite high. If, however, the tendons are draped, as shown in Figure 15.7, it is

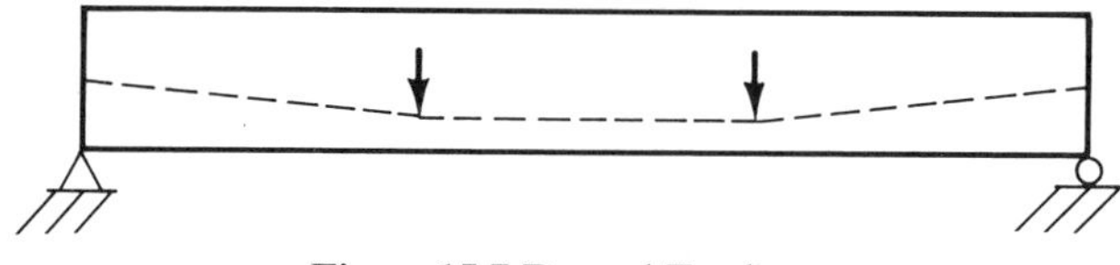

Figure 15.7 Draped Tendons

possible to reduce or even eliminate the tensile stresses. Out in the span the centroid of the strands may be below the lower kern point, but if at the ends of the beam, where there is no stress due to dead load moment, it is below the kern point, tensile stresses in the top will be the result. If the tendons are draped so that at the ends they are located at or above this point, tension will not occur in the top of the beam.

In posttensioning the sleeve or conduit is placed in the forms in the curved position desired. The tendons in pretensioned members can be placed at or below the lower kern points and then can be pushed down to the desired depth at the center line or at other points. In Figure 15.7 the tendons are shown held down at the one-third points. Two alternatives to draped tendons that have been used are straight tendons, located below the lower kern point but which are encased in tubes at their ends, or have their ends greased. Both methods are used to prevent the development of negative moments at the beam ends.

Example 15.2 shows the calculations necessary to locate the kern point for the beam of Example 15.1. In addition, the stresses at the top and bottom of the beam ends are computed. It will be noted that according to these calculations the kern point is 4 in. below the middepth of the beam, and it would thus appear that the prestress tendons should be located at the kern point at the beam ends and pushed down to the desired depth further out in the beam. Actually, however, the tendons do not have to be as high as the kern points because the ACI Code (18.4.1) permits some tension in the top of the beam when the tendons are cut. This value is $3\sqrt{f_{ci}'}$, where f_{ci}' is the strength of the concrete at the time the tendons are cut, as determined by testing concrete cylinders. This permissible value equals about 40% of the cracking strength or modulus of rupture of the concrete ($7.5\sqrt{f_{ci}'}$) at that time. The stress at the bottom of the beam, which is compressive, is permitted to go as high as $0.60f_{ci}'$.

The Code actually permits tensile stresses at the ends of simple beams to go as high as $6\sqrt{f_{ci}'}$. These allowable tensile values are applicable to the stresses which occur immediately after the transfer of the prestressing forces and after the losses occur due to elastic shortening of the concrete and relaxation of the tendons and anchorage seats. It is further assumed that the time-dependent losses of creep and shrinkage have not occurred. A discussion of these various losses is presented in Section 15.7 of this chapter. If the calculated tensile stresses are greater than the permissible values it is necessary to use some additional bonded reinforcing (prestressed or unprestressed) to resist the total tensile force in the concrete computed on the basis of an uncracked section.

Section 18.4.2 of the Code gives allowable stresses at service loads after all prestress losses have occurred. The compression in the concrete may be $0.45f_c'$ while the allowable stress in the precompressed tensile zone for ordinary prestressed beams is $6\sqrt{f_c'}$. If, however, certain conditions are met such as additional cover and satisfactory deflections a tensile stress of $12\sqrt{f_c'}$ is permitted. This higher value which is not applicable to two-way slabs provides improved performance under service loads particularly where live loads are of a transient nature.

EXAMPLE 15.2

Determine the location of the lower kern point at the ends of the beam of Example 15.1. Calculate the stresses at the top and bottom of the beam ends assuming the tendons are placed at the kern point.

SOLUTION

Locating the kern point

$$f_{\text{top}} = -\frac{P}{A} + \frac{Pec}{I} = 0$$

$$-\frac{250}{288} + \frac{(250)(e)(12)}{13{,}824} = 0$$

$$-0.868 + 0.217e = 0$$

$$e = 4''$$

Computing stresses

$$f_{\text{top}} = -\frac{P}{A} - \frac{Pec}{I} = -\frac{250}{288} + \frac{(250)(4)(12)}{13{,}824}$$

$$= -0.868 + 0.868 = 0$$

$$f_{\text{bottom}} = -\frac{P}{A} - \frac{Pec}{I} = -0.868 - 0.868 = -1.736 \text{ ksi}$$

15.6 SHAPES OF PRESTRESSED SECTIONS

For simplicity in introducing prestressing theory, rectangular sections are used for most of the examples of this chapter. From the viewpoint of formwork alone, rectangular sections are the most economical, but more complicated shapes, such as Is and Ts, will require smaller quantities of concrete and prestressing steel to carry the same loads and, as a result, they frequently have the lowest overall costs.

If a member is to be made only one time, a cross section requiring simple formwork (thus often rectangular) will probably be used. For instance, simple formwork is essential for most cast-in-place work. Should, however, the forms be used a large number of times to make many identical members, more complicated cross sections, such as Is and Ts, channels, or boxes, will be used. For such sections the cost of the formwork as a percentage of each member's total cost will be much reduced. Several types of commonly used prestressed sections are shown in Figure 15.8. The same general theory used for the determination of stresses and flexural strengths applied to shapes such as these as it does to rectangular sections.

The usefulness of a particular section depends on the simplicity and reusability of the formwork, the appearance of the sections, the degree of difficulty of placing the concrete, as well as the theoretical properties of the cross section. The greater the amount of concrete located near the extreme fibers of a beam, the greater will be the lever arm between the C and T forces and thus the greater the resisting moment. Of course, there are some limitations on the widths and thicknesses of the flanges. In addition, the webs must be sufficiently large to resist shear and to allow the proper placement of the concrete and at the same time be sufficiently thick to avoid buckling.

A prestressed T such as the one shown in Figure 15.8(a) is often a very economical section because a large proportion of the concrete is placed in the compression flange, where it is quite effective in resisting compressive forces. The double T shown in Figure 15.8(b) is used for schools, office buildings, stores, and so on and is probably the most used prestressed section in the United States today. The total width of the flange provided by a double T is in the range of about 5 to 8 ft and spans of 30 to 50 ft are common. You can see that a floor or roof system can easily and quickly be erected by placing a series of precast double Ts side by side TT TT TT the sections providing both the beams and slabs for the floor or roof system. Single Ts are normally used for heavier loads and longer spans up to as high as 100 or 120 ft. Double Ts for such spans would be very heavy and difficult to handle.

The I and box sections, shown in parts (c) and (d) of Figure 15.8, have a large proportion of their concrete placed in their flanges, with the result that larger moments of inertia are possible (as compared to rectangular sections with the same amounts of concrete and prestressing tendons). The formwork, however, is complicated and the placing of concrete is difficult. Box girders are frequently used for intermediate bridge spans. Their properties are the same as for I sections. Unsymmetrical Is [Figure 15.8(e)], with large bottom flanges to contain the tendons and small top flanges, may be economical for certain composite sections where they are used together with a slab poured in place to provide the compression flange. A similar situation in shown in Figure 15.8(f), where an inverted T is used with a cast-in-place slab.

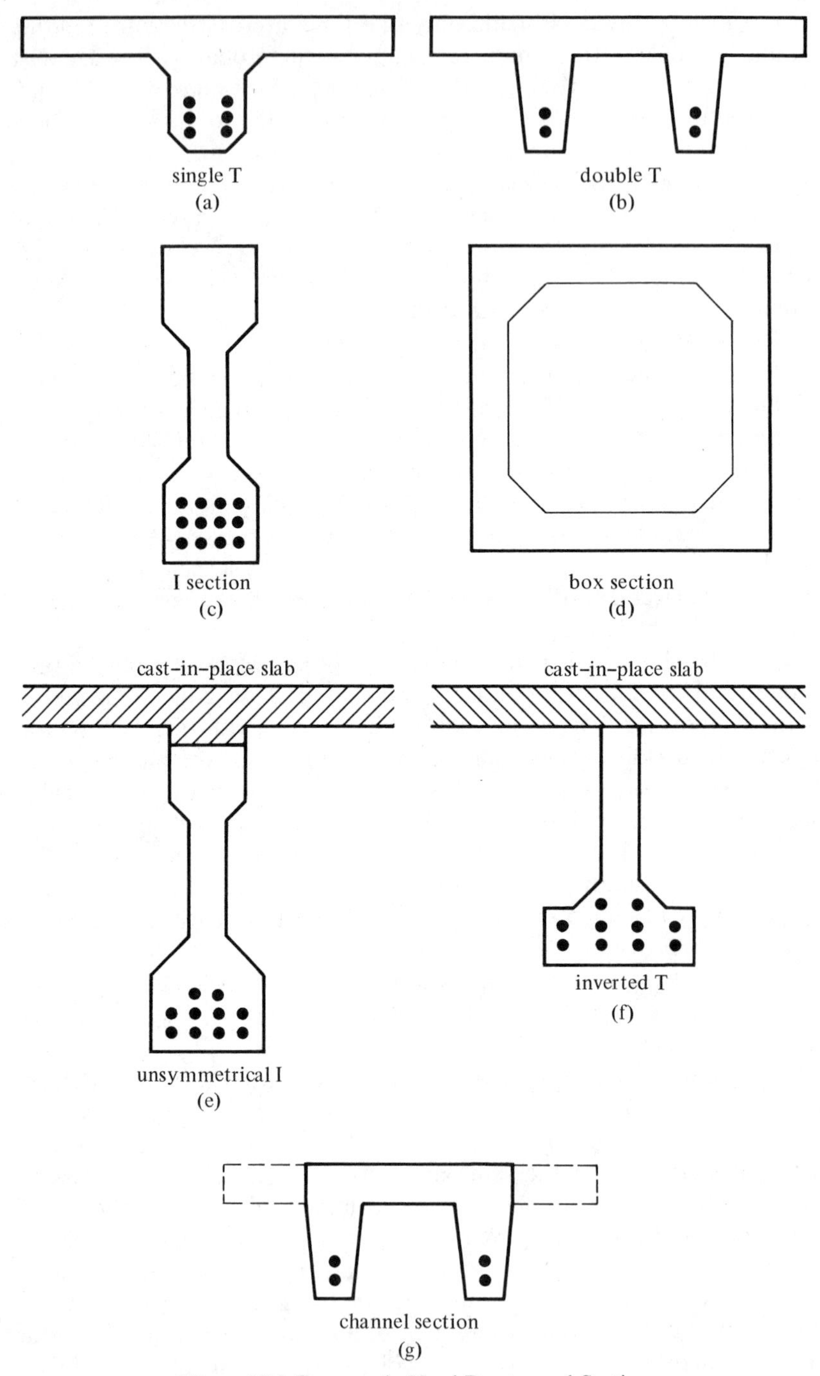

Figure 15.8 Commonly Used Prestressed Sections

Many variations of these sections are used, such as the channel section shown in Figure 15.8(g). Such a section might be made by blocking out the flanges of a double T form as shown and the resulting members used for stadium seats or similar applications.

15.7 PRESTRESS LOSSES

The flexural stresses calculated for the beams of Examples 15.1 and 15.2 were based on initial stresses in the prestress tendons. The stresses, however, become smaller with time due to several factors. These factors, which are discussed in the paragraphs to follow, include

1. elastic shortening of the concrete
2. shrinkage and creep of the concrete
3. relaxation or creep in the tendons
4. slippage in posttensioning and anchorage systems
5. friction along the ducts used in posttensioning.

Elastic Shortening of the Concrete

When the tendons are cut for a pretensioned member, the prestress force is transferred to the concrete, with the result that the concrete is put in compression and shortens, thus permitting some relaxation or shortening of the tendons. The stress in the concrete adjacent to the tendons can be computed as described in the preceding examples. The strain in the concrete ε_c, which equals f_c/E_c, is assumed due to bond to equal the steel strain ε_s. Thus the loss in prestress can be computed as $\varepsilon_s E_s$. An average value of prestress loss in pretensioned members due to elastic shortening is about 3% of the initial value.

For posttensioned members the situation is a little more involved because it is rather common to stress a few of the strands at a time and tie them to the end plates. As a result, the losses vary, with the greatest losses occurring in the first strands stressed and the least losses occurring in the last strands stressed. For this reason an average loss may be calculated for the different strands. Losses due to elastic shortening average about $1\frac{1}{2}$% for posttensioned members. It is, by the way, often possible to calculate the expected losses in each set of tendons and overstress them by that amount so the net losses will be close to zero.

Shrinkage and Creep of the Concrete

The losses in prestressing due to the shrinkage and creep in the concrete are quite variable. For one thing the amount of shrinkage that occurs in concrete varies from almost zero up to about 0.0005 in./in. (depending on dampness and on the age of the concrete when it is loaded), with an average value of about 0.0003 in./in. being the usual approximation. The shrinkage loss is taken as approximately 7% in pretensioned sections and 6% in posttensioned ones.

The amount of creep in the concrete depends on several factors, which have been previously discussed in this text and can vary from 1 to 5 times the instantaneous elastic shortening. Prestress forces are usually applied to pretensioned members much earlier in the age of the concrete than for posttensioned members. Pretensioned members are normally cast in a bed at the prestress yard where the speed of production of members is an important economic matter. The owner wants to tension the steel, place the concrete, and take the members out of the prestress bed as quickly as the concrete gains sufficient strength so he can start on the next set of members. As a result, creep and shrinkage are larger, as are the resulting losses. Average losses used for pretensioned members are about 6% and about 5% for posttensioned members.

Relaxation or Creep in the Tendons

The plastic flow or relaxation of steel tendons is quite small when the stresses are low, but the percentage of relaxation increases as stresses become higher. In general, the estimated losses run from about 2% to 3% of the initial stresses. The amount of these losses actually varies quite a bit for different steels and should be determined from test data available from the steel manufacturer in question.

Slippage in Posttensioning and Anchorage Systems

When the jacks are released and the prestress forces transferred to the end anchorage system, a little slippage of the tendons occurs. The amount of the slippage depends on the system used and tends to vary from about 0.10 in. to 0.20 in. Such deformations are quite important if the members and thus the tendons are short, but if they are long, the percentage is much less important.

Friction Along the Ducts Used in Posttensioning

There are losses in posttensioning due to friction between the tendons and the surrounding ducts. In other words, the stress in the tendons gradually falls off as the distance from the tension points increases due to friction between the tendons and the surrounding material. These losses are due to the so-called length and curvature effects.

The *length effect* is the friction that would have existed if the cable had been straight and not curved. Actually it is impossible to have a perfectly straight duct in posttensioned construction, and the result is friction, called the length effect or sometimes the *wobble effect*. The magnitude of this friction is dependent on the stress in the tendons, their length, the workmanship for the particular member in question, and the coefficient of friction between the materials.

The *curvature effect* is the amount of friction that occurs in addition to the unplanned wobble effect. The resulting loss is due to the coefficient of friction between the materials caused by the pressure on the concrete from the tendons, which is dependent on the stress and the angle change in the curved tendons.

It is possible to substantially reduce frictional losses in prestressing by several methods. These include jacking from both ends, overstressing the tendons initially, and lubricating unbonded cables.

The ACI Code (18.6.2) requires that frictional losses for posttensioned members be computed with wobble and curvature coefficients experimentally obtained and verified during the prestressing operation. Furthermore, the Code gives equations for making the calculations.

Total Prestress Losses

Table 15.1 shows a summary of the estimated losses for all the items except slippage in end anchorage systems and friction along ducts in posttensioning.

It is quite common to estimate total prestress losses rather than calculating each one separately, as described in the preceding paragraphs. Values in the range of 15% to 20% of the initial steel stresses are frequently used. The ACI-ASCE Joint Committee 325[3] recommended that the total losses in the tendons, not including friction, be estimated as 35,000 psi for pretensioned members and 25,000 psi for posttensioned ones. These same values are also recommended by the ACI Commentary.

Table 15.1 Prestress Losses

Item	Pretensioning	Posttensioning
Elastic shortening of concrete	3%	$1\frac{1}{2}$%
Creep of concrete	7%	6%
Shrinkage of concrete	6%	5%
Relaxation of steel	2%	3%
Total	18%	$15\frac{1}{2}$%

15.8 ULTIMATE STRENGTH OF PRESTRESSED SECTIONS

Considerable emphasis is given to the ultimate strength of prestressed sections, the objective being to obtain a satisfactory factor of safety against collapse. You might wonder why it is necessary in prestress work to

[3] ACI-ASCE Joint Committee 325, 1958, "Tentative Recommendation for Prestressed Concrete," *Journal ACI*, 54 (January), pp. 545–578.

consider *both* working stress and ultimate strength situations. The answer lies in the tremendous change that occurs in a prestressed member's behavior after tensile cracks occur. Before the cracks begin to form, the entire cross section of a prestressed member is effective in resisting forces, but after the tensile cracks begin to develop, the cracked part is not effective in resisting tensile forces. Cracking is usually assumed to occur when calculated tensile stresses equal the modulus of rupture of the concrete (about $7.5\sqrt{f_c'}$).

Another question that might enter your mind at this time is this: "What effect do the prestress forces have on the ultimate strength of a section?" The answer to the question is quite simple. An ultimate or strength analysis is based on the assumption that the prestressing strands are stressed above their yield point. If the strands have yielded, the tensile side of the section has cracked and the theoretical ultimate resisting moment is the same as for a nonprestressed beam constructed with the same concrete and reinforcing.

The theoretical calculation of ultimate capacities for prestressed sections is not such a routine thing as it is for ordinary reinforced concrete members. The high-strength steels from which prestress tendons are manufactured do not have distinct yield points. Despite this factor the strength method for determining the ultimate moment capacities of sections checks rather well with load tests as long as the steel percentage is sufficiently small as to insure a tensile failure and as long as bonded strands are being considered.

In the expressions used here, f_{ps} is the average stress in the prestressing steel at the design load. This stress is used in the calculations because the prestressing steels usually used in prestressed beams do not have well-defined yield points [thus the flat portions which are common to stress-strain curves for ordinary structural steels.] Unless the yield points of these steels are determined from detailed studies, their values are normally specified. For instance, the ACI Code (18.7.2) states that the following approximate expression may be used for calculating f_{ps}. In this expression f_{pu} is the ultimate strength of the prestressing steel, ρ_p is the percentage of prestress reinforcing A_{ps}/bd, and f_{se} is the effective stress in the prestressing steel after losses. If more accurate stress values are available, they may be used instead of the specified values. In no case may the resulting values be taken as more than the specified yield strength f_{py} nor f_{se} + 60,000. For bonded members:

$$f_{ps} = f_{pu}\left(1 - 0.5\rho_p \frac{f_{pu}}{f_c'}\right)$$

For unbonded members:

$$f_{ps} = f_{se} + 10{,}000 + \frac{f_c'}{100\rho_p}$$

As in reinforced concrete members, the amount of steel in prestressed sections is limited to ensure tensile failures. The limitation rarely presents a problem except in members with very small amounts of prestressing or in members that have not only prestress strands but also some regular reinforcing bars.

In Section 18.8.1 of the Code ω, the ratio of prestressed and nonprestressed reinforcement used for calculating moment strength, is limited to a maximum value of 0.30 to ensure an underreinforced member. If this value is exceeded the theoretical ultimate strength equations do not give results which correspond well with test results. For such over-reinforced members the design moment may not exceed the moment resistance calculated using the compression part of the internal resisting couple. For rectangular and flanged sections with $\omega > 0.3$ the ACI Commentary (18.8.2) recommends the following values.

For rectangular or flanged sections where the neutral axis falls within the flange, $\omega > 0.3$.

$$M_u = \phi(0.25f_c'bd^2)$$

For flanged sections where the neutral axis falls outside the flange, $\omega > 0.3$

$$M_u = \phi[0.25f_c'b_wd^2 + 0.85f_c'(b - b_w)h_f(d - 0.5h_f)]$$

Example 15.3 illustrates the calculations involved in determining the permissible ultimate capacity of a rectangular prestressed beam. Some important comments about the solution and about ultimate moment calculations in general are made at the end of the problem.

EXAMPLE 15.3

Determine the permissible ultimate moment capacity of the prestressed bonded beam of Figure 15.9 if f_{pu} is 275,000 psi and f_c' is 5,000 psi.

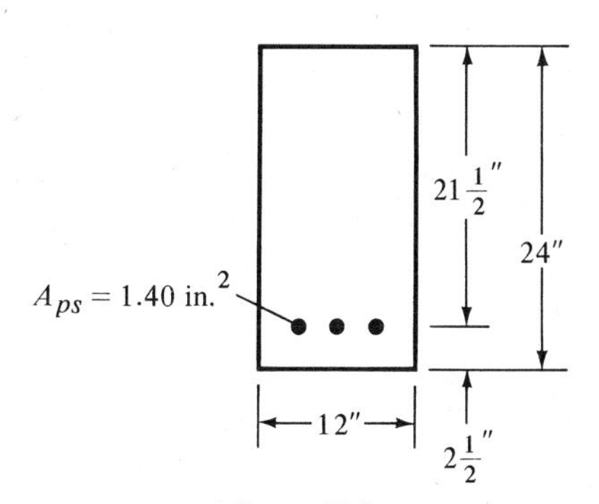

Figure 15.9

SOLUTION

Approximate value of f_{ps} from ACI Code

$$\rho_p = \frac{A_{ps}}{bd} = \frac{1.40}{(12)(24)} = 0.00486$$

$$f_{ps} = f_{pu}\left(1 - 0.5\rho_p \frac{f_{pu}}{f'_c}\right)$$

$$= 275[1 - (0.5)(0.00486)(\tfrac{275}{5})] = 238.2 \text{ ksi}$$

Computing value of reinforcing ratio (Code 18.8.1)

$$w_p = \frac{\rho_p f_{ps}}{f'_c} = \frac{(0.00486)(238.2)}{5} = 0.232 < 0.30$$

tensile failure occurs

Moment capacity

$$a = \frac{A_{ps} f_{ps}}{0.85 f'_c b} = \frac{(1.40)(238.2)}{(0.85)(5)(12)} = 6.54''$$

$$M_u = \phi A_{ps} f_{ps}\left(d - \frac{a}{2}\right) = (0.9)(1.40)(238.2)\left(21.5 - \frac{6.54}{2}\right)$$

$$= 5471''^{k} = \underline{456 \text{ ft-k}}$$

Discussion

The approximate value of f_{ps} obtained by the ACI formula is very satisfactory for all practical purposes. Actually a slightly more accurate value of f_{ps} and thus of the moment capacity of the section can be obtained by calculating the strain in the prestress strands due to the prestress and adding to it the strain due to the ultimate moment. This latter strain can be determined from the values of a and the strain diagram as frequently used in earlier chapters for checking to see if tensile failures control in reinforced concrete beams. With the total strain a more accurate cable stress can be obtained by referring to the stress-strain curve for the prestressing steel being used. Such a curve is shown in Figure 15.10.

The analysis described herein is satisfactory for pretensioned beams or for bonded posttensioned beams but is not so good for unbonded posttensioned members. In these latter beams the steel can slip with respect to the concrete and, as a result, the steel stress is almost constant throughout the member. The calculations for M_u for such members are less accurate than for bonded members. Unless some ordinary reinforcing bars are added to these members, large cracks may form in the members, which are not attractive and which can lead to some corrosion of the prestress strands.

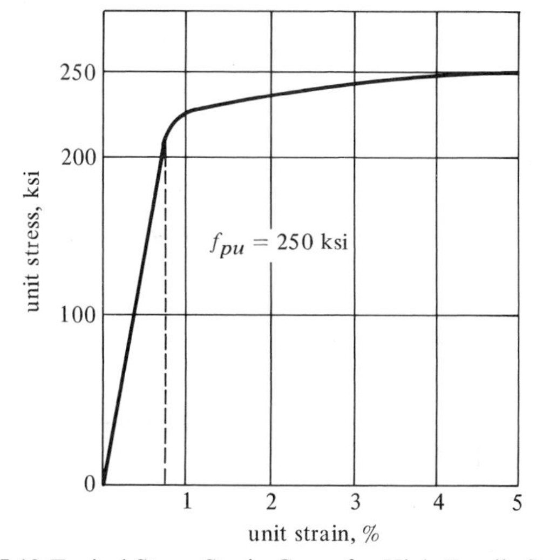

Figure 15.10 Typical Stress-Strain Curve for High-Tensile Steel Wire

Should a prestressed beam be satisfactorily designed based on service loads, then checked by strength methods and found to have insufficient strength to resist the factored loads ($M_u = 1.4M_D + 1.7M_L$), nonprestressed reinforcement may be added to increase the factor of safety. The increase in T due to these bars is assumed to equal $A_s f_y$ (Code 18.7.3). The Code (18.8.3) further states that the total amount of prestressed and nonprestressed reinforcement shall be sufficient to develop an ultimate moment equal to at least 1.2 times the cracking moment of the section calculated with the modulus of rupture of the concrete. This additional steel also will serve to reduce cracks.

15.9 DEFLECTIONS

The deflections of prestressed concrete beams must be calculated very carefully. Some members which are completely satisfactory in all other respects are not satisfactory for practical use because of the magnitudes of their deflections. The actual deflection calculations are made as they are for members made of other materials, such as structural steel, reinforced concrete, and so on. There is, however, the same problem that exists for reinforced concrete members and that is the difficulty of determining the modulus of elasticity to be used in the calculations. The modulus varies with age, with different stress levels, and with other factors. Usually the gross moments of inertia are used for deflection calculations (ACI Code 9.5.4.1), although the transformed I may on occasion be used if there is a large percentage of reinforcing steel in the members or if the members

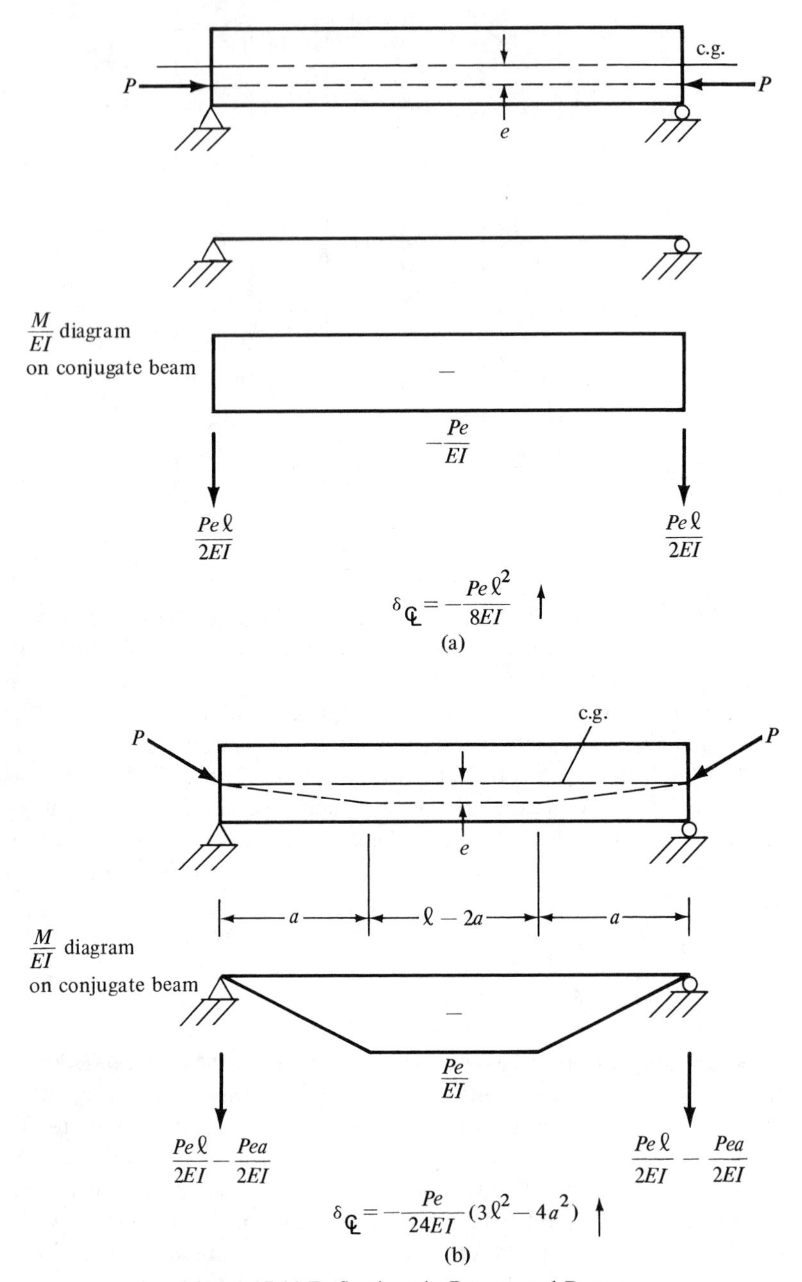

Figure 15.11 Deflections in Prestressed Beams

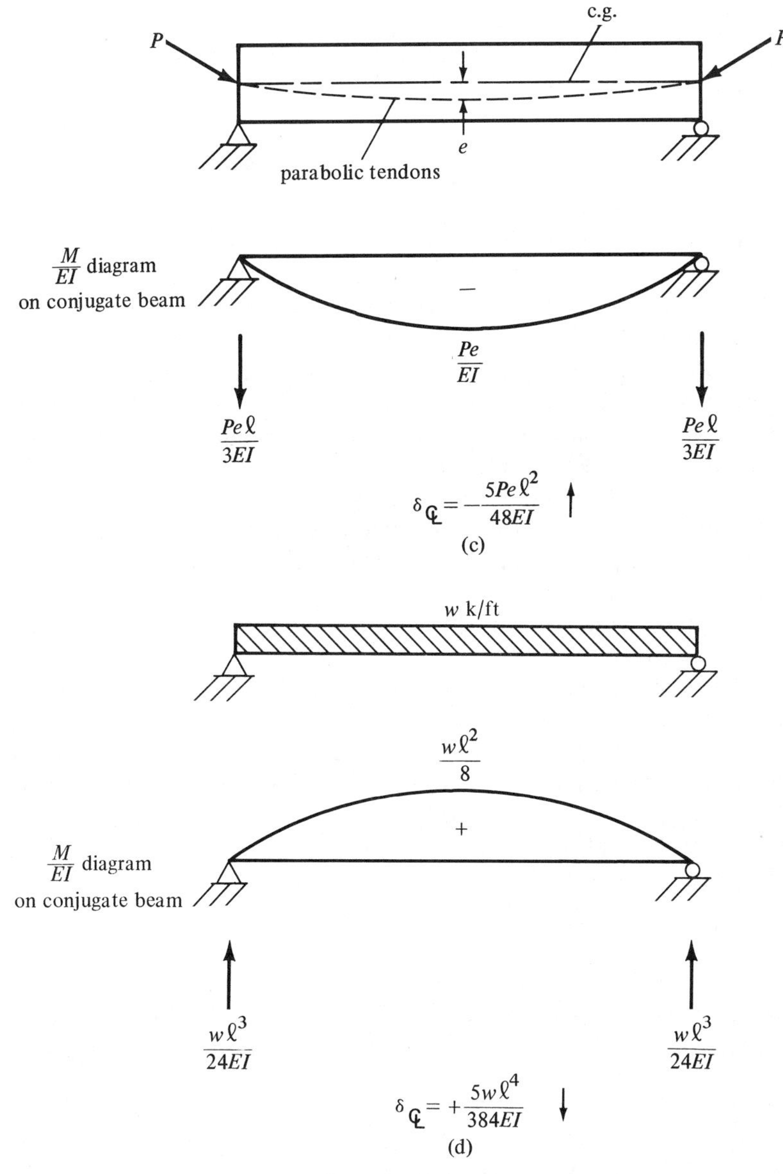

Figure 15.11 *(cont.)*

have a relatively high tensile stress as permitted in Section 18.4.2(c) of the Code.

The deflection due to the force in a set of straight tendons is considered first in this section, with reference being made to Figure 15.11(a). The prestress forces cause a negative moment equal to *Pe* and thus an upward deflection or camber of the beam. This deflection can be calculated by taking moments at the point desired when the conjugate beam is loaded with the M/EI diagram. It equals

$$-\left(\frac{Pel}{EI}\right)\left(\frac{l}{2}-\frac{l}{4}\right)=-\frac{Pel^2}{8EI}$$

upwards for the straight cables shown.

Should the cables not be straight, the deflection will be different due to the different negative moment diagram produced by the cable force. If the cables are bent down or curved, as shown in parts (b) and (c) of Figure 15.11, the conjugate beam can again be applied to compute the deflections. The resulting values are shown in the figure.

The deflections due to the tendon stresses will change with time. First of all, the losses in stress in the prestress tendons will reduce the negative moments they produce and thus the upward deflections. On the other hand, however, the long term compressive stresses in the bottom of the beam due to the prestress negative moments will cause creep and therefore increase the upward deflections.

In addition to the deflections caused by the tendon stresses, there are deflections due to the beam's own weight and due to the additional dead and live loads subsequently applied to the beam. These deflections can be computed and superimposed on the ones caused by the tendons. Figure 15.11(d) shows the ℄ deflection of a uniformly loaded simple beam obtained by taking moments at the ℄ when the beam is loaded with the M/EI diagram.

Example 15.4 shows the initial and long-term deflection calculations for a rectangular pretensioned beam.

EXAMPLE 15.4

The pretensioned rectangular beam shown in Figure 15.12 has straight cables with initial stresses of 175 ksi and final stresses after losses of 140 ksi. Determine the deflection at the beam ℄ immediately after the cables are

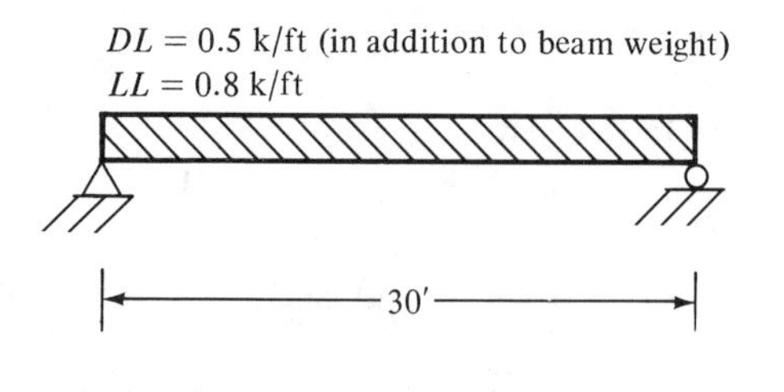

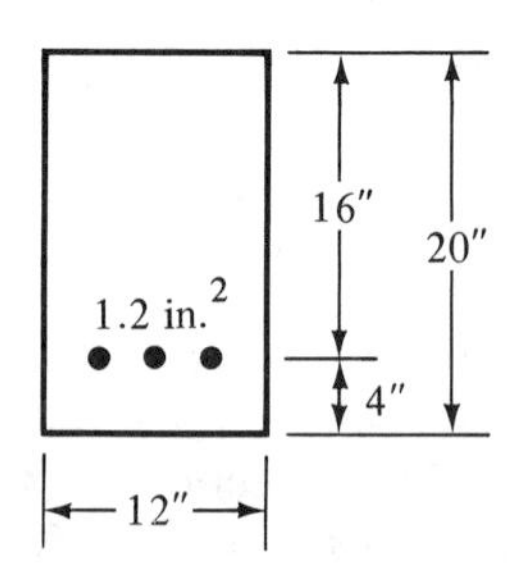

Figure 15.12

cut and six months later under full dead and live loads. Assume a creep coefficient of 2.0 for this latter case. $E = 4 \times 10^6$ psi.

SOLUTION

$$I_g = (\tfrac{1}{12})(12)(20)^3 = 8{,}000 \text{ in.}^4$$
$$e = 6''$$
$$Bm \text{ weight} = \frac{(12)(20)}{144}(150) = 250 \ \#/\text{ft}$$

Deflection immediately after cables cut

$$\delta \text{ due to cable} = -\frac{Pel^2}{8EI} = \frac{(1.2 \times 175{,}000)(6)(12 \times 30)^2}{(8)(4 \times 10^6)(8{,}000)} = -0.638''\uparrow$$

$$\delta \text{ due to beam weight} = +\frac{5wl^4}{384EI} = \frac{(5)\left(\frac{250}{12}\right)(12 \times 30)^4}{(384)(4 \times 10^6)(8{,}000)} = +0.142''\downarrow$$

$$\text{total deflection} = -0.496''\uparrow$$

Deflection at six months under full load

$$\delta \text{ due to cable} = \left(\frac{140{,}000}{175{,}000}\right)(-0.638)(2) = -1.021''$$

$$\delta \text{ due to beam weight} + \text{additional } DL = +\frac{5wl^4}{384EI}\,C_c$$

$$= \frac{(5)\left(\frac{750}{12}\right)(12 \times 30)^4}{(384)(4 \times 10^6)(8{,}000)}(2.0) = +0.852''\downarrow$$

$$\delta \text{ due to } LL = \left(\frac{800}{250}\right)(+0.142) = +0.454''\downarrow$$

$$\text{total deflection} = +0.285''\downarrow$$

Additional Deflection Comments

From the preceding example it can be seen that, not counting external loads, the beam is initially cambered upwards by 0.496 in. and as time goes by this camber increases due to creep in the concrete. Such a camber is often advantageous in offsetting deflections caused by the superimposed loads. In some members, however, the camber can be quite large, particularly for long spans and where lightweight aggregates are used.

To illustrate one problem that can occur, it is assumed that the roof of a school is being constructed by placing 50 ft double Ts, made with a lightweight aggregate, side by side over a classroom. The resulting cambers may be rather large and, worse, they may not be equal in the different sections. It then becomes necessary to force the different sections to the same deflection and tie them together in some fashion so that a smooth surface is provided for roofing. Once the surface is even, the members may be connected by welding together metal inserts, such as angles that were cast in the edges of the different sections for this purpose.

Both reinforced concrete members and prestressed members with overhanging or cantilevered ends will often have rather large deflections. The total deflections at the free ends of these members are due to the sum of the normal deflections plus the effect of support rotations. This latter effect may frequently be the larger of the two and, as a result, the sum of the two deflections may be so large as to detrimentally affect the appearance of the structure. As a result, many designers try to avoid cantilevered members in prestressed construction.

15.10 SHEAR IN PRESTRESSED SECTIONS

Web reinforcement for prestressed sections is handled in a manner similar to that used for a conventional reinforced concrete beam. In the expressions which follow, b_w is the web width or the diameter of a circular section and d is the distance from the extreme fiber in compression to the centroid of the tensile reinforcement. Should the reaction introduce compression into the end region of a prestressed member, sections of the beam located at distances less than $h/2$ from the face of the support may be designed for the shear computed at $h/2$, where h is the overall thickness of the member.

$$v_u = \frac{V_u}{\phi b_w d}$$

The Code (11.4.1) provides two methods for estimating the shear strength that the concrete of a prestressed section can resist. There is an approximate method, which can only be used when the effective prestress force is equal to at least 40% of the tensile strength of the flexural reinforcement f_{pu}, and a more detailed analysis, which can be used regardless of the magnitude of the effective prestress force. These methods are discussed in the paragraphs to follow.

Approximate Method

With this method the nominal shear capacity of a prestressed section can be taken as

$$V_c = \left(0.6\sqrt{f_c'} + 700\frac{V_u d}{M_u}\right) b_w d$$

The Code (11.4.1) states that regardless of the value given by this equation, V_c need not be taken as less than $2\sqrt{f_c'}b_w d$ nor may it be larger than $5\sqrt{f_c'}b_w d$. In this expression V_u is the maximum design shear at the section being considered and M_u is the design moment at the same section occurring simultaneously with V_u and d is the distance from the extreme compression fiber to the centroid of the prestressed tendons. The value of $V_u d/M_u$ is limited to a maximum value of 1.0.

More Detailed Analysis

If a more detailed analysis is desired (it will have to be used if the effective prestressing force is less than 40% of the tensile strength of the flexural reinforcement), the nominal shear force carried by the concrete is considered to equal the smaller of V_{ci} and V_{cw}, to be defined here. The shear force at diagonal cracking due to all design loads when the cracking is due to flexural shear is called V_{ci}. The shear force when cracking is due to the principal tensile stress is referred to as V_{cw}. In both expressions to follow d is the distance from the extreme compression fiber to the centroid of the prestressed tendons or $0.8h$, whichever is greater (Code 11.4.2.3).

The estimated shear capacity V_{ci} can be computed by the following expression, given by the ACI Code (11.4.2):

$$V_{ci} = 0.6\sqrt{f_c'}b_w d + V_d + \frac{V_i M_{cr}}{M_{max}}$$

$$\text{but not less than } 1.7\sqrt{f_c'}b_w d$$

In this expression V_d is the shear at the section in question due to service dead load, M_{max} is the maximum bending moment at the section due to externally applied design loads, V_i is the shear that occurs simultaneously with M_{max}, and M_{cr} is the cracking moment, which is to be determined from the expression to follow:

$$M_{cr} = \left(\frac{I}{y_t}\right)(6\sqrt{f_c'} + f_{pe} - f_d)$$

in which

I = the moment of inertia of the section that resists the externally applied loads

y_t = the distance from the centroidal axis of the gross section (neglecting the reinforcing) to the extreme fiber in tension

f_{pe} = the compressive stress in the concrete due to prestress after all losses at the extreme fiber of the section where the applied loads cause tension

f_d = the stress due to dead load at the extreme fiber where the applied loads cause tension.

From a somewhat simplified principal tension theory, the shear capacity of a beam is equal to the value given by the following expression but need not be less than $1.7\sqrt{f_c'}b_wd$.

$$V_{cw} = (3.5\sqrt{f_c'} + 0.3f_{pc})b_wd + V_p$$

In this expression f_{pc} is the calculated compressive stress, in pounds per square inch, in the concrete at the centroid of the section resisting the applied loads due to the effective prestress after all losses have occurred. (Should the centroid be in the flange, f_{pc} is to be computed at the junction of the web and flange.) V_p is the vertical component of the effective prestress at the section under consideration. Alternately, the Code states that V_{cw} may be taken as the shear force that corresponds to a multiple of dead load plus live load, which results in a calculated principal tensile stress equal to $4\sqrt{f_c'}$ at the centroid of the member or at the intersection of the flange and web if the centroid falls in the web.

A further comment should be made here about the computation of f_{pc} for pretensioned members, as it is affected by the transfer length. The Code (11.4.3) states that the transfer length can be taken as 50 diameters for strand tendons and 100 diameters for wire tendons. The prestress force may be assumed to vary linearly from zero at the end of the tendon to a maximum at the aforesaid transfer distance. If the value of $h/2$ is less than the transfer length, it is necessary to consider the produced prestress when V_{cw} is calculated.

15.11 DESIGN OF SHEAR REINFORCEMENT

Should the computed value of V_u exceed ϕV_c, the area of vertical stirrups (the Code not permitting inclined stirrups or bent-up bars in prestressed members) must not be less than A_v as determined by the following expression from the Code (11.5.6.2):

$$V_s = \frac{A_v f_y d}{s}$$

As in conventional reinforced concrete design, a minimum area of shear reinforcing is required at all points where V_u is greater than $\frac{1}{2}\phi V_c$. This minimum area is to be determined from the expression to follow if the effective prestress is less than 40% of the tensile strength of the flexural reinforcement (ACI Code 11.5.5.3):

$$A_v = 50\frac{b_w s}{f_y}$$

If the effective prestress is equal to or greater than 40% of the tensile strength of the flexural reinforcement, the following expression, in which A_{ps} is the area of prestressed reinforcement in the tensile zone, is to be used to calculated A_v:

$$A_v = \left(\frac{A_{ps}}{80}\right)\left(\frac{f_{pu}}{f_y}\right)\left(\frac{s}{d}\right)\sqrt{\left(\frac{d}{b_w}\right)}$$

Section 11.5.4.1 of the ACI Code states that in no case may the maximum spacing exceed $0.75h$ nor 24 in. Examples 15.5 and 15.6 illustrate the calculations necessary for determining the shear strength and for selecting the stirrups for a prestressed beam.

EXAMPLE 15.5

Calculate the shearing strength of the section shown in Figure 15.13 at 4 ft from the supports using both the approximate method and the more detailed method allowed by the ACI Code. Assume the area of the prestressing steel is 1.0 in.^{2} and the effective prestress force is 250^k. $f_c' =$ 4000 psi.

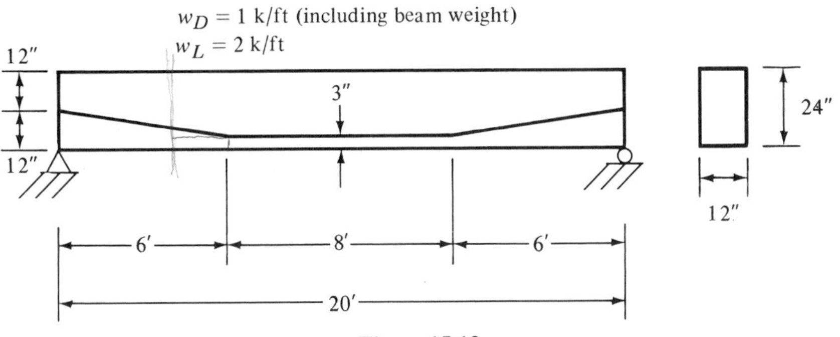

Figure 15.13

SOLUTION

1. Approximate method:

$$w_u = (1.4)(1.0) + (1.7)(2.0) = 4.8 \text{ k/ft}$$

$$V_u = (10)(4.8) - (4)(4.8) = 28.8^k$$

$$M_u = (10)(4.8)(4) - (4)(4.8)(2) = 153.6 \text{ ft-k}$$

$$\frac{V_u d}{M_u} = \frac{(28.8)(24-3-3)}{(12)(153.6)} = 0.281 < 1.0 \qquad \underline{\text{ok}}$$

$$V_c = \left(0.6\sqrt{f_c'} + 700\frac{V_u d}{M_u}\right)b_w d$$

$$= [0.6\sqrt{4000} + (700)(0.281)](12)(18) = 50{,}684\#$$

$$\text{minimum } V_c = (2\sqrt{4000})(12)(18) = 27{,}322\# < 50{,}684\#$$

$$\text{maximum } V_c = (5\sqrt{4000})(12)(18) = 68{,}305\# > 50{,}684\#$$

$$V_c = 50{,}684\#$$

2. More detailed method:

$$I = (\tfrac{1}{12})(12)(24)^3 = 13{,}824 \text{ in.}^4$$

$$y_t = 12''$$

f_{pe} = compressive stress in concrete due to prestress after all losses

$$= \frac{P}{A} + \frac{Pec}{I}$$

$$f_{pe} = \frac{250{,}000}{(12)(24)} + \frac{(250{,}000)(6)(12)}{13{,}824} = 2{,}170 \text{ psi}$$

$$M_d = \text{dead load moment at } 4' \text{ point} = (10)(1)(4) - (4)(1)(2)$$
$$= 32 \text{ ft-k}$$

$$f_d = \text{stress due to the dead load moment} = \frac{(12)(32{,}000)(12)}{13{,}824}$$

$$= 333 \text{ psi}$$

$$M_{cr} = \text{cracking moment} = \left(\frac{I}{y_t}\right)(6\sqrt{f_c'} + f_{pe} - f_d)$$

$$= \left(\frac{13{,}824}{12}\right)(6\sqrt{4{,}000} + 2{,}170 - 333) = 2{,}553{,}373''\#$$

$$= 212{,}780 \text{ ft-lb}$$

$$\text{beam weight} = \frac{(12)(24)}{144}(150) = 300 \ \#/\text{ft}$$

$$w_u \text{ not counting bm wt} = (1.4)(1.0 - 0.3) + (1.7)(2) = 4.38 \text{ k/ft}$$

$$M_{\max} = (10)(4.38)(4) - (4)(4.38)(2) = 140.2 \text{ ft-k} = 140{,}200 \text{ ft-lb}$$

$$V_i \text{ due to } w_u \text{ occurring same time as } M_{\max} = (10)(4.38) - (4)(4.38)$$
$$= 26.28 \text{ k} = 26{,}280\#$$

$$V_d = \text{dead-load shear} = (10)(1.0) - (4)(1.0) = 6^k = 6000\#$$

$$d = 24 - 3 - 3 = 18'' \quad \text{or} \quad (0.8)(24) = \underline{19.2''}$$

$$V_{ci} = 0.6\sqrt{f_c'}b_w d + V_d + \frac{V_i M_{cr}}{M_{\max}}$$

$$= (0.6\sqrt{4000})(12)(19.2) + 6000 + \frac{(26{,}280)(212{,}780)}{140{,}200} = 54{,}628\#$$

$$> (1.7\sqrt{4000})(12)(19.2) = 24{,}772\#$$

Computing V_{cw}

f_{pc} = calculated compressive stress in psi at the centroid of the concrete due to the effective prestress

$$= \frac{250{,}000}{(12)(24)} = 868 \text{ psi}$$

V_p = vertical component of effective prestress at section

$$= \left(\frac{9}{72.56}\right)(250{,}000) = 31{,}009\,\#$$

$$V_{cw} = (3.5\sqrt{f_c'} + 0.3f_{pc})b_w d + V_p$$
$$= (3.5\sqrt{4000} + 0.3 \times 868)(12)(19.2) + 31{,}009 = 142{,}006\,\#$$

Using lesser of V_{ci} or V_{cw}

$$V_c = 54{,}628\,\#$$

EXAMPLE 15.6

Determine the spacing of #3 ⊔ stirrups required for the beam of Example 15.5 at 4 ft from the end support if f_{pu} is 250 ksi for the prestressing steel and f_y for the stirrups is 40 ksi. Use the value of V_c obtained by the approximate method, 50,684#

SOLUTION

$$w_u = (1.4)(1.0) + (1.7)(2.0) = 4.8\text{k/ft}$$
$$V_u = (10)(4.8) - (4)(4.8) = 28.8^{\text{k}}$$
$$\phi V_c = (0.85)(50{,}684) = 43{,}081\,\#$$
$$> V_u = 28{,}800\,\#$$
$$V_u > \frac{\phi V_c}{2} = 21{,}541\,\# < \phi V_c$$

A minimum amount of reinforcement is needed.

Since effective prestress is greater than 40% of tensile strength of reinforcing,

$$A_v = \left(\frac{A_{ps}}{80}\right)\left(\frac{f_{pu}}{f_y}\right)\left(\frac{s}{d}\right)\sqrt{\left(\frac{d}{b_v}\right)}$$

$$(2)(0.11) = \left(\frac{1.0}{80}\right)\left(\frac{250}{40}\right)\left(\frac{s}{18}\right)\sqrt{\frac{18}{12}}$$

$s = 41.38''$ but maximum $s = (\frac{3}{4})(24) = 18''$ <u>use 18″</u>

15.12 ADDITIONAL TOPICS

This chapter has presented a brief discussion of prestressed concrete. There are quite a few other important topics that have been omitted from this introductory material. Several of these items are briefly mentioned in the paragraphs to follow.

Stresses in End Blocks

The part of a prestressed member around the end anchorages of the steel tendons is called the *end block*. In this region the prestress forces are transferred from very concentrated areas out into the whole beam cross section. It has been found that the length of transfer for posttensioned members is less than the height of the beam and in fact is probably much less.

For posttensioned members there is direct-bearing compression at the end anchorage and, as a result, solid end blocks are usually used there to spread out the concentrated prestress forces. To prevent bursting of the block, either wire mesh or a grid of vertical and horizontal reinforcing bars is placed near the end face of the beam. In addition, both vertical and horizontal reinforcing is placed throughout the block.

For pretensioned members where the prestress is transferred to the concrete by bond over a distance approximately equal to the beam depth, a solid end block is probably not necessary but spaced stirrups are needed. A great deal of information on the subject of end block stresses for posttensioned and pretensioned members is available.[4]

Composite Construction

Precast prestressed sections are frequently used in buildings and bridges in combination with cast-in-place concrete. Should such members be properly designed for shear transfer so the two parts will act together as a unit, they are called *composite sections*. Examples of such members were previously shown in parts (e) and (f) of Figure 15.8. In composite construction the parts that are difficult to form and that contain most of the reinforcing are precast, while the slabs and perhaps the top of the beams, which are relatively easy to form, are cast in place.

The precast sections are normally designed to support their own weights plus the green cast-in-place concrete in the slabs plus any other loads applied during construction. The dead and live loads applied after the slab hardens are supported by the composite section. The combination of the two parts will yield a composite section that has a very

[4] Guyon, Y., 1960, *Prestressed Concrete* (New York: Wiley), pp. 127–174.

large moment of inertia and thus a very large resisting moment. It is usually quite economical to use a precast prestressed beam made with a high-strength concrete and a slab made with an ordinary grade of concrete. If this practice is followed, it will be necessary to account for the different moduli of elasticity of the two materials in calculating the composite properties (thus it becomes a transformed area problem).

Continuous Members

Continuous prestress sections may be cast in place completely with their tendons running continuously from one end to the other. It should be realized for such members that where the service loads tend to cause positive moments, the tendons should produce negative moments and vice versa. This means that the tendons should be below the member's center of gravity in normally positive moment regions and above the center of gravity in normally negative moment regions. To produce the desired stress distributions, it is possible to use curved tendons and members of constant cross section or straight tendons with members of variable cross section. In Figure 15.14 several continuous beams of these types are shown.

Another type of continuous section that has been very successfully used in the United States, particularly for bridge construction, involves the use of precast prestressed members made into continuous sections

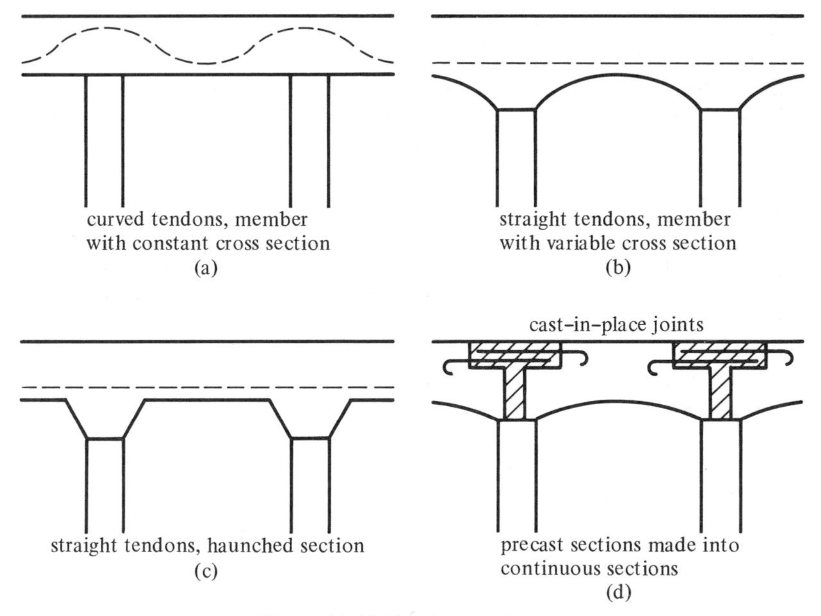

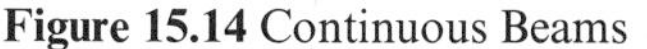
Figure 15.14 Continuous Beams

with cast-in-place concrete and regular reinforcing steel. Figure 15.14(d) shows such a case. For such construction the precast section resists a portion of the dead load, while the live load and the dead load that is applied after the cast-in-place concrete hardens are resisted by the continuous member.

Partial Prestressing

During the early days of prestressed concrete, the objective of the designer was to proportion members that could never be subject to tension when service loads were applied. Such members are said to be *fully prestressed.* Subsequent investigations of fully prestressed members have shown they often have an appreciable amount of extra strength. As a result, many designers now believe that certain amounts of tensile stresses can be permitted under service loads. Members that are permitted to have some tensile stresses are said to be *partially prestressed.*

A major advantage of a partially prestressed beam is a decrease in camber. This is particularly important when the beam load or the dead load is quite low when compared to the total design load.

To provide additional safety for partially prestressed beams, it is common practice to add some conventional reinforcement. This reinforcement will increase the ultimate flexural strength of the members as well as help to carry the tensile stresses in the beam.[5]

PROBLEMS

15.1 The beam shown in the accompanying illustration has an effective prestress of 240^k. Calculate the fiber stresses in the top and bottom of the beam at the ends and ℄. The tendons are assumed to be straight.

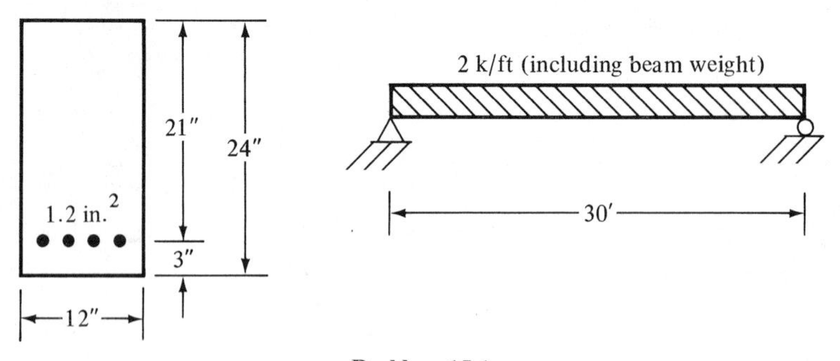

Problem 15.1

[5] Lin, T. Y., 1963, *Design of Prestressed Concrete Structures*, 2d ed. (New York: Wiley), pp. 280–299.

15.2 Compute the stresses in the top and bottom of the beam shown at the ends and center line immediately after the cables are cut. Assume straight cables. Initial prestress is 170 ksi.

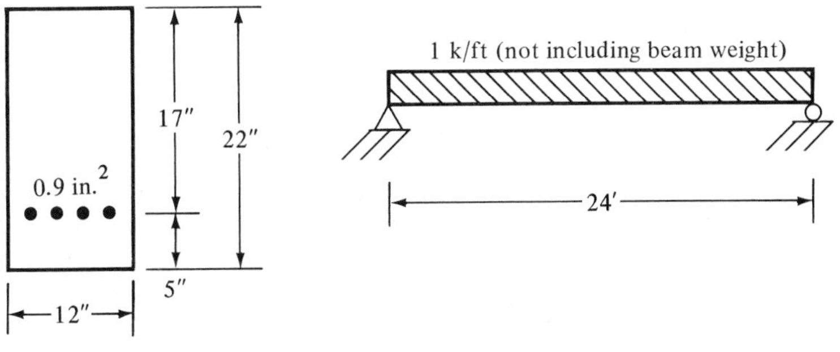

Problem 15.2

15.3 The beam shown has a 30-ft simple span; $f_c' = 5{,}000$ psi, $f_{pu} = 250{,}000$ psi and the initial prestress is 160,000 psi.

(a) Calculate the concrete stresses in the top and bottom of the beam at midspan immediately after the tendons are cut.

(b) Recalculate the stresses at midspan after assumed prestress losses in the tendons of 20%.

(c) What maximum service live load can this beam support if allowable stresses of $0.45f_c'$ in compression and $6\sqrt{f_c'}$ in tension are permitted?

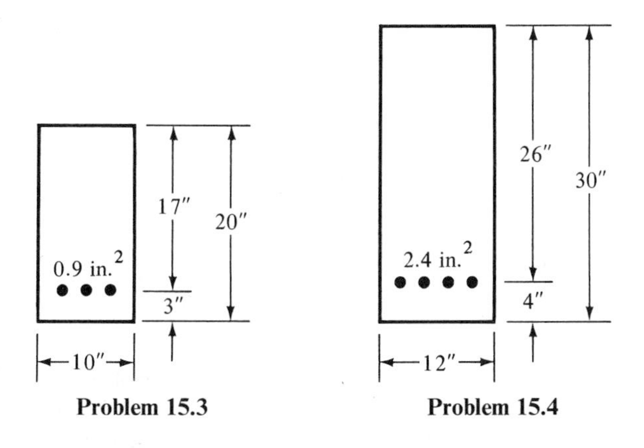

Problem 15.3 **Problem 15.4**

15.4 Using the same allowable stresses permitted in Problem 15.3, what total uniform load can the beam shown in the accompanying illustration support for a 50-ft simple span?

15.5 Compute the cracking moment and the permissible ultimate moment capacity of the beam of Problem 15.3. Assume dead load (not including beam weight) is 500 #/ft.

15.6 Compute the stresses in the top and bottom at the ℄ of the beam of Problem 15.1 if it is picked up at the ℄. Assume 100% impact. Concrete weighs 150 #/ft^3.

15.7 Determine the stresses at the one-third points of the beam of Problem 15.6 if the beam is picked up at those points.

15.8 Calculate the permissible ultimate moment capacity of a 12-in. × 20-in. pretensioned beam that is prestressed with 1.2 in.^{2} of steel tendons stressed to an initial stress of 160 ksi. The center of gravity of the tendons is 3 in. above the bottom of the beam $f_c' = 5{,}000$ psi and $f_{pu} = 250{,}000$ psi.

15.9 Compute the ultimate capacity of the pretensioned beam of Problem 15.2 if $f_c' = 5{,}000$ psi and $f_{pu} = 225$ ksi.

15.10 Compute the permissible ultimate moment capacity of the T beam shown in the accompanying illustration. $f_c' = 5{,}000$ psi, $f_{pu} = 250{,}000$ psi, and the initial stress in the cables is 160,000 psi.

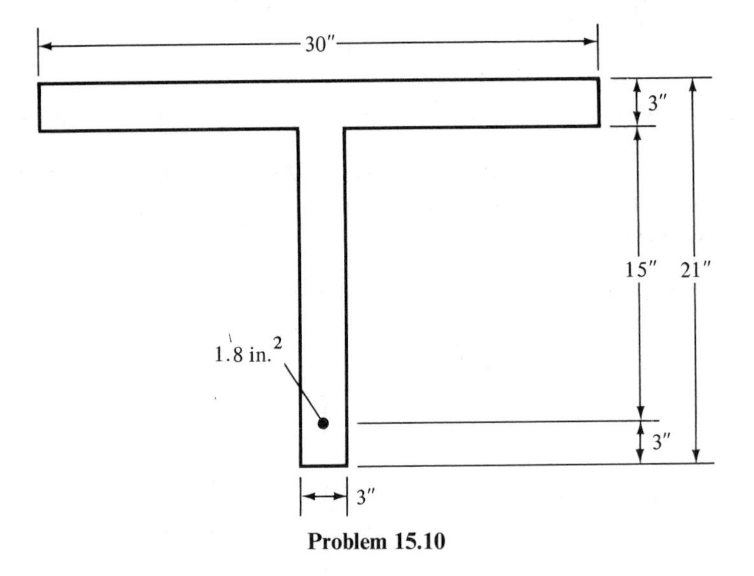

Problem 15.10

15.11 Calculate the deflection at the ℄ of the beam of Problem 15.3 immediately after the cables are cut assuming cable stress = 160,000 psi. Calculate the same deflection two years later if a live load of 2 klf is considered and a creep coefficient of 1.8 is used. Assume a 20% prestress loss.

15.12 Calculate the deflection at the ℄ of the beam of Problem 15.1 immediately after the cables are cut assuming P initial = 240^k and P after losses = 190^k. Repeat the calculation two years later if 20^k concentrated live loads are located at the one-third points of the beam and a creep coefficient of 2.0 is used. $f_c' = 5{,}000$ psi.

15.13 Determine the shearing strength of the beam shown in the accompanying illustration 3 ft from the supports using the approximate method allowed by the ACI Code. Determine the required spacing of #3 ⊔ stirrups at the same section. $f_c' = 5{,}000$ psi, $f_{pu} = 250{,}000$ psi, and f_y for stirrups = 50,000 psi. $f_{se} = 200{,}000$ psi

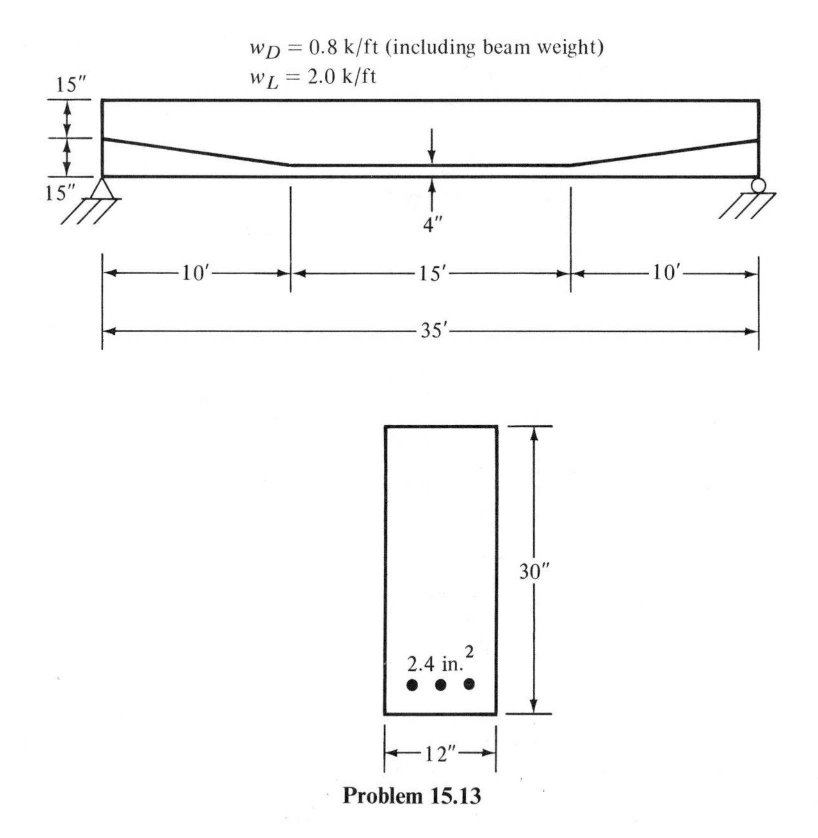

Problem 15.13

PROBLEMS WITH SI UNITS

15.14 Immediately after cutting the cables in the beam shown, they have an effective prestress of 1.260 GPa. Determine the stresses at the top and bottom of the beam at the ends and center line. The concrete weighs 23.5 kN/m³. Cables are straight. $E_c = 27\,924$ MPa.

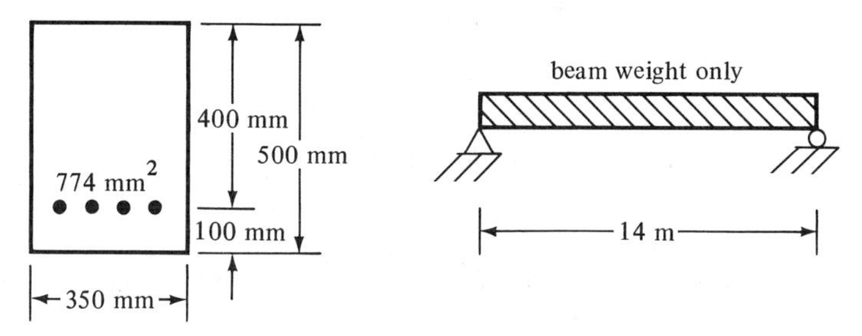

Problem 15.14

15.15 The beam shown has a 12-m simple span; $f_c' = 34.5$ MPa, $f_{pu} = 1.725$ GPa, and the initial prestress is 1.10 GPa.

(a) Calculate the concrete stresses in the top and bottom of the beam at midspan immediately after the tendons are cut.

(b) Recalculate the stresses after assumed losses in the tendons of 18%.

(c) What maximum service uniform live load can the beam support in addition to its own weight if allowable stresses of $0.45f_c'$ in compression and $0.5\sqrt{f_c'}$ in tension are permitted?

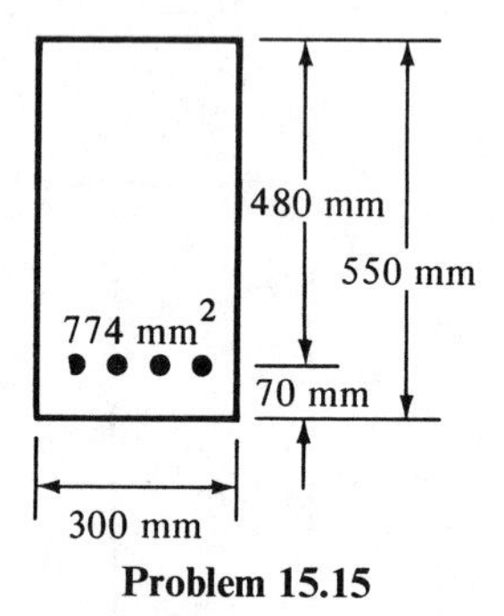

Problem 15.15

15.16 Using the same allowable stresses permitted and cable stresses in Problem 15.15, what total uniform load including beam weight can the beam shown in the accompanying illustration support for a 15-m simple span?

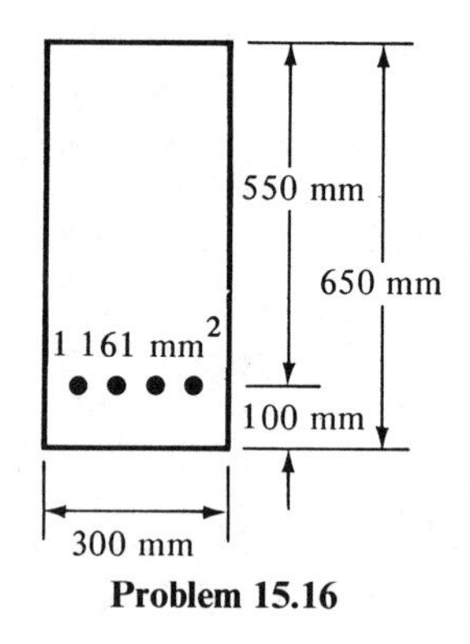

Problem 15.16

15.17 Compute the cracking moment and the permissible ultimate moment capacity of the beam of Problem 15.15.

15.18 Compute the stresses in the top and bottom of the beam of Problem 15.14 if it is picked up at its one-third points. Assume an impact of 100%.

15.19 Compute the permissible ultimate moment capacity of the T beam shown in the accompanying illustration. $f_c' = 34.5$ MPa, $f_{pu} = 1.725$ GPa, and the initial stress in the cables is 1.100 GPa.

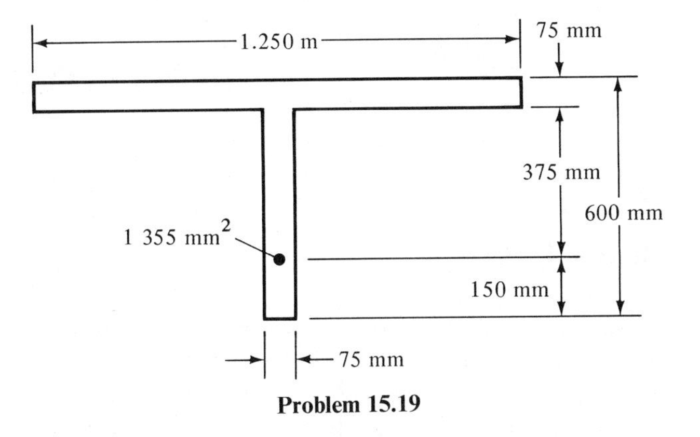

Problem 15.19

15.20 Calculate the deflection at the ℄ of the beam of Problem 15.14 immediately after the cables are cut. Repeat the calculation for a time two years later if 80-kN concentrated live loads are placed at the one-third points of the beam and a creep coefficient of 2.0 is used.

CHAPTER 16

Formwork

16.1 INTRODUCTION

Concrete forms are molds into which semiliquid concrete is placed. The molds need to be sufficiently strong to hold the concrete in the desired size and shape until the concrete hardens. Forms are structures and they should be carefully and economically designed to support the imposed loads using the methods required for the design of other engineering structures.

Safety is a major concern in formwork because a rather large percentage of the accidents that occur during the construction of concrete structures is due to formwork failures. Normally formwork failures are not caused by the application of excessive gravity loading. Although such failures do occasionally occur, the usual failures are due to lateral forces that cause the supporting members to be displaced. These lateral forces may be caused by wind, by moving equipment on the forms, by vibration from passing traffic, or by the lateral pressure of freshly placed and vibrated concrete. A rather sad thing is that most of these failures could have been prevented if only a little additional lateral bracing had been used. There are, of course, other causes of failure, such as stripping the forms too early and improper control of the placement rate of the concrete.

Though you might think that shape, finish, and safety are the most important items in concrete formwork, you should realize that economy is also a major consideration. The cost of the formwork, which can range from one-third to almost two-thirds of the total cost of a concrete structure, is often more than the cost of both the concrete and the reinforcing steel. For the average concrete structure, the formwork is considered to represent about 50% of the total cost. *From this discussion it is obvious that any efforts made to improve the economy of concrete structures should be primarily concentrated on reducing formwork costs.* Formwork must be treated as an integral part of the overall job plan and the low bidder will be the contractor who has planned the most economical forming job.

16.2 RESPONSIBILITY FOR FORMWORK DESIGN

Normally the structural designer is responsible for the design of reinforced concrete structures, while the design of the formwork and its construction is the responsibility of the contractor. For some structures, however, where the formwork is very complex, such as for shells, folded plates, and arches, it is desirable for the designer to take responsibility for the formwork design. There does seem to be an increasing trend, particularly on larger jobs, to employ structural designers to design and detail the formwork.

The contract documents should clearly specify the responsibility of each party for the work. Normally it is wise to allow the contractor as much room as possible to plan his own formwork and construction details, as more economical bids will often be the result.

It is to be remembered in contract law that if the contract documents describe exactly how something is to be done and at the same time spell out the results to be accomplished, it is not generally enforceable. In other

Wastewater Treatment Plant, Santa Rosa, California
(Courtesy Simpson Timber Company).

words, if the exact form of the construction details are specified in the contract and followed by the contractor, and the formwork fails, he cannot be held responsible.

16.3 MATERIALS USED FOR FORMWORK

Several decades ago lumber was the universal material used for constructing concrete forms. The forms were constructed, used one time, torn down, and there was little salvage other than perhaps a board or timber here or there. Such a practice still exists in some parts of the world where labor costs are very low. In a high-labor cost area, such as the United States, the trend for several decades has been toward reusable forms made from many different materials. Wood is still the most widely used material, however, and no matter what types of material are used, some wood will still probably be required, whether it be lumber or plywood.

The lumber usually comes from the softwood varieties, with pine, fir and western hemlock being the most common varieties. These woods are relatively light and are commonly available. Form lumber should be only partly seasoned because if it is too green, it shrinks and warps in hot dry weather. If it is too dry, it swells a great deal when it becomes wet. The lumber should be planed if it is going to be in contact with the concrete, but it is possible to use rough lumber for braces and shoring.

The introduction of plywood made great advances in formwork possible. Large sheets of plywood save a great deal of labor in the construction of the forms and at the same time result in large areas of joint-free concrete surfaces with corresponding reductions in costs of finishing and rubbing of the exposed concrete. Plywood also has considerable resistance to changes in shape when it becomes wet, can withstand rather rough usage, and can be bent to a certain degree, making it possible to construct curved surfaces.

Steel forms are extremely important for today's concrete construction. Not only are steel panel systems important for buildings but steel bracing and framing are important where wood or plywood forms may be used. If steel forms are properly maintained, they may be reused many times. In addition, steel forms with their great strength may be used in places where other materials are not feasible, as in forming long spans. Forms for some other types of structures are frequently made with steel simply as a matter of expediency. Falling into this class are the forms for round columns, tunnels, and so on.

The Sonotube paper-fiber forms are another type of form often used for round columns. These patented forms are lightweight one-piece units that can be sawed to fit beams, utility outlets, and other structures. They have built-in moisture barriers that aid in curing. As they are disposable, there are no cleaning, reshipping, or inventory costs. They can be quickly stripped with electric saws or hand tools.

Two other materials often used for concrete forms are glass-fiber-reinforced plastic and insulation boards. The glass-fiber-reinforced plastic can be sprayed over wooden base forms or can be used to fabricate special forms such as the pans for the waffle-type floors. Quite a few types of insulation board used as form liners are on the market. They are fastened to the forms, and when the forms are removed, the boards are left in place either bonded to the concrete or held in place by some type of clips.

16.4 FURNISHING OF FORMWORK

There are companies that specialize in formwork and from whom forms, shoring, and needed accessories can be rented or purchased. The contractor may very well take bids from these types of companies for forms for columns, floors, and walls. If structures are designed with standard dimensions, the use of rented forms may provide the most economical solution. Such a process makes the contractor's job a little easier. He can regulate and smooth out the work load of his company by assigning the responsibility and risk involved in the formwork to a company that specializes in that particular area. Renting formwork permits the contractor to reduce his investment in the many stock items, such as spacers, fasteners, and so forth, that would be necessary if he did all of the formwork himself.

If forms of the desired sizes are not available for rent or purchase, it is possible for the contractor to make his own forms or to have them made by one of the companies that specialize in form manufacture. If this latter course is chosen, the result is usually a very high quality of formwork.

Finally, the forms can be constructed in place. Though the quality may not be as high as for the last two methods mentioned, it may be the only option available for complicated structures for which the forms cannot be reused. Of course, for many jobs a combination of rented or built-in-place or shop-built forms may be used for best economy.

16.5 ECONOMY IN FORMWORK

The first opportunities for form economy occur during the design of the structure. The concrete designer can often save the owner a great deal of money if he will take into account those factors that permit formwork economy as he designs the structure. Among the items he can do in this regard are the following:

1. Attempt to coordinate his design with the architectural design, such as varying room sizes a little to accomodate standard forms.
2. Keep the sizes of beams, slabs and columns, and story heights the same for as many floors as possible to permit the reuse of forms without change from floor to floor. Without a doubt the most important item affecting the total cost of formwork is the number of times the forms can be used.
3. Remember in the design of columns for multistory buildings that their sizes may be kept constant for a good many floors by changing the percentage of steel used from floor to floor, starting at the top with approximately 1% steel and increasing it floor by floor until the maximum percentage is reached.
4. Design structures that permit the use of commercial forms, such as metal pans or corrugated metal deck sheets for floor or roof slabs.
5. Use beam and column shapes that do not have complicated haunches, offsets, and cut outs.
6. Allow the contractor to use his own methods in constructing the forms, holding him responsible only for their adequacy.

16.6 FORM MAINTENANCE

To be able to reuse forms many times, it is obvious that they should be properly removed, maintained, and stored. One particular thing the contractor can do to appreciably lengthen form lives is to require the same crews to do both erection and stripping. If while stripping a set of forms, the men know that they are going to have to erect those same forms for the next job, they will be much more careful in their handling and maintenance. (Using the same crews for both jobs may not be possible in some areas because of labor contracts.)

Metal bars or prys should not be used for stripping plywood forms as they may easily cause splitting. Wooden wedges, which are gradually tapped to separate the forms from the hardened concrete, should be used. After the forms are removed, they should be thoroughly cleaned and oiled. Any concrete that has adhered to metal parts of the forms should be thoroughly scraped off. The holes in wooden form faces caused by nails, ties (to be described in Section 16.7), and other items should be carefully filled with metal plates, cork, or plastic materials. After the cleaning and repairs are completed, the forms should be coated with oil or other preservative. Steel forms should be coated both front and back to keep them from rusting and to prevent spilled concrete from sticking to them.

Plywood surfaces should be lightly oiled and the forms stored in a flat position with the oiled faces together. The stored panels should be kept out of the sun and rain or should be lightly covered so as to permit air circulation and prevent heat buildup. In addition, wood strips may be placed between forms to increase this ventilation. Although some plywood forms have reportedly been used successfully for 200 or more times, it is normal with proper maintenance to be able to use them approximately 30 or 40 times. After plywood panels are deemed to be no longer suitable for formwork, they can usually be used by the wise contractor for subflooring or wall or roof sheathing.

Before forms are used, their surfaces should be wetted and oiled or coated with some type of material that will not stain or soften the concrete. The coating is primarily used to keep the concrete from sticking to the forms. There are a large number of various compounds on the market that will reduce sticking and will at the same time serve as sealers or protective coatings to substantially reduce the absorption of water by the forms from the concrete.

Formwork for Water Tower Place, Chicago, Illinois (Courtesy Symons Corporation).

Various types of oil or oil compounds have been the most commonly used form-release agents used in the past for both wood and metal surfaces. There are many form-release and sealer agents available, but before he tries a new one on a large job, the contractor would be wise to experiment with a small application. No material should be used that will form a coating on the concrete which will, after stripping, interfere with the wetting of the concrete for curing purposes. In the same manner, if the concrete is later to be plastered or painted, the coating should not be one that will leave a waxy or oily surface that will interfere with the sticking of paint or plaster.

There are various kinds of products used to coat the plywood during its manufacture. Some manufacturers of these products say that the oiling of forms coated with their products is not necessary in the field. Glass-fiber-reinforced plastic forms, which are used a great deal for precast concrete and for architectural concrete because of the very excellent surfaces produced, need only be oiled very lightly before use. In fact, they have been satisfactorily used by some contractors with no oiling whatsoever.

16.7 DEFINITIONS

In this section several terms relating to formwork, with which you need to be familiar, are introduced. They are not listed alphabetically but rather are given in the order in which the author thinks they will help the reader understand the material in subsequent sections of this chapter.

Double-headed nails: A tremendous number of devices have been developed to simplify the erection and stripping of forms. For instance, there are the very simple double-headed nails. These nails can be easily removed, permitting the quick removal of braces and dismantling of forms.

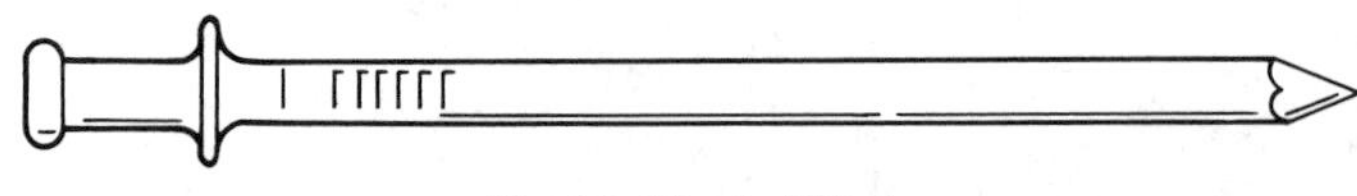

Double-Headed Nail

Sheathing: The material that forms the contact face of the forms to the concrete is called sheathing or lagging or sheeting. (See Figure 16.1.)

Joists: The usually horizontal beams that support the sheathing for floor and roof slabs are called joists. (See Figure 16.1.)

Stringers: The usually horizontal beams that support the joists and that rest on the vertical supports are called stringers. (See Figure 16.1.)

Shores: The temporary members (probably wood or metal) that are used for vertical support for formwork holding up fresh concrete are called shores. (See Figure 16.1.) Other names given to these members

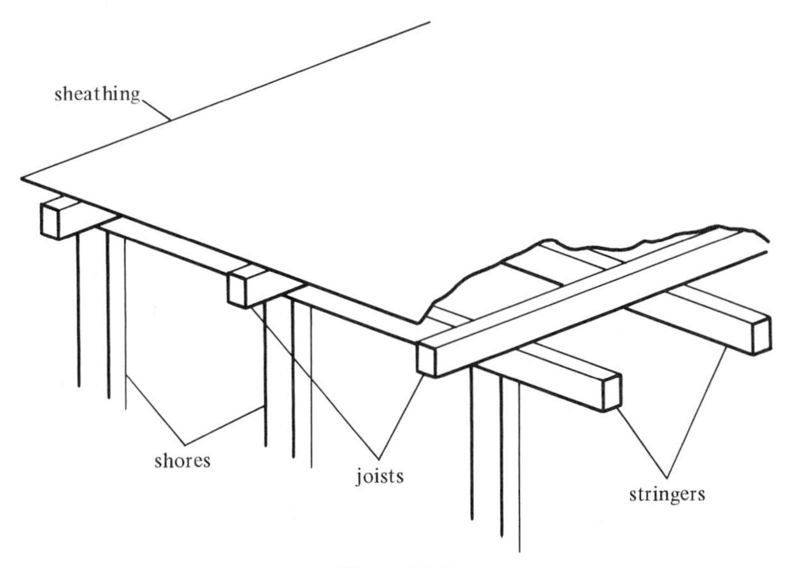

Figure 16.1

are struts, posts, or props. The whole system of vertical supports for a particular structure is called the shoring.

Ties are those devices used to support both sides of wall forms against the lateral pressure of the concrete. They pass through the concrete and are fastened on each side. (See Figure 16.2.) The lateral pressure caused by the fresh concrete, which tends to push out against the forms, is resisted in tension by the ties. They can, when correctly used, eliminate most of the external braces otherwise required for wall forms. They may or may not have a provision for holding the forms the desired distance apart.

Spreaders: Braces, usually made of wood, that are placed in a form to keep its two sides from being drawn together are called spreaders. They should be removed when the concrete is placed so they are not cast in the concrete.

Snap ties: Rather than using spreaders separate from the ties, it is more common to use a type of spreader called a snap tie which is combined with the spreader. The ends of these patented devices can be twisted or snapped off a little distance from the face of the concrete. After the forms are removed, the small holes resulting can easily be repaired with cement mortar.

Studs: The vertical members that support sheathing for wall forms are called studs. (See Figure 16.2.)

Wales: The long horizontal members that support the studs are called wales. (See Figure 16.2.) Notice that each wale usually consists of two timbers with the ties passing in between so that holes do not have to be drilled.

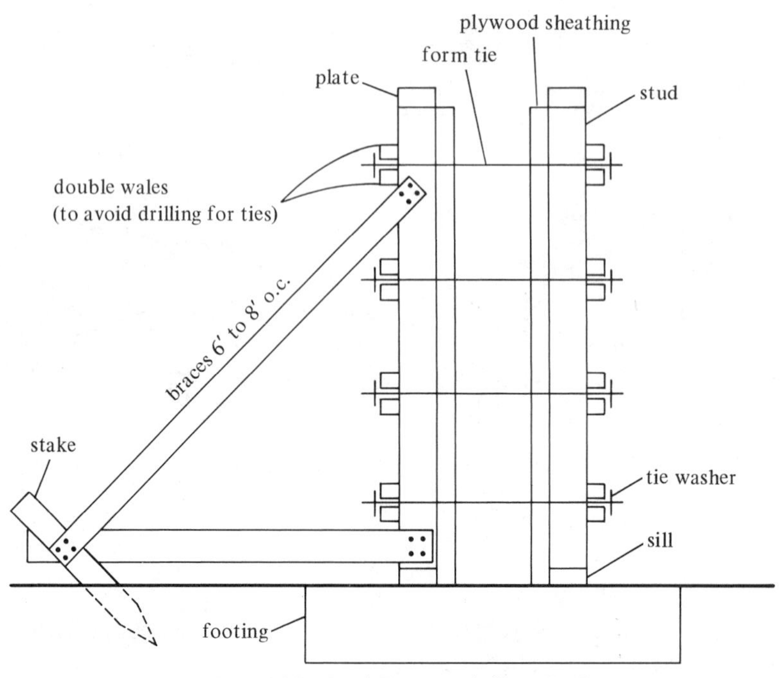

Figure 16.2 Wood Formwork for a Wall

16.8 FORCES APPLIED TO CONCRETE FORMS

Formwork must be designed to resist all vertical and lateral loads applied to it until the concrete gains sufficient strength to do the job itself. The vertical loads applied to the forms include the weights of the concrete, the reinforcing steel, and the forms themselves, as well as the construction live loads. The lateral loads include the liquid pressure of freshly placed concrete, wind forces, and the lateral forces caused by the movement of the construction equipment. These forces are discussed in the paragraphs to follow.

Vertical Loads

The vertical dead loads that must be supported by the formwork in addition to its own weight include the weight of the concrete and the reinforcing bars. The weight of ordinary concrete including the reinforcing is taken as 150 $\#/\text{ft}^3$. The weight of the formwork, which is frequently neglected in the design calculations, may vary from 3 or 4 psf up to 12 to 15 psf.

The vertical live load to be supported includes the weight of the workmen, the construction equipment, and the storage of materials on freshly hardened slabs. A minimum load of 50 psf of horizontal projection is recommended by ACI Committee 347.[1] This figure includes the weight of the workmen, equipment, runways, and impact. When powered concrete buggies are used, this value should be increased to a minimum of at least 75 psf. Furthermore, to make allowance for the storage of materials on freshly hardened slabs (a very likely circumstance), it is common in practice to design slab forms for vertical live loads as high as 150 psf.

Lateral Loads

For walls and columns the loads are different from those of slabs. The semiliquid concrete is placed in the form and exerts lateral pressures against the form as does any liquid. The amount of this pressure is dependent on the concrete weight per cubic foot, the rate of placing, and the method of placing.

The exact pressures applied to forms by freshly placed concrete are difficult to estimate due to several factors. These factors include the following:

1. The rate of placing the concrete. Obviously, the faster the concrete is placed, the higher the pressure will be.

2. The temperature of the concrete. The liquid pressure on the forms varies quite a bit with different temperatures. For instance, the pressure when the concrete is at 50°F is appreciably higher than when the concrete is at 70°F. The reason for this is that concrete placed at 50°F hardens at a slower rate than does concrete deposited at 70°F. As a result, the colder concrete remains in a semiliquid state for a longer time and thus to a greater depth, and pressures may be as much as 25% or more higher. Because of this fact forms for winter-placed concrete should be designed for much higher lateral pressures than those designed for summer-placed concrete.

3. The method of placing the concrete. If high-frequency vibration is used, the concrete is kept in a liquid state to a fairly high depth, and it acts very much like a liquid with a weight per cubic foot equal to that of the concrete. Vibration may increase lateral pressures by as much as 20% over pressures caused by spading.

4. Size and shape of the forms and the consistency and proportions of the concrete. These items do affect to some degree lateral pressures but for the usual building are felt to be negligible.

[1] ACI, Committee 301, 1972, *Specifications for Structural Concrete for Buildings* (ACI 301-72) (Detroit: American Concrete Institute), p. 156.

The pressures that occur in formwork due to the semiliquid concrete are often of such magnitude that other people do not believe it when the designer says the lateral pressure is 1,500 psf or 1,800 psf or more. ACI Committee 347 has published recommended formulas for calculating lateral concrete pressures for different temperatures and rates of placing the concrete. In these expressions, which follow, p is the maximum equivalent liquid pressure, in pounds per square foot, at any elevation in the form, R is the vertical rate of concrete placement, in feet per hour, T is the temperature of the concrete in the forms, and h is the maximum height of fresh concrete above the point being considered. These expressions are to be used for walls for which concrete with a slump no greater than 4 in. is placed at a rate not exceeding 7 ft/h. Vibration depth is limited to 4 ft below the concrete surface.[2]

In walls with *R* not exceeding 7 ft/h

$$p = 150 + \frac{9{,}000R}{T} \text{(maximum 2,000 psf or 150}h\text{, whichever is less)}$$

In walls with *R* greater than 7 ft/h but not greater than 10 ft/h

$$p = 150 + \frac{43{,}400}{T} + \frac{2{,}800R}{T} \text{(maximum 2,000 psf or 150}h\text{, whichever is less)}$$

Column forms are often filled very quickly. They may in fact be completely filled in less time than is required for the bottom concrete to set. Furthermore, vibration will frequently extend for the full depth of the form. As a result, greater lateral pressures are produced than would be expected for the wall conditions just considered. ACI Committee 347 presents the following equation for estimating the maximum pressures for column form design:

Columns with maximum horizontal dimension of 6 ft or less

$$p = 150 + \frac{9{,}000R}{T} \text{(maximum 3,000 psf or 150}h\text{, whichever is less)}$$

In Figure 16.3 a comparison of the pressures calculated with the preceding expressions is presented for temperatures ranging from 30°F to 100°F.

In addition to the lateral pressures on the forms caused by the fresh concrete, it is necessary for the braces and shoring to be designed to

[2] ACI Committee 301, 1972, *Specifications for Structural Concrete for Buildings* (ACI 301-72) (Detroit: American Concrete Institute), pp. 156–157.

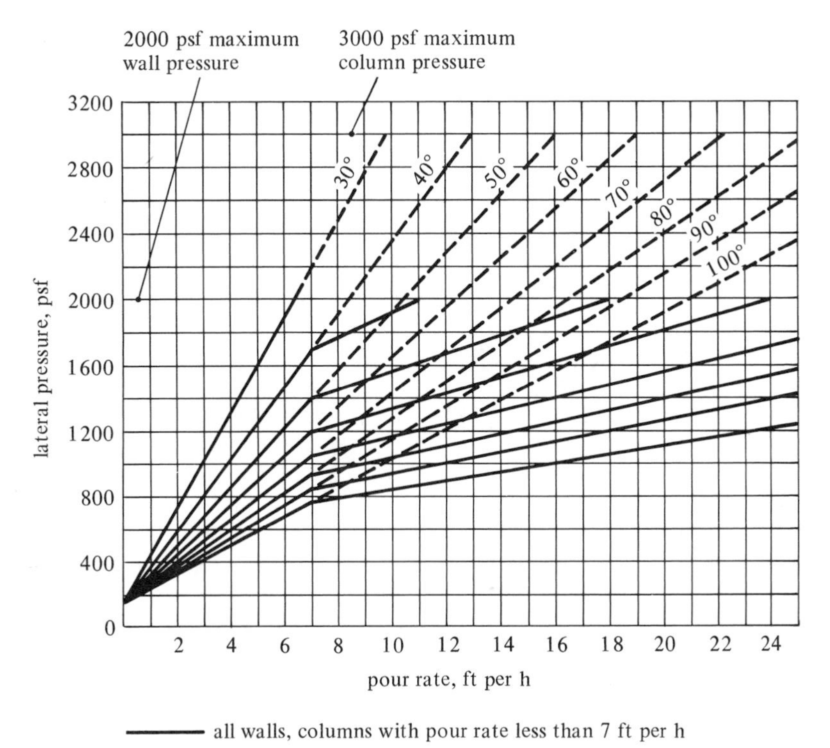

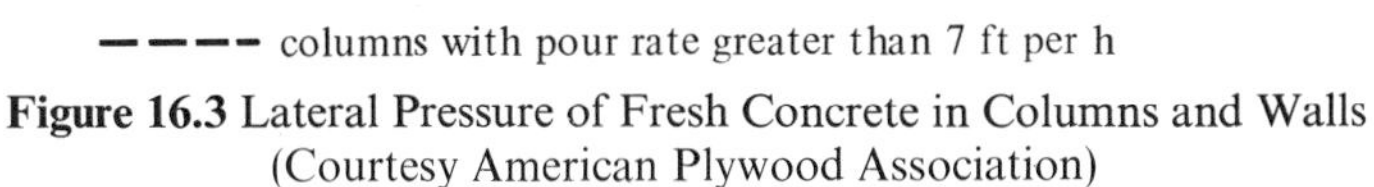

Figure 16.3 Lateral Pressure of Fresh Concrete in Columns and Walls (Courtesy American Plywood Association)

resist all other possible lateral forces, such as wind, dumping of concrete, movement of equipment, and other forces.

16.9 ANALYSIS OF FORMWORK FOR FLOOR AND ROOF SLABS

This section is devoted to the calculations needed for analyzing formwork for concrete floor and roof slabs; its design is presented in the following section. As wood formwork is the most common type faced in practice, the discussion and examples of this chapter are primarily concerned with wood.

Wood members have a very useful property with which you should be familiar and that is their ability to support excessive loads for short periods of time. As a result of this characteristic, it is a common practice to allow them a 25% increase over their normal allowable stresses if the applied loads are of short duration.

Formwork is usually considered to be a temporary type of structure because it remains in place only a short time. Furthermore, the loads supported by the forms reach a peak during pouring activity and then rapidly fall off as the concrete hardens around the reinforcing and begins to support the load. As a result, the formwork designer will want to use increased allowable stresses where possible. The ability of wood to take overloads is based on the total time of application of the loads. In other words if forms are to be used many times the amount of overload which they can resist is not as large as if they are to be used only one time.

In the paragraphs to follow some discussion is presented regarding flexure, shear, and deflections in formwork.

Flexure

Normally sheathing, joists, and stringers are continuous over several spans. For such cases it seems reasonable to assume that the maximum moment is equal to the maximum moment that would occur in a uniformly loaded span continuous over three or more spans. This value is approximately equal to

$$M = \frac{wl^2}{10}$$

Shear

The horizontal and vertical shear stresses at any one point in a beam are equal. For materials for which strengths are the same in every direction, no distinction is made between shear values acting in different directions. Materials such as wood, however, have entirely different shear strengths in the different directions. Wood tends to split or shear between its fibers (usually parallel to the beam axis). Since horizontal shear is rather critical for wood members, it is common in talking about wood formwork to use the term "horizontal shear."

The horizontal shearing stress in a rectangular wooden member can be calculated by the usual formula as follows:

$$f_v = \frac{VQ}{Ib} = \frac{(V)[b \times (h/2) \times (h/4)]}{(\frac{1}{12}bh^3)(b)} = \frac{3V}{2bh} = \frac{3V}{2A}$$

In this expression V is equal to $wl/2$ for uniformly loaded simple spans. As in earlier chapters relating to reinforced concrete, it is permissible to calculate the shearing stress at a distance h from the face of the support. If h is given in inches and if w is the uniform load per foot, the external shear at a distance h from the support can be calculated as follows:

$$V = 0.5wl - \frac{h}{12}w$$

$$V = 0.5w\left(l - \frac{2h}{12}\right)$$

The sheathing, joists, stringers, wales, and so on for formwork are normally continuous. For uniformly loaded beams continuous over three or more spans, V is approximately equal to $0.6wl$. At a distance h from the support, it is *assumed* to equal the following value:

$$V = 0.6w\left(l - \frac{2h}{12}\right)$$

The allowable shear stresses for construction grades of lumber are quite low and the designer is not really out of line if he increases the given allowable shear stresses by something more than 25% if the forms in question are to be used only once. Such increases are included in the allowable stresses given later in this chapter. It is not usually necessary to check the shearing stresses in sheathing where they are normally very low.

Deflections

Formwork must be designed to limit deflections to certain maximum values. If deflections are not controlled, the end results will be unsightly looking concrete, with bulges and perhaps cracks marring its appearance. The amounts of deflection permitted are dependent on the desired finish of the concrete as well as its location. For instance, where a rough finish is used, small deflections might not be very obvious, but where smooth surfaces are desired, small deflections are easily seen and are quite objectionable.

For some concrete forms deflection limitations of $l/240$ or $l/270$ are considered to be satisfactory, while for others, particularly for horizontal surfaces, a limitation of $l/360$ may be required. For some cases, particularly for architectural exposure, even tighter limitations may be required, perhaps $l/480$.

It should also be realized that perfectly straight beams and girders seem to the eye of a person below to sag downwards; consequently, just a little downward deflection seems to be very large and quite detrimental to the appearance of the structure. For this reason it is considered good practice to camber the forms of such members so they do not appear to be deflecting. A camber frequently used is about $\frac{1}{4}$ in. for each 10 ft of length.

For uniformly loaded equal-span beams continuous over three or more spans, the maximum center-line deflections can be approximately determined with the following expression:

$$\delta = \frac{wl^4}{145EI}$$

In using deflection expressions it is to be noted that the deflections that occur in the formwork are going to be permanent in the completed structure, and thus the modulus of elasticity E should not be increased even if the forms are to be used just one time. As a matter of fact, it may be necessary to reduce E rather than increase it for some parts of the formwork. When wood becomes wet, it becomes more flexible, with the result that larger deflections occur. You can see that wet formwork, particularly the sheathing, is the normal situation. As a result it is considered wise to use a smaller E than normally permitted and a reduction factor of $\frac{10}{11}$ is commonly specified.

Three tables useful for both the analysis and design of formwork are presented in this section. The first of these, Table 16.1, gives the properties of several sizes of lumber that are commonly used for formwork. These properties include the dressed dimensions, cross-sectional areas, moments of inertia, and section moduli.

Plywood is the standard material used in practice for floor and wall sheathing for formwork and it is so used for the examples presented in this chapter. Almost all exterior plywoods manufactured with waterproof glue can be satisfactorily used, but the industry produces a particular type called plyform intended especially for formwork. It is this product that is referred to in the remainder of this chapter.

Plyform is manufactured by gluing together an odd number of sheets of wood with the grain of each sheet being perpendicular to the grain of the sheet adjoining it. The alternating of the grains of the wood sheets greatly reduces the shrinkage and warping of the resulting panels. Plyform can be obtained in thicknesses from $\frac{1}{8}$ to $1\frac{1}{8}$ in. Actually, however, many suppliers only carry the $\frac{5}{8}$- and $\frac{3}{4}$-in. sizes and the others may have to be specially ordered.

The various types of wood used for manufacturing plywood (fir, spruce, larch, redwood, cedar, etc.) have different strengths and stiffnesses. Those types of woods having similar properties are assigned to a particular species group. To simplify design calculations the plywood industry has accounted for the different species and the effects of gluing the wood sheets together with alternating grains in developing the properties given in Table 16.2. As a result, the designer needs only to consider the allowable stresses for the plyform and the plyform properties given in the table.

Table 16.1 Properties of American Standard Board, Plank, Dimension, and Timber Sizes Commonly Used for Form Construction (Courtesy American Concrete Institute)

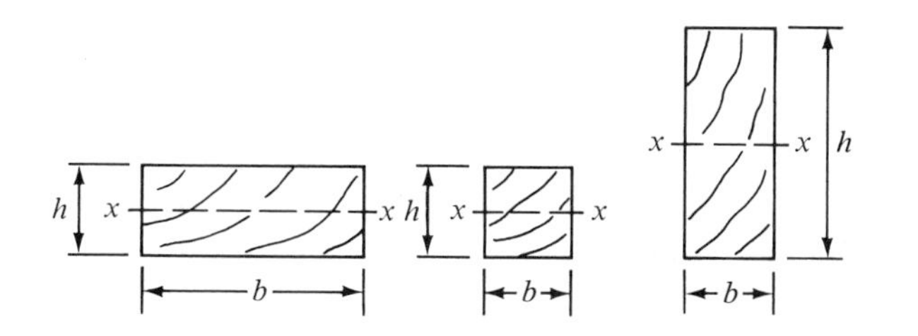

Nominal Size in Inches, bxh	American Standard Size in Inches, bxh S4S* 19% Maximum Moisture	Area of Section $A = bh$, sq in.		Moment of Inertia, in.^{4} $I = \frac{bh^3}{12}$		Section Modulus, in.^{3} $S = \frac{bh^2}{6}$		Board Feet per Linear Foot of Piece
		Rough	S4S	Rough	S4S	Rough	S4S	
4×1	$3\frac{1}{2} \times \frac{3}{4}$	3.17	2.62	0.20	0.12	0.46	0.33	$\frac{1}{3}$
6×1	$5\frac{1}{2} \times \frac{3}{4}$	4.92	4.12	0.31	0.19	0.72	0.52	$\frac{1}{2}$
8×1	$7\frac{1}{4} \times \frac{3}{4}$	6.45	5.44	0.41	0.25	0.94	0.68	$\frac{2}{3}$
10×1	$9\frac{1}{4} \times \frac{3}{4}$	8.20	6.94	0.52	0.32	1.20	0.87	$\frac{5}{6}$
12×1	$11\frac{1}{4} \times \frac{3}{4}$	9.95	8.44	0.63	0.39	1.45	1.05	1
$4 \times 1\frac{1}{4}$	$3\frac{1}{2} \times 1$	4.08	3.50	0.43	0.29	0.76	0.58	$\frac{5}{12}$
$6 \times 1\frac{1}{4}$	$5\frac{1}{2} \times 1$	6.33	5.50	0.68	0.46	1.19	0.92	$\frac{5}{8}$
$8 \times 1\frac{1}{4}$	$7\frac{1}{4} \times 1$	8.30	7.25	0.87	0.60	1.56	1.21	$\frac{5}{6}$
$10 \times 1\frac{1}{4}$	$9\frac{1}{4} \times 1$	10.55	9.25	1.11	0.77	1.98	1.54	$1\frac{1}{24}$
$12 \times 1\frac{1}{4}$	$11\frac{1}{4} \times 1$	12.80	11.25	1.35	0.94	2.40	1.87	$1\frac{1}{4}$
$4 \times 1\frac{1}{2}$	$3\frac{1}{2} \times 1\frac{1}{4}$	4.98	4.37	0.78	0.57	1.14	0.91	$\frac{1}{2}$
$6 \times 1\frac{1}{2}$	$5\frac{1}{2} \times 1\frac{1}{4}$	7.73	6.87	1.22	0.89	1.77	1.43	$\frac{3}{4}$
$8 \times 1\frac{1}{2}$	$7\frac{1}{4} \times 1\frac{1}{4}$	10.14	9.06	1.60	1.18	2.32	1.89	1
$10 \times 1\frac{1}{2}$	$9\frac{1}{4} \times 1\frac{1}{4}$	12.89	11.56	2.03	1.50	2.95	2.41	$1\frac{1}{4}$
$12 \times 1\frac{1}{2}$	$11\frac{1}{4} \times 1\frac{1}{4}$	15.64	14.06	2.46	1.83	3.58	2.93	$1\frac{1}{2}$
4×2	$3\frac{1}{2} \times 1\frac{1}{2}$	5.89	5.25	1.30	0.98	1.60	1.31	$\frac{2}{3}$
6×2	$5\frac{1}{2} \times 1\frac{1}{2}$	9.14	8.25	2.01	1.55	2.48	2.06	1
8×2	$7\frac{1}{4} \times 1\frac{1}{2}$	11.98	10.87	2.64	2.04	3.25	2.72	$1\frac{1}{3}$
10×2	$9\frac{1}{4} \times 1\frac{1}{2}$	15.23	13.87	3.35	2.60	4.13	3.47	$1\frac{2}{3}$
12×2	$11\frac{1}{4} \times 1\frac{1}{2}$	18.48	16.87	4.07	3.16	5.01	4.21	2

Table 16.1 *(cont.)*

Nominal Size in Inches, bxh	American Standard Size in Inches, bxh S4S* 19% Maximum Moisture	Area of Section $A = bh$, sq in.		Moment of Inertia, in.^{4} $I = \frac{bh^3}{12}$		Section Modulus, in.^{3} $S = \frac{bh^2}{6}$		Board Feet per Linear Foot of Piece
		Rough	S4S	Rough	S4S	Rough	S4S	
2 × 4	$1\frac{1}{2} \times 3\frac{1}{2}$	5.89	5.25	6.45	5.36	3.56	3.06	$\frac{2}{3}$
2 × 6	$1\frac{1}{2} \times 5\frac{1}{2}$	9.14	8.25	24.10	20.80	8.57	7.56	1
2 × 8	$1\frac{1}{2} \times 7\frac{1}{4}$	11.98	10.87	54.32	47.63	14.73	13.14	$1\frac{1}{3}$
2 × 10	$1\frac{1}{2} \times 9\frac{1}{4}$	15.23	13.87	111.58	98.93	23.80	21.39	$1\frac{2}{3}$
2 × 12	$1\frac{1}{2} \times 11\frac{1}{4}$	18.48	16.87	199.31	177.97	35.04	31.64	2
3 × 4	$2\frac{1}{2} \times 3\frac{1}{2}$	9.52	8.75	10.42	8.93	5.75	5.10	1
3 × 6	$2\frac{1}{2} \times 5\frac{1}{2}$	14.77	13.75	38.93	34.66	13.84	12.60	$1\frac{1}{2}$
3 × 8	$2\frac{1}{2} \times 7\frac{1}{4}$	19.36	18.12	87.74	79.39	23.80	21.90	2
3 × 10	$2\frac{1}{2} \times 9\frac{1}{4}$	24.61	23.12	180.24	164.89	38.45	35.65	$2\frac{1}{2}$
3 × 12	$2\frac{1}{2} \times 11\frac{1}{4}$	29.86	28.12	321.96	296.63	56.61	52.73	3
4 × 4	$3\frac{1}{2} \times 3\frac{1}{2}$	13.14	12.25	14.39	12.50	7.94	7.15	$1\frac{1}{3}$
4 × 6	$3\frac{1}{2} \times 5\frac{1}{2}$	20.39	19.25	53.76	48.53	19.12	17.65	2
4 × 8	$3\frac{1}{2} \times 7\frac{1}{4}$	26.73	25.38	121.17	111.15	32.86	30.66	$2\frac{2}{3}$
4 × 10	$3\frac{1}{2} \times 9\frac{1}{4}$	33.98	32.38	248.91	230.84	53.10	49.91	$3\frac{1}{3}$
6 × 3	$5\frac{1}{2} \times 2\frac{1}{2}$	14.77	13.75	8.48	7.16	6.46	5.73	$1\frac{1}{2}$
6 × 4	$5\frac{1}{2} \times 3\frac{1}{2}$	20.39	19.25	22.33	19.65	12.32	11.23	2
6 × 6	$5\frac{1}{2} \times 5\frac{1}{2}$	31.64	30.25	83.43	76.26	29.66	27.73	3
6 × 8	$5\frac{1}{2} \times 7\frac{1}{2}$	42.89	41.25	207.81	193.36	54.51	51.56	4
8 × 8	$7\frac{1}{2} \times 7\frac{1}{2}$	58.14	56.25	281.69	263.67	73.89	70.31	$5\frac{1}{3}$

* Rough dry sizes are $\frac{1}{8}$ in. larger, both dimensions.

It will be noted that section properties are given in Table 16.2 for stresses applied both parallel to the face grain and perpendicular to it. Should the face grain be placed parallel to the direction of bending, the section is stronger; it is weaker if placed in the opposite direction. This situation is shown in Figure 16.4.

In plyform the shearing stresses between the plys or along the glue lines are calculated with the usual expression $H = f_v = VQ/Ib$ and are referred to as the *rolling shear stresses*. For convenience in making the calculations, the values of Ib/Q are presented in the table for the different plyform sizes and H can be simply determined for a particular case by dividing V by the appropriate Ib/Q value from the table.

Table 16.2 Section Properties and Allowable Stresses for Certain Sizes of Plyform (Courtesy American Plywood Association)

Thickness (inches)	Approx. Weight (psf)	Properties for Stress Applied Parallel with Face Grain			Properties for Stress Applied Perpendicular to Face Grain		
		Moment of inertia I (in.^{4}/ft.)	Effective section modulus KS (in.^{3}/ft.)	Rolling shear constant Ib/Q (in.^{2}/ft.)	Moment of inertia I (in.^{4}/ft.)	Effective section modulus KS (in.^{3}/ft.)	Rolling shear constant Ib/Q (in.^{2}/ft.)
CLASS I							
$\frac{1}{2}$	1.5	0.077	0.268	5.127	0.035	0.167	2.919
$\frac{5}{8}$	1.8	0.130	0.358	6.427	0.064	0.250	3.692
$\frac{3}{4}$	2.2	0.199	0.455	7.854	0.136	0.415	4.565
$\frac{7}{8}$	2.6	0.280	0.553	8.204	0.230	0.581	5.418
1	3.0	0.427	0.737	8.871	0.373	0.798	7.242
$1\frac{1}{8}$	3.3	0.554	0.849	9.872	0.530	0.986	8.566
CLASS II							
$\frac{1}{2}$	1.5	0.075	0.267	4.868	0.029	0.182	2.671
$\frac{5}{8}$	1.8	0.130	0.357	6.463	0.053	0.320	3.890
$\frac{3}{4}$	2.2	0.198	0.455	7.892	0.111	0.530	4.814
$\frac{7}{8}$	2.6	0.280	0.553	8.031	0.186	0.742	5.716
1	3.0	0.421	0.754	8.614	0.301	1.020	7.645
$1\frac{1}{8}$	3.3	0.566	0.869	9.571	0.429	1.260	9.032
STRUCTURAL I							
$\frac{1}{2}$	1.5	0.078	0.271	5.166	0.042	0.229	3.076
$\frac{5}{8}$	1.8	0.131	0.361	6.526	0.077	0.343	3.887
$\frac{3}{4}$	2.2	0.202	0.464	7.926	0.162	0.570	4.812
$\frac{7}{8}$	2.6	0.288	0.569	7.539	0.275	0.798	6.242
1	3.0	0.479	0.827	7.978	0.445	1.098	7.639
$1\frac{1}{8}$	3.3	0.623	0.955	8.841	0.634	1.356	9.031

All Properties Adjusted to Account for Reduced Effectiveness of Plies with Grain Perpendicular to Applied Stress

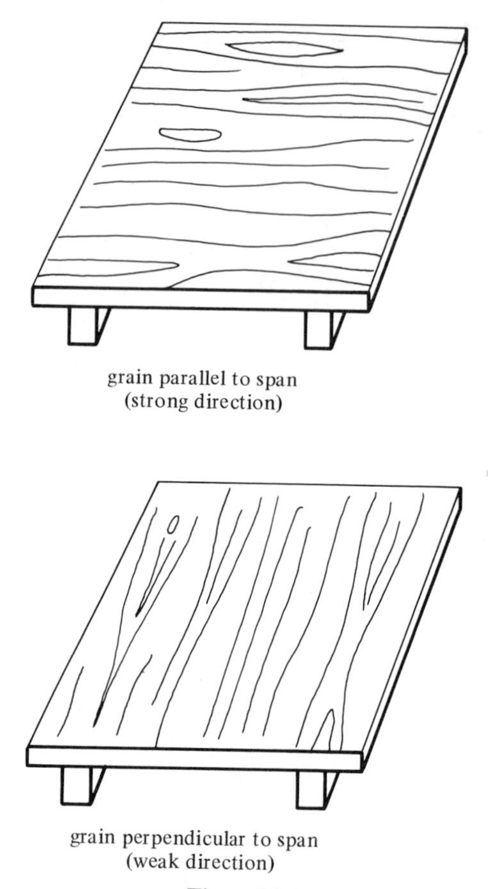

grain parallel to span
(strong direction)

grain perpendicular to span
(weak direction)

Figure 16.4

The normal framing applications of plyform are for wet conditions, and adjustments in allowable stresses should be made for those conditions as well as for duration of load and other experience factors. As a result of these items, the following allowable stresses are normally given for plyform:

	Plyform Class I	Plyform Class II	Structural I Plyform
modulus of elasticity (psi)	1,650,000	1,430,000	1,650,000
bending stress (psi)	1,930	1,330	1,930
rolling shear stress (psi)	80	72	102

Table 16.3 shows allowable bending stresses, allowable compression stresses perpendicular and parallel to the grain, allowable shearing stresses, and moduli of elasticity for several types of lumber often used in formwork.

Table 16.3 Suggested Working Stresses for Design of Wood Formwork. Derived from "National Design Specification for Stress-Grade Lumber and Its Fastenings," 1960 Edition, as Amended, and from Recommendations of the Douglas Fir Plywood Association

	Class I Formwork, Single Use or Light Construction					Class II Formwork, Multiple Use or Heavy Construction				
	Allowable unit stress, psi				Modulus of elasticity, psi	Allowable unit stress, psi				Modulus of elasticity, psi
Species and Grade of Form Framing Lumber	Extreme fiber bending	Compression ⊥ to grain	Compression ‖ to grain	Horizontal shear		Extreme fiber bending	Compression ⊥ to grain	Compression ‖ to grain	Horizontal shear	
Douglas fir, coastal, construction	1,875	490	1,500	180	1,760,000	1,500	390	1,200	145	1,760,000
Douglas fir, inland, common structural	1,815	475	1,565	180	1,760,000	1,450	380	1,250	145	1,760,000
Hemlock, west coast, construction	1,875	455	1,380	150	1,540,000	1,500	365	1,100	120	1,540,000
Larch, common structural	1,815	490	1,660	180	1,650,000	1,450	390	1,325	145	1,650,000
Pine, southern No. 1 SR	1,875	490	1,625–1,875	220	1,760,000	1,500	390	1,300–1,500	175	1,760,000
Redwood, heart structural	1,625	400	1,380	135	1,320,000	1,300	320	1,100	110	1,320,000
Spruce, eastern, 1450 f structural	1,815	375	1,310	160	1,320,000	1,450	300	1,050	130	1,320,000

Table 16.3 (*cont.*)

SHEATHING wet or damp location assumed

Species and Grade of Form Framing Lumber	Class I Formwork, Single Use or Light Construction					Class II Formwork, Multiple Use or Heavy Construction				
	Allowable unit stress, psi				Modulus of elasticity, psi	Allowable unit stress, psi				Modulus of elasticity, psi
	Extreme fiber bending	Compression ⊥ to grain	Compression ‖ to grain	Horizontal shear		Extreme fiber bending	Compression ⊥ to grain	Compression ‖ to grain	Horizontal shear	
Plywood, Douglas fir, concrete form B-B	2000	(Bearing on face) 435	1500	*	1,600,000	1500	(Bearing on face) 325	1100	*	1,600,000
Board sheathing, all species	Use 100% of value for species, shown above	Use 67% of value for species, shown above	Use 90% of value for species, shown above	Use 100% of value for species, shown above	Use $\frac{10}{11}$ of value for species, shown above	Use 100% of value for species, shown above	Use 67% of value for species, shown above	Use 90% of value for species, shown above	Use 100% of value for species, shown above	Use $\frac{10}{11}$ of value for species, shown above

* Shear is not a governing design consideration for plywood form panels except in the case of very short, heavily loaded spans. If 1-in. or $1\frac{1}{8}$-in. panels are being used, check for rolling shear according to plywood manufacturer's suggestions.

Example 16.1 illustrates the calculations necessary to determine the bending and shear stresses and the maximum deflections that occur in the sheathing, joists, and stringers used as the formwork for a certain concrete slab and live loading. In an actual design problem, it may be necessary to consider several different arrangements of the joists, stringers, and so on and the different grades of lumber available with their different allowable stresses before a final design is selected.

EXAMPLE 16.1

The formwork for a 6-in. thick reinforced concrete floor slab of normal weight consists of $\frac{3}{4}$-in. class II plyform sheathing supported by 2 × 6-in. joists spaced 2 ft 0 in. on center, which in turn are supported by 2 × 8-in. stringers spaced 4 ft 0 in. on centers. The stringers are themselves supported by shores spaced 4 ft 0 in. on centers.

The joists and stringers are constructed with Douglas fir, coastal construction, and are to be used many times. The allowable flexural stress in the fir members is 1,875 psi, the allowable horizontal shear stress is 180 psi, and the modulus of elasticity is 1,760,000 psi. For the plyform the corresponding values are 1,330 psi, 72 psi, and 1,430,000 psi, with the stress applied parallel with the face grain.

Using a live load of 150 psf, check the bending and shear stresses in the sheathing, joists, and stringers. Check deflections assuming the maximum permissible deflection of each of these elements is $l/360$ for all loads.

SOLUTION

Sheathing

Properties of 12-in.-wide piece of $\frac{3}{4}$ plyform from Table 16.2:

$$S = 0.455 \text{ in.}^3,\ I = 0.198 \text{ in.}^4,\ \frac{Ib}{Q} = 7.892 \text{ in.}^2$$

Loads (neglecting weight of forms):

$$\begin{aligned} \text{concrete} &= (\tfrac{6}{12})(150) = 75 \text{ psf} \\ LL &= 150 \\ \text{total load} &= 225 \text{ psf} \end{aligned}$$

Bending stresses (with reference made to Figure 16.5):

$$M = \frac{wl^2}{10} = \frac{(225)(2)^2}{10} = 90 \text{ ft-lb}$$

$$f_b = \frac{(12)(90)}{0.445} = 2{,}373 \text{ psi} > 1{,}330 \text{ psi} \qquad \underline{\text{no good}}$$

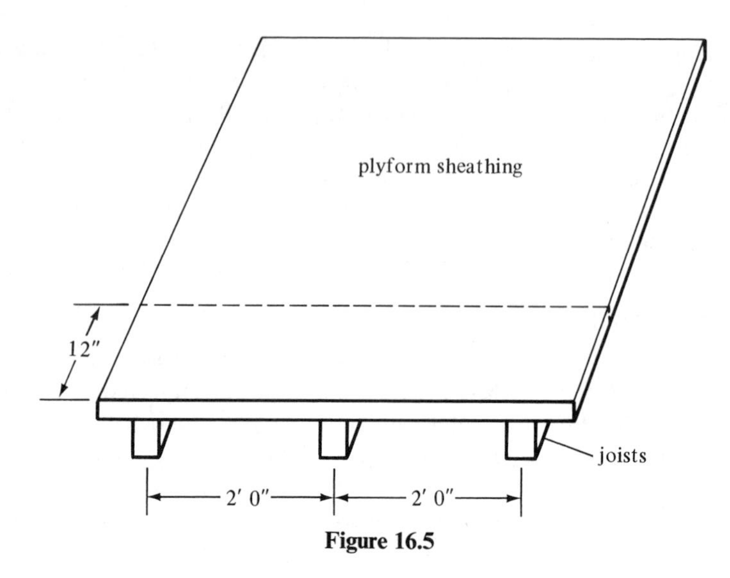

Figure 16.5

Shear stresses (not usually checked in sheathing):

$$V = 0.6w\left(l - \frac{2h}{12}\right) = (0.6)(225)\left(2 - \frac{2 \times \frac{3}{4}}{12}\right) = 253\ \#$$

$$f_v = \frac{VQ}{Ib} = \frac{V}{Ib/Q} = \frac{253}{7.892} = 32 \text{ psi} < 72 \text{ psi} \qquad \underline{\text{ok}}$$

Deflection:

$$\text{permissible deflection} = \left(\tfrac{1}{360}\right)(12 \times 2) = 0.067''$$

$$\text{actual } \delta = \frac{wl^4}{125EI} = \frac{\left(\frac{225}{12}\right)(12 \times 2)^4}{(145)(1.43 \times 10^6)(0.198)} = 0.0064 < 0.067'' \qquad \underline{\text{ok}}$$

Joists

Properties of 2 × 6 joist with finished dimensions $1\frac{1}{2} \times 5\frac{1}{2}$:

$$A = 8.250 \text{ in.}^2,\ I = 20.80 \text{ in.}^4,\ S = 7.56 \text{ in.}^3$$

$$\text{load/ft} = (2)(225) = 450\ \#/\text{ft}$$

Bending and shear stresses:

$$M = \frac{wl^2}{10} = \frac{(450)(4)^2}{10} = 720 \text{ ft-lb}$$

$$f_b = \frac{(12)(720)}{7.56} = 1{,}143 \text{ psi} < 1{,}875 \text{ psi} \qquad \underline{\text{ok}}$$

$$V = 0.6w\left(l - \frac{2h}{12}\right) = (0.6)(450)\left(4 - \frac{2 \times 5.50}{12}\right) = 832.5\ \#$$

$$f_v = \frac{3V}{2A} = \frac{(3)(832.5)}{(2)(8.25)} = 151 \text{ psi} < 180 \text{ psi} \qquad \underline{\text{ok}}$$

Deflection:

$$\text{permissible deflection} = (\tfrac{1}{360})(12 \times 4) = 0.133''$$

$$\text{actual } \delta = \frac{wl^4}{145EI} = \frac{(\frac{450}{12})(12 \times 4)^4}{(145)(1.76 \times 10^6)(20.80)} = 0.0375 < 0.133'' \qquad \underline{\text{ok}}$$

Stringers

Properties of 2 × 8 stringer with finished dimensions $1\frac{1}{2} \times 7\frac{1}{4}$

$$A = 10.87 \text{ in.}^2,\ I = 47.63 \text{ in.}^4,\ S = 13.14 \text{ in.}^3$$

$$\text{load/ft} = (4)(225) = 900\ \#/\text{ft}$$

Bending and shear stresses:

$$M = \frac{wl^2}{10} = \frac{(900)(4)^2}{10} = 1{,}440 \text{ ft-lb}$$

$$f_b = \frac{(12)(1{,}440)}{13.14} = 1{,}315 \text{ psi} < 1{,}875 \text{ psi} \qquad \underline{\text{ok}}$$

$$V = 0.6w\left(l - \frac{2h}{12}\right) = (0.6)(900)\left(4 - \frac{2 \times 7.25}{12}\right) = 1{,}507\ \#$$

$$f_v = \frac{(3)(1{,}507)}{(2)(10.87)} = 208 > 180 \text{ psi} \qquad \underline{\text{no good}}$$

Deflection:

$$\text{permissible deflection} = (\tfrac{1}{360})(12 \times 4) = 0.133''$$

$$\text{actual } \delta = \frac{(\frac{900}{12})(12 \times 4)^4}{(145)(1.76 \times 10^6)(47.63)} = 0.0328'' < 0.133'' \qquad \underline{\text{ok}}$$

16.10 DESIGN OF FORMWORK FOR FLOOR AND ROOF SLABS

From the same principles used for calculating the flexural and shear stresses and the deflections in the preceding section, it is possible to calculate maximum permissible spans for certain sizes of sheathing, joists, or stringers. As an illustration it is assumed that a rectangular section is used to support a total uniform load of w lb/ft over three or more continuous equal spans. The term l used in the equations to follow is the span, center to center, of supports in inches, while f is the allowable flexural stress.

Moment

The bending moment can be calculated in inch pounds and equated to the resisting moment also in inch pounds. The resulting expression can be solved for l, the maximum permissible span, in inches.

$$M = \frac{wl^2}{10} \text{ in foot pounds} = \frac{wl^2}{120} \text{ in inch pounds}$$

$$M_{\text{res}} = fS \text{ in inch pounds}$$

Equating and solving for l,

$$\frac{wl^2}{120} = fS$$

$$l = 10.95\sqrt{\frac{fS}{w}}$$

Shear

In the same fashion, an expression can be written for the maximum permissible span from the standpoint of horizontal shear. In the expression to follow, H is the allowable shearing stress in pounds per square inch while V is the maximum external shear applied to the member at a distance h from the support previously assumed to equal $0.6w[l - (2h/12)]$.

$$H = \frac{3V}{2A} = \frac{(3)(0.6w)[l - (2h/12)]}{2bh}$$

The value of l used in the preceding expression was given in feet; therefore, when the expression is solved for l, it must be multiplied by 12 to give an answer in inches to coincide with the units of the moment and deflection expressions.

$$l = 12\left(\frac{Hbh}{0.9w} + \frac{2h}{12}\right)$$

Deflection

If the maximum center-line deflection is equated to the maximum permissible deflection, the resulting equation can be solved for l. For this case the maximum permissible deflection is assumed to equal $l/360$.

$$\delta = \frac{wl^4}{145EI} = \frac{(w/12)(l^4)}{145EI} = \frac{l}{360}$$

$$l = 1.69 \sqrt[3]{\frac{EI}{w}}$$

EXAMPLE 16.2

It is desired to support a total uniform load of 200 psf for a floor slab with $\frac{3}{4}$-in. Class II plyform. If the forms are to be reused many times and the maximum permissible deflection is $l/360$, determine the maximum permissible span of the sheathing, center to center, of the joists if $f =$ 1,330 psi and $E =$ 1,430,000 psi. Neglect shear in the sheathing. Make similar calculations but consider shear values for 2 × 4-in. joists spaced 2 ft 0 in. on center if $f =$ 1,500 psi, $H =$ 180 psi, and $E =$ 1,700,000 psi.

SOLUTION

Sheathing

Assuming a 12-in. wide piece of plywood:

$$S = 0.455 \text{ in.}^3, I = 0.198 \text{ in.}^4$$

Moment:

$$l = 10.95 \sqrt{\frac{fs}{w}} = 10.95 \sqrt{\frac{(1{,}330)(0.455)}{200}} = 19.05''$$

Deflection:

$$l = 1.69 \sqrt[3]{\frac{EI}{w}} = 1.69 \sqrt[3]{\frac{(1{,}430{,}000)(0.198)}{200}} = \underline{18.98''}$$

Joists

Properties of 2 × 4 joists with finished dimensions $1\frac{1}{2} \times 3\frac{1}{2}$:

$$A = 5.250 \text{ in.}^2, I = 5.36 \text{ in.}^4, S = 3.06 \text{ in.}^3$$

load per foot if joists spaced 1′ 6″ on center = (1.5)(200) = 300 #/ft

Moment:

$$l = 10.95 \sqrt{\frac{(1{,}500)(3.063)}{300}} = \underline{42.85''}$$

Table 16.4 Safe Spacing, in., of Supports for Joists, Studs (or Other Beam Components of Formwork), Continuous over Three or More Spans (Courtesy American Concrete Institute)

$f = 1{,}875$ psi $E = 1{,}700{,}000$ psi $H = 225$ psi

Uniform Load, lb per Lineal ft (Equals Uniform Load on Forms Times Spacing between Joists or Studs, ft)	Nominal size of S4S lumber																		
	2 × 4	2 × 6	2 × 8	2 × 10	2 × 12	3 × 4	3 × 6	3 × 8	3 × 10	4 × 2	4 × 4	4 × 6	4 × 8	6 × 2	6 × 4	6 × 6	6 × 8	8 × 2	8 × 8	10 × 2
100	76	111	137	164	190	90	126	156	187	43	98	138	169	50	110	154	195	55	210	60
200	59	92	115	138	160	72	106	131	157	34	80	116	142	40	92	130	164	44	177	47
300	48	75	99	125	145	62	96	118	142	30	70	105	129	35	81	117	148	38	160	41
400	41	65	86	110	133	54	84	110	132	27	63	97	120	32	74	109	138	35	149	38
500	37	58	77	98	119	48	75	99	125	24	57	89	113	29	69	103	130	32	141	35
600	33	52	69	88	107	44	69	91	116	22	52	81	107	28	65	98	124	30	134	33
700	29	46	61	78	95	40	64	84	107	20	48	75	99	26	60	94	120	29	129	31
800	27	42	55	70	86	38	59	78	100	19	45	70	93	24	56	88	116	28	125	30
900	24	38	51	65	79	36	56	74	94	18	42	66	87	23	53	83	112	26	121	29
1,000	23	36	47	60	73	33	52	69	88	17	40	63	83	21	50	79	109	25	118	28
1,100	21	33	44	56	68	31	48	64	82	16	38	60	79	20	48	75	103	24	115	27
1,200	20	32	42	53	65	29	45	60	76	16	37	57	76	20	46	72	99	23	113	25
1,300	19	30	40	50	61	27	43	56	72	15	35	55	73	19	44	69	94	22	110	24
1,400	18	29	38	48	59	26	40	53	68	14	33	52	69	18	42	67	91	21	106	24
1,500	17	27	36	46	56	24	38	51	65	13	31	49	65	18	41	64	88	20	103	23

←Spacing→←Spacing→←Spacing→

1,600	17	26	35	44	54	23	37	48	62	13	30	47	62	17	40	62	85	19	100	22
1,700	16	26	34	43	52	22	35	46	59	12	29	45	59	16	38	61	83	19	97	21
1,800	16	25	33	42	51	22	34	45	57	12	27	43	57	16	37	59	80	18	94	21
1,900	15	24	32	40	49	21	33	43	55	11	26	41	55	16	36	57	78	18	91	20
2,000	15	23	31	39	48	20	32	42	53	11	25	40	53	15	35	56	76	17	89	20
2,100	14	23	30	38	47	19	31	40	51	10	24	38	51	15	34	54	74	17	87	19
2,200	14	22	29	37	45	19	30	39	50	10	24	37	49	14	33	52	71	17	85	19
2,300	14	22	29	37	44	18	29	38	49	10	23	36	48	14	32	50	69	16	83	18
2,400	14	21	28	36	44	18	28	37	47	10	22	35	46	13	31	49	66	16	81	18
2,500	13	21	27	35	43	17	27	36	46	9	22	34	45	13	30	47	63	16	80	18
2,600	13	20	27	34	42	17	27	35	45	9	21	33	44	12	29	46	61	15	78	17
2,700	13	20	27	34	41	17	26	35	44	9	20	32	43	12	28	45	60	15	77	17
2,800	13	20	26	33	41	16	26	34	43	9	20	32	42	12	28	43	59	15	75	17
2,900	12	19	26	32	40	16	25	33	42	8	20	31	41	11	27	42	58	14	73	16
3,000	12	19	25	32	39	16	25	32	41	8	19	30	40	11	26	41	56	14	71	16
3,200	12	19	25	31	38	15	24	31	40	8	18	29	38	11	25	39	54	13	68	16
3,400	12	18	24	31	37	15	23	30	39	8	18	28	37	10	24	38	51	13	65	15
3,600	11	18	24	30	37	14	22	30	38	7	17	27	36	10	23	36	49	12	62	15
3,800	11	17	23	30	36	14	22	29	37	7	17	26	34	9	22	35	48	12	59	14
4,000	11	17	23	29	35	14	21	28	36	7	16	25	33	9	21	34	46	11	57	13
4,500	10	16	22	28	34	13	20	27	34	6	15	24	31	8	20	31	43	10	53	12
5,000	10	16	21	27	33	12	19	25	32	6	14	22	30	8	18	29	40	9	49	11

NOTE: Span values above the solid line are governed by deflection. Values within dashed line box are spans governed by shear.
$\Delta_{max} = 1/360$, but not to exceed $\frac{1}{4}$ in.

Shear:

$$l = (12)\left(\frac{Hbh}{0.9w} + \frac{2h}{12}\right) = (12)\left(\frac{180 \times 1.50 \times 3.50}{0.9 \times 300} + \frac{2 \times 3.50}{12}\right) = 49.00''$$

Deflection:

$$l = 1.69\sqrt[3]{\frac{1{,}760{,}000 \times 5.36}{300}} = 53.34''$$

For practical design work many tables are available to simplify and expedite the design calculations. Table 16.4, which shows the maximum permissible spacings, center to center, of supports for joists, stringers, and other beams continuous over four or more supports for one particular grade of lumber, is one such example. From this table the maximum permissible spacing of the joists of Example 16.2 can be found to be 43 in., which corresponds with the 42.85 in. obtained in the example.

16.11 DESIGN OF SHORING

Wood shores are usually designed as simply supported columns using a modified form of the Euler equation. If the Euler equation is divided by a factor of safety of 3, and if r is replaced with 0.3 times d (the least lateral dimension of square or rectangular shores), as follows;

$$\frac{P}{A} = \frac{\pi^2 E}{3(l/0.3d)^2}$$

the so-called National Forest Products Association formula results:

$$\frac{P}{A} = \frac{0.3E}{(l/d)^2}$$

The maximum l/d value normally specified is 50. If a round shore is being used, it may be replaced for calculation purposes with a square shore having the same cross-sectional area. Should a shore be braced at different points laterally so that it has different unsupported lengths along its different faces, it will be necessary to calculate the l/d ratios in each direction and use the largest one to determine the allowable stress.

The allowable stress used may not be greater than the value obtained from this equation nor the allowable unit stress in compression parallel to the grain for the grade and type of lumber in question. If the formwork is to be used only once, the allowable stress in the column can reasonably be increased by 25%. Example 16.3 shows the calculation of the permissible column load for a particular shore. Tables are readily available for making these calculations for shores as they are for sheathing and beam members. For example, Table 16.5 gives the allowable column loads as determined here for a set of simple shores.

Table 16.5 Allowable Load in Pounds on Simple Wood Shores,* for Lumber of the Indicated Strength, Based on Unsupported Length (Courtesy American Concrete Institute)

c ‖ to grain = 1,150 psi $E = 1,400,000$ psi $\frac{l}{d_{max}} = 50$ $\frac{P}{A_{max}} = \frac{0.30E}{(l/d)^2}$

Nominal Lumber Size, in.	2 × 4		3 × 4		4 × 4		4 × 2		4 × 3		4 × 6		6 × 6	
	R**	S4S**	R	S4S	R	S4S	R	S4S	R	S4S	R	S4S	R	S4S
Unsupported length, ft							Bracing needed†							
4	2,800	2,200	10,900	10,000	15,100	15,000	6,800	6,000	10,900	10,100	23,400	22,100	36,400	34,800
5	1,800	1,400	7,700	6,400	15,100	15,000	6,800	6,000	10,900	10,100	23,400	22,100	36,400	34,800
6	1,300	1,000	5,300	4,400	14,000	12,200	6,300	5,200	10,100	8,700	21,700	19,100	36,400	34,800
7			3,900	3,300	10,300	8,900	4,600	3,800	7,400	6,400	15,900	14,000	36,400	34,800
8			3,000	2,500	7,870	6,800	3,500	2,900	5,700	4,900	12,200	10,700	36,400	34,800
9			2,400	2,000	6,200	5,400	2,800	2,300	4,500	3,900	9,700	8,500	36,200	32,900
10			1,900	1,600	5,000	4,400	2,300	1,900	3,600	3,100	7,800	6,900	29,200	26,700
11			—	—	4,200	3,600	1,900	1,500	3,000	2,600	6,500	5,700	24,100	22,100
12					3,500	3,000	1,600	1,300	2,500	2,200	5,400	4,800	20,300	18,500
13					3,000	2,600	1,300	1,100	2,200	1,800	4,600	4,100	17,300	15,800
14					2,600	2,200	1,200	1,000	1,900	1,600	4,000	3,500	14,900	13,600
15					2,200	—	1,000	—	1,600	—	3,500	—	13,000	11,900
16					—		—		—				11,400	10,400
17													10,100	9,200
18													9,000	8,200
19													8,100	7,100
20													7,300	6,700

* Calculated to nearest 100 lb. **R indicates rough lumber; S4S indicates lumber finished on all four sides.

† The dimension used in determining l/d is that shown first in the size column. Where this is the larger dimension, the column must be braced in the other direction so that l/d is equal to or less than that used in arriving at the loads shown. For 4 × 2's bracing in the plane of the 2-in. dimension must be at intervals not greater than 0.4 times the unsupported length. For 4 × 3's bracing in the plane of the 3-in. dimension must be at intervals not more than 0.7 times the unsupported length.

Table 16.5 *(cont.)*

c ‖ to grain = 1,000 psi					$E = 1{,}600{,}000$ psi		$\frac{l}{d_{max}} = 50$				$\frac{P}{A_{max}} = \frac{0.30E}{(l/d)^2}$			
Nominal Lumber Size, in.	2 × 4		3 × 4		4 × 4		4 × 2		4 × 3		4 × 6		6 × 6	
	R**	S4S**	R	S4S	R	S4S	R	S4S	R	S4S	R	S4S	R	S4S
Unsupported length, ft							Bracing needed †							
4	3,200	2,500	9,500	8,700	13,100	12,200	5,900	5,200	9,500	8,700	20,400	19,200	31,600	30,200
5	2,100	1,600	8,700	7,300	13,100	12,200	5,900	5,200	9,500	8,700	20,400	19,200	31,600	30,200
6	1,400	1,100	6,100	5,100	13,100	12,200	5,900	5,200	9,500	8,700	20,400	19,200	31,600	30,200
7	—	—	4,500	3,700	11,700	10,200	5,300	4,400	8,500	7,300	18,200	16,000	31,600	30,200
8			3,400	2,800	9,000	7,800	4,000	3,300	6,500	5,600	13,900	12,300	31,600	30,200
9			2,700	2,200	7,100	6,200	3,200	2,600	5,100	4,400	11,000	9,700	31,600	30,200
10			2,200	1,800	5,800	5,000	2,600	2,100	4,200	3,600	8,900	7,900	31,600	30,200
11			—	—	4,800	4,100	2,100	1,800	3,400	2,900	7,400	6,500	27,600	25,200
12					4,000	3,500	1,800	1,500	2,900	2,500	6,200	5,500	23,200	21,200
13					3,400	3,000	1,500	1,300	2,500	2,100	5,300	4,600	19,700	18,000
14					2,900	2,500	1,300	1,100	2,100	1,800	4,600	4,000	17,000	15,600
15					2,600	—	1,100	—	1,800	—	4,000	—	14,800	13,600
16					—		—		—		—		13,000	11,900
17													11,500	10,500
18													10,300	9,400
19													9,200	8,400
20													8,300	7,600

c ‖ to grain = 1,000 psi $\quad E = 1{,}700{,}000$ psi $\quad \frac{l}{d_{max}} = 50 \quad \frac{P}{A_{max}} = \frac{0.30E}{(l/d)^2}$

Nominal Lumber Size, in.	2 × 4		3 × 4		4 × 4		4 × 2		4 × 3		4 × 6		6 × 6	
	R**	S4S**	R	S4S	R	S4S	R	S4S	R	S4S	R	S4S	R	S4S
Unsupported length, ft							Bracing needed†							
4	3,400	2,600	9,500	8,800	13,100	12,300	5,900	5,300	9,500	8,800	20,400	19,200	31,600	30,200
5	2,200	1,700	9,300	7,700	13,100	12,300	5,900	5,300	9,500	8,800	20,400	19,200	31,600	30,200
6	1,500	1,200	6,500	5,400	13,100	12,300	5,900	5,300	9,500	8,800	20,400	19,200	31,600	30,200
7	—	—	4,700	4,000	12,500	10,800	5,600	4,600	9,000	7,700	19,400	17,000	31,600	30,200
8			3,600	3,000	9,600	8,300	4,300	3,600	6,900	5,900	14,800	13,100	31,600	30,200
9			2,900	2,400	7,600	6,600	3,400	2,800	5,500	4,700	11,700	10,300	31,600	30,200
10			2,300	1,900	6,100	5,300	2,700	2,300	4,400	3,800	9,500	8,400	31,600	30,200
11			—	—	5,100	4,400	2,300	1,900	3,700	3,100	7,900	6,900	29,300	28,000
12					4,200	3,700	1,900	1,600	3,100	2,600	6,600	5,800	24,600	22,500
13					3,600	3,100	1,600	1,300	2,600	2,200	5,600	4,900	21,000	19,200
14					3,100	2,700	1,400	1,200	2,300	1,900	4,800	4,300	18,100	16,500
15					2,700	—	1,200	—	2,000	—	4,200	—	15,800	14,400
16					—		—		—		—		13,800	12,600
17													12,300	11,200
18													10,900	10,000
19													9,800	9,000
20													8,900	8,100

* Calculated to nearest 100 lb. **R indicates rough lumber; S4S indicates lumber finished on all four sides.

† The dimension used in determining l/d is that shown first in the size column. Where this is the larger dimension, the column must be braced in the other direction so that l/d is equal to or less than that used in arriving at the loads shown. For 4 × 2's bracing in the plane of the 2-in. dimension must be at intervals not greater than 0.4 times the unsupported length. For 4 × 3's bracing in the plane of the 3-in. dimension must be at intervals not more than 0.7 times the unsupported length.

EXAMPLE 16.3

Are 4 × 6-in. shores (*S4S*) 10 ft long and 4 ft on centers satisfactory for supporting the floor system of Example 16.1? Assume the allowable compression stress parallel to the grain is 1,000 psi and $E = 1.70 \times 10^6$ psi.

SOLUTION

Using 4 × 6 shores (dressed dimensions $3\frac{1}{2} \times 5\frac{1}{2}$, $A = 19.250$ in.^{2})

load applied to each shore = (4)(4)(225) = 3,600 #

$$\frac{l}{d} = \frac{(12)(10)}{3.5} = 34.29 < 50 \qquad \underline{\text{ok}}$$

$$\text{allowable}\,\frac{P}{A} = \frac{(0.3)(1.70 \times 10^6)}{(34.29)^2} = 434 \text{ psi} < 1{,}800 \text{ psi} \qquad \underline{\text{ok}}$$

$$\text{allowable } P = (434)(19.250) = 8{,}354 > 3{,}600\ \# \qquad \underline{\text{ok}}$$

Wood shores may consist of different pieces of lumber that are nailed or bolted together to form larger built-up columns or shores. Such members are usually given lower allowable stresses than those available for solid-sawn or round columns. These built-up shores may be spaced columns consisting of two pieces of lumber with spacer blocks in between.

Spaced shores are very important for formwork because of economic factors. Logs can be efficiently sawed into 2 × material (2 × 4's, 2 × 6's, etc.). The cost of larger pieces, such as 4 ×, 6 ×, and so on, does not make such efficient use of the logs and prices are appreciably higher. As a result, built-up or spaced shores are frequently used. The allowable stresses, maximum l/d ratios, and other factors are somewhat different from single shores.

There are a number of adjustable shoring systems available. The simplest type is made by overlapping two wood members. The workmen use a portable jacking tool to make vertical adjustments. Various hardware is available for joining the shores to the stringers with a minimum amount of nailing. Another type of adjustable shore consists of dimension lumber combined with a steel column section and a jacking device. Several manufacturers have available all-metal shores, called *jack shores*, which are adjustable for heights from 4 to 16 ft.[3]

16.12 BEARING STRESSES

The bearing stresses produced when one member rests upon another may be critical in the design of formwork. These stresses need to be

[3] Hurd, M. K., 1973, *Formwork for Concrete* (Detroit: American Concrete Institute), pp. 68–69.

carefully checked where joists rest on stringers, where stringers rest on shores, and where studs rest on wales, as well as where form ties bear on the wales through brackets or washers. In each of the cases mentioned, the bearing forces applied to the horizontal timber members cause compression stresses perpendicular to the grain. The allowable stresses for compression perpendicular to the grain are much less than they are for compression parallel to the grain and, in fact, will often be less than the allowable compression stresses in shores as determined by the National Forest Products Association formula.

As a first illustration the bearing stresses in stringers resting on shores are considered. One particular point that should be noted is that the bearing area may well be less than the full cross-sectional area of the shore. For instance, in Examples 16.1 and 16.3 the 2 × 8-in. stringers are supported by 4 × 6-in. shores, as shown in Figure 16.6. Obviously, the bearing area does not equal the full cross-sectional area of the 4 × 6 (3.50 × 5.50 = 19.25 in.^{2}) but rather equals the smaller crosshatched area (1.50 × 5.50 = 8.25 in.^{2}).

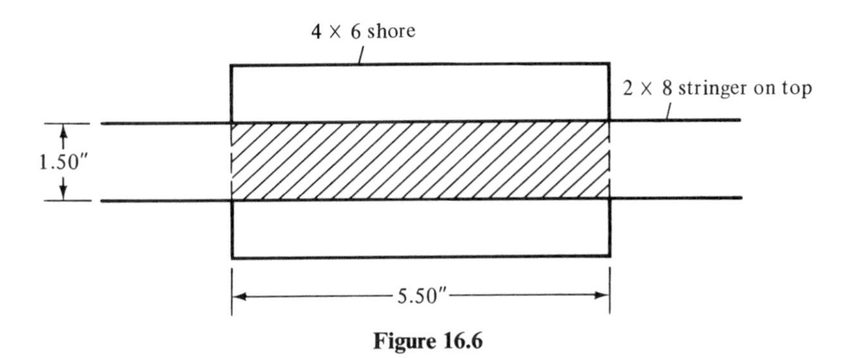

Figure 16.6

The bearing stress can be calculated by dividing the total load applied to the shore (3,600 # from Example 16.3) by the crosshatched area.

$$\text{bearing stress} = \frac{3{,}600}{8.25} = 436 \text{ psi}$$

From Table 16.3 the allowable bearing stress perpendicular to the grain in the stringer is 490 psi for the Doublas fir, coastal-construction lumber used in the examples of this chapter. The values given in this table for compression perpendicular to the grain are actually applicable to bearing spread over any length at the end of a member and for bearing at 6 in. or more in length at interior locations.

Should the bearing be located more than 3 in. from the end of a beam and be less than 6 in. long, the allowable compressive stresses perpendicular to the grain may be safely increased by multiplying them by the following factor, in which l is the length of bearing measured along the grain. Should the bearing area be circular, as for a washer, the length of bearing is considered to equal the diameter of the circle.

$$\text{multiplier} = \frac{l + \frac{3}{8}}{l}$$

If the shore is bearing at an interior point of the stringer in Figure 16.6, the allowable compression stress perpendicular to the grain equals

$$\left(\frac{l + \frac{3}{8}}{l}\right)(490) = \left(\frac{5.50 + 0.375}{5.50}\right)(490) = 523 \text{ psi} > 436 \text{ psi} \qquad \underline{\text{ok}}$$

If the calculated bearing stress exceeds the allowable stress, it is necessary to spread the load out, as by using larger members or by using a bearing plate or a hardwood cap on top of the shore.[4]

Another bearing stress situation is presented in Example 16.4, where the bearing stress caused in the wales of a wall form by the form ties is considered.

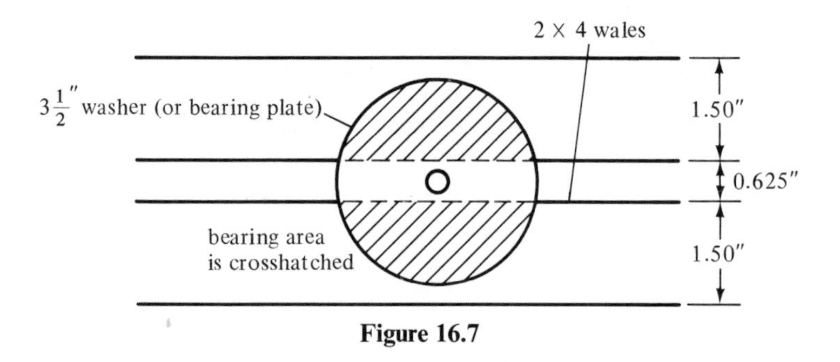

Figure 16.7

EXAMPLE 16.4

Form ties are subjected to an estimated force of 5,000 # from the fresh concrete in a wall form. The load is transferred to double 2 × 4-in. wales through $3\frac{1}{2}$ inch brackets or washers as shown in Figure 16.7. Are the calculated bearing stresses within the permissible values if the wales are Douglas fir, coastal construction, and the bearing locations are at interior points in the wales?

[4] Hurd, M. K., 1973, *Formwork for Concrete* (Detroit: American Concrete Institute), pp. 89–90.

SOLUTION

$$\text{bearing area} = \frac{(\pi)(3.50)^2}{4} - (0.625)(3.50) = 7.43 \text{ in.}^2$$

$$\text{bearing stress} = \frac{5{,}000}{7.43} = 673 \text{ psi}$$

$$\text{allowable compression} \perp \text{to grain} = \left(\frac{l + \frac{3}{8}}{l}\right)(490) = \left(\frac{3.50 + 0.375}{3.50}\right)(490)$$

$$= 543 \text{ psi} < 673 \text{ psi} \qquad \underline{\text{no good}}$$

Larger washers or smaller spacing of ties is needed as allowable bearing is not satisfactory.

16.13 DESIGN OF FORMWORK FOR WALLS

In Section 16.8 the lateral pressures on wall and column forms were discussed. These pressures increase from a minimum at the top to a maximum at the bottom of the semiliquid concrete. Occasionally in design this varying pressure is recognized, but most of the time it is practical to consider that the maximum pressure exists for the entire height. For convenience in construction, the sheathing, studs, wales, and ties are usually kept at the same size and spacings throughout the entire height of a wall, as are the wales and ties. In effect, therefore, uniform maximum pressures are assumed.

The sheathing and studs for a wall are designed as they are for the sheathing and joists for roof and floor slabs. In addition, the wales are designed as they are for the stringers for slabs, the spans being the center-to-center spacing of the ties.

If the wales are framed on both sides, the ties will extend through the formwork and be attached to the wales on each side, as was shown in Figure 16.2. The tensile force applied to each tie is then dependent on the pressure from the liquid concrete and on the horizontal and vertical spacing of the ties. If the spacing of ties for a certain wall is 2 ft vertically and 3 ft horizontally, and the calculated wall pressure is 800 psf, the total tensile force on each tie equals $2 \times 3 \times 800 = 4{,}800$ #, assuming a uniform pressure.

Should concrete be placed at 50°F at a rate of 4 ft/h in a 12-ft-high wall, the lateral pressure for the semiliquid concrete could be calculated from the ACI expression as follows:

$$p = 150 + \frac{9{,}000R}{T} = 150 + \frac{(9{,}000)(4)}{50} = 870 \text{ psf}$$

Column Forms (Courtesy Burke Concrete Accessories, Inc.).

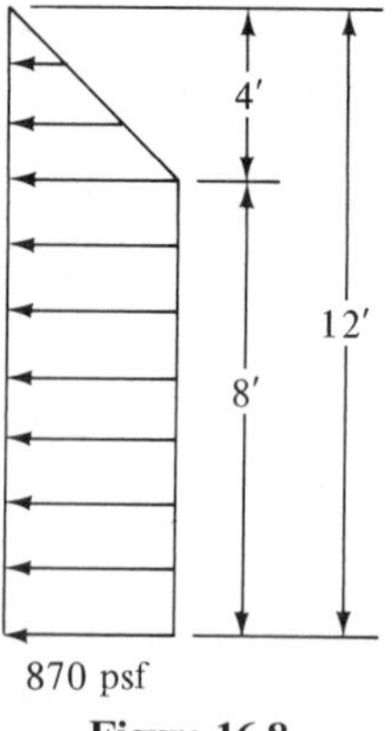

Figure 16.8

The resulting pressure diagram for the wall would be as shown in Figure 16.8.

Assuming a uniform pressure of 870 psf, the spacing of the studs could be selected as was the spacing for the joists in the floor slab formwork. Example 16.5 shows the design procedure for the sheathing, studs, wales, and ties for a wall form.

EXAMPLE 16.5

Design the forms for a 12-ft-high concrete wall for which the concrete is to be placed at a rate of 4 ft/hr at a temperature of 50°F. (Refer to Figure 16.8.) Use the following data:

1. Sheathing is to be $\frac{3}{4}$-inch Class I plyform. $f = 1{,}930$ psi and $E = 1{,}650{,}000$ psi
2. Studs and wales are to consist of Douglas fir, coastal construction. $f = 1{,}875$ psi, $H = 180$ psi, and $E = 1{,}760{,}000$ psi. Allowable compression perpendicular to the grain is 490 psi.
3. Ties can carry 5,000 # each and the tie washers are the same size as the one shown in Figure 16.7.
4. Double wales are used to avoid drilling for ties.
5. Maximum deflection in any form component is $\frac{1}{360}$ of the span.

SOLUTION

Design of Sheathing

Properties of 12-in. wide piece of $\frac{3}{4}$ plyform:

$$S = 0.455 \text{ in.}^3,\ I = 0.198 \text{ in.}^4$$

$$\text{load} = 870\ \#/\text{ft}$$

Moment

$$l = 10.95\sqrt{\frac{fS}{w}} = 10.95\sqrt{\frac{(1{,}930)(0.455)}{870}} = \underline{11.00''}$$

Deflection

$$l = 1.62\sqrt[3]{\frac{EI}{w}} = 1.62\sqrt[3]{\frac{(1{,}650{,}000)(0.198)}{870}} = 11.69''$$

<u>use studs at 0′ 11″ on center</u>

Design of studs (spaced 0 ft 11 in. on center)

Assuming 2 × 4 studs ($1\frac{1}{2} \times 3\frac{1}{2}$)

$$I = 5.36 \text{ in.}^4,\ S = 3.06 \text{ in.}^3$$

$$\text{load} = (\tfrac{11}{12})(870) = 797.5\ \#/\text{ft}$$

Moment

$$l = 10.95\sqrt{\frac{(1{,}875)(3.06)}{797.5}} = 29.37''$$

Deflection

$$l = 1.62\sqrt[3]{\frac{(1{,}760{,}000)(5.36)}{797.5}} = 36.91''$$

Shear

$$l = 12\left(\frac{Hbh}{0.9w} + \frac{2h}{12}\right) = (12)\left(\frac{180 \times 1.50 \times 3.50}{0.9 \times 797.5} + \frac{2 \times 3.50}{12}\right)$$

$$= 22.80''$$

use 1′ 10″ spacing of wales on center

$$\text{load on wales} = (\tfrac{22}{12})(797.5) = 1{,}462\ \#/\text{ft}$$

$$\text{maximum tie spacing} = \frac{5{,}000}{1{,}462} = 3.42'$$

Design of wales

Assuming two 2 × 4 wales ($I = 2 \times 5.36 = 10.72$ in.^{4}, $S = 2 \times 3.06 = 6.12$ in.^{3})

Moment

$$l = 10.95\sqrt{\frac{(1{,}875)(6.12)}{1{,}462}} = 30.68''$$

Deflection

$$l = 1.62\sqrt[3]{\frac{(1{,}760{,}000)(6.12)}{1{,}462}} = 31.52''$$

Shear

$$l = (12)\left(\frac{180 \times 3.00 \times 3.50}{0.9 \times 1{,}462} + \frac{2 \times 3.50}{12}\right) = 24.24''$$

use ties at every other stud 1′ 10″ on center

Check bearing for studs to wales

$$\text{maximum load transferred} = \frac{(22)(11)}{144}(870) = 1462\ \#$$

$$\text{bearing area (see Figure 16.9)} = (2)(1.50)(1.50) = 4.50\ \text{in.}^2$$

$$\text{bearing stress} = \frac{1{,}462}{4.50} = 325\ \text{psi} < 490\ \text{psi} \qquad \underline{\text{ok}}$$

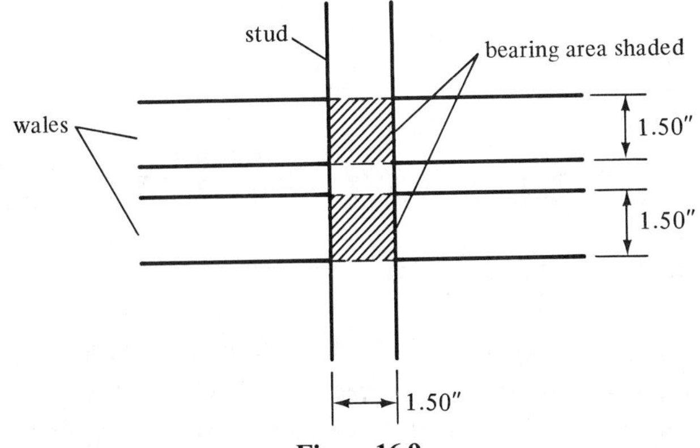

Figure 16.9

Check bearing from ties to wedges

Assume same dimensions as in Figure 16.7.

$$\text{bearing load} = \frac{(22)(22)}{144}(870) = 2924\ \#$$

$$\text{bearing stress} = \frac{2{,}924}{7.43} = 394 \text{ psi} < 543 \text{ psi}$$

as determined in Example 16.4 ok

Conclusion of Textbook

The author has attempted to include in this textbook only the elementary phases of reinforced concrete design. One of his main purposes has been to try to interest you in the subject as he feels that this is the first and most important item in gaining proficiency in any subject.

This book has only begun to present the knowledge now available concerning reinforced concrete. The amount of concrete research being conducted around the world far exceeds that being done at any time in the past, and new projects are being initiated at an ever-increasing rate. As a result, a person going into reinforced concrete design will need to study continually to keep abreast of the latest developments. In an effort to assist such a person in further study, the author recommends the following excellent references that will provide more detailed information in certain areas. For convenience an attempt has been made to divide these references into areas where they are particularly useful.

Basic Textbooks

Ferguson, P. M. 1973. *Reinforced Concrete Fundamentals*, 3d ed. New York: Wiley.

Wang, C. K., and Salmon, C. G. 1978. *Reinforced Concrete Design*, 3d ed. New York: Harper & Row.

Winter, G., and Nilson, A. H. 1972. *Design of Concrete Structures*, 8th ed. New York: McGraw-Hill.

Design Applications

CRSI. 1972. *CRSI Handbook.* Chicago: Concrete Reinforcing Steel Institute.

Fintel, M., ed. 1974. *Handbook of Concrete Engineering.* New York: Van Nostrand Reinhold.

Rice, P. F., and Hoffman, E. S. 1972. *Structural Design Guide to the ACI Building Code.* New York: Van Nostrand Reinhold.

Material Behavior

Bresler, B., ed. 1974. *Reinforced Concrete Engineering.* New York: Wiley.

Park, R., and Paulay, T. 1975. *Reinforced Concrete Structures.* New York: Wiley.

Appendix

Table A.1 Values of Modulus of Elasticity for Normal-Weight Concrete

Customary Units		SI Units	
f_c' (psi)	E_c (psi)	f_c' (MPa)	E_c (MPa)
3,000	3,140,000	20.7	21 650
3,500	3,390,000	24.1	23 373
4,000	3,620,000	27.6	24 959
4,500	3,850,000	31.0	26 545
5,000	4,050,000	34.5	27 924

Source Notes: Tables A-4, A-6, A-7, A-15, A-16, A-17, and Graphs 1–9 are reprinted from *Design of Concrete Structures* by Winter and Nilson. Copyright © 1972 by McGraw-Hill, Inc. and with permission of the McGraw-Hill Book Company.

Tables A-3(a) and A-3(b) are reprinted from *Manual of Standard Practice*, 22nd Ed., 1976, second printing, Concrete Reinforcing Steel Institute, Chicago, Ill.

Table A-5 is reprinted from *Reinforced Concrete Design*, 2nd Ed, by Wang and Salmon (IEP, New York, 1973).

Table A-8 is reprinted from *Notes on ACI 318-71 Building Code Requirements with Design Applications*, dated 1972. Portland Cement Association, Skokie, Ill.

Table A.2 Designations, Areas, Perimeters, and Weights of Standard Bars

Bar No.	Customary Units Diameter (in.)	Cross-sectional Area (in.^{2})	Unit Weight (lb/ft)	SI Units Diameter (mm)	Cross-sectional Area (mm^2)	Unit Weight (kg/m)
3	0.375	0.11	0.376	9.52	71	0.560
4	0.500	0.20	0.668	12.70	129	0.994
5	0.625	0.31	1.043	15.88	200	1.552
6	0.750	0.44	1.502	19.05	284	2.235
7	0.875	0.60	2.044	22.22	387	3.042
8	1.000	0.79	2.670	25.40	510	3.973
9	1.128	1.00	3.400	28.65	645	5.060
10	1.270	1.27	4.303	32.26	819	6.404
11	1.410	1.56	5.313	35.81	1006	7.907
14	1.693	2.25	7.650	43.00	1452	11.384
18	2.257	4.00	13.600	57.33	2581	20.238

Table A.3(a) Common Stock Styles of Welded Wire Fabric

Style Designation	Steel Area sq. in. per ft. Longit.	Steel Area sq. in. per ft. Transv.	Weight Approx. lbs. per 100 sq. ft.
Rolls			
6 × 6—W1.4 × W1.4	0.03	0.03	21
6 × 6—W2 × W2	0.04	0.04	29
6 × 6—W2.9 × W2.9	0.06	0.06	42
6 × 6—W4 × W4	0.08	0.08	58
4 × 4—W1.4 × W1.4	0.04	0.04	31
4 × 4—W2 × W2	0.06	0.06	43
4 × 4—W2.9 × W2.9	0.09	0.09	62
4 × 4—W4 × W4	0.12	0.12	86
Sheets			
6 × 6—W2.9 × W2.9	0.06	0.06	42
6 × 6—W4 × W4	0.08	0.08	58
6 × 6—W5.5 × W5.5	0.11	0.11	80
4 × 4—W4 × W4	0.12	0.12	86

Table A.3(b) Sectional Area and Weight of Welded Wire Fabric

[a]Wire Size Number		Nominal Diameter inches	Nominal Weight lbs./lin. ft.	Area in sq. in. per ft. of Width for Various Spacings						
				Center-To-Center Spacing						
Smooth	Deformed			2″	3″	4″	6″	8″	10″	12″
W31	D31	0.628	1.054		1.24	0.93	0.62	0.465	0.372	0.31
W28	D28	0.597	0.952		1.12	0.84	0.56	0.42	0.336	0.28
W26	D26	0.575	0.934		1.04	0.78	0.52	0.39	0.312	0.26
W24	D24	0.553	0.816		0.96	0.72	0.48	0.36	0.288	0.24
W22	D22	0.529	0.748		0.88	0.66	0.44	0.33	0.264	0.22
W20	D20	0.505	0.680	1.20	0.80	0.60	0.40	0.30	0.24	0.20
W18	D18	0.479	0.612	1.08	0.72	0.54	0.36	0.27	0.216	0.18
W16	D16	0.451	0.544	0.96	0.64	0.48	0.32	0.24	0.192	0.16
W14	D14	0.422	0.476	0.84	0.56	0.42	0.28	0.21	0.168	0.14
W12	D12	0.391	0.408	0.72	0.48	0.36	0.24	0.18	0.144	0.12
W11	D11	0.374	0.374	0.66	0.44	0.33	0.22	0.165	0.132	0.11
W10	D10	0.357	0.340	0.60	0.40	0.30	0.20	0.15	0.12	0.10
W9.5		0.348	0.323	0.57	0.38	0.285	0.19	0.142	0.114	0.095
W9	D9	0.339	0.306	0.54	0.36	0.27	0.18	0.135	0.108	0.09
W8.5		0.329	0.289	0.51	0.34	0.255	0.17	0.127	0.102	0.085

Table A.3(b) (*cont.*)

[a]Wire Size Number		Nominal Diameter	Nominal Weight	Area in sq. in. per ft. of Width for Various Spacings						
				Center-To-Center Spacing						
Smooth	Deformed	inches	lbs./lin. ft.	2″	3″	4″	6″	8″	10″	12″
W8	D8	0.319	0.272	0.48	0.32	0.24	0.16	0.12	0.096	0.08
W7.5		0.309	0.255	0.45	0.30	0.225	0.15	0.112	0.09	0.075
W7	D7	0.299	0.238	0.42	0.28	0.21	0.14	0.105	0.084	0.07
W6.5		0.288	0.221	0.39	0.26	0.195	0.13	0.097	0.078	0.065
W6	D6	0.276	0.204	0.36	0.24	0.18	0.12	0.09	0.072	0.06
W5.5		0.265	0.187	0.33	0.22	0.165	0.11	0.082	0.066	0.055
W5	D5	0.252	0.170	0.30	0.20	0.15	0.10	0.075	0.06	0.05
W4.5		0.239	0.153	0.27	0.18	0.135	0.09	0.067	0.054	0.045
W4	D4	0.226	0.136	0.24	0.16	0.12	0.08	0.06	0.048	0.04
W3.5		0.211	0.119	0.21	0.14	0.105	0.07	0.052	0.042	0.035
W2.9		0.192	0.099	0.174	0.116	0.087	0.058	0.043	0.035	0.029
W2.5		0.178	0.085	0.15	0.10	0.075	0.05	0.037	0.03	0.025
W2		0.160	0.068	0.12	0.08	0.06	0.04	0.03	0.024	0.02
W1.4		0.134	0.048	0.084	0.056	0.042	0.028	0.021	0.017	0.014

Note: The above listing of smooth and deformed wire sizes represents wires normally selected to manufacture welded wire fabric styles to specific areas of reinforcement. Wire sizes and spacings other than those listed above may be produced provided the quantity required is sufficient to justify manufacture.

[a] The number following the prefix W or the prefix D identifies the cross-sectional area of the wire in hundredths of a square inch.

The nominal diameter of a deformed wire is equivalent to the diameter of a smooth wire having the same weight per foot as the deformed wire.

Table A.4 Areas of Groups of StandardBars (in.^{2})

Bar No.	Number of Bars												
	2	3	4	5	6	7	8	9	10	11	12	13	14
4	0.39	0.58	0.78	0.98	1.18	1.37	1.57	1.77	1.96	2.16	2.36	2.55	2.75
5	0.61	0.91	1.23	1.53	1.84	2.15	2.45	2.76	3.07	3.37	3.68	3.99	4.30
6	0.88	1.32	1.77	2.21	2.65	3.09	3.53	3.98	4.42	4.86	5.30	5.74	6.19
7	1.20	1.80	2.41	3.01	3.61	4.21	4.81	5.41	6.01	6.61	7.22	7.82	8.42
8	1.57	2.35	3.14	3.93	4.71	5.50	6.28	7.07	7.85	8.64	9.43	10.21	11.00
9	2.00	3.00	4.00	5.00	6.00	7.00	8.00	9.00	10.00	11.00	12.00	13.00	14.00
10	2.53	3.79	5.06	6.33	7.59	8.86	10.12	11.39	12.66	13.92	15.19	16.45	17.72
11	3.12	4.68	6.25	7.81	9.37	10.94	12.50	14.06	15.62	17.19	18.75	20.31	21.87
14	4.50	6.75	9.00	11.25	13.50	15.75	18.00	20.25	22.50	24.75	27.00	29.25	31.50
18	8.00	12.00	16.00	20.00	24.00	28.00	32.00	36.00	40.00	44.00	48.00	52.00	56.00

Table A.5 Minimum Beam Width (in.) According to the 1977 ACI Code[a]

Size of Bars	Number of Bars in Single Layer of Reinforcement							Add for Each Added Bar
	2	3	4	5	6	7	8	
#4	6.1	7.6	9.1	10.6	12.1	13,6	15.1	1.50
#5	6.3	7.9	9.6	11.2	12.8	14.4	16.1	1.63
#6	6.5	8.3	10.0	11.8	13.5	15.3	17.0	1.75
#7	6.7	8.6	10.5	12.4	14.2	16.1	18.0	1.88
#8	6.9	8.9	10.9	12.9	14.9	16.9	18.9	2.00
#9	7.3	9.5	11.8	14.0	16.3	18.6	20.8	2.26
#10	7.7	10.2	12.8	15.3	17.8	20.4	22.9	2.54
#11	8.0	10.8	13.7	16.5	19.3	22.1	24.9	2.82
#14	8.9	12.3	15.6	19.0	22.4	25.8	29.2	3.39
#18	10.5	15.0	19.5	24.0	28.6	33.1	37.6	4.51

Table shows minimum beam widths when stirrups are used.
For additional bars, add dimension in last column for each added bar.
For bars of different size, determine from table the beam width for smaller size bars and then add last column figure for each larger bar used.
[a] Assumes maximum aggregate size does not exceed three-fourths of the clear space between bars.

Table A.6 Areas of Bars in Slabs (in.^{2}/ft)

Spacing, in.	Bar No.								
	3	4	5	6	7	8	9	10	11
3	0.44	0.78	1.23	1.77	2.40	3.14	4.00	5.06	6.25
$3\frac{1}{2}$	0.38	0.67	1.05	1.51	2.06	2.69	3.43	4.34	5.36
4	0.33	0.59	0.92	1.32	1.80	2.36	3.00	3.80	4.68
$4\frac{1}{2}$	0.29	0.52	0.82	1.18	1.60	2.09	2.67	3.37	4.17
5	0.26	0.47	0.74	1.06	1.44	1.88	2.40	3.04	3.75
$5\frac{1}{2}$	0.24	0.43	0.67	0.96	1.31	1.71	2.18	2.76	3.41
6	0.22	0.39	0.61	0.88	1.20	1.57	2.00	2.53	3.12
$6\frac{1}{2}$	0.20	0.36	0.57	0.82	1.11	1.45	1.85	2.34	2.89
7	0.19	0.34	0.53	0.76	1.03	1.35	1.71	2.17	2.68
$7\frac{1}{2}$	0.18	0.31	0.49	0.71	0.96	1.26	1.60	2.02	2.50
8	0.17	0.29	0.46	0.66	0.90	1.18	1.50	1.89	2.34
9	0.15	0.26	0.41	0.59	0.80	1.05	1.33	1.69	2.08
10	0.13	0.24	0.37	0.53	0.72	0.94	1.20	1.52	1.87
12	0.11	0.20	0.31	0.44	0.60	0.78	1.00	1.27	1.56

Table A.7 Design of Rectangular Beams and Slabs

n and f_c'	f_s	f_c	k	j	ρ_s	K
10 (2,500)	18,000	1,000	0.357	0.881	0.0099	157
		1,125	0.385	0.872	0.0120	189
	20,000	1,000	0.333	0.889	0.0083	148
		1,125	0.360	0.880	0.0101	178
	24,000	1,000	0.294	0.902	0.0061	133
		1,125	0.319	0.894	0.0075	160
9 (3,000)	18,000	1,200	0.375	0.875	0.0125	197
		1,350	0.403	0.866	0.0151	235
	20,000	1,200	0.351	0.883	0.0105	186
		1,350	0.377	0.874	0.0128	223
	24,000	1,200	0.310	0.897	0.0078	167
		1,350	0.336	0.888	0.0095	201
8 (4,000)	18,000	1,600	0.416	0.861	0.0185	286
		1,800	0.444	0.852	0.0222	341
	20,000	1,600	0.390	0.870	0.0156	272
		1,800	0.419	0.860	0.0188	324
	24,000	1,600	0.345	0.884	0.0116	246
		1,800	0.375	0.875	0.0141	295
7 (5,000)	18,000	2,000	0.438	0.854	0.0243	374
		2,250	0.467	0.844	0.0292	443
	20,000	2,000	0.412	0.863	0.0206	355
		2,250	0.441	0.853	0.0248	423
	24,000	2,000	0.368	0.877	0.0154	323
		2,250	0.396	0.868	0.0186	387

Table A.8 Balanced Ratio of Reinforcement ρ_b for Rectangular Sections with Tension Reinforcement Only

f_y \ f_c'		2,500 psi (17.2 MPa) $\beta_1 = 0.85$	3,000 psi (20.7 MPa) $\beta_1 = 0.85$	4,000 psi (27.6 MPa) $\beta_1 = 0.85$	5,000 psi (34.5 MPa) $\beta_1 = 0.80$	6,000 psi (41.4 MPa) $\beta_1 = 0.75$
Grade 40	ρ_b	0.0309	0.0371	0.0495	0.0582	0.0655
40,000 psi	$0.75\rho_b$	0.0232	0.0278	0.0371	0.0437	0.0492
(275.8 MPa)	$0.50\rho_b$	0.0155	0.0186	0.0247	0.0291	0.0328
Grade 50	ρ_b	0.0229	0.0275	0.0367	0.0432	0.0486
50,000 psi	$0.75\rho_b$	0.0172	0.0206	0.0275	0.0324	0.0365
(344.8 MPa)	$0.50\rho_b$	0.0115	0.0138	0.0184	0.0216	0.0243
Grade 60	ρ_b	0.0178	0.0214	0.0285	0.0335	0.0377
60,000 psi	$0.75\rho_b$	0.0134	0.0161	0.0214	0.0252	0.0283
(413.7 MPa)	$0.50\rho_b$	0.0089	0.0107	0.0143	0.0168	0.0189
Grade 75	ρ_b	0.0129	0.0155	0.0207	0.0243	0.0274
75,000 psi	$0.75\rho_b$	0.0097	0.0116	0.0155	0.0182	0.0205
(517.1 MPa)	$0.50\rho_b$	0.0065	0.0078	0.0104	0.0122	0.0137

Table A.9 $f_y = 40{,}000$ psi (275.8 MPa); $f_c' = 3{,}000$ psi (20.7 MPa)

	ρ	$\dfrac{M_u}{\phi bd^2}$	ρ	$\dfrac{M_u}{\phi bd^2}$	ρ	$\dfrac{M_u}{\phi bd^2}$		ρ	$\dfrac{M_u}{\phi bd^2}$
ρ_{min}	0.0050	192.1	0.0107	392.0	0.0164	571.4		0.0222	732.9
	0.0051	195.8	0.0108	395.3	0.0165	574.3		0.0223	735.5
	0.0052	199.5	0.0109	398.6	0.0166	577.3		0.0224	738.1
	0.0053	203.2	0.0110	401.9	0.0167	580.2		0.0225	740.7
	0.0054	206.8	0.0111	405.2	0.0168	583.2		0.0226	743.3
	0.0055	210.5	0.0112	408.5	0.0169	586.1		0.0227	745.8
	0.0056	214.1	0.0113	411.8	0.0170	589.1		0.0228	748.4
	0.0057	217.8	0.0114	415.1	0.0171	592.0		0.0229	751.0
	0.0058	221.4	0.0115	418.4	0.0172	594.9		0.0230	753.5
	0.0059	225.0	0.0116	421.7	0.0173	597.8		0.0231	756.1
	0.0060	228.7	0.0117	424.9	0.0174	600.7		0.0232	758.6
	0.0061	232.3	0.0118	428.2	0.0175	603.6		0.0233	761.2
	0.0062	235.9	0.0119	431.4	0.0176	606.5		0.0234	763.7
	0.0063	239.5	0.0120	434.7	0.0177	609.4		0.0235	766.2
	0.0064	243.1	0.0121	437.9	0.0178	612.3		0.0236	768.7
	0.0065	246.7	0.0122	441.2	0.0179	615.2		0.0237	771.2
	0.0066	250.3	0.0123	444.4	0.0180	618.0		0.0238	773.7
	0.0067	253.9	0.0124	447.6	0.0181	620.9		0.0239	776.2
	0.0068	257.4	0.0125	450.8	0.0182	623.8		0.0240	778.7
	0.0069	261.0	0.0126	454.0	0.0183	626.6		0.0241	781.2
	0.0070	264.6	0.0127	457.2	0.0184	629.5		0.0242	783.7
	0.0071	268.1	0.0128	460.4	0.0185	632.3		0.0243	786.2
	0.0072	271.7	0 0129	463.6	0.0186	635.1		0.0244	788.6
	0.0073	275.2	0.0130	466.8	0.0187	638.0		0.0245	791.1
	0.0074	278.8	0.0131	470.0	0.0188	640.8		0.0246	793.6
	0.0075	282.3	0.0132	473.2	0.0189	643.6		0.0247	796.0
	0.0076	285.8	0.0133	476.3	0.0190	646.4		0.0248	798.5
	0.0077	289.3	0.0134	479.5	0.0191	649.2		0.0249	800.9
	0.0078	292.9	0.0135	482.6	0.0192	652.0		0.0250	803.3
	0.0079	296.4	0.0136	485.8	0.0193	654.8		0.0251	805.7
	0.0080	299.9	0.0137	488.9	0.0194	657.6		0.0252	808.2
	0.0081	303.4	0.0138	492.1	0.0195	660.3		0.0253	810.6
	0.0082	306.8	0.0139	495.2	0.0196	663.1		0.0254	813.0
	0.0083	310.3	0.0140	498.3	0.0197	665.9		0.0255	815.4
	0.0084	313.8	0.0141	501.4	0.0198	668.6		0.0256	817.8
	0.0085	317.3	0.0142	504.5	0.0199	671.4		0.0257	820.2
	0.0086	320.7	0.0143	507.6	0.0200	674.1		0.0258	822.5
	0.0087	324.2	0.0144	510.7	0.0201	676.9		0.0259	824.9
	0.0088	327.6	0.0145	513.8	0.0202	679.6		0.0260	827.3
	0.0089	331.1	0.0146	516.9	0.0203	682.3		0.0261	829.6
	0.0090	334.5	0.0147	520.0	0.0204	685.0		0.0262	832.0
	0.0091	337.9	0.0148	523.1	0.0205	687.8		0.0263	834.3
	0.0092	341.4	0.0149	526.1	0.0206	690.5		0.0264	836.7
	0.0093	344.8	0.0150	529.2	0.0207	693.2		0.0265	839.0
	0.0094	348.2	0.0151	532.2	0.0208	695.9		0.0266	841.3
	0.0095	351.6	0.0152	535.3	0.0209	698.5		0.0267	843.7
			0.0153	538.3	0.0210	701.2		0.0268	846.0
	0.0096	355.0	0.0154	541.4	0.0211	703.9		0.0269	848.3
	0.0097	358.4	0.0155	544.4	0.0212	706.6		0.0270	850.6
	0.0098	361.8	0.0156	547.4	0.0213	709.2		0.0271	852.9
	0.0099	365.2	0.0157	550.4	0.0214	711.9		0.0272	855.2
	0.0100	368.5	0.0158	553.4	0.0215	714.5		0.0273	857.5
	0.0101	371.9	0.0159	556.4	0.0216	717.2		0.0274	859.7
	0.0102	375.3	0.0160	559.4	0.0217	719.8		0.0275	862.0
	0.0103	378.6	0.0161	562.4	0.0218	722.4		0.0276	864.3
	0.0104	382.0			0.0219	725.1		0.0277	866.5
	0.0105	385.3	0.0162	565.4	0.0220	727.7	ρ_{max}	0.0278	868.8
	0.0106	388.6	0.0163	568.4	0.0221	730.3			

For SI units multiply $M_u/\phi bd^2$ by 0.006 895 for Tables A.9 through A.14.

Table A.10 $f_y = 40{,}000$ psi (275.8 MPa); $f_c' = 4000$ psi (27.6 MPa)

	ρ	$\frac{M_u}{\phi bd^2}$	ρ	$\frac{M_u}{\phi bd^2}$	ρ	$\frac{M_u}{\phi bd^2}$	ρ	$\frac{M_u}{\phi bd^2}$
ρ_{min}	0.0050	194.1	0.0091	344.5	0.0131	483.5	0.0171	615.0
	0.0051	197.9	0.0092	348.0	0.0132	486.9	0.0172	618.2
	0.0052	201.6	0.0093	351.6	0.0133	490.2	0.0173	621.4
	0.0053	205.4	0.0094	355.1	0.0134	493.6	0.0174	624.5
	0.0054	209.1	0.0095	358.7	0.0135	497.0	0.0175	627.7
	0.0055	212.9			0.0136	500.3	0.0176	630.9
	0.0056	216.6	0.0096	362.2	0.0137	503.7	0.0177	634.1
	0.0057	220.3	0.0097	365.8	0.0138	507.0	0.0178	637.2
	0.0058	224.1	0.0098	369.3	0.0139	510.4	0.0179	640.4
	0.0059	227.8	0.0099	372.9	0.0140	513.7	0.0180	643.5
	0.0060	231.5	0.0100	376.4	0.0141	517.1	0.0181	646.7
	0.0061	235.2	0.0101	379.9	0.0142	520.4	0.0182	649.8
	0.0062	238.9	0.0102	383.4	0.0143	523.7	0.0183	653.0
	0.0063	242.6	0.0103	387.0	0.0144	527.1	0.0184	656.1
	0.0064	246.3	0.0104	390.5	0.0145	530.4	0.0185	659.2
	0.0065	250.0	0.0105	394.0	0.0146	533.7	0.0186	662.3
	0.0066	253.7	0.0106	397.5	0.0147	537.0	0.0187	665.5
	0.0067	257.4	0.0107	401.0	0.0148	540.3	0.0188	668.6
	0.0068	261.1	0.0108	404.5	0.0149	543.6	0.0189	671.7
	0.0069	264.8	0.0109	408.0	0.0150	546.9	0.0190	674.8
	0.0070	268.4	0.0110	411.4	0.0151	550.2	0.0191	677.9
	0.0071	272.1	0.0111	414.9	0.0152	553.5	0.0192	681.0
	0.0072	275.8	0.0112	418.4	0.0153	556.7	0.0193	684.1
	0.0073	279.4	0.0113	421.9	0.0154	560.0	0.0194	687.2
	0.0074	283.1	0.0114	425.3	0.0155	563.3	0.0195	690.3
	0.0075	286.7	0.0115	428.8	0.0156	566.6	0.0196	693.3
	0.0076	290.4	0.0116	432.2	0.0157	569.8	0.0197	696.4
	0.0077	294.0	0.0117	435.7	0.0158	573.1	0.0198	699.5
	0.0078	297.6	0.0118	439.1	0.0159	576.3	0.0199	702.5
	0.0079	301.3	0.0119	442.6	0.0160	579.6	0.0200	705.6
	0.0080	304.9	0.0120	446.0	0.0161	582.8	0.0201	708.6
	0.0081	308.5	0.0121	449.4			0.0202	711.7
	0.0082	312.1	0.0122	452.9	0.0162	586.1	0.0203	714.7
	0.0083	315.7	0.0123	456.3	0.0163	589.3	0.0204	717.8
	0.0084	319.3	0.0124	459.7	0.0164	592.5	0.0205	720.8
	0.0085	322.9	0.0125	463.1	0.0165	595.7	0.0206	723.8
	0.0086	326.5	0.0126	466.5	0.0166	599.0	0.0207	726.9
	0.0087	330.1	0.0127	469.9	0.0167	602.2	0.0208	729.9
	0.0088	333.7	0.0128	473.3	0.0168	605.4	0.0209	732.9
	0.0089	337.3	0.0129	476.7	0.0169	608.6	0.0210	735.9
	0.0090	340.9	0.0130	480.1	0.0170	611.8	0.0211	738.9

Table A.10 *(cont.)*

ρ	$\frac{M_u}{\phi bd^2}$	ρ	$\frac{M_u}{\phi bd^2}$	ρ	$\frac{M_u}{\phi bd^2}$		ρ	$\frac{M_u}{\phi bd^2}$
0.0212	741.9	0.0252	858.1	0.0293	969.4		0.0332	1067.9
0.0213	744.9	0.0253	860.9				0.0333	1070.3
0.0214	747.9	0.0254	863.7	0.0294	972.0		0.0334	1072.7
0.0215	750.9	0.0255	866.5	0.0295	974.6		0.0335	1075.1
0.0216	753.9	0.0256	869.3	0.0296	977.2		0.0336	1077.6
0.0217	756.9	0.0257	872.1	0.0297	979.8		0.0337	1080.0
0.0218	759.8	0.0258	874.9	0.0298	982.4		0.0338	1082.4
0.0219	762.8	0.0259	877.7	0.0299	985.0		0.0339	1084.8
0.0220	765.8	0.0260	880.5	0.0300	987.6		0.0340	1087.2
0.0221	768.7	0.0261	883.2	0.0301	990.2		0.0341	1089.6
0.0222	771.7	0.0262	886.0	0.0302	992.7		0.0342	1091.9
0.0223	774.6	0.0263	888.7	0.0303	995.3		0.0343	1094.3
0.0224	777.6	0.0264	891.5	0.0304	997.9		0.0344	1096.7
0.0225	780.5	0.0265	894.3	0.0305	1000.4		0.0345	1099.1
0.0226	783.4	0.0266	897.0	0.0306	1003.0		0.0346	1101.5
0.0227	786.4	0.0267	899.7	0.0307	1005.6		0.0347	1103.8
		0.0268	902.5	0.0308	1008.1		0.0348	1106.2
0.0228	789.3	0.0269	905.2	0.0309	1010.7		0.0349	1108.5
0.0229	792.2	0.0270	907.9	0.0310	1013.2		0.0350	1110.9
0.0230	795.1	0.0271	910.7	0.0311	1015.7		0.0351	1113.2
0.0231	798.1	0.0272	913.4	0.0312	1018.3		0.0352	1115.6
0.0232	801.0	0.0273	916.1	0.0313	1020.8		0.0353	1117.9
0.0233	803.9	0.0274	918.8	0.0314	1023.3		0.0354	1120.2
0.0234	806.8	0.0275	921.5	0.0315	1025.8		0.0355	1122.6
0.0235	809.7	0.0276	924.2	0.0316	1028.3		0.0356	1124.9
0.0236	812.5	0.0277	926.9	0.0317	1030.8		0.0357	1127.2
0.0237	815.4	0.0278	929.6	0.0318	1033.3		0.0358	1129.5
0.0238	818.3	0.0279	932.3	0.0319	1035.8		0.0359	1131.8
0.0239	821.2	0.0280	935.0	0.0320	1038.3		0.0360	1134.1
0.0240	824.1	0.0281	937.6	0.0321	1040.8		0.0361	1136.4
0.0241	826.9	0.0282	940.3	0.0322	1043.3		0.0362	1138.7
0.0242	829.8	0.0283	943.0	0.0323	1045.8		0.0363	1141.0
0.0243	832.6	0.0284	945.6	0.0324	1048.2		0.0364	1143.3
0.0244	835.5	0.0285	948.3	0.0325	1050.7		0.0365	1145.6
0.0245	838.3	0.0286	950.9	0.0326	1053.2		0.0366	1147.8
0.0246	841.2	0.0287	953.6	0.0327	1055.6		0.0367	1150.1
0.0247	844.0	0.0288	956.2	0.0328	1058.1		0.0368	1152.4
0.0248	846.8	0.0289	958.9	0.0329	1060.5		0.0369	1154.6
0.0249	849.7	0.0290	961.5	0.0330	1063.0		0.0370	1156.9
0.0250	852.5	0.0291	964.1	0.0331	1065.4	ρ_{max}	0.0371	1159.2
0.0251	855.3	0.0292	966.8					

Table A.11 $f_y = 50{,}000$ psi (344.8 MPa); $f_c' = 3{,}000$ psi (20.7 MPa)

	ρ	$\frac{M_u}{\phi bd^2}$	ρ	$\frac{M_u}{\phi bd^2}$	ρ	$\frac{M_u}{\phi bd^2}$		ρ	$\frac{M_u}{\phi bd^2}$
$\rho_{\min}$	0.0040	192.1	0.0082	376.9	0.0124	544.4		0.0166	694.5
	0.0041	196.7	0.0083	381.1	0.0125	548.2		0.0167	697.9
	0.0042	201.3	0.0084	385.3	0.0126	551.9		0.0168	701.2
	0.0043	205.9	0.0085	389.5	0.0127	555.7		0.0169	704.6
	0.0044	210.5	0.0086	393.6	0.0128	559.4		0.0170	707.9
	0.0045	215.0	0.0087	397.8	0.0129	563.2		0.0171	711.2
	0.0046	219.6	0.0088	401.9	0.0130	566.9		0.0172	714.5
	0.0047	224.1	0.0089	406.1	0.0131	570.6		0.0173	717.8
	0.0048	228.7	0.0090	410.2	0.0132	574.3		0.0174	721.1
	0.0049	233.2	0.0091	414.3	0.0133	578.0		0.0175	724.4
	0.0050	237.7	0.0092	418.4	0.0134	581.7		0.0176	727.7
	0.0051	242.2	0.0093	422.5	0.0135	585.4		0.0177	731.0
	0.0052	246.7	0.0094	426.6	0.0136	589.1		0.0178	734.2
	0.0053	251.2	0.0095	430.6	0.0137	592.7		0.0179	737.5
	0.0054	255.7	0.0096	434.7	0.0138	596.4		0.0180	740.7
	0.0055	260.1	0.0097	438.7	0.0139	600.0		0.0181	743.9
	0.0056	264.6	0.0098	442.8	0.0140	603.6		0.0182	747.1
	0.0057	269.0	0.0099	446.8	0.0141	607.2		0.0183	750.3
	0.0058	273.5	0.0100	450.8	0.0142	610.9		0.0184	753.5
	0.0059	277.9	0.0101	454.8	0.0143	614.5		0.0185	756.7
	0.0060	282.3	0.0102	458.8	0.0144	618.0		0.0186	759.9
	0.0061	286.7	0.0103	462.8	0.0145	621.6		0.0187	763.1
	0.0062	291.1	0.0104	466.8	0.0146	625.2		0.0188	766.2
	0.0063	295.5	0.0105	470.8	0.0147	628.7		0.0189	769.4
	0.0064	299.9	0.0106	474.8	0.0148	632.3		0.0190	772.5
	0.0065	304.2	0.0107	478.7	0.0149	635.8		0.0191	775.6
	0.0066	308.6	0.0108	482.6	0.0150	639.4		0.0192	778.7
	0.0067	312.9	0.0109	486.6	0.0151	642.9		0.0193	781.8
	0.0068	317.3	0.0110	490.5	0.0152	646.4		0.0194	784.9
	0.0069	321.6	0.0111	494.4	0.0153	649.9		0.0195	788.0
	0.0070	325.9	0.0112	498.3	0.0154	653.4		0.0196	791.1
	0.0071	330.2	0.0113	502.2	0.0155	656.9		0.0197	794.2
	0.0072	334.5	0.0114	506.1	0.0156	660.3		0.0198	797.2
	0.0073	338.8	0.0115	510.0	0.0157	663.8		0.0199	800.3
	0.0074	343.1	0.0116	513.8	0.0158	667.3		0.0200	803.3
	0.0075	347.3	0.0117	517.7	0.0159	670.7		0.0201	806.4
	0.0076	351.6	0.0118	521.5	0.0160	674.1		0.0202	809.4
	0.0077	355.8	0.0119	525.4	0.0161	677.5		0.0203	812.4
	0.0078	360.1	0.0120	529.2	0.0162	681.0		0.0204	815.4
	0.0079	364.3	0.0121	533.0	0.0163	684.4		0.0205	818.4
	0.0080	368.5	0.0122	536.8	0.0164	687.8	$\rho_{\max}$	0.0206	821.3
	0.0081	372.7	0.0123	540.6	0.0165	691.1			

Table A.12 $f_y = 50{,}000$ psi (344.8 MPa); $f_c' = 4{,}000$ psi (27.6 MPa)

	ρ	$\frac{M_u}{\phi bd^2}$	ρ	$\frac{M_u}{\phi bd^2}$	ρ	$\frac{M_u}{\phi bd^2}$	ρ	$\frac{M_u}{\phi bd^2}$		ρ	$\frac{M_u}{\phi bd^2}$
			0.0082	385.2	0.0130	587.7	0.0178	773.2		0.0227	945.0
			0.0083	389.6	0.0131	591.7	0.0179	776.8		0.0228	948.3
			0.0084	394.0	0.0132	595.7	0.0180	780.5		0.0229	951.6
			0.0085	398.4	0.0133	599.8	0.0181	784.2		0.0230	954.9
			0.0086	402.7	0.0134	603.8	0.0182	787.8		0.0231	958.2
			0.0087	407.1	0.0135	607.8	0.0183	791.5		0.0232	961.5
			0.0088	411.4	0.0136	611.8	0.0184	795.1		0.0233	964.8
ρ_{min}	0.0040	194.1	0.0089	415.8	0.0137	615.8	0.0185	798.8		0.0234	968.1
	0.0041	198.8	0.0090	420.1	0.0138	619.8	0.0186	802.4		0.0235	971.3
	0.0042	203.5	0.0091	424.5	0.0139	623.7	0.0187	806.0		0.0236	974.6
	0.0043	208.2	0.0092	428.8	0.0140	627.7	0.0188	809.7		0.0237	977.9
	0.0044	212.9	0.0093	433.1	0.0141	631.7	0.0189	813.3		0.0238	981.1
	0.0045	217.5	0.0094	437.4	0.0142	635.6	0.0190	816.9		0.0239	984.4
	0.0046	222.2	0.0095	441.7	0.0143	639.6	0.0191	820.5		0.0240	987.6
	0.0047	226.9			0.0144	643.5	0.0192	824.1		0.0241	990.8
	0.0048	231.5	0.0096	446.0	0.0145	647.5	0.0193	827.6		0.0242	994.0
	0.0049	236.1	0.0097	450.3	0.0146	651.4	0.0194	831.2		0.0243	997.2
	0.0050	240.8	0.0098	454.6	0.0147	655.3	0.0195	834.8		0.0244	1000.4
	0.0051	245.4	0.0099	458.9	0.0148	659.2	0.0196	838.3		0.0245	1003.6
	0.0052	250.0	0.0100	463.1	0.0149	663.1	0.0197	841.9		0.0246	1006.8
	0.0053	254.6	0.0101	467.4	0.0150	667.0	0.0198	845.4		0.0247	1010.0
	0.0054	259.2	0.0102	471.6	0.0151	670.9	0.0199	849.0		0.0248	1013.2
	0.0055	263.8	0.0103	475.9	0.0152	674.8	0.0200	852.5		0.0249	1016.4
	0.0056	268.4	0.0104	480.1	0.0153	678.7	0.0201	856.0		0.0250	1019.5
	0.0057	273.0	0.0105	484.3	0.0154	682.5	0.0202	859.5		0.0251	1022.7
	0.0058	277.6	0.0106	488.6	0.0155	686.4	0.0203	863.0		0.0252	1025.8
	0.0059	282.2	0.0107	492.8	0.0156	690.3	0.0204	866.5		0.0253	1029.0
	0.0060	286.7	0.0108	497.0	0.0157	694.1	0.0205	870.0		0.0254	1032.1
	0.0061	291.3	0.0109	501.2	0.0158	697.9	0.0206	873.5		0.0255	1035.2
	0.0062	295.8	0.0110	505.4	0.0159	701.8	0.0207	877.0		0.0256	1038.3
	0.0063	300.4	0.0111	509.6	0.0160	705.6	0.0208	880.5		0.0257	1041.4
	0.0064	304.9	0.0112	513.7	0.0161	709.4	0.0209	883.9		0.0258	1044.5
	0.0065	309.4	0.0113	517.9			0.0210	887.4		0.0259	1047.6
	0.0066	313.9	0.0114	522.1	0.0162	713.2	0.0211	890.8		0.0260	1050.7
	0.0067	318.4	0.0115	526.2	0.0163	717.0	0.0212	894.3		0.0261	1053.8
	0.0068	322.9	0.0116	530.4	0.0164	720.8	0.0213	897.7		0.0262	1056.9
	0.0069	327.4	0.0117	534.5	0.0165	724.6	0.0214	901.1		0.0263	1059.9
	0.0070	331.9	0.0118	538.6	0.0166	728.4	0.0215	904.5		0.0264	1063.0
	0.0071	336.4	0.0119	542.8	0.0167	732.1	0.0216	907.9		0.0265	1066.0
	0.0072	340.9	0.0120	546.9	0.0168	735.9	0.0217	911.3		0.0266	1069.1
	0.0073	345.3	0.0121	551.0	0.0169	739.7	0.0218	914.7		0.0267	1072.1
	0.0074	349.8	0.0122	555.1	0.0170	743.4	0.0219	918.1		0.0268	1075.1
	0.0075	354.3	0.0123	559.2	0.0171	747.2	0.0220	921.5		0.0269	1078.2
	0.0076	358.7	0.0124	563.3	0.0172	750.9	0.0221	924.9		0.0270	1081.2
	0.0077	363.1	0.0125	567.4	0.0173	754.6	0.0222	928.3		0.0271	1084.2
	0.0078	367.6	0.0126	571.4	0.0174	758.3	0.0223	931.6		0.0272	1087.2
	0.0079	372.0	0.0127	575.5	0.0175	762.1	0.0224	935.0		0.0273	1090.2
	0.0080	376.4	0.0128	579.6	0.0176	765.8	0.0225	938.3		0.0274	1093.1
	0.0081	380.8	0.0129	583.6	0.0177	769.5	0.0226	941.6	ρ_{max}	0.0275	1096.1

Table A.13 $f_y = 60{,}000$ psi (413.7 MPa); $f_c' = 3{,}000$ psi (20.7 MPa)

	ρ	$\frac{M_u}{\phi bd^2}$	ρ	$\frac{M_u}{\phi bd^2}$		ρ	$\frac{M_u}{\phi bd^2}$
ρ_{min}	0.0033	190.3	0.0077	420.0		0.0119	613.7
	0.0034	195.8	0.0078	424.9		0.0120	618.0
	0.0035	201.3	0.0079	429.8		0.0121	622.3
	0.0036	206.8				0.0122	626.6
	0.0037	212.3	0.0080	434.7		0.0123	630.9
	0.0038	217.8	0.0081	439.5		0.0124	635.1
	0.0039	223.2	0.0082	444.4		0.0125	639.4
	0.0040	228.7	0.0083	449.2		0.0126	643.6
	0.0041	234.1	0.0084	454.0		0.0127	647.8
	0.0042	239.5	0.0085	458.8		0.0128	652.0
	0.0043	244.9	0.0086	463.6		0.0129	656.2
	0.0044	250.3	0.0087	468.4		0.0130	660.3
	0.0045	255.7	0.0088	473.2		0.0131	664.5
	0.0046	261.0	0.0089	477.9		0.0132	668.6
	0.0047	266.4	0.0090	482.6		0.0133	672.8
	0.0048	271.7	0.0091	487.4		0.0134	676.9
	0.0049	277.0	0.0092	492.1		0.0135	681.0
	0.0050	282.3	0.0093	496.8		0.0136	685.0
	0.0051	287.6	0.0094	501.4		0.0137	689.1
	0.0052	292.9	0.0095	506.1		0.0138	693.2
	0.0053	298.1				0.0139	697.2
	0.0054	303.4	0.0096	510.7		0.0140	701.2
	0.0055	308.6	0.0097	515.4		0.0141	705.2
	0.0056	313.8	0.0098	520.0		0.0142	709.2
	0.0057	319.0	0.0099	524.6		0.0143	713.2
	0.0058	324.2	0.0100	529.2		0.0144	717.2
	0.0059	329.4	0.0101	533.8		0.0145	721.1
	0.0060	334.5	0.0102	538.3		0.0146	725.1
	0.0061	339.7	0.0103	542.9		0.0147	729.0
	0.0062	344.8	0.0104	547.4		0.0148	732.9
	0.0063	349.9	0.0105	551.9		0.0149	736.8
	0.0064	355.0	0.0106	556.4		0.0150	740.7
	0.0065	360.1	0.0107	560.9		0.0151	744.6
	0.0066	365.2	0.0108	565.4		0.0152	748.4
	0.0067	370.2	0.0109	569.9		0.0153	752.3
	0.0068	375.3	0.0110	574.3		0.0154	756.1
	0.0069	380.3	0.0111	578.8		0.0155	759.9
	0.0070	385.3	0.0112	583.2		0.0156	763.7
	0.0071	390.3	0.0113	587.6		0.0157	767.5
	0.0072	395.3	0.0114	592.0		0.0158	771.2
	0.0073	400.3	0.0115	596.4		0.0159	775.0
	0.0074	405.2	0.0116	600.7		0.0160	778.7
	0.0075	410.2	0.0117	605.1	ρ_{max}	0.0161	782.5
	0.0076	415.1	0.0118	609.4			

Table A.14 $f_y = 60{,}000$ psi (413.7 MPa); $f_c' = 4{,}000$ psi (27.6 MPa)

	ρ	$\frac{M_u}{\phi bd^2}$	ρ	$\frac{M_u}{\phi bd^2}$	ρ	$\frac{M_u}{\phi bd^2}$		ρ	$\frac{M_u}{\phi bd^2}$
ρ_{min}	0.0033	192.2	0.0079	440.9	0.0124	662.3		0.0169	862.3
	0.0034	197.9	0.0080	446.0	0.0125	667.0		0.0170	866.5
	0.0035	203.5	0.0081	451.2	0.0126	671.7		0.0171	870.7
	0.0036	209.1	0.0082	456.3	0.0127	676.3		0.0172	874.9
	0.0037	214.7	0.0083	461.4	0.0128	681.0		0.0173	879.1
	0.0038	220.3	0.0084	466.5	0.0129	685.6		0.0174	883.2
	0.0039	225.9	0.0085	471.6	0.0130	690.3		0.0175	887.4
	0.0040	231.5	0.0086	476.7	0.0131	694.9		0.0176	891.5
	0.0041	237.1	0.0087	481.8	0.0132	699.5		0.0177	895.6
	0.0042	242.6	0.0088	486.9	0.0133	704.1		0.0178	899.7
	0.0043	248.2	0.0089	491.9	0.0134	708.6		0.0179	903.9
	0.0044	253.7	0.0090	497.0	0.0135	713.2		0.0180	907.9
	0.0045	259.2	0.0091	502.0	0.0136	717.8		0.0181	912.0
	0.0046	264.8	0.0092	507.1	0.0137	722.3		0.0182	916.1
	0.0047	270.3	0.0093	512.1	0.0138	726.9		0.0183	920.2
	0.0048	275.8	0.0094	517.1	0.0139	731.4		0.0184	924.2
	0.0049	281.2	0.0095	522.1	0.0140	735.9		0.0185	928.3
	0.0050	286.7			0.0141	740.4		0.0186	932.3
	0.0051	292.2	0.0096	527.1	0.0142	744.9		0.0187	936.3
	0.0052	297.6	0.0097	532.0	0.0143	749.4		0.0188	940.3
	0.0053	303.1	0.0098	537.0	0.0144	753.9		0.0189	944.3
	0.0054	308.5	0.0099	542.0	0.0145	758.3		0.0190	948.3
	0.0055	313.9	0.0100	546.9	0.0146	762.8		0.0191	952.3
	0.0056	319.3	0.0101	551.8	0.0147	767.2		0.0192	956.2
	0.0057	324.7	0.0102	556.7	0.0148	771.7		0.0193	960.2
	0.0058	330.1	0.0103	561.7	0.0149	776.1		0.0194	964.1
	0.0059	335.5	0.0104	566.6	0.0150	780.5		0.0195	968.1
	0.0060	340.9	0.0105	571.5	0.0151	784.9		0.0196	972.0
	0.0061	346.2	0.0106	576.3	0.0152	789.3		0.0197	975.9
	0.0062	351.6	0.0107	581.2	0.0153	793.7		0.0198	979.8
	0.0063	356.9	0.0108	586.1	0.0154	798.1		0.0199	983.7
	0.0064	362.2	0.0109	590.9	0.0155	802.4		0.0200	987.6
	0.0065	367.6	0.0110	595.7	0.0156	806.8		0.0201	991.5
	0.0066	372.9	0.0111	600.6	0.0157	811.1		0.0202	995.3
	0.0067	378.2	0.0112	605.4	0.0158	815.4		0.0203	999.2
	0.0068	383.4	0.0113	610.2	0.0159	819.7		0.0204	1003.0
	0.0069	388.7	0.0114	615.0	0.0160	824.1		0.0205	1006.8
	0.0070	394.0	0.0115	619.8	0.0161	828.3		0.0206	1010.7
	0.0071	399.2	0.0116	624.5				0.0207	1014.5
	0.0072	404.5	0.0117	629.3	0.0162	832.6		0.0208	1018.3
	0.0073	409.7	0.0118	634.1	0.0163	836.9		0.0209	1022.0
	0.0074	414.9	0.0119	638.8	0.0164	841.2		0.0210	1025.8
	0.0075	420.1	0.0120	643.5	0.0165	845.4		0.0211	1029.6
	0.0076	425.3	0.0121	648.2	0.0166	849.7		0.0212	1033.3
	0.0077	430.5	0.0122	653.0	0.0167	853.9		0.0213	1037.1
	0.0078	435.7	0.0123	657.7	0.0168	858.1	ρ_{max}	0.0214	1040.8

Table A.15 Development Lengths (in.)

		f_c'							
		3000		4000		5000		6000	
Bar No.	f_y	Basic l_d	Top Bars $1.4l_d$	Basic l_d	Top Bars $1.4l_d$	Basic l_d	Top Bars $1.4l_d$	Basic l_d	Top Bars $1.4l_d$
2	40	12	12	12	12	12	12	12	12
	50	12	12	12	12	12	12	12	12
	60	12	12	12	12	12	12	12	12
3	40	12	12	12	12	12	12	12	12
	50	12	12	12	12	12	12	12	12
	60	12	13	12	13	12	13	12	13
4	40	12	12	12	12	12	12	12	12
	50	12	14	12	14	12	14	12	14
	60	12	17	12	17	12	17	12	17
5	40	12	14	12	14	12	14	12	14
	50	13	18	13	18	13	18	13	18
	60	15	21	15	21	15	21	15	21
6	40	13	18	12	17	12	17	12	17
	50	16	22	15	21	15	21	15	21
	60	19	27	18	25	18	25	18	25
7	40	18	25	15	21	14	20	14	20
	50	22	31	19	27	18	25	18	25
	60	26	37	23	32	21	29	21	29
8	40	23	32	20	28	18	25	16	23
	50	29	40	25	35	22	31	20	29
	60	35	48	30	42	27	38	24	34
9	40	29	41	25	35	23	32	21	29
	50	37	51	32	44	28	40	26	36
	60	44	61	38	53	34	48	31	43
10	40	37	52	32	45	29	40	26	37
	50	46	65	40	56	36	50	33	46
	60	56	78	48	67	43	60	39	55
11	40	46	64	39	55	35	49	32	45
	50	57	80	49	69	44	62	40	56
	60	68	96	59	83	53	74	48	68
14	40	62	87	54	75	48	67	44	61
	50	78	109	67	94	60	84	55	77
	60	93	130	81	113	72	101	66	92
18	40	80	113	70	97	62	87	57	80
	50	100	141	87	122	78	109	71	99
	60	121	169	104	146	93	131	85	119

Table A.16 Size and Pitch of Spirals, ACI Code

Diameter of Column (in.)	Out to Out of Spiral (in.)	f_c'			
		2500	3000	4000	5000
$f_y = 40{,}000$:					
14, 15	11, 12	$\frac{3}{8}$–2	$\frac{3}{8}$–$1\frac{3}{4}$	$\frac{1}{2}$–$2\frac{1}{2}$	$\frac{1}{2}$–$1\frac{3}{4}$
16	13	$\frac{3}{8}$–2	$\frac{3}{8}$–$1\frac{3}{4}$	$\frac{1}{2}$–$2\frac{1}{2}$	$\frac{1}{2}$–2
17–19	14–16	$\frac{3}{8}$–$2\frac{1}{4}$	$\frac{3}{8}$–$1\frac{3}{4}$	$\frac{1}{2}$–$2\frac{1}{2}$	$\frac{1}{2}$–2
20–23	17–20	$\frac{3}{8}$–$2\frac{1}{4}$	$\frac{3}{8}$–$1\frac{3}{4}$	$\frac{1}{2}$–$2\frac{1}{2}$	$\frac{1}{2}$–2
24–30	21–27	$\frac{3}{8}$–$2\frac{1}{4}$	$\frac{3}{8}$–2	$\frac{1}{2}$–$2\frac{1}{2}$	$\frac{1}{2}$–2
$f_y = 60{,}000$:					
14, 15	11, 12	$\frac{1}{4}$–$1\frac{3}{4}$	$\frac{3}{8}$–$2\frac{3}{4}$	$\frac{3}{8}$–2	$\frac{1}{2}$–$2\frac{3}{4}$
16–23	13–20	$\frac{1}{4}$–$1\frac{3}{4}$	$\frac{3}{8}$–$2\frac{3}{4}$	$\frac{3}{8}$–2	$\frac{1}{2}$–3
24–29	21–26	$\frac{1}{4}$–$1\frac{3}{4}$	$\frac{3}{8}$–3	$\frac{3}{8}$–$2\frac{1}{4}$	$\frac{1}{2}$–3
30	27	$\frac{1}{4}$–$1\frac{3}{4}$	$\frac{3}{8}$–3	$\frac{3}{8}$–$2\frac{1}{4}$	$\frac{1}{2}$–$3\frac{1}{4}$

Table A.17 Weights, Areas, and Moments of Inertia of Circular Columns and Moments of Inertia of Column Verticals Arranged in a Circle 5 in. Less Than the Diameter of Column

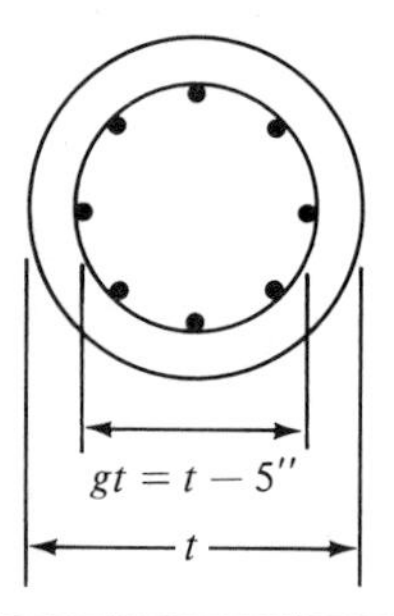

Diameter of Column h (in.)	Weight per Foot (lb)	Area (in.^{2})	I (in.^{4})	A_s Where $\rho_g = 0.01$*	I_s (in.^{4})†
12	118	113	1,018	1.13	6.92
13	138	133	1,402	1.33	10.64
14	160	154	1,886	1.54	15.59
15	184	177	2,485	1.77	22.13
16	210	201	3,217	2.01	30.40
17	237	227	4,100	2.27	40.86
18	265	255	5,153	2.55	53.87
19	295	284	6,397	2.84	69.58
20	327	314	7,854	3.14	88.31
21	361	346	9,547	3.46	110.7
22	396	380	11,500	3.80	137.2
23	433	416	13,740	4.16	168.4
24	471	452	16,290	4.52	203.9
25	511	491	19,170	4.91	245.5
26	553	531	22,430	5.31	292.7
27	597	573	26,090	5.73	346.7
28	642	616	30,170	6.16	407.3
29	688	661	34,720	6.61	475.9
30	736	707	39,760	7.07	552.3

* For other values of ρ_g, multiply the value in the table by $100\rho_g$.

† The bars are assumed transformed into a thin-walled cylinder having the same sectional area as the bars. Then $I_s = A_s(\gamma t)^2/8$.

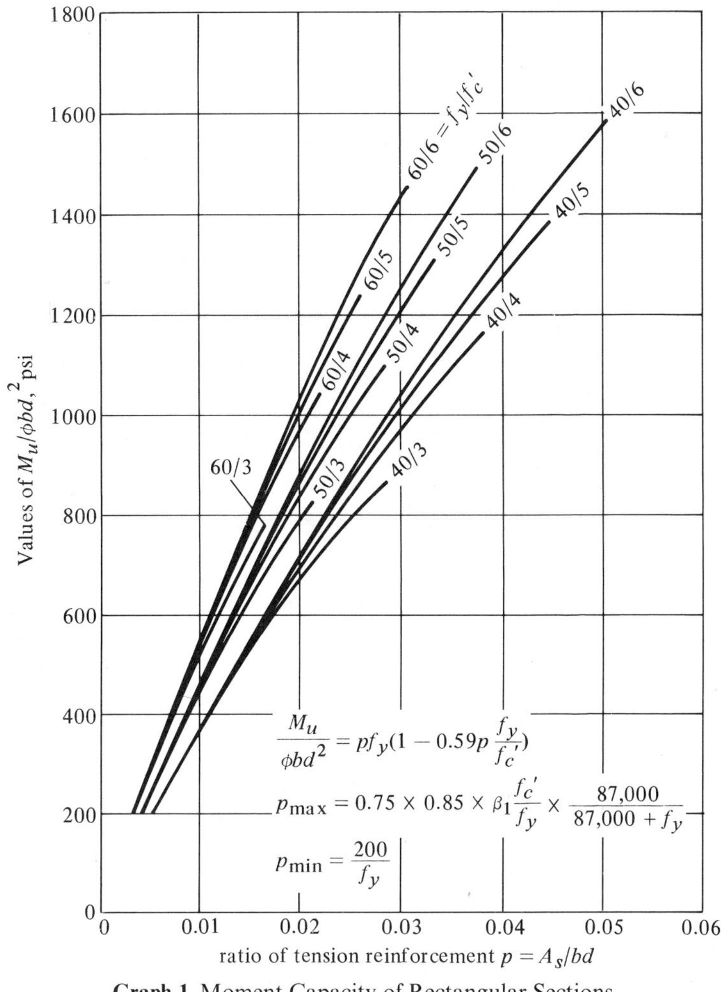

Graph 1 Moment Capacity of Rectangular Sections

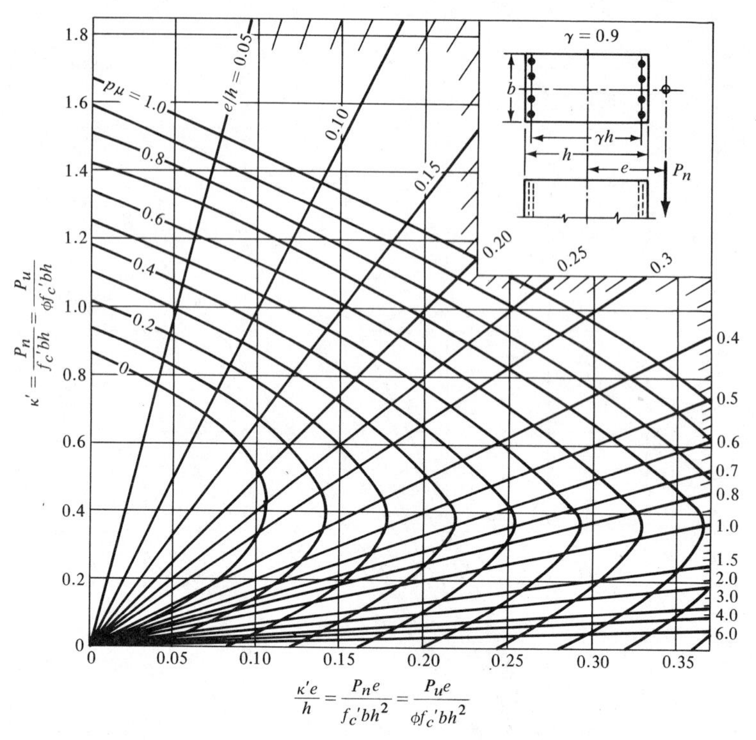

Graph 2 Column Interaction Diagram—Rectangular Section: $\gamma = 0.9$, $f_c' = 4$ ksi, $f_y = 60$ ksi, $\mu = f_y/0.85f_c'$, $P = P_t = A_{s\,\text{total}}/bh$

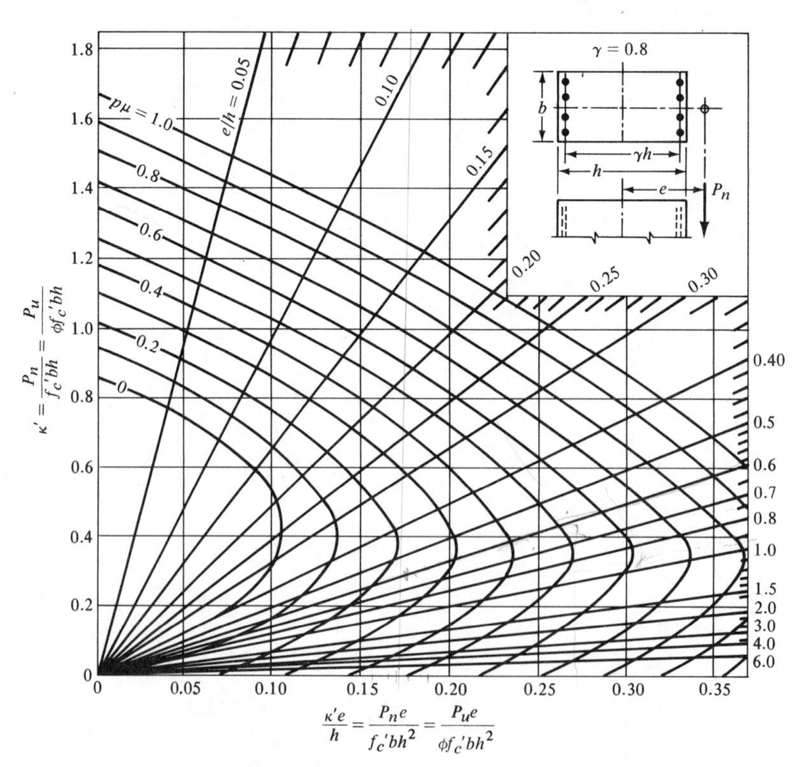

Graph 3 Column Interaction Diagram—Rectangular Section $\gamma = 0.8$, $f_c' = 4$ ksi, $f_y = 60$ ksi

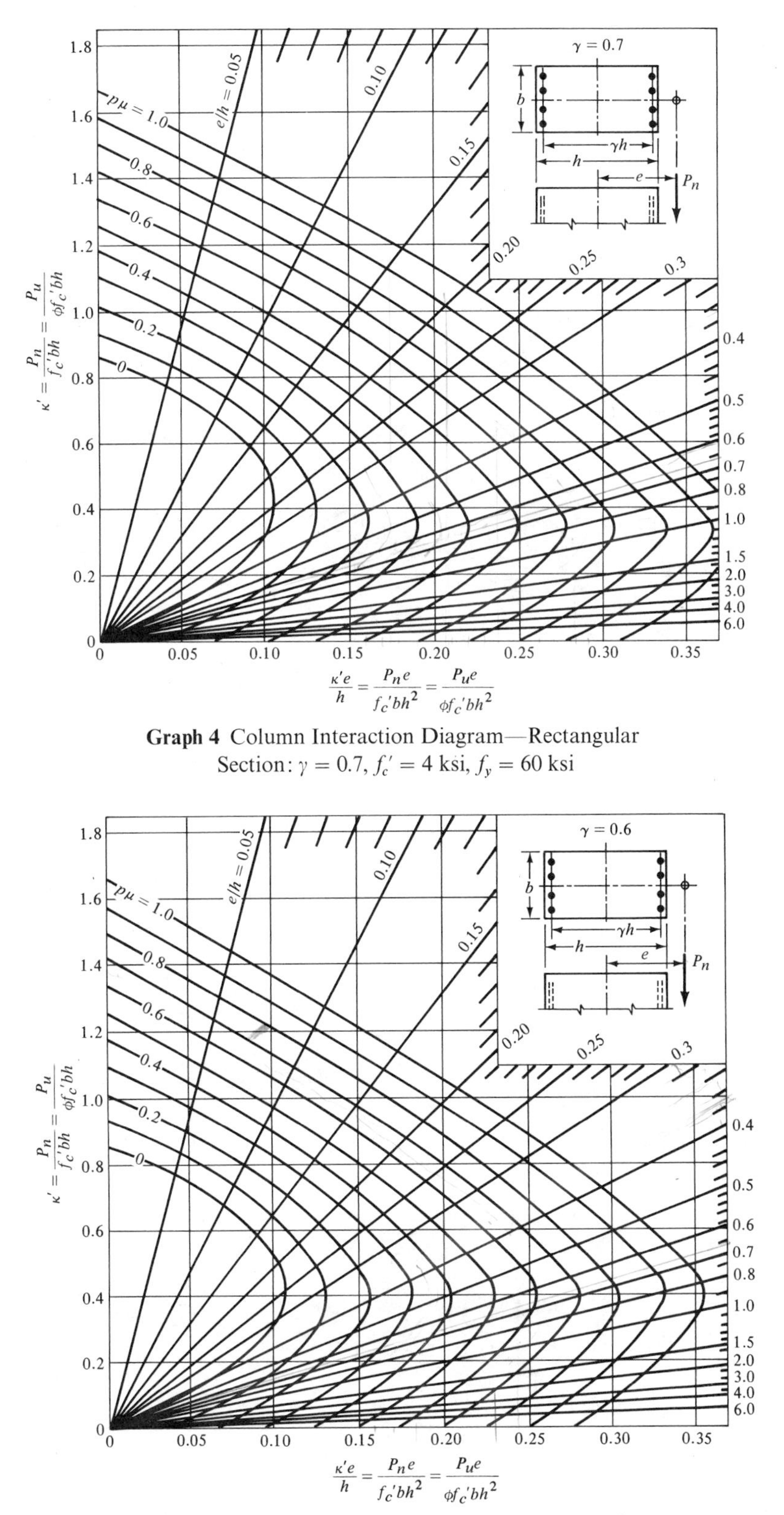

Graph 4 Column Interaction Diagram—Rectangular Section: $\gamma = 0.7$, $f_c' = 4$ ksi, $f_y = 60$ ksi

Graph 5 Column Interaction Diagram—Rectangular Section: $\gamma = 0.6$, $f_c' = 4$ ksi, $f_y = 60$ ksi

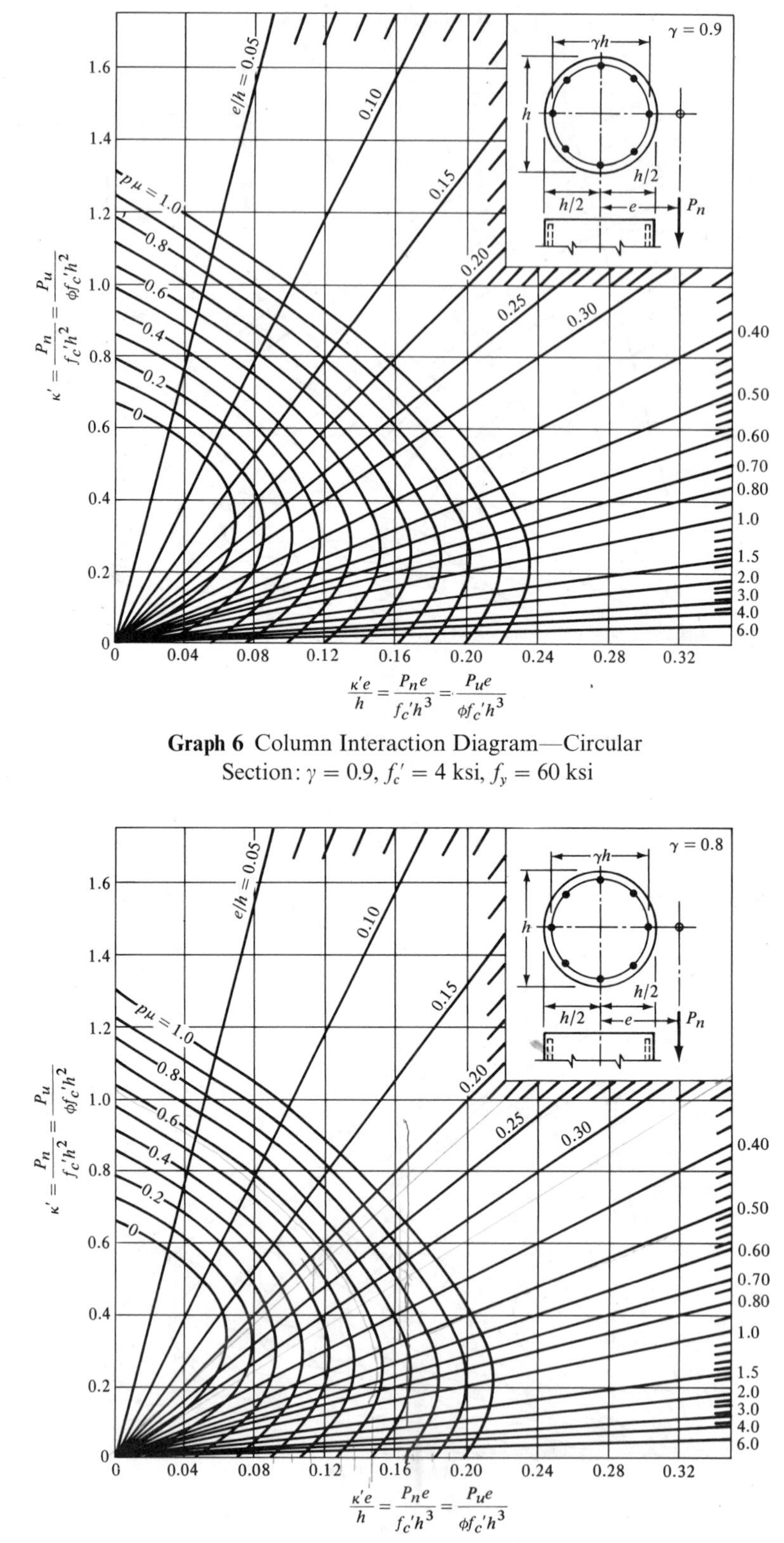

Graph 6 Column Interaction Diagram—Circular
Section: $\gamma = 0.9$, $f_c' = 4$ ksi, $f_y = 60$ ksi

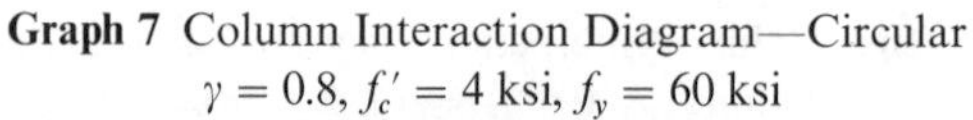

Graph 7 Column Interaction Diagram—Circular
$\gamma = 0.8$, $f_c' = 4$ ksi, $f_y = 60$ ksi

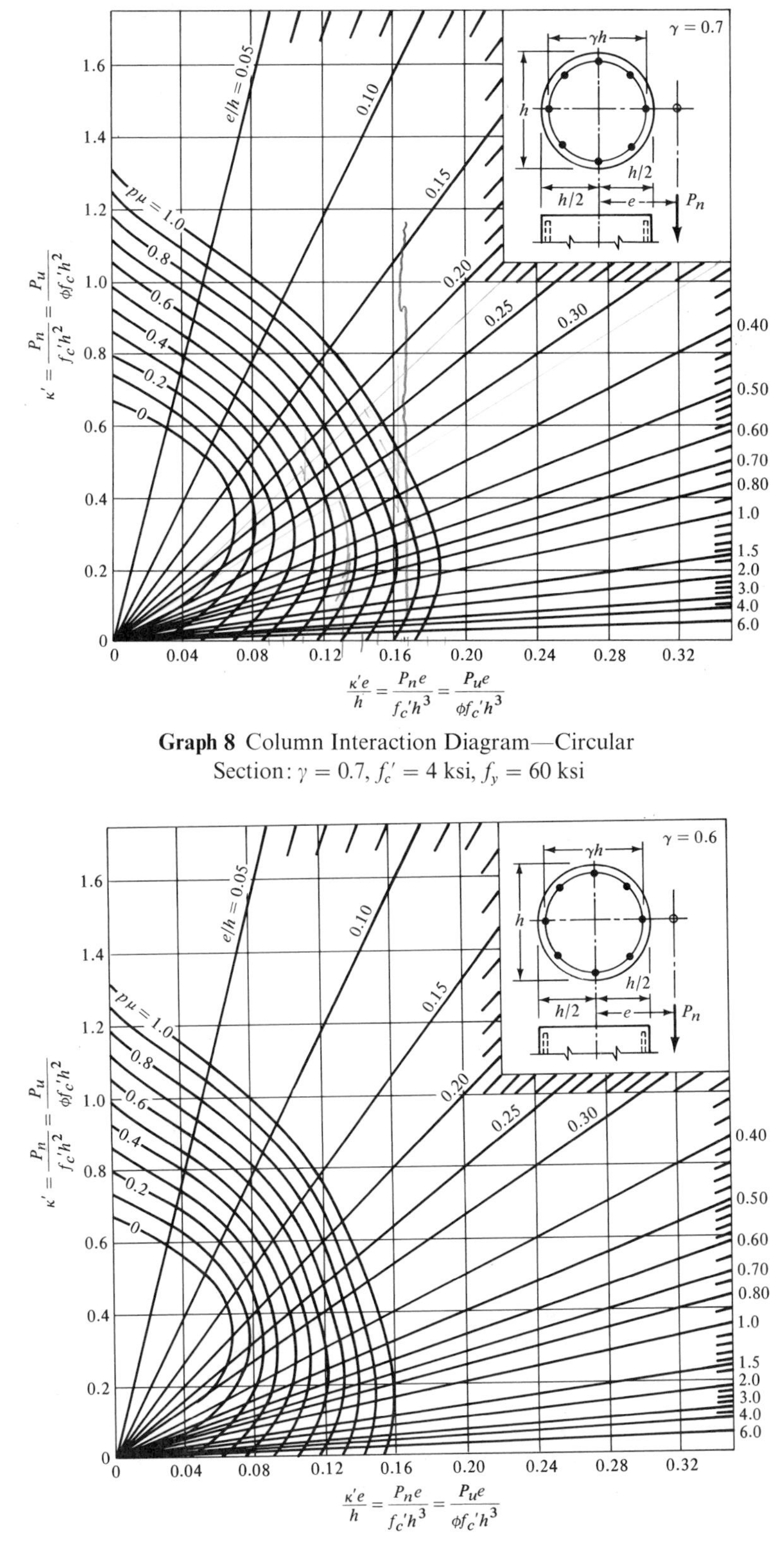

Graph 8 Column Interaction Diagram—Circular Section: $\gamma = 0.7$, $f_c' = 4$ ksi, $f_y = 60$ ksi

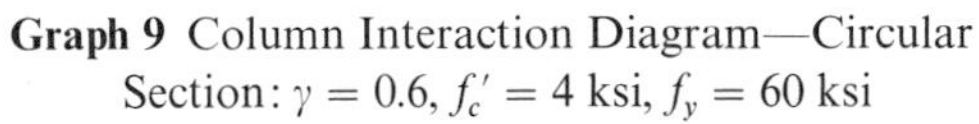

Graph 9 Column Interaction Diagram—Circular Section: $\gamma = 0.6$, $f_c' = 4$ ksi, $f_y = 60$ ksi

INDEX

78 79 80 81 82 9 8 7 6 5 4 3 2 1